Turbulent Shear Flows 5

Selected Papers from the Fifth International
Symposium on Turbulent Shear Flows,
Cornell University, Ithaca, New York, USA,
August 7–9, 1985

Editors:

F. Durst B. E. Launder J. L. Lumley
F. W. Schmidt J. H. Whitelaw

With 260 Figures

Springer-Verlag Berlin Heidelberg New York
London Paris Tokyo

Franz Durst
Lehrstuhl für Strömungsmechanik, Technische Fakultät
Friedrich-Alexander-Universität, Egerlandstraße 13,
D-8520 Erlangen, Fed. Rep. of Germany

Brian E. Launder
Department of Mechanical Engineering, University of Manchester,
Institute of Science and Technology, PO Box 88,
Manchester M60 1QD, England

John L. Lumley
Sibley School of Mechanical Aero Engineering,
Cornell University, 238 Upson Hall,
Ithaca, NY 14853, USA

Frank W. Schmidt
Mechanical Engineering Department, The Pennsylvania State University,
University Park, PA 16802, USA

James H. Whitelaw
Department of Mechanical Engineering, Imperial College of Science and Technology, Exhibition Road, London SW7 2BX, England

ISBN-13: 978-3-642-71437-5 e-ISBN-13: 978-3-642-71435-1
DOI: 10.1007/978-3-642-71435-1

Library of Congress Cataloging-in-Publication Data.
International Symposium on Turbulent Shear Flows (5th : 1985 : Cornell University)
Selected papers from the Fifth International Symposium on Turbulent Shear Flows, Cornell University, Ithaca, New York, USA, August 7–9, 1985.
1. Shear flow – Congresses. 2. Turbulence – Congresses. I. Durst, F. II. Title.
TA357.I59 1985 620.1'064 86-22076

Softcover reprint of the hardcover 1st edition 1987

Typesetting: Daten- und Lichtsatz-Service, Würzburg;

2153/3020-543210

Preface

The first four symposia in the series on turbulent shear flows have been held alternately in the United States and Europe with the first and third being held at universities in eastern and western States, respectively. Continuing this pattern, the Fifth Symposium on Turbulent Shear Flows was held at Cornell University, Ithaca, New York, in August 1985. The meeting brought together more than 250 participants from around the world to present the results of new research on turbulent shear flows. It also provided a forum for lively discussions on the implications (practical or academic) of some of the papers. Nearly 100 formal papers and about 20 shorter communications in open forums were presented. In all the areas covered, the meeting helped to underline the vitality of current research into turbulent shear flows whether in experimental, theoretical or numerical studies.

The present volume contains 25 of the original symposium presentations. All have been further reviewed and edited and several have been considerably extended since their first presentation. The editors believe that the selection provides papers of archival value that, at the same time, give a representative statement of current research in the four areas covered by this book:

- Homogeneous and Simple Flows
- Free Flows
- Wall Flows
- Reacting Flows

Each of these sections begins with an introductory article by a distinguished worker in the field. These articles provide both a thumbnail sketch of the contributions made by the different papers and sets new contributions against the background of the research in the field. It is the editors' hope that in this way the book provides an up-to-date collection of high quality papers for the expert and, at the same time, offers sufficient signposts to help the newcomer (in conjunction with the earlier volumes in this series) to gain an appreciation of the present preoccupations in turbulent flow research.

The papers in this volume are arranged according to their complexity. The accounts of studies of homogeneous and simple flows include descriptions of modelling of pressure terms of the scalar fluxes and considerations on inhomogeneous turbulence with applications to boundary layer flows. Interactions of turbulent scales are studied and structural considerations of homogeneous turbulence are presented. Wake shear layer interactions are described and mixing layer and jet measurements presented. Particular attention is given to wall boundary layer flows providing new experimental and numerical results. The last section concentrates on experimental and numerical studies of combustion flows.

Financial support for the Fifth Symposium was generously contributed by the Boeing Aerospace Company, and the ASME-Heat Transfer Division also kindly provided material assistance. The success of the Cornell meeting depended strongly on the efforts of many individuals in various aspects of the pre-conference organization as well as in the highly visible contributions during the Symposium itself.

As with earlier symposia responsibility for setting the technical programme rested with a *Papers Committee* which for the Fifth Symposium was composed of J. L. Lumley (Chairman), B. E. Launder, W. C. Reynolds and J. H. Whitelaw. Each of the nearly two hundred 1000-word abstracts offered for presentation at the Symposium were reviewed by two members of the Advisory Committee. Many Advisory Committee members later served as session chairmen at the Symposium and have throughout been valuable sources of advice and helpful criticism. The Advisory Committee consisted of:

R. J. Adrian
J.-C. Andre
L. H. Back
R. W. Bilger
R. Blackwelder
R. Borghi
R. Briley
D. Bushnell
M. Coantic
S. Corrsin
J. Cousteix
J. K. Eaton
H. Fiedler
I. Gartshore
M. M. Gibson
V. W. Goldschmidt
K. Hanjalic
T. J. Hanratty
J. R. Herring
M. Hino
M. Hirata
J. A. C. Humphrey
A. K. M. F. Hussain
J. P. Johnston
P. N. Joubert
E. Krause
J. C. LaRue
P. A. Libby
O. Martynenko
J. Mathieu
S. V. Patankar
G. Raithby
W. Rodi
U. Schumann
R. L. Simpson
K. Suzuki
W. Wyngaard

The editors of this book are also very thankful to the staff of Springer-Verlag who were actively involved and helped greatly in the completion of the present book.

As this volume was going to press, the turbulent shear flows community was saddened by the death of Stanley Corrsin after a long illness. Stan, who had contributed so much over forty years to the measurement and understanding of turbulence, served as a member of the Advisory Committee for all five TSF Symposia. To his memory, therefore, this volume is affectionately dedicated.

Erlangen, November 1986

The Editors

Contents

Part I **Homogeneous and Simple Flows**

Introductory Remarks. By A. E. Perry 1

On Modelling the Pressure Terms of the Scalar Flux Equations. By T. Dakos and M. M. Gibson 7

A Simplified Spectral Closure for Inhomogeneous Turbulence: Application to the Boundary Layer. By J.-P. Bertoglio and D. Jeandel (With 4 Figures) 19

The Interaction of Two Distinct Turbulent Velocity Scales in the Absence of Mean Shear. By S. Veeravalli and Z. Warhaft (With 11 Figures) 31

The Mixing Layer Between Turbulent Fields of Different Scales. By S. B. Pope and D. C. Haworth (With 7 Figures) 44

On the Structure of Homogeneous Turbulence. By M. J. Lee and W. C. Reynolds (With 12 Figures) 54

Turbulence in a Stably Stratified Shear Flow: A Progress Report. By J. J. Rohr, K. N. Helland, E. C. Itsweire and C. W. Van Atta (With 8 Figures) 67

Part II **Free Shear Flows**

Introduction. By R. J. Adrian 79

Wake-Shear Layer Interaction and the Onset of Turbulence Behind a Circular Cylinder. By A. Kourta, H. C. Boisson, M. Braza, P. Chassaing and H. Ha Minh (With 17 Figures) 82

Investigation of the Detailed Spread Characteristics of Plane Turbulent Wakes. By P. R. Louchez, J. G. Kawall and J. F. Keffer (With 11 Figures) 98

Coherent Structures in a Turbulent Mixing Layer: A Comparison Between Direct Numerical Simulations and Experiments. By R. W. Metcalfe, A. K. M. F. Hussain, S. Menon and M. Hayakawa (With 8 Figures) 110

Structure of the Vorticity Field in a Plane Free Shear-Layer. By J. C. Lasheras and T. Maxworthy (With 9 Figures) 124

Coherent Structures in the Axisymmetric Turbulent Jet Mixing Layer. M. N. Glauser, St. J. Leib, W. K. George (With 12 Figures) 134

Laser Doppler Velocimeter Measurements in 3-D Impinging Twin-Jet Fountain Flows. By K. R. Saripalli (With 24 Figures) 146

Part III **Wall Flows**

Introduction. By P. Bradshaw 171

Turbulent Flow Over Mesh Roughness. By M. Acharya and M. P. Escudier (With 7 Figures) 176

Low Reynolds Number Turbulent Boundary Layers on a Smooth Flat Surface in a Zero Pressure Gradient. By L. P. Erm, A. J. Smits and P. N. Joubert (With 9 Figures) 186

Probability Density Distribution in Turbulent Wall Boundary Layer Flows. By F. Durst, J. Jovanovic and Lj. Kanevce (With 19 Figures) 197

Evolution of a Vortical Structure Associated with the Bursting Event in a Channel Flow. By J. Kim (With 12 Figures) 221

Direct Numerical Simulation of Equilibrium Turbulent Boundary Layers. By P. R. Spalart and A. Leonard (With 16 Figures) 234

Calculation of Longitudinal Vortices Imbedded in a Turbulent Boundary Layer. By J. Liandrat, B. Aupoix and J. Cousteix (With 5 Figures) 253

Interaction Between a Vortex and a Turbulent Boundary Layer in a Streamwise Pressure Gradient. By R. V. Westphal, J. K. Eaton and W. R. Pauley (With 9 Figures) 266

The Pulsed Wire Skin-Friction Measurement Technique. By I. P. Castro, M. Dianat and L. J. S. Bradbury (With 12 Figures) 278

Part IV **Reacting Flows**

Introduction: Turbulence Modeling of Reacting Flows: The State of the Art. By N. Peters 293

Multi-Scalar *pdf* Transport Equations for Turbulent Diffusion Flames. By W. P. Jones and W. Kollmann (With 12 Figures) 296

Modelling the Flow and Combustion in a Production Gas Turbine Combustor. By J. Coupland and C. H. Priddin (With 9 Figures) 310

Connection with a First-Order Chemically Reactive Passive Scalar. By J. R. Herring and J. C. Wyngaard (With 9 Figures) 324

Experimental and Numerical Study of a Turbulent Recirculation Zone with Combustion. By P. Moreau, J. Labbé, F. Dupoirieux, R. Borghi (With 6 Figures) 337

Experimental Study of the Interaction Between a Premixed Confined Laminar Flame and Coherent Structures. By D. Escudié and G. Charnay (With 12 Figures) 347

Index of Contributors 361

Part I

Homogeneous and Simple Flows

Introductory Remarks

A. E. Perry

Department of Mechanical Engineering, University of Melbourne, Parkville, 3052, Australia

The papers selected for this chapter are all concerned either directly or indirectly with the central problem of turbulence theory, which is to obtain realistic statistical solutions of the Navier-Stokes and scalar transport equations and apply them to practical situations. Before dealing with the papers in detail, it is important to remind ourselves of why this extremely complex topic developed the way it did, take stock of its achievements and review the direction it is heading.

Most efforts of the past have focused on homogeneous turbulence. Because of its relative simplicity, isotropic homogeneous turbulence has been studied the most. Between the years of 1930 and 1960 most of the basic formulations of the problem were made and the state of knowledge for that period is well summarized in books by Batchelor (1971), Townsend (1976) and Hinze (1975). It has not been possible even for the simple isotropic case to obtain general complete solutions for the differential equations involved except for the final period of decay of grid turbulence. The essential difficulty is that the Navier-Stokes equations when time or ensemble averaged do not yield a closed set of differential equations for the velocity covariance. To quote Kraichnan (1959) "the equation of motion for this covariance contain third-order moments of the velocity field, the equations of motion for the third-order moments contain fourth-order moments and so fourth *ad infinitum.* A central goal of turbulence theory is the closing of this infinite chain of coupled equations into a determinate set containing only moments below some finite order". During the 70's and up to the present day a world wide industry has sprung up devoted to this closure problem. The practical enigneer may well ask why we torture ourselves attempting what has proved to be a most difficult if not impossible task and why concentrate on homogeneous turbulence. Surely with modern large computers we could solve the complete time dependent Navier-Stokes equations by specifying the appropriate initial and boundary conditions and allow the solution to run its course.

As Leslie (1973) states (and it is still true today) "The short answer is that large though they are, present day computers are not large enough". A mesh must be sufficiently fine to resolve the dissipating eddies. The so called "Full Direct Simulations" or "Full Turbulence Simulations" mentioned in some of the papers here, attempt to do this. However, they are limited to low Reynolds numbers. For channel flow, Leslie quotes 10^8 mesh points are required for (Re) $= 10^4$ and 10^{13} for (Re) $= 10^6$ where (Re) is the channel flow Reynolds number. Also the number of time steps required for numerical stability rises rapidly with (Re). See also Corrsin (1961). For one of the papers here, the numbers of mesh points used is of order 10^6. Then there is the question of running time and cost. A simulation needs to be run many times to obtain ensemble averages with stable statistics although if homogeneous turbulence is considered we could average along the homogeneous direction (Lumley and Bejan Khajeh-Nouri, 1974). It has been the experience of the writer that using "nature's own computer" – namely, the wind tunnel, 40,000 data samples are needed for convergence of quantities like Reynolds shear stress at a point with 1% repeatability.

The reason why there has been such an interest in homogeneous isotropic turbulence is that this is the case we know most about. In more general flow situations, perturbations about this "basic" flow are made and most papers considered in this chapter are concerned with various types of perturbations of this basic state. Many early schemes were based on the so called two-point closure methods where the equations for the two-point correlations were often solved in wavenumber space. Even when truncated to fourth order moments, the computational power required for such simulations can often rival that needed for the Full-Direct Simulations as is pointed out in the paper by Bertoglio and Jeandal given later in this chapter. Unless someone can come up with a better idea, it appears that the only hope we have at present of producing practical computational methods is to develop closure models for what Lumley (1978) calls "second order modelling" or "invariant modelling" or what the French call "one-point closure". The $k - \varepsilon$ model is an example of this. Inspired by some early ideas of Donaldson (1968), workers developed a general tensor notation and expressed everything as far as possible in an invariant form. All unknown correlations are expressed in terms of the second order correlations. With this invariant form, models developed for one flow geometry, with all of the associated empirically determined constants, have the greatest likelihood of being applicable in another flow geometry. However, since we need to truncate to a finite probability moment (usually the fourth) and because of the necessary over simplications made in the modelling details, there will probably never be a single "law of turbulence" based on this formulation for all cases. The method is basically an expansion about the homogeneous stationary case and Lumley (1978) sums up the underlying philosophy when explaining why it can be successfully used for flows with appreciable inhomogeneity; "by following a rational procedure we have created a physically possible phenomenon, not quite real turbulence perhaps, but one which conserves momentum and energy; transports the right amount of everything budgeted (momentum, energy, Reynolds stress, heat flux, etc.) although not by quite the right mechanism; satisfies realizability (so that non-negative quantities are never negative, Schwarz' inequality is always satisfied, etc.); behaves correctly for both large and small Reynolds numbers; and reduces to real turbulence in one limit (weak inhomogeneity and unsteadiness.) Probably any mechanism that satisfied all of these restrictions would behave about the same".

Many of the papers here are concerned with this modelling although there are also hybrid models. Let us consider the papers in detail.

The first paper is by T. Dakos and M. M. Gibson "*On Modelling the Pressure Terms of the Scalar Flux Equations*". The emphasis is placed on modelling the pressure and scalar gradient correlation $\left\langle \frac{p}{\varrho} \frac{\partial \theta}{\partial x_i} \right\rangle$ which appears in the equation for the scalar flux $\langle u_i \theta \rangle$. The method assumes nearly homogeneous undirectional flow with a weak mean scalar gradient. Part of the expression for $\left\langle \frac{p}{\varrho} \frac{\partial \theta}{\partial x_i} \right\rangle$ has been derived from theoretical considerations using formal solutions in wavenumber space. The remaining part is modelled using dimensional considerations and conventional ideas about the dependence on second moments and the anisotropy tensor. Various coefficients have been evaluated from the data of Sirivat and Warhaft (1982), Dakos (1985) and Tavoularis and Corrsin (1981). The authors hope in the near future to see if the model will correctly predict inhomogeneous flow data.

The second paper is by J. P. Bertoglio and D. Jeandel "*Simplified Spectral Closure for Homogeneous Turbulence: Application to the Boundary Layer*". The authors consider this as an "attempt to show that two-point closures can provide practical tools for computing complex flows". They point out that the application of the two-point closure scheme, particularly the Eddy Damped Quasi-Normal Markovian theory by Orszag (1970) to inhomogeneous turbulence in its full form, would probably require as much computational power as a Large Eddy Simulation or even a Fully Direct Simulation. However, the authors suggest that by means of certain crude assumptions perhaps some aspects of the two-point modelling (for the energy transfer between wavenumbers) could be combined with single point modelling (eg. the $k - \varepsilon$ model) for the inhomogeneous transport terms. Based on this, the authors claim to have produced a method of calculation which is "one step further in sophistication than the standard $k - \varepsilon$ model" and does not suffer from many of the disadvantages of that model. Comparison with the data of Klebanoff (1955) look promising, although these days there is an abundance of more extensive and detailed data which the authors could have chosen. Perhaps this data will be used in future work.

The third paper to be considered here is "*The Interaction of Two Distinct Turbulent Velocity Scales in the Absence of Mean Shear*" by S. Veeravalli and Z. Warhaft. Here we have a novel perturbation to homogeneous turbulence which is sufficiently simple to give insights into many aspects of scale interactions and transport phenomena. A *shearless* mixing layer bounded by two homogeneous fields of different properties is considered. The first experiment along these lines was carried out by Gilbert (1980). However, here a greater scale ratio of the two streams is considered and this reveals vital effects unobserved by Gilbert. The authors point out that the important point about this work is that this "second order" mixing layer focuses particular attention on the third and fourth moments and has a distinctly different character to the "first order" mixing layer where production of kinetic energy and shear stress play an important role. Furthermore with the shearless layer, simple gradient diffusion models are quite inadequate. The experimental results are preliminary but the facility could readily be used for investigating the transport of scalar quantities in the near future.

"*The Mixing Layer Between Turbulent Fields of Different Scales*" by S. B. Pope and C. D. Howarth follows on from the previous paper. Here a model is proposed which fits the data of Veeravalli and Warhaft rather well. The model is based on a transport equation for the joint probability density function of the three components of velocity as formulated by Pope (1981). A novel feature of the modelling is the use of a Lagrangian formulation for the dissipation and a simple relaxation equation is used to account for the fact that packets of fluid can traverse the layer from the large scale side (say) to the small scale side more rapidly than the turbulence in the packet can respond to its changed surroundings. As discussed by Veeravalli and Warhaft, the most revealing statistics are in the third and fourth moments.

In the next paper "*on the Structure of Homogeneous Turbulence*" by J. Lee and William C. Reynolds a Full Turbulence Simulation is being used. Here homogeneous turbulence is subjected to irrotational strains. A code developed by Rogallo (1981) is used to provide "data" which might be useful for the formulation of closure hypotheses in the "single point" or "second order" modelling methods as summarized by Lumley (1978). If the computations can be believed, considerable insights are provided. Anisotropy invariant maps are produced which have the boundaries identified by Lumley.

Turbulence which is initially isotropic is subjected to various irrotational strain rates. It is found that the vorticity field is more anisotropic than the velocity field, likely due to definition and Reynolds number. Most models including those of Launder et al. (1975) and Lumley (1978) assume that the "return to isotropy" tensor ϕ_{ij} is correlated with the anisotropy tensor b_{ij} whereas these simulations show that ϕ_{ij} correlates better with the dissipation rate anisotropy d_{ij}. A simple relationship is given which not only agrees with the runs given in this paper but also with the shear runs of Rogallo (1981). As he points out, these simulations are limited to a small range of eddy scales. Nevertheless, "data" can be provided concerning those aspects of turbulence which are most difficult to resolve in wind tunnel experiments. Hence such simulations might well compliment experiments for the formulation of more efficient closure schemes.

The final paper of this selection is "*Turbulence in a stably stratified shear Flow: A Progress Report*" by J. J. Rohr, K. N. Helland, E. C. Itsweire and W. C. Van Atta. This describes work of an experimental nature where sheared homogeneous turbulence is studied in a salt statified water channel. It is found that the results exhibit trends similar to ocean data in regions where the turbulence is growing under the combined influence of shear and buoyancy but not in regions close to the grid where buoyancy is relatively unimportant. It appears that the facility developed will be most relevant to the interpretation of oceanic microstructure data and for the modelling of turbulence in stratified fluids.

References

Batchelor, K. G. (1971): *The Theory of Homogeneous Turbulence* (2nd ed). (Cambridge University Press, London)

Corrsin, S. (1961): Turbulent Flow. Am. Sci. 49, 300

Dakos, T. (1985): Fundamental Heat-Transfer Studies in Grid Generated Homogeneous Turbulence. Ph.D. Thesis Imperial College, London

Donaldson, C. (1968): see Kline, S. J., Morkovin, M. J., Sovran, G., Cockrell, J. J. (eds). Proc. Comput. Turbulent Boundary Layers 1968 AFOSR-IFP-Stanford. Thermosci. Div. Stanford University, Stanford, California, 114–118

Gilbert, B. (1980): Diffusion mixing in grid turbulence without mean shear. J. Fluid Mech. **100**, 349–365

Hinze, J. O. (1975): *Turbulence* 2nd ed. (McGraw-Hill, New York)

Klebanoff, P. S. (1955): Characteristics of turbulence in a boundary layer with zero pressure gradient. NACA Report 1247

Kraichnan, R. H. (1959): The structure of isotropic turbulence at very high Reynolds numbers. J. Fluid Mech. **15**, 497–543

Launder, B. E., Reece, G. J., and Rodi, W. (1975): Progress in the development of a Reynolds stress turbulent closure: J. Fluid Mech. **68**, 537–566

Leslie, D. C. (1973): *Developments in the Theory of Turbulence* (Clarendon Press, Oxford)

Lumley, J. L., Bejan Khajeh-Nouri (1974): Computational modeling of turbulent transport. Adv. Geophys. **18A**, 169–192

Lumley, J. L. (1978): Computational modeling of turbulent flows. Adv. Appl. Mech. **18**, 23–176

Orszag, S. A. (1970): Analytical theories of turbulence. J. Fluid Mech. **41**, 363–386

Pope, S. B. (1981): Transport equation for the joint PDF of velocity and scalars in turbulent flow. Phys. Fluids **24**, 588–596

Rogallo, R. S. (1981): Numerical experiments in homogeneous turbulence. NASA T. M. 81315

Sirivat, A., Warhaft, Z. (1982): The effect of a passive cross-stream temperature gradient on the evolution of temperature variance and heat flux in grid turbulence. J. Fluid Mech. **120**, 475

Townsend, A. A. (1976): *The Structure of Turbulent Shear Flow*, 2nd ed. (Cambridge University Press)

Travoularis, S., Corrsin, S. (1981): Experiments in nearly homogeneous turbulence with a uniform mean temperature gradient. J. Fluid Mech. **104**, 311

On Modelling the Pressure Terms of the Scalar Flux Equations

T. Dakos and M. M. Gibson

Fluids Section, Mechanical Engineering Department, Imperial College of Science & Technology, London SW7 2BX, England

Abstract

The paper is concerned with the problem of modelling the ensemble averaged product of the pressure and scalar gradient that appears in the equations for the scalar flux $\langle u_i \theta \rangle$. The method (for nearly homogeneous unidirectional flow with weak gradients) is to derive nonlinear expressions for the fluctuating parts of the pressure and the scalar by formal solution (in wave vector space) of the Navier-Stokes and the scalar equations. Substitution in the Fourier transform of the pressure correlation shows that this quantity is the sum of four terms, one of which contains the mean scalar gradient. A fourth-order cumulant discard approximation allows this term to be expressed in terms of the single point double products and the turbulent energy and stress spectra. A numerical calculation is made when the spectra are represented by simple functions. Finally, a tentative model is proposed in which the remaining terms of the pressure correlation are evaluated by reference to experimental data.

Nomenclature

b_{ij}	Anisotropy tensor $2\langle u_i u_j \rangle / q^2 - 2\delta_{ij}/3$
$C_{\theta 1} - C_{\theta 5}$	Turbulence model constants
$I(\boldsymbol{r}, t)$	Quantity defined by Eq. (7)
$G_0(\boldsymbol{r}, t; \boldsymbol{r}_1, t_1)$	Green's function, Eq. (10)
$\boldsymbol{k}$	Wave vector
k_{ij}	j component of k_i vector
$N(\boldsymbol{k}, t)$	Fourier transform defined by Eq. (5)
P	Turbulent energy production rate
p	Fluctuating part of the pressure
Pr_t	Turbulent Prandtl number
$\boldsymbol{r}$	Position vector
$S_{ij}(\boldsymbol{k}, t, t_1)$	Two-time energy spectrum tensor
$S_{ij}(\boldsymbol{k}, t)$	One-time energy spectrum tensor
T	Mean value of scalar quantity
t	Time
U	Mean velocity
$\boldsymbol{u}$	Velocity vector
$\langle u_i u_j \rangle$	Reynolds stress
$\langle u_i \theta \rangle$	Scalar flux
V	Volume
v_0^2	$\frac{1}{3} q^2$
$q^2 = \langle u_i u_i \rangle$	$2 \times$ turbulent kinetic energy
x_i $(i = 1, 3)$	Cartesian coordinates
α, β, γ	Constants in spectrum models
δ	Dirac delta function
δ_{ij}	Kronecker delta
ε	Turbulent energy dissipation rate
θ	Fluctuating part of scalar quantity
λ	Diffusivity
ϱ	Fluid density
$\Phi_{i\theta}$	Pressure-scalar-gradient correlation

Subscripts and Superscripts

i, j, k	Tensor indices
$*$	Complex conjugate

Introduction

In order to close the equations for the turbulent stresses and scalar fluxes it is necessary to model the correlations that contain the fluctuating part of the pressure. Because the pressure interactions are responsible for the distribution of turbulent kinetic energy between components, it has been convenient to rearrange the pressure term in the stress equations as the sum of a divergence (which is added to the turbulent diffusion) and a traceless distributive term (the pressure-strain correlation). Rotta [1] appears to have been the first to recognize the value of further splitting the pressure-strain correlation into a non-linear turbulence part,

Turbulent Shear Flows 5

and a "rapid" part containing the mean velocity gradient, so that the two parts could be modelled separately. The first step in the analogous treatment of the pressure terms in the scalar-flux equations is to introduce the fluctuating scalar gradient by writing

$$\left\langle \frac{\theta}{\varrho} \frac{\partial p}{\partial x_i} \right\rangle = \delta_{ik} \left\langle \frac{\partial}{\partial x_k} \left(\frac{p\theta}{\varrho} \right) \right\rangle - \left\langle \frac{p}{\varrho} \frac{\partial \theta}{\partial x_i} \right\rangle, \tag{1}$$

where the first part may be added to the transport terms as "pressure diffusion". The second step is to take the divergence of the Navier-Stokes equations and multiply the solution of the resulting Poisson equation for p by $d\theta/dx_i$. The well-known result [2]

$$\left\langle \frac{p}{\varrho} \frac{\partial \theta}{\partial x_i} \right\rangle = \frac{1}{4\pi} \int \left\{ \left\langle \left(\frac{\partial^2 u_i u_j}{\partial x_i \partial x_j} \right)' \frac{\partial \theta}{\partial x_i} \right\rangle + 2 \left(\frac{\partial U_i}{\partial x_j} \right)' \left\langle \left(\frac{\partial u_j}{\partial x_i} \right)' \frac{\partial \theta}{\partial x_j} \right\rangle \right\} \frac{dV}{r} \tag{2}$$

(where the primes mean that quantities so marked are evaluated at $\boldsymbol{x} + \boldsymbol{r}$) indicates that the pressure-scalar-gradient correlation also consists of turbulence and velocity-gradient components. The mean scalar gradient does not appear explicitly here and it is also usually omitted from models of the correlation, but not by Jones and Musonge [3]. These authors argue that, because both terms in the integral depend on the mean field, there is no point in modelling them separately. They show that the inclusion of a scalar-gradient term in the model, and the use of a scalar time scale in the turbulence-only part, produces good results for the strongly sheared, nearly-homogeneous, free shear flow realized by Tavoularis and Corrsin [4], where appreciable departures from equilibrium provide a demanding test for modelling techniques.

We now approach the problem from a different direction, following closely the lead of Weinstock [5, 6], who has used the Fourier-transform method to calculate the pressure-strain terms in the stress equations for a simple turbulent shear flow at high Reynolds numbers. We now use this method to derive nonlinear expressions for the fluctuating parts of the pressure and the scalar by formal solution (in wave-vector space) of the Navier-Stokes and the scalar equations. These expressions are substituted in the Fourier transform of the pressure-scalar-gradient correlation which is then expressed as the sum of four terms in turbulence quantities only, the mean velocity gradient, the mean scalar gradient, and the product of the two mean gradients. Although we have, for simplicity, limited the analysis to nearly homogeneous unidirectional flow with weak gradients, the results are still too complicated for immediate application to practical flow calculation. It is, however, possible to express the scalar-gradient term in terms of the single-point velocity products and the turbulent energy and stress spectrum functions. The method (like Weinstock's) is based on the neglect of the cumulant of the two-time fourth-order correlation that is basic to the direct-interaction approximation but here, as in Weinstock's analysis, the goal is not to calculate the spectra but to derive part of the pressure correlation in terms of the spectra. An approximate calculation of the term is made when the spectra are represented by simple functions. Finally, a tentative model is proposed in which the remaining terms are evaluated by reference to measurements in equilibrium free flow and to the data of Tavoularis and Corrsin [4]. The analysis is straightforward but lengthy, particularly for the second part concerning the scalar-gradient term. For some details the reader is referred to the two papers by Weinstock (as a guide to the method) and to the thesis by Dakos [7].

Derivation of the Pressure Term

The Fluctuating Part of the Pressure

We consider, for simplicity, a homogeneous turbulent incompressible flow field at high Reynolds and Peclet numbers in which the mean flow is in the x_1 direction and gradients of the mean velocity and mean scalar are uniform in the x_2 direction. The flux and variance of a passive scalar are described by equations that are uncoupled from the equations of the turbulent velocity field. The starting point is the equation for the fluctuating part of the pressure that is obtained by taking the divergence of the equation for the fluctuating velocity

$$-\frac{1}{\varrho} \nabla^2 p = 2 \frac{\partial u_m}{\partial x_i} \frac{\partial U_i}{\partial x_m} + \frac{\partial u_m}{\partial x_i} \frac{\partial u_i}{\partial x_m}. \tag{3}$$

Instead of solving (3) in physical space we follow Weinstock [5] in taking the Fourier transform to obtain, with the aid of Parseval's theorem and the assumption that $\partial U_1/\partial x_2$ is uniform,

$$\varrho^{-1} p(\boldsymbol{k}, t) = N(\boldsymbol{k}, t) + \frac{2ik_1}{k^2} u_2(\boldsymbol{k}, t) \frac{\partial U_1}{\partial x_2}, \tag{4}$$

where the Fourier transform of the nonlinear term is

$$N(\boldsymbol{k}, t) = -\int d\boldsymbol{k}_1 \frac{k_i k_l}{k^2} \{u_i(\boldsymbol{k}_1)\, u_l(\boldsymbol{k} - \boldsymbol{k}_1)\}. \tag{5}$$

The Fluctuating Part of the Scalar

The fluctuating part of the scalar, θ, is obtained by solving the equation obtained from the transport equation for $\tilde{\theta} = T + \theta$ by subtracting the mean and using the homogeneity condition

$$\frac{D\theta}{Dt} = \frac{\partial \theta}{\partial t} + U_m \frac{\partial \theta}{\partial x_m} = I(\boldsymbol{r}, t), \tag{6}$$

where

$$I(\boldsymbol{r}, t) = -u_m \frac{\partial T}{\partial x_m} - u_m \frac{\partial \theta}{\partial x_m} + \lambda \frac{\partial^2 \theta}{\partial x_m^2}. \tag{7}$$

A formal solution of (6) is

$$\theta(\boldsymbol{r}, t) = \int d\boldsymbol{r}_1\, G_0(\boldsymbol{r}, t; \boldsymbol{r}_1, 0)\, \theta(\boldsymbol{r}_1, 0) + \int_0^t dt_1 \int d\boldsymbol{r}_1\, G_0(\boldsymbol{r}, t; \boldsymbol{r}_1, t_1)\, I(\boldsymbol{r}_1, t_1), \tag{8}$$

where $G_0(\boldsymbol{r}, t; \boldsymbol{r}_1, t_1)$ is a Green's function chosen to satisfy the conditions

$$\frac{\partial G_0}{\partial t_1} + U_m \frac{\partial G_0}{\partial x_m} = 0, \qquad G_0(\boldsymbol{r}, t; \boldsymbol{r}_1, t_1) = \delta(\boldsymbol{r} - \boldsymbol{r}_1) \tag{9}$$

or,

$$G_0(\boldsymbol{r}, t; \boldsymbol{r}_1, t_1) = \delta(\boldsymbol{r} - U_1(t - t_1) - \boldsymbol{r}_1). \tag{10}$$

The Green's function acts as a propagator of information to relate $\theta(\boldsymbol{r}, t)$ at the point under consideration to values at all points in space and time. When the Peclet number is sufficiently high the dissipation term in (7) can be discarded so that when $I(\boldsymbol{r}, t)$ from (7) is substituted in (8) we have

$$\begin{aligned}\theta(\boldsymbol{r}, t) &= \int d\boldsymbol{r}_1\, G_0(\boldsymbol{r}, t; \boldsymbol{r}_1, 0)\, \theta(\boldsymbol{r}_1, 0)\\ &\quad + \int_0^t dt_1 \int d\boldsymbol{r}_1\, G_0(\boldsymbol{r}, t; \boldsymbol{r}_1, t_1) \left[- u_m \frac{\partial T}{\partial x_m} - u_m \frac{\partial \theta}{\partial x_m} \right]_{\boldsymbol{r}_1, t_1},\end{aligned} \tag{11}$$

where I is to be evaluated at r_1 and t_1. We now assume that the mean scalar gradient is uniform in space and take the Fourier transform of (11) to obtain

$$\begin{aligned}\theta(\boldsymbol{k}, t) &= G_0(\boldsymbol{k}, t)\, \theta(\boldsymbol{k}, 0)\\ &\quad - \int_0^t dt_1\, G_0(\boldsymbol{k}, t - t_1) \left[\left(u_m \frac{\partial \theta}{\partial x_m} \right)_{\boldsymbol{k}, t_1} + u_2(\boldsymbol{k}, t_1) \frac{\partial T}{\partial x_2} \right],\end{aligned} \tag{12}$$

where $(u_m \partial\theta/\partial x_m)_{\boldsymbol{k}, t_1}$ is the Fourier transform of $u_m \partial\theta/\partial x_m$ evaluated at k and t_1, and $G_0(\boldsymbol{k}, t - t_1)$ is given by

$$G_0(\boldsymbol{k}, t - t_1) = \exp[-i\boldsymbol{k} \cdot \boldsymbol{U}(t - t_1)]. \tag{13}$$

This result (12) is similar to the expression obtained by Weinstock [5] for the fluctuating velocity component u_2.

The Pressure-Scalar-Gradient Correlation

The pressure-scalar-gradient correlation is obtained with the aid of Parseval's theorem and the Fourier transform pairs for the fluctuating parts of the pressure and scalar

$$\Phi_{i\theta} = \left\langle \frac{p}{\varrho} \frac{\partial \theta}{\partial x_i} \right\rangle = \iint i k_{1i} \langle \theta(\boldsymbol{k}_1)\, p(\boldsymbol{k} - \boldsymbol{k}_1)\, \varrho^{-1} \exp(i\boldsymbol{k} \cdot \boldsymbol{r}) \rangle\, d\boldsymbol{k}\, d\boldsymbol{k}_1. \tag{14}$$

The integral is carried out over all space to obtain

$$\left\langle \frac{p}{\varrho} \frac{\partial \theta}{\partial x_i} \right\rangle = -\frac{(2\pi)^3}{V} \int i k_i \langle p(\boldsymbol{k}, t)\, \varrho^{-1} \theta^*(\boldsymbol{k}, t) \rangle\, d\boldsymbol{k}, \tag{15}$$

where V is the volume of the system (which will cancel out later on) and $\theta^*(\boldsymbol{k}, t)$ is the complex conjugate of the Fourier transform of the fluctuating scalar. This equation is the desired result; it remains only to substitute from (4) and (12) to obtain

$$\Phi_{i\theta} = \left\langle \frac{p}{\varrho} \frac{\partial \theta}{\partial x_i} \right\rangle = \Phi_{i\theta}^{(1)} + \Phi_{i\theta}^{(2)} + \Phi_{i\theta}^{(3)} + \Phi_{i\theta}^{(4)}, \tag{16}$$

where

$$\Phi_{i\theta}^{(1)} = -\frac{(2\pi)^3}{V} \int i k_i$$

$$\times \left\langle N(\boldsymbol{k}, t) \left[G_0^*(\boldsymbol{k}, t)\, \theta^*(\boldsymbol{k}, 0) - \int_0^t dt_1\, G_0^*(\boldsymbol{k}, t - t_1) \left(u_m \frac{\partial \theta}{\partial x_m} \right)^*_{\boldsymbol{k}, t_1} \right] \right\rangle d\boldsymbol{k}, \tag{17}$$

$$\Phi_{i\theta}^{(2)} = \frac{(2\pi)^3}{V} \int_0^t dt_1 \int i k_i \langle N(\boldsymbol{k}, t)\, G_0^*(\boldsymbol{k}, t - t_1)\, u_2^*(\boldsymbol{k}, t_1)\rangle \frac{\partial T}{\partial x_2}\, d\boldsymbol{k}, \tag{18}$$

$$\Phi_{i\theta}^{(3)} = -\frac{(2\pi)^3}{V} \int i k_i$$

$$\times \left\langle \frac{2ik_1}{k^2} u_2(\boldsymbol{k}, t) \left[G_0^*(\boldsymbol{k}, t)\, \theta^*(\boldsymbol{k}, 0) - \int_0^t dt_1\, G_0^*(\boldsymbol{k}, t - t_1) \left(u_m \frac{\partial \theta}{\partial x_m} \right)^*_{\boldsymbol{k}, t_1} \right] \right\rangle \frac{\partial U_1}{\partial x_2}\, d\boldsymbol{k}, \tag{19}$$

$$\Phi_{i\theta}^{(4)} = \frac{(2\pi)^3}{V} \int_0^t dt_1 \int i k_i \left\langle \frac{2ik_1}{k^2} u_2(\boldsymbol{k}, t)\, u_2^*(\boldsymbol{k}, t_1)\, G_0^*(\boldsymbol{k}, t - t_1) \right\rangle \frac{\partial T}{\partial x_2} \frac{\partial U_1}{\partial x_2}\, d\boldsymbol{k}. \tag{20}$$

The first term in (16) contains only turbulence quantities, the second term accounts for the interaction of the turbulent velocity field (expressed here by the third-order tensor $\langle N(\boldsymbol{k}, t)\, u_2^*(\boldsymbol{k}, t)\rangle$) with the mean scalar gradient, the third term is the product of turbulence correlations and the mean velocity gradient [which apppears because the Poisson equation (3) has been used to determine p] and the fourth term contains the product of the mean velocity and scalar gradients. Comparison with the terms in the volume integral (2) shows that the sum of the first two terms in (16) corresponds to the turbulence-only part of the correlation and the sum of the third and fourth terms corresponds to the mean-velocity-gradient part. The mean scalar gradient appears in two terms because, instead of simply multiplying p obtained as the solution of (2) by the fluctuating scalar gradient, we have obtained in (8) a formal solution for θ (which necessarily contains grad T) before forming the pressure-scalar-gradient correlation.

The terms (17–20) in (16) are too complicated for immediate use in closing the system of equations. It is, however, possible to continue the analysis to obtain a usable form of the second (scalar gradient) term (18) but before doing this we examine the problems in modelling the other three terms. In his treatment of the stress equations Weinstock [5, 6] was able to show that, to first order in anisotropy, the nonlinear turbulence part of the pressure-stain-rate correlation corresponding to (17) reduced to Rotta's linear return-to-isotropy model, and that it was possible to evaluate the Rotta constant with the aid of assumed turbulent energy and shear-stress spectrum functions. The obstacle to applying the same method to the turbulence-only part of (16) is that, although the fourth-order correlation in (17) may be treated in the same way as the fourth-order correlation in the equivalent pressure-strain-rate term (by making use of a two-time cumulant discard), the spectrum of $\langle u_i \theta\rangle$ cannot be split into isotropic and non-isotropic parts as could the energy spectrum in Weinstock's calculation. It does not, then, seem to be possible to use this method to derive the Monin model equivalent to that of Rotta for the turbulence part of (16), or to evaluate the model constant. The same difficulty also prevents the derivation of a simple expression for the velocity-gradient term (19). The fourth term (20) is more amenable to treatment; it is possible to introduce the energy spectrum tensor and to perform the integration with respect to time to obtain, subject to the restrictions imposed by the homogeneity condition,

$$\Phi_{i\theta}^{(4)} = -2\left(\frac{3\pi}{2}\right)^{1/2} \iint \frac{k_i k_1}{k^2} S_{22}(\boldsymbol{k}, t) \frac{1}{kq} \, d\boldsymbol{k} \frac{\partial U_1}{\partial x_2} \frac{\partial T}{\partial x_2}. \tag{21}$$

Unfortunately, not enough is known about the spectrum function $S_{22}(\boldsymbol{k})$ to allow further progress in this direction. Attention is therefore concentrated on the scalar-gradient term (18).

The Mean Scalar-Gradient Part

Outline of the Calculation

In this calculation of the scalar-gradient part of the pressure term the method is

(a) to derive a general expression for (18) in terms of turbulent energy and stress spectra,
(b) to show that, to first order in anisotropy, this is proportional to the anisotropy tensor b_{ij} and,
(c) to use modelled spectrum functions to evaluate the coefficient of proportionality.

We make use of the expression for the fluctuating velocity component u_2 obtained by Weinstock, (21) in [5], from a formal solution of the transport equation

$$u_2(\boldsymbol{k}, t) = G_0(\boldsymbol{k}, t)\, u_2(\boldsymbol{k}, 0) - \int_0^t dt_1\, G_0(\boldsymbol{k}, t - t_1)$$
$$\times \left[\left(u_m \frac{\partial u_2}{\partial x_m}\right)_{\boldsymbol{k}, t_1} + i k_2 N(\boldsymbol{k}, t_1) - \frac{2 k_1 k_2 u_2}{u^2}(\boldsymbol{k}, t_1) \frac{\partial U_1}{\partial x_2}\right]. \tag{22}$$

Calculation of the Scalar-Gradient Term

When the fluctuating part of the velocity given by (22) is substituted in (18) a term containing the product of the mean velocity and scalar gradients appears in the new expression for $\Phi_{i\theta}^{(2)}$ which can be absorbed in the fourth part of the correlation (20). We consider only the remaining, scalar-gradient, part

$$\Phi_{i\theta}^{(2)} = \frac{(2\pi)^3}{V} \int_0^t dt_1 \int i k_{0i} \langle N(\boldsymbol{k}_0, t)\, G_0^*(\boldsymbol{k}_0, t - t_1)\rangle \Big\{ G_0^*(\boldsymbol{k}_0, t_1)\, u_2^*(\boldsymbol{k}_0, 0)$$
$$- \int_0^{t_2} dt_2\, G_0^*(\boldsymbol{k}_0, t_1 - t_2) \left[\left(u_m \frac{\partial u_2}{\partial x_m}\right)^*_{\boldsymbol{k}_0, t_2} - i k_{0\,2} N^*(\boldsymbol{k}_0, t_2)\right]\Big\} \frac{\partial T}{\partial x_2} \, d\boldsymbol{k}_0. \tag{23}$$

For time separations t much larger than the integral time scale the two-time third-order tensor $\langle N(\boldsymbol{k}_0, t)\, u_2^*(\boldsymbol{k}_0, 0)\rangle$ on the right-hand side of (23) vanishes so that, after rearrangement, we obtain

$$\Phi_{i\theta}^{(2)} = -\frac{(2\pi)^3}{V} \int_0^t dt_1 \int_0^{t_1} dt_2 \int i k_{0i}\, G_0^*(\boldsymbol{k}_0, t - t_1)\, G_0^*(\boldsymbol{k}_0, t_1 - t_2)$$
$$\times \left\langle N(\boldsymbol{k}_0, t) \left[\left(u_m \frac{\partial u_2}{\partial x_m}\right)^*_{\boldsymbol{k}_0, t_2} - i k_{0\,2} N^*(\boldsymbol{k}_0, t_2)\right]\right\rangle \frac{\partial T}{\partial x_2} \, d\boldsymbol{k}_0. \tag{24}$$

When the Fourier transform of the nonlinear term $u_m \partial u_2/\partial x_m$, and $N(\boldsymbol{k}_0, t)$ from (5), are substituted in (24) the statistical average of

$$N(\boldsymbol{k}_0, t)\left[\left(u_m \frac{\partial u_2}{\partial x_m}\right)^*_{\boldsymbol{k}_0, t_2} - i k_{02} N^*(\boldsymbol{k}_0, t_2)\right],$$

hereafter denoted by B, becomes

$$B = \iint i \frac{k_{0n} k_{0j} k_{0m}}{k_0^2}\left(\delta_{2l} - \frac{k_{02} k_{0l}}{k_0^2}\right) \times \langle u_n(\boldsymbol{k}_1, t)\, u_j(\boldsymbol{k}_0 - \boldsymbol{k}_1, t)\, u_m^*(\boldsymbol{k}_2, t_2)\, u_l^*(\boldsymbol{k}_0 - \boldsymbol{k}_2, t_2)\rangle\, d\boldsymbol{k}_1\, d\boldsymbol{k}_2. \tag{25}$$

The two-time fourth-order tensor on the right-hand side of (25) is then expanded in cumulants, as in Appendix A of [5], and the fourth-order cumulant of the correlation is discarded. This cumulant is very small for large time separations; the main approximation is to neglect it for any time separation. Weinstock [5] has pointed out that such a cumulant neglect is also basic to the direct-interaction approximation and is much weaker than the neglect of single-time fourth-order cumulants in quasi-normal theory. Although Kraichnan [8] has objected to this cumulant-discard approximation on the grounds that it violates the principle of energy conservation, Meecham [9] has argued that this objection is invalid, that the failure to conserve energy is due to unsymmetrical application of the hypothesis and that the theory can be constructed so that energy is conserved. Whether the present approximation is inconsistent with energy conservation for uniformly-sheared homogeneous turbulence remains to be shown. However, our purpose is not to calculate the energy spectra but to obtain an expression for this part of the pressure correlation in terms of the stress spectra.

The two-time fourth-order tensor is now written as

$$\begin{aligned}
&\langle u_n(\boldsymbol{k}_1, t)\, u_j(\boldsymbol{k}_0 - \boldsymbol{k}_1, t)\, u_m^*(\boldsymbol{k}_2, t_2)\, u_l^*(\boldsymbol{k}_0 - \boldsymbol{k}_2, t_2)\rangle \\
&\quad = \langle u_n(\boldsymbol{k}_1, t)\, u_m^*(\boldsymbol{k}_2, t_2)\rangle \cdot \langle u_j(\boldsymbol{k}_0 - \boldsymbol{k}_1, t)\, u_l^*(\boldsymbol{k}_0 - \boldsymbol{k}_2, t_2)\rangle \\
&\qquad + \langle u_n(\boldsymbol{k}_1, t)\, u_l^*(\boldsymbol{k}_0 - \boldsymbol{k}_2, t_2)\rangle \, \langle u_j(\boldsymbol{k}_0 - \boldsymbol{k}_1, t)\, u_m^*(\boldsymbol{k}_2, t_2)\rangle
\end{aligned} \tag{26}$$

with the two-time velocity covariances

$$\langle u_i(\boldsymbol{k}_1, t_1)\, u_j^*(\boldsymbol{k}_2, t_2)\rangle = S_{ij}(\boldsymbol{k}_1, t_1, t_2)\, \delta(\boldsymbol{k}_1 - \boldsymbol{k}_2) \tag{27}$$

$$\langle u_i(\boldsymbol{k}_1, t_1)\, u_j^*(\boldsymbol{k}_1, t_2)\rangle = \frac{V}{(2\pi)^3} S_{ij}(\boldsymbol{k}_1, t_1, t_2) \tag{28}$$

and the two-time spectrum tensor defined by

$$\int S_{ij}(\boldsymbol{k}_1, t_1, t_2)\, d\boldsymbol{k}_1 = \langle u_i(\boldsymbol{r}, t_1)\, u_j(\boldsymbol{r}, t_2)\rangle. \tag{29}$$

Substitution in (25) and rearrangement then produces

$$B = \frac{2V}{(2\pi)^3} \int i \frac{k_{0n} k_{0j} k_{0m}}{k_0^2}\left(\delta_{2l} - \frac{k_{02} k_{0l}}{k_0^2}\right) S_{nm}(\boldsymbol{k}_1, t, t_2)\, S_{jl}(\boldsymbol{k}_0 - \boldsymbol{k}_1, t, t_2)\, d\boldsymbol{k}_1. \tag{30}$$

The next step is to express the two-time spectra in terms of single-time spectra at time t. Here we make use of Weinstock's result ([5], Appendix A) to obtain

$$B = \frac{2V}{(2\pi)^3} \int i \frac{k_{0n}k_{0j}k_{0m}}{k_0^2} \left(\delta_{2l} - \frac{k_{02}k_{0l}}{k_0^2}\right) S_{nm}(\boldsymbol{k}_1, t)\, S_{jl}(\boldsymbol{k}_0 - \boldsymbol{k}_1, t)$$
$$\times \exp\left\{-\frac{1}{2}[k_{1p}k_{1q} + (k_{0p} - k_{1p})]\,\langle u_p u_q\rangle\,(t - t_2)^2 - i\boldsymbol{k}_0\boldsymbol{u}(t - t_2)\right\} \tag{31}$$

and, consequently, upon substitution of the G_0^* functions from (13)

$$\Phi_{i\theta}^{(2)} = 2 \iint d\boldsymbol{k}_0\, d\boldsymbol{k}_1 \frac{k_{0i}k_{0j}k_{0n}k_{0m}}{k_0^2} \left(\delta_{2l} - \frac{k_{02}k_{0l}}{k_0^2}\right) S_{nm}(\boldsymbol{k}_1, t)\, S_{jl}(\boldsymbol{k}_0 - \boldsymbol{k}_l, t)$$
$$\times \int_0^t dt_1 \int_0^{t_1} dt_2 \exp\left\{-\frac{1}{2}[k_{1p}k_{1q} + (k_{0p} - k_{1p})\,(k_{0q} - k_{1q})]\,\langle u_p u_q\rangle\,(t - t_2)^2\right\} \frac{\partial T}{\partial x_2}. \tag{32}$$

The time integrations are now performed, first with respect to t_2 from 0 to t_1, then with respect to t_1 from 0 to the present time t. We then put $\boldsymbol{k}_0 = \boldsymbol{k}_1 + \boldsymbol{k}_2$ and use the incompressibility condition to obtain

$$\Phi_{i\theta}^{(2)} = 2 \frac{\partial T}{\partial x_2} \iint d\boldsymbol{k}_1\, d\boldsymbol{k}_2 \frac{(k_{1i} + k_{2i})\, k_{1j}k_{2m}k_{2n}}{(\boldsymbol{k}_1 + \boldsymbol{k}_2)^2} \left[\delta_{2l} - \frac{(k_{12} + k_{22})\, k_{1l}}{(\boldsymbol{k}_1 + \boldsymbol{k}_2)^2}\right]$$
$$\times \frac{S_{nm}(\boldsymbol{k}_1, t)\, S_{jl}(\boldsymbol{k}_2, t)}{(k_{1p}k_{1q} + k_{2p}k_{2q})\,\langle u_p u_q\rangle}. \tag{33}$$

This is the desired relationship between the turbulence-only part of the pressure-scalar-gradient correlation and the single-time spectrum functions $S_{ij}(\boldsymbol{k}, t)$. No assumption has been made regarding the anisotropy of the flow and the relationship is correct to all orders of anisotropy.

For isotropic turbulence and $i = 1$, $\Phi_{i\theta}^{(2)}$ is zero because the integrand in (33) is an odd function of $k_{11} + k_{21}$. The results of numerical integration using the approximations described below show that the integrand is very small for $i = 2$. These findings, and dimensional arguments, suggest that this part of the pressure correlation might be modelled to first order in anisotropy as

$$\Phi_{i\theta}^{(2)} = C_{\theta 2} \frac{\overline{q^2}}{2} b_{i2} \frac{\partial T}{\partial x_2}, \tag{34}$$

where $C_{\theta 2}$ is a coefficient which can be evaluated theoretically with the aid of modelled spectra.

$\Phi_{i\theta}^{(2)}$ to First Order in Anisotropy

The right-hand side of (33) is expanded in powers of anisotropy, retaining only the first-order terms, and the energy-spectrum and stress tensors are split into isotropic and anisotropic parts

$$S_{ij}(\boldsymbol{k}) = S_{ij}^{(I)}(\boldsymbol{k}) + S_{ij}^{(A)}(\boldsymbol{k}), \qquad \langle u_i u_j\rangle = \tfrac{1}{3}\,\delta_{ij} q^2 + \langle u_i u_j\rangle^{(A)}. \tag{35}$$

A Taylor series expansion of the denominator of (33) for small anisotropy

$$[(k_{1p}k_{1q} + k_{2p}k_{2q}) \langle u_p u_q \rangle]^{-1} = \frac{(k_1^2 + k_2^2)^{-1}}{v_0^2} \left[1 - \frac{\langle u_p u_q \rangle^A (k_{1p}k_{1q} + k_{2p}k_{2q})}{(k_1^2 + k_2^2)\, v_0^2} \right]^{-1} \tag{36}$$

gives the desired expression

$$\Phi_{i\theta}^{(2)} = 2 \frac{\partial T}{\partial x_2} \iint d\boldsymbol{k}_1 d\boldsymbol{k}_2 \frac{(k_{1i} + k_{2i})\, k_{ij} k_{2m} k_{2n}}{(\boldsymbol{k}_1 + \boldsymbol{k}_2)^2 (k_1^2 + k_2^2)\, v_0^2} \left[\delta_{2l} - \frac{(k_{12} + k_{22})\, k_{1l}}{(\boldsymbol{k}_1 + \boldsymbol{k}_2)^2} \right]$$
$$\times \left[S_{nm}^{(I)}(\boldsymbol{k}_1, t)\, S_{jl}(\boldsymbol{k}_2, t) + S_{nm}(\boldsymbol{k}_1, t)\, S_{jl}^{(I)}(\boldsymbol{k}_2, t) \right.$$
$$\left. - S_{nm}^{(I)}(\boldsymbol{k}_1, t)\, S_{jl}^{(I)}(\boldsymbol{k}_2, t) \frac{(k_{1p}k_{1q} + k_{2p}k_{2q}) \langle u_p u_q \rangle^{(A)}}{(k_1^2 + k_2^2)\, v_0^2} \right], \tag{37}$$

in which only the first-order terms are retained. It remains only to substitute the isotropic part of the spectrum tensor

$$S_{ij}^{(I)}(\boldsymbol{k}) = \left(\delta_{ij} - \frac{k_i k_j}{k^2} \right) \frac{E(k)}{4\pi k^2} \tag{38}$$

and to introduce

$$E_{ij}(k) = k^2 S_{ij}(\boldsymbol{k}) \int_0^\pi \sin\theta\, d\theta \int_0^{2\pi} d\phi. \tag{39}$$

The volume integrals of k_1 and k_2 are written in spherical coordinates and the continuity condition is used to express $\Phi_{i\theta}^{(2)}$ in terms of the anisotropy

$$\Phi_{1\theta}^{(2)} = 0.63 \langle u_1 u_2 \rangle \int_0^\infty \int_0^\infty \frac{k_1^2 k_2^2}{(k_1^2 + k_2^2)^2} \frac{E_{12}(k_1)}{\langle u_1 u_2 \rangle} \frac{E(k_2)}{3/2 v_0^2}\, dk_1 dk_2 \frac{\partial T}{\partial x_2}, \tag{40}$$

$$\Phi_{2\theta}^{(2)} = 0.36 \left(\langle u_2^2 \rangle - \frac{1}{3} q^2 \right) \int_0^\infty \int_0^\infty \frac{k_1^2 k_2^2 E_{22}(k_1)\, E(k_2)}{(k_2^2 + k_2^2)^2 \langle u_2^2 \rangle\, v_0^2}\, dk_1 dk_2 \frac{\partial T}{\partial x_2}. \tag{41}$$

The integrals on the right-hand side of this expression can be evaluated when the shapes of the spectra are known. For an approximate theoretical estimate of the coefficient $C_{\theta 2}$ in (34), and in the absence of spectral details in anisotropic turbulence, simple models will suffice.

Models of the Spectra

We again follow Weinstock [5] in adopting the two-range model of $E(k, t)$ proposed by Comte-Bellot and Corrsin [10] in their study of decaying grid turbulence. In the low wavenumber range containing the largest eddies E is assumed to vary as k^m and, at high Reynolds numbers, an inertial subrange exists in which E varies as $k^{-5/3}$. The intersection of the two curves defines the wavenumber k_L (roughly the reciprocal of the integral scale) at which $E(k, t)$ is a maximum. The spectrum is then defined by

$$\begin{aligned} E(k, t) &= \alpha \varepsilon^{2/3} (k_L^{-5/3-m})\, k^m \quad && \text{for } 0 < k \leqq k_L, \\ E(k, t) &= \alpha \varepsilon^{2/3} k^{-5/3} && \text{for } k \geqq k_L, \end{aligned} \tag{42}$$

where α is the Kolmogorov constant, assumed here to be 1.5. The exponent m has values of 2 or 4 according to different theories [11]. It turns out (as Weinstock also found) that the results of the calculation are insensitive to the precise value and we have assumed $m = 2$.

The corresponding scalar spectrum of the shear stress due to Kaimal et al. [12] is

$$\begin{aligned} E_{12}(k, t) &= \beta(k_L^{-7/5-m})\, k^m \quad \text{for } 0 < k \leqq k_L, \\ E_{12}(k, t) &= \beta k^{-7/3} \quad \text{for } k \geqq k_L, \end{aligned} \tag{43}$$

where $\beta = \langle u_1 u_2 \rangle\, k_L^{-4/3}(m+1)/[1 + 0.75(m+1)]$ is obtained from the condition $\int E_{12}(k)\, dk = \langle u_1 u_2 \rangle$. Finally, for the spectra of the diagonal components of the Reynolds stress tensor we assume

$$\begin{aligned} E_{ii}(k, t) &= \gamma(k_L^{-5/3-m})\, k^m \quad \text{for } 0 < k \leqq k_L, \\ E_{ii}(k, t) &= \gamma k^{-5/3} \quad \text{for } k \geqq k_L, \end{aligned} \tag{44}$$

where $\int E_{ii}(k)\, dk = \langle u_i u_i \rangle$.

Calculation of $C_{\theta 2}$

Substitution of these modelled spectra in (40) and (41) permits these equations to be integrated numerically and reduced to the precise form of (34) for the streamwise ($i = 1$) and cross-stream ($i = 2$) components of $\Phi_{i\theta}^{(2)}$. For $m = 2$, $C_{\theta 2}$ is evaluated as 0.079 for $i = 1$, and 0.124 for $i = 2$; the corresponding values for $m = 4$ are 0.078 and 0.120.

That different values of $C_{\theta 2}$ are obtained for the streamwise and cross-stream components is not altogether surprising because the wave-number exponents in the inertial subrange parts of the assumed spectra are different. A more fundamental reason is that the scalar-gradient part of the pressure correlation is not linear in the anisotropy as assumed in (34) and the subsequent analysis, but is more probably a function of the higher invariants as postulated by Lumley [2]. The linear model can only be regarded as a relatively crude, but potentially useful, approximation.

Considered in the context of current modelling practice the difference in the two values of $C_{\theta 2}$ cannot be regarded as very important. The main point is that the coefficient in the assumed linear model is (a) positive and (b) of order 0.1. This finding may be contrasted with estimate of $C_{\theta 2}$ as -0.23 made by Jones and Musonge [3].

Practical Modelling Considerations

In the last Section we have dealt only with the scalar-gradient part of the pressure-scalar-gradient correlation, the other parts being too intractable for this sort of theoretical calculation. We now combine the principal result (34) with models of the other terms suggested by dimensional considerations and the basic idea of linear dependence on the second moments and the anisotropy tensor. Equation (16) is then expressed in modelled form as

$$\begin{aligned} \left\langle \frac{p}{\varrho} \frac{\partial \theta}{\partial x_i} \right\rangle = &- C_{\theta 1} \frac{2\varepsilon}{q^2} \langle u_i \theta \rangle + C_{\theta 2} \left(\langle u_i u_k \rangle - \frac{1}{3} \delta_{ik} q^2 \right) \frac{\partial T}{\partial x_2} \\ &+ C_{\theta 3} \langle u_k \theta \rangle \frac{\partial U_i}{\partial x_k} + C_{\theta 4} \langle u_k \theta \rangle \frac{\partial U_k}{\partial x_i} + C_{\theta 5} \frac{q^2}{2\varepsilon} \left(\langle u_i u_l \rangle - \frac{1}{3} \delta_{il} q^2 \right) \frac{\partial U_l}{\partial x_k} \frac{\partial T}{\partial x_k} . \end{aligned} \tag{45}$$

The task is now to estimate the values of the unknown coefficients from the limited amount of experimental data currently available.

The first term on the right-hand side of (45) is analogous to Rotta's model of the turbulence-only part of the pressure-strain-rate correlation. In principle, the constant $C_{\theta 1}$ can be evaluated directly from measurements in grid turbulence with an imposed linear variation of the mean scalar (temperature) across the flow. The data of Sirivat and Warhaft [13] at two Reynolds numbers can be used to evalauted $C_{\theta 1}$ as approximately 2.5 while the measurements of Dakos [7], at a higher Reynolds number but with a weaker temperature gradient, suggest a value of 1.44. On the other hand, values of 3.0 and over have been used previously in calculation methods [14], probably because the effects of the extra scalar-gradient term (and in some cases the velocity-gradient term as well) have been absorbed. Jones and Musonge [3] do include a scalar-gradient term in a model which has a constant of 1.9 associated with the thermal time scale. This value appears to be consistent with $C_{\theta 1} = 2$ approximately when the thermal/mechanical time-scale ratio is taken as roughly 0.5, when it is noted that the Jones-Musonge formulation apparently involves another factor of two.

In order to estimate the other constants we turn to local-equilibrium flow data and the measurements of Tavoularis and Corrsin [4] in a nearly homogeneous turbulence with uniform mean velocity and temperature gradients. In the absence of significant turbulent transport the scalar-flux equations can be written in the dimensionless form

$$\frac{q^2}{2\varepsilon}\frac{U_1}{\langle u_2\theta\rangle}\frac{\partial\langle u_1\theta\rangle}{\partial x_1} = -C_{\theta 1}\frac{\langle u_1\theta\rangle}{\langle u_2\theta\rangle} - C_{\theta 2}\frac{q^2}{2\langle u_1u_2\rangle}\frac{P}{\varepsilon}\mathrm{Pr}_t - C_{\theta 3}\frac{q^2}{2\langle u_1u_2\rangle}\frac{P}{\varepsilon}$$
$$+ C_5 b_{11}\left(\frac{q^2}{2\langle u_1u_2\rangle}\right)^3\left(\frac{P}{\varepsilon}\right)^2\mathrm{Pr}_t + \frac{q^2}{2\langle u_1u_2\rangle}\frac{P}{\varepsilon}(1+\mathrm{Pr}_t), \tag{46}$$

$$\frac{q^2}{2\varepsilon}\frac{U_1}{\langle u_2\theta\rangle}\frac{\partial\langle u_2\theta\rangle}{\partial x_1} = -C_{\theta 1} - C_{\theta 2}b_{22}\left(\frac{q^2}{2\langle u_1u_2\rangle}\right)^2\frac{P}{\varepsilon}\mathrm{Pr}_t - C_{\theta 4}\frac{\langle u_1\theta\rangle}{\langle u_2\theta\rangle}\frac{P}{\varepsilon}$$
$$+ C_{\theta 5}\left(\frac{q^2}{2\langle u_1u_2\rangle}\right)^2\left(\frac{P}{\varepsilon}\right)^2\mathrm{Pr}_t + \frac{2\langle u_2^2\rangle}{q^2}\left(\frac{q^2}{2\langle u_1u_2\rangle}\right)^2\frac{P}{\varepsilon}\mathrm{Pr}_t, \tag{47}$$

in which P/ε is the ratio of turbulent energy production to dissipation, and Pr_t is the turbulent Prandtl number. If, as a result of the preceeding analysis, $C_{\theta 2}$ is given an average value of 0.1, the remaining four constants may, in principle, be found from the two sets of equations corresponding to $P/\varepsilon = 1.0$ for local equilibrium conditions and $P/\varepsilon = 1.72$ in the Tavoularis-Corrsin flow. It turns out, however, that the values so obtained are highly sensitive to the input data, particularly to Pr_t, whose value in local equilibrium turbulence is, at best, uncertain. Moreover, solutions for a range of plausible values of the quantities in (46) and (47) show that the contribution from the $C_{\theta 5}$ term is fairly insignificant and that there is little point in retaining it as an unecessary complication. With $C_{\theta 5} = 0$ and $C_{\theta 1}$ taken as about 2.0 in line with the discussion above, it is found that values of the constants

$$(C_{\theta 1}, C_{\theta 2}, C_{\theta 3}, C_{\theta 4}, C_{\theta 5}) = (2.1, 0.1, 0.96, 0.5, 0)$$

are consistent with the experimental data presented in Table 1. For local-equilibrium conditions the anisotropies are estimated from the measurements of Champagne et al. [15] in weakly-sheared homogeneous turbulence; the scalar-flux ratio and turbulent Prandtl number are taken from Launder's [14] survey of heat-transfer measurements.

Table 1. Values of turbulence quantities used to determine the model constants from (46) and (47)

Turbulence Quantity	Local Equilibrium [14, 15]	Tavoularis-Corrsin [4]
P/ε	1.0	1.72
b_{11}	0.30	0.40
b_{22}	− 0.18	− 0.295
b_{12}	− 0.32	− 0.284
$\langle u_1 \theta \rangle / \langle u_2 \theta \rangle$	− 1.10	− 2.23
Pr_t	0.775	1.12

It has been noted that the first two terms in (16) together make up the turbulence-only part of the volume integral in (2). With the above constants the effect of the scalar-gradient component is to reduce the relative (positive) contribution of the combined term in the $\langle u\theta \rangle$ equation by (by 17% and 35% for $P/\varepsilon = 1.0$ and 1.72 respectively) and to increase the relative (negative) contribution in the $\langle v\theta \rangle$ equation (by 8% and 38%).

Acknowledgement. We gratefully acknowledge financial support provided by the Science and Engineering Reasearch Council through Grant No. GR/B83674.

References

1. Rotta, J. C.: Statistische Theorie nichthomogener Turbulenz. Z. Phys. **129**, 547 (1951)
2. Lumley, J. L.: Computational modelling of turbulent flows. Adv. Appl. Mech. **18**, 123 (1978)
3. Jones, W. P., Musonge, P.: Modelling of scalar transport in homogeneous turbulent flows, Proceedings of the 4th Turbulent Shear Flows Symposium, Karlsruhe (1983)
4. Tavoularis, S., Corrsin, S.: Experiments in nearly homogeneous turbulence with a unitform mean temperature gradient. J. Fluid Mech. **104**, 311 (1981)
5. Weinstock, J.: Theory of the pressure strain rate correlation for Reynolds-stress turbulence closures. Part 1. Off-diagonal element. J. Fluid Mech. **105**, 369 (1981)
6. Weinstock, J.: Theory of the pressure-strain rate. Part 2. Diagonal elements. J. Fluid Mech. **116**, 1 (1982)
7. Dakos, T.: Fundamental heat-transfer studies in grid-generated homogeneous turbulence. Ph.D. Thesis, Imperial College, London (in preparation)
8. Kraichnan, R.: The structure of isotropic turbulence at high Reynolds numbers. J. Fluid Mech. **5**, 497 (1959)
9. Meecham, W. C.: Turbulence-energy principles for quasi-normal and Weiner-Hermite expansions. Phys. Fluids **8**, 1738 (1965)
10. Comte-Bellot, G., Corrsin, S.: The use of a contraction to improve the isotropy of grid-generated turbulence. J. Fluid Mech. **25**, 657 (1966)
11. Hinze, J. O.: *Turbulence,* 2nd ed. (McGraw-Hill, New York 1975)
12. Kaimal, J. C., Wyngaard, J. C., Izumi, Y., Cote, O. R.: Spectral characteristics of surface-layer turbulence. Q. J. Roy. Met. Soc. **98**, 563 (1972)
13. Sirivat, A., Warhaft, Z.: The effect of a passive cross-stream temperature gradient on the evolution of temperature variance and heat flux in grid turbulence. J. Fluid Mech. **120**, 475 (1982)
14. Launder, B. E.: Heat and Mass Transport in Turbulence, in *Turbulence*, ed. by P. Bradshaw, Topics in Appl. Phys., Vol. 12, 2nd ed. (Springer, Berlin, Heidelberg 1978) Chap. 6
15. Champagne, F. H., Harris, V. B., Corrsin, S.: "Experiments on nearly homogeneous turbulent shear flow". J. Fluid Mech. **41**, 81 (1970)

A Simplified Spectral Closure for Inhomogeneous Turbulence: Application to the Boundary Layer

Jean-Pierre Bertoglio and Denis Jeandel

Laboratoire de Mécanique des Fluides, Ecole Centrale de Lyon 36, Avenue Guy de Collongue, F 69131 Ecully Cedex, France

Abstract

A turbulence model is proposed for the prediction of inhomogeneous flows. A spectral closure is used to predict the energy transfer to the dissipative scales. The inhomogeneous transport terms are evaluated within the framework of one-point closures. Rough approximations are made concerning the anisotropy of turbulence. In the case of a flat plate boundary layer, the model is found to compare favorably with the experimental data. Possible improvements are discussed.

Nomenclature

$\langle\ \rangle$	Statistical average
a_{KPQ}	$=\frac{1}{2}(b_{KPQ}+b_{KQP})$
b	Constant
b_{KPQ}	Coefficient depending on the geometry of a triad in wave space
$D(K, X, t)$	Transport term in the energy spectrum equation
$E(K, X, t)$	Energy spectrum
$g(K, K_{\max})$	Modeled spectral shape for the transport term D
$\boldsymbol{K}$	Wave-vector
K	Wave-number, $=\lvert\boldsymbol{K}\rvert$
$K_{\min}$	Infrared cutoff for the spectral computation
$K_{\max}$	Wave-number corresponding to the maximum of the energy spectrum
k	Turbulent kinetic energy
$\mathbb{K}$	Constant
L	Integral length scale
$\langle q^2\rangle$	Twice the turbulent kinetic energy
r	Constant
$T(K, X, t)$	Energy transfer
$U_i(X, t)$	Velocity field
u^*	Shear stress velocity
δ	Boundary layer thickness
σ_1, σ_2	Constants
$\Phi_{ij}(\boldsymbol{K}, X, t)$	Spectral tensor, Fourier transform of the double velocity correlation at x and x', with $X=\frac{x+x'}{2}$
$\varphi_{ij}(k, X, t)$	Integral of Φ_{ij} over a spherical shell of radius K
$\phi_{ij,l}$	Fourier transform of the triple correlation
λ	Constant
ω	Dimensionless stream function, $=\frac{\psi-\psi_I}{\psi_E-\psi_I}$
π_i	Fourier transform of the pressure velocity correlation
ψ	Stream function
μ	Constant
η	Damping factor
ν	Kinematic viscosity
θ_{KPQ}	E.D.Q.N.M. characteristic time

Introduction

During the last few years, two-point closures and particularly the Eddy Damped Quasi-Normal Markovian theory (Orszag, 1970), have been applied to the study of homogeneous anisotropic turbulence subjected to uniform mean velocity gradients (see for example Cambon et al., 1981 a, Bertoglio, 1982 or Weinstock, 1981 and 1982). The results of such methods were compared with experimental data, particularly at the 1981 Stanford Conference (Cambon et al., 1981 b) and the agreement was found to be reasonably good. Extensions to inhomogeneous and wall-bounded fields are now a challenging problem since, at the moment, one-point closures, such as $k - \varepsilon$ models, are still regarded as the only efficient tools for computing complex turbulent flows.

As a matter of fact, the complete set of equations resulting from the inhomogeneous formulation of two-point closures is very complicated (see for example Kraichnan, 1972). It has been numerically integrated only in the simplest case, in which non-linear effects are neglected (Rapid Distortion Theory: Hunt, 1978, Goldstein and Durbin, 1980 and Hamadiche, 1981). A non-linear computation directly based on the complete set of equations would probably require as much computational effort as a Large Eddy Simulation or even a Full Direct Simulation. Therefore simplified models are to be found in order to reduce the numerical task to an acceptable level.

Simplifications can be introduced at many different levels. One can attempt to introduce only slight approximations; this is, for example, the route chosen by Dannevik (1984) in the case of thermal convection (see also Herring, 1985). In the present paper we adopt a completely different point of view: it consists in making rough assumptions and devising a model which would be only one step further in sophistication than the standard $k - \varepsilon$ model. The model only retains from the two-point closures their description of the spectral energy cascade. Any other effects are approximated in the framework of one-point models. The E.D.Q.N.M. closure for homogeneous and isotropic turbulence is thus used to predict the energy transfer between wavenumbers, whereas inhomogenous transport terms are evaluated in physical space, with the help of the classical one-point formulations ($k - \varepsilon$ model, or more precisely a $k - kL$ model in the case of the present work).

The present model must be considered as an attempt to show that two-point closures can provide practical tools for computing complex flows. Obviously many of the assumptions introduced are very questionable. One could argue for example that, due to the high degree of organization existing in wall-bounded flows, the mechanisms of energy transfer in inhomogeneous flows are totally different from what they are in freely decaying turbulence, and that they cannot be subsumed under an isotropic approximation. It can be reasonably answered that the same sort of approximation exists in one point models: the terms accounting for the energy transfer are modelled with reference to the isotropic decay of turbulence.

It is hoped that the present model would be as efficient as the $k - \varepsilon$ model in the case of strong inhomogeneities, and as accurate as the E.D.Q.N.M. closure for nearly homogeneous and isotropic flows. The model does not involve an equation for the turbulent dissipation of kinetic energy (or an equation for a length-scale). This is a major advantage since it is known that this kind of equation is the weakest point in the $k - \varepsilon$ technique. In particular the turbulent field is no longer assumed to have a spectrum of universal form that could be characterized completely by the scales of the energy containing range (see Lumley, 1983). A different time scale is introduced at each wavenumber.

The model is applied here only to the simplest case of a flat plate boundary layer and very crude approximations are introduced concerning the field anisotropy. Many improvements of the method could be devised. They are briefly discussed in the last section.

Equations and Approximations

By writing the equation for the double velocity correlation at two points $\boldsymbol{x}$ and $\boldsymbol{x}'$, and by Fourier transforming with respect to $\boldsymbol{r} = (\boldsymbol{x}' - \boldsymbol{x})/2$, the equation for the double velocity correlation sepctrum is obtained. Using Taylor developments of the two-point correlations in the vicinity of $X = (x + x')/2$, we obtain (Ménoret, 1982)

$$
\begin{array}{lll}
\multicolumn{3}{l}{\dfrac{\partial}{\partial t}\,\Phi_{ij}(\boldsymbol{K}, \boldsymbol{X}, t)} \\[2ex]
\quad = -2\nu K^2 \Phi_{ij} \qquad\qquad (a) & + 2\nu \dfrac{\partial 2 \Phi_{ij}}{4\partial X_1^2} \qquad (b) & \\[2ex]
\quad - \left(\dfrac{\partial \langle U_i \rangle}{\partial X_1} \Phi_{1j} + \dfrac{\partial \langle U_j \rangle}{\partial X_1} \Phi_{i1} \right) \quad (c) & - \langle U_1 \rangle \dfrac{\partial \Phi_{ij}}{\partial X_1} \qquad (d) & \\[2ex]
\quad + \dfrac{\partial \langle U_1 \rangle}{\partial X_m} K_1 \dfrac{\partial \Phi_{ij}}{\partial K_m} \qquad (e) & + \dfrac{\partial^2 \langle U_1 \rangle}{\partial X_m \partial X_n} \left\{ \dfrac{1}{8} \dfrac{\partial}{\partial X_1} \dfrac{\partial^2 \Phi_{mj}}{\partial K_m \partial K_n} \right. & \text{Linear terms} \\[2ex]
 & \left. + \cdots + \dfrac{i}{2} \dfrac{\partial}{\partial K_n} (\delta_{i1} \Phi_{mj} - \delta_{j1} \Phi_{mi}) \right\} \quad (f) & \\[2ex]
\quad + i K_1 (\phi_{i1,j} - \phi^*_{j1,i}) \qquad (g) & - \dfrac{1}{2} \dfrac{\partial}{\partial X_1} (\phi_{i1,j} + \phi^*_{j1,i}) \qquad (h) & \text{Non Linear terms} \\[2ex]
\quad + i K_1 (\delta_{i1} \pi_j - \delta_{j1} \pi_i^*) \qquad (i) & - \dfrac{\partial}{2\partial X_1} (\delta_{i1} \pi_j + \delta_{j1} \pi_i^*) \qquad (j) & \text{Pressure terms} \\[1ex]
\quad \text{Homogeneous terms} & \text{Inhomogeneous terms} &
\end{array}
\qquad (1)
$$

in which the $\phi_{ij,k}$ and π_i are unknown terms respectively related to triple velocity correlations and pressure velocity correlations. $\langle U_i \rangle$ denotes the mean velocity. All the quantities are functions of the wave-vector $\boldsymbol{K}$, of time t, and of a position variable $\boldsymbol{X}$ which takes into account the turbulent field inhomogeneity.

A first simplification of (1) consists in integrating over the directions of the wave-vector $\boldsymbol{K}$. An average over spherical shells in wave space was thus introduced by Ménoret (1982), who derived the corresponding equation for

$$\varphi_{ij}(K, X, t) = \int_{\Sigma(K)} \Phi_{ij}(\boldsymbol{K}, \boldsymbol{X}, t)\, d\Sigma,$$

in which $\Sigma(K)$ is the spherical shell of radius K.

The route suggested by Ménoret was to close the resulting equation using the E.D.Q.N.M. theory. Although this kind of approach was successful for homogeneous anisotropic turbulence (Cambon et al., 1981 a), it led to some disillusionment in the case of wall-bounded flows, due to the excessive number of terms to be modeled.

A semi-empirical but straightforward way out of this difficulty is proposed in the present paper. We suggest to work with the equation for the turbulent kinetic energy spectrum

$$E(K, X, t) = \tfrac{1}{2}\, \varphi_{ii}(K, X, t),$$

written in the following simplified form:

$$\left\{ \frac{\partial}{\partial t} + \langle U_m \rangle \frac{\partial}{\partial X_m} \right\} E(K, X, t) = -2\nu K^2 E(K, X, t) - \frac{\partial \langle U_i \rangle}{\partial X_m} \varphi_{im}(K, X, t) + T(K, X, t) + D(K, X, t). \qquad (2)$$

In (2), the advection term $\langle U_m \rangle\ \partial E/\partial X_m$; the dissipation term $-2\nu K^2 E$, and the production term $\partial\langle U_i\rangle/\partial X_m \varphi_{im}$ are exact terms, respectively corresponding to terms (d), (a) and (c) in (1).

The linear transfer associated with term (e) in (1) has been discarded, as well as the term involving second order derivatives of the mean velocity profile (f).

T and D are modeled forms corresponding to the remaining terms of (1).

T is the transfer term accounting for the energy cascade. It comes from term (g).

The basic idea is to evaluate T as if the turbulent field were homogeneous and isotropic. It is known that, for this type of field, the E.D.Q.N.M. expression for T is (Orszag, 1970)

$$T(K, X, t) = \frac{K^2}{2\pi} \int d^3P\, d^3Q\, \theta_{KPQ}\, \delta(-\boldsymbol{K} + \boldsymbol{P} + \boldsymbol{Q})$$
$$\times \left(a_{KPQ} \left(\frac{K}{PQ}\right)^2 E(P, X, t)\, E(Q, X, t) - b_{KPQ}\, \frac{1}{Q^2}\, E(Q, X, t)\, E(K, X, t) \right), \quad (3)$$

in which

$$b_{KPQ} = \frac{P}{K}(xy + z^3)$$

and

$$a_{KPQ} = \tfrac{1}{2}(b_{KPQ} + b_{KQP}).$$

x, y and z are the cosines of the interior angles oposite K, P and Q in the $\boldsymbol{K}$, $\boldsymbol{P}$, $\boldsymbol{Q}$ triangle. θ_{KPQ} is a characteristic time for the triple correlations

$$\theta_{KPQ}^{-1} = \eta(P) + \eta(Q) + \eta(K) \quad (4)$$

with

$$\eta(K) = \lambda \left(\int_0^K R^2 E(R, X, t)\, dR \right)^{1/2} + \nu K^2, \quad (5)$$

λ being a constant.

In order to account for the effect of a solid body rotation on the triple correlations, Cambon et al. (1981 b) have proposed a modified expression for η in which the rotational part of the mean field is included:

$$\eta(K) = \lambda \left[\int_0^K R^2 E(R, X, t)\, dR + \frac{1}{4} \left(\frac{\partial\langle U_i\rangle}{\partial X_j} - \frac{\partial\langle U_j\rangle}{\partial X_i} \right) \left(\frac{\partial\langle U_i\rangle}{\partial X_j} - \frac{\partial\langle U_j\rangle}{\partial X_i} \right) \right]^{1/2} + \nu K^2. \quad (5')$$

Both expressions (5) and (5′) will be tested in the present method.

The last term in (2), $D(K, X, t)$, is a typically inhomogeneous transport term. It involves pressure (j) and triple velocity correlation (h) effects, as well as a viscous contribution. The complete expression for D is rather complex. A crude approximation is retained here. Let us first remark that the equation governing the turbulent kinetic energy,

$$k = \tfrac{1}{2}\langle q^2 \rangle = \int_0^\infty E(K, X, t)\, dK,$$

is obtained by integrating (2) over the wave-number K. Therefore identification with the equation for k, closed in the framework of classical one-point models, allows the integral of

D to be evaluated. Using the models of Ng and Spalding (1976) or Jeandel (1976) (see also Jeandel et al., 1978) leads to

$$\int_0^\infty D(K, \boldsymbol{X}, t)\, dK = \frac{\partial}{\partial X_1}\left\{\left(\frac{(\langle q^2\rangle/2)^{1/2} L}{\sigma_1} + \nu\right) \frac{\partial \langle q^2\rangle/2}{\partial X_1}\right\}, \tag{6}$$

where σ_1 is a constant and L an integral length scale defined as

$$L = \frac{3\pi}{8} \int_0^\infty K^{-1} E(K, \boldsymbol{X}, t)\, dK \Big/ \int_0^\infty E(K, \boldsymbol{X}, t)\, dK.$$

A similar relation can be obtained by dividing (2) by K before integrating and identifying with the equation for the energy-length product $\langle q^2\rangle L$. We obtain

$$\frac{3\pi}{8} \int_0^\infty \frac{1}{K} D(K, \boldsymbol{X}, t)\, dK = \frac{\partial}{\partial X_1}\left\{\left(\frac{(\langle q^2\rangle/2)^{1/2} L}{\sigma_2} + \nu\right) \frac{\partial \langle q^2\rangle L/2}{\partial X_1}\right\}, \tag{7}$$

where σ_2 is another constant.

This suggests that we may model $D(K, \boldsymbol{X}, t)$ in the following way: a plausible spectral shape is first assumed for D and then we impose the two conditions (6) and (7). A simple form for D is:

$$D(K, \boldsymbol{X}, t) = \alpha(\boldsymbol{X}, t)\, g(K, K_{max}), \tag{8}$$

where K_{max} is the wave-number corresponding to the maximum of $E(K, \boldsymbol{X}, t)$ and where

$$\begin{aligned} g(K, K_{max}) &= K^2 && \text{if } K \leqq r K_{max}, \\ g(K, K_{max}) &= (r K_{max})^2 \left(\frac{K}{r K_{max}}\right)^{-b} && \text{if } K > r K_{max}. \end{aligned} \tag{9}$$

The quantities r and b are adjustable parameters. A value of b large enough to preserve numerical stability in the large wave-number range is adopted: $b = 2$. r is supposed to be equal to 1 for the sake of simplicity. At each value of $\boldsymbol{X}$ and t, application of the constraint (6) permits determination of α.

A more elaborate form is needed if (7) has also to be satisfied. The following shape is finally adopted:

$$D(K, \boldsymbol{X}, t) = \alpha(\boldsymbol{X}, t)\, g(K, K_{max}) + A(\boldsymbol{X}, t) \frac{\partial E(K, \boldsymbol{X}, t)}{\partial K}, \tag{10}$$

$A(\boldsymbol{X}, t)$ being determined by (7).

The only unknown term in (2) is now the production term

$$\frac{\partial \langle U_i \rangle}{\partial X_m} \varphi_{im}(K, \boldsymbol{X}, t);$$

it has also to be modeled since only the energy spectrum is known, but not the whole tensor φ_{ij}.

For shear dominated flows, such as boundary layers on flat plates, where

$$\frac{\partial \langle U_i \rangle}{\partial X_m} = \frac{\partial \langle U_1 \rangle}{\partial X_2} \delta_{i1} \delta_{m2},$$

φ_{12} is the only component appearing in the production term. To model this component, we simply assume the Reynolds stress to be directly proportional to the local turbulent kinetic energy:

$$\langle u_1 u_2 \rangle = \int_0^\infty \varphi_{12}(K, X, t)\, dK = -0.15 \langle q^2 \rangle. \tag{11}$$

We furthermore assume that the Reynolds stress spectrum has the following shape

$$\begin{aligned} \varphi_{12}(K, X, t) &= B && \text{for } K \leqq K_{\max}, \\ \varphi_{12}(K, X, t) &= B\left(\frac{K}{K_{\max}}\right)^{-7/3} && \text{for } K > K_{\max}. \end{aligned} \tag{12}$$

The $K^{-7/3}$ behaviour is supported by the results of a previous study (Bertoglio, 1981). The value of B is deduced from (11).

Such a modeling of the anisotropy of the turbulent spectrum certainly constitutes a poor approximation. Possible improvements and extensions to other types of mean velocity gradients will shortly be discussed in the last section.

At this point, all the terms appearing in (2) have been modeled. (2), together with the mean velocity equations, form a closed set, since the quantities $\langle q^2 \rangle$ and $\langle q^2 \rangle L$ can be determined by integrating $E(K, t)$. However there is still one effect which has not been taken into account. This is the limitation of the large scale growth imposed by the presence of the wall. This effect is taken into account by the pressure term (j) in (1), but certainly not by the corresponding modeled term in (2) since a diffusive form is used for D.

We suggest to model this external scaling in a straightforward way, by simply not allowing energy to be transferred to wavenumbers smaller than some limit. This limit is assumed to be inversely proportional to the distance from the wall X_2:

$$K_{\min} = \mu / X_2.$$

Such a condition can easily be implemented in an E.D.Q.N.M. computational code by simply discarding triadic interactions involving wavenumbers smaller than $K_{\min}$. In other words, we impose the largest possible scale of the turbulent field as a function of X_2. This kind of "infrared cutoff" is an important condition, since if it were not imposed, the spectrum would behave like a homogeneous field submitted to a uniform shear flow: energy would eventually always grow and the balance between production and dissipation would never be reached.

The value of μ can be determined by assuming that production equals dissipation in the logarithmic zone of the boundary layer and that a Kolmogorov inertial range exists in the spectrum. We obtain

$$\mu = 0.9.$$

The constants appearing in the model are σ_1, σ_2, b, r, μ and the E.D.Q.N.M. constant λ, the usual value of which is 0.355. We already saw that b and r are arbitrarily determined.

Table 1. The typical values of the constants

Constants	σ_1	σ_2	μ	λ	b	r
Values	1	1	0.8	0.355	2	1

Typical values for the diffusive constants are chosen equal to

$$\sigma_1 = 1$$

and

$$\sigma_2 = 1$$

since this figure seems to correspond to the order of magnitude of the values usually introduced in one-point closures (Rodi and Spalding, 1970; Ng and Spalding, 1972 and 1976; Jeandel et al., 1978). However, one must be careful when selecting the values of the diffusive constants: they are not necessarily equal to those used in $k - kL$ models since L is now defined as a true integral length scale (at least for isotropic turbulence), which was not the case for earlier models (see Jeandel 1976). The influence of the choice of σ_1 and σ_2 will be discussed in the next section.

The retained set of constant values is tabulated in Table 1.

The effects of a variation of μ between 0.6 and 1.1 will also be investigated.

Boundary-Layer Application

The model is applied to the computation of a two-dimensional boundary layer. The numerical method is the one proposed by Jeandel (1976). An $X_1 - \omega$ coordinate system is used, ω being the dimensionless stream function. ω is defined as follows (Patankar and Spalding, 1967):

$$\omega = \frac{\psi - \psi_I}{\psi_E - \psi_I},$$

where ψ_I and ψ_E are respectively the values of the stream function at the wall and at the external boundary of the layer. The solution procedure employed to solve the mean velocity equation and to evaluate the inhomogeneous turbulent terms is a finite-difference method.

The spectral equation is integrated along the iso-ω curves. In order to simplify the numerical method, $K_{\min}$ is assumed to be constant on these lines, an approximation justified only if the boundary layer thickness does not vary too much during the computation.

The free-stream velocity and the initial data which are chosen correspond to the experiment of Klebanoff (1955). Computations are run between $X_{1_i} = 3.2$ m and $X_{1_f} = 3.6$ m, X_1 being the distance from the leading edge of the boundary layer.

Initial Profiles

The initial data for $\langle q^2 \rangle/2$ correspond to Klebanoff's measurements. An initial profile for the length scale is also estimated to fit with Klebanoff's data. Initial spectra are thereafter assumed to have the following shape:

$$E(K, X_{1_i}, X_2) = 0 \qquad \text{for} \quad K \leqq K_{\min},$$

$$E(K, X_{1_i}, X_2) \sim K^s \qquad \text{for } K_{\min} < K \leqq K_{\max},$$

$$E(K, X_{1_i}, X_2) = C\varepsilon^{2/3} K^{-5/3} \qquad \text{for } K_{\max} < K \leqq K_\eta,$$

$$E(K, X_{1_i}, X_2)\text{: strongly decreasing with } K \qquad \text{for} \quad K > K_\eta,$$

with $2 \leqq s \leqq 4$; $C = 1.5$ denotes the Kolmogoroff constant and ε is the dissipation (values for ε are defined by writing $\varepsilon = C_D \dfrac{\langle q^2/2\rangle^{3/2}}{L}$, with $C_D = 0.155$, a value consistent with the chosen shape for the spectra). $K_{\max}$ and K_η are then determined at each point of the initial profile by the relations

$$\int_0^\infty E(K, X_{1_i}, X_2)\, dK = \langle q^2(X_{1_i}, X_2)\rangle/2,$$

$$2\nu \int_0^\infty K^2 E(K, X_{1_i}, X_2)\, dK = \varepsilon(X_{1_i}, X_2).$$

Boundary conditions

The same kind of spectrum is used for the near wall boundary conditions. The first point being situated in the logarithmic region, we furthermore assume

$$\frac{\langle U_1\rangle}{u^*} = \frac{1}{\mathbb{K}} \operatorname{Log} \frac{X_2 u^*}{\nu} + F$$

$$-\langle u_1 u_2\rangle = 0.15\,\langle q^2\rangle \simeq u^{*2},$$

$$L = \chi X_2,$$

with $\chi = \dfrac{\mathbb{K} C_D}{(0.3)^{3/2}}$ and $\mathbb{K} = 0.41$.

At the free stream edge of the boundary layer the turbulent kinetic energy is supposed to be equal to zero. Of course, one could also work with finite values of $\langle q^2\rangle$, corresponding to a finite value of the free-stream turbulence. In this case a spectrum corresponding to the external turbulence would have to be specified. The mesh size in the X_1 direction is chosen to be $\Delta X_1 = 0.03\,\delta$ where δ is the boundary layer thickness. The results of predictions presented in the following section were obtained with 16 cross-stream grid points. Wave-numbers were discretized using a number of spectral points increasing from 9, in the vicinity of the wall, to 21, at the free stream boundary.

It appeared that it was not necessary to compute $T(K, X)$ at each section, except in the near wall region. Hence, in the outer part of the layer, $T(K, X)$ was assumed to be constant on $3\,\Delta X_1$ steps.

Results

Results concerning the energy spectra are plotted in Fig. 1. They were obtained with the constant values presented in Table 1. It can be seen that the wave-number corresponding to

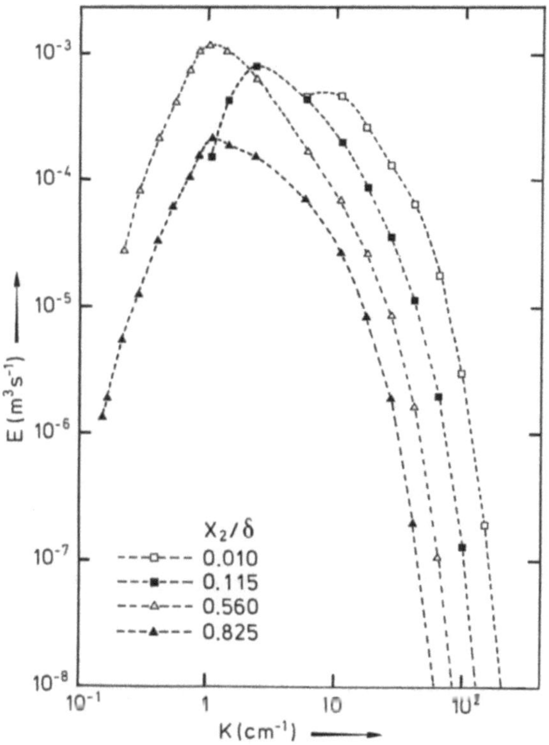

Fig. 1. Kinetic energy spectrum at different X_2/δ across the boundary layer. Results obtained with the values of constants appearing in Table 1

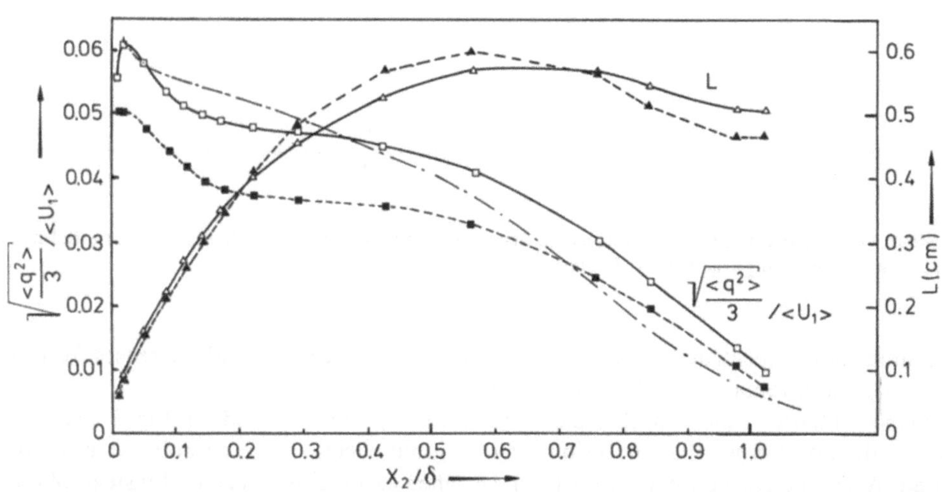

Fig. 2. Turbulent intensity and integral length scale profiles; model modified to account for mean rotational effects: –□–, –△–; and model not modified: --■--, –▲–; at $X_1 = X_{1f}$. Results corresponding to the values of the constants appearing in Table 1. Comparison with Klebanoff's experiment: — · —

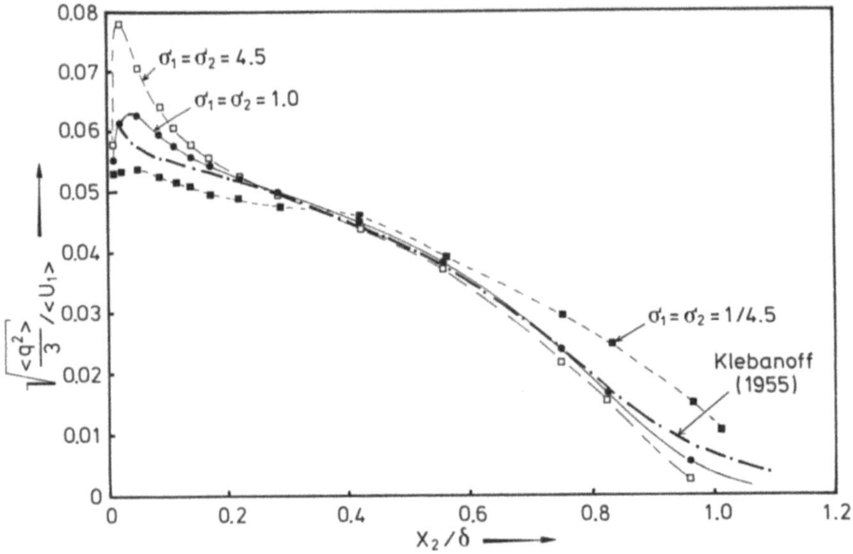

Fig. 3. Influence of the values of the diffusive constants σ_1 and σ_2; Turbulence intensities at $X_1 = 3.38$ m. Comparison with Klebanoff's data: —·—

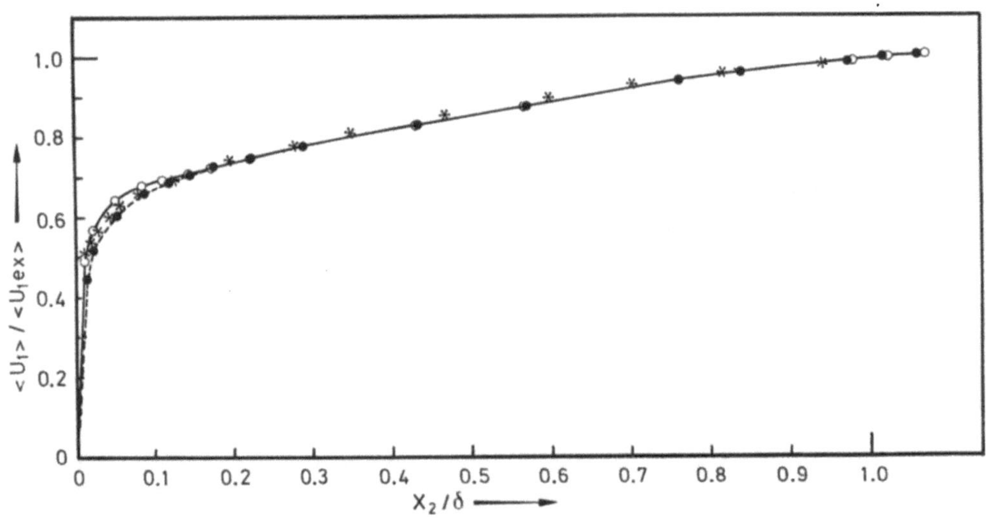

Fig. 4. Mean velocity profile; model modified to account for mean rotational effects: —o—; model not modified: --●--; and comparison with Klebanoff's data: *

the maximum of the spectrum decreases when the distance from the wall increases from 0 to $X_2/\delta = 0.56$ and remains nearly constant for larger distances.

Turbulence intensity and length-scale profiles at X_{1f} are plotted in Fig. 2. Results obtained with and without the correction for rotational effects in the damping term η are compared. When the correction is introduced, i.e. when expression (5′) is used instead of (5), larger values for the turbulent kinetic energy are obtained. Such a behaviour can be related to the fact that the energy transfer is then reduced. Agreement with Klebanoff's measurements seems to be improved when (5′) is used.

When the rotational effects are accounted for, the mean velocity profile is slightly modified: on Fig. 3 it appears that $\langle U_1 \rangle$ is affected only in the inner part of the boundary layer.

The effects of a change of the values of the diffusive constants σ_1 and σ_2 has briefly been investigated. Values corresponding to Jeandel et al. (1978): $\sigma_1 = \sigma_2 = 1/4.5$ lead to overestimation of the diffusive effect (Fig. 4) whereas the choice of $\sigma_1 = \sigma_2 = 4.5$ results in insufficient diffusional transport.

The choice of the value of the constant μ affects the behaviour of turbulence only in the inner part of the layer.

From Fig. 2–4 one can infer that agreement with experimental data seems to be reasonably good, especially if we consider that no real adjustment of the constants has been performed and that there is probably still room for improvement.

Discussion

The model has been applied only to flat plate boundary layers. Encouraging results were obtained. Now the method must be extended to other types of flows, and in particular should be used to predict situations in which a spectral approach is likely to provide better results than the $k - \varepsilon$ technique. Let us for instance mention flows in which peaks occur in the spectra, or flows involving non constant viscosities, due for example to the presence of polymers or to the action of a compression (Wu et al., 1985) or of combustion.

One of the weakest points in the present method is probably the crude modeling of anisotropy. Improvements can be proposed. One could suggest the introduction of an equation for $\langle u_1 u_2 \rangle / \langle q^2 \rangle$, or for the Reynolds stress tensor

$$b_{ij} = \langle u_i u_j \rangle / \langle q^2 \rangle - \delta_{ij}/3$$

and the use of one-point closure or algebraic stress type expressions. It seems more convenient to use a spectral approximation. This could be done by introducing an equation for φ_{ij} and by using the spherically averaged model of Cambon et al. (1981 a). An alternative procedure is to use Rapid Distortion Theory together with the concept of effective strain (see Maxey, 1982). The corresponding equations are now being implemented in the numerical code. Since the energy spectrum is explicitly predicted by the present model, a spectral characteristic time is now available and it becomes possible to extend Maxey's method by introducing different effective strains corresponding to the different wavenumbers.

One of the advantages of a spectral determination of the anisotropy is that it will provide a way to evaluate the linear transfer term (e) which has been neglected in the present version of the model.

A possible extension of the model to temperature fluctuations could permit discussion of Prandtl number effects. The information provided by the present model could also be useful for subgrid models in Large Eddy Simulation.

References

Bertoglio, J.-P. (1982): A model of three-dimensional transfer in non-isotropic homogeneous turbulence, in *Turbulent Shear Flows 3*, ed. by L. J. S. Bradbury, F. Durst, B. E. Launder, F. W. Schmidt, J. H. Whitelaw (Springer, Berlin, Heidelberg) p. 253

Cambon, C., Jeandel, D., Mathieu, J. (1981 a): Spectral modeling of homogeneous non-isotropic Turbulence. J. Fluid Mech. **104**, pp. 247–262

Cambon, C., Bertoglio, J.-P., Jeandel, D., (1981 b); Spectral closure of homogeneous turbulence, AFOSR-HTTM-Stanford Conf. on Complex Turb. Flows, Sept. 1981
Dannevik, W., (1984): Two-point closure study of covariance budgets for turbulent Rayleigh-Benard convection, Ph. D. Thesis, St. Louis Univ., MO
Goldstein, M. E., Durbin, P. A. (1980): The effect of finite turbulence spatial scale on the amplification of turbulence by contracting stream. J. Fluid Mech. **98**, (3), 473–508
Hamadiche, M. (1981): Modélisation spectrale des mécanismes linéaires dans les écoulements turbulents non homogènes. Thèse Univ. Cl. Bernard, Lyon I, Déc. 1981
Herring, J. R. (1985): Some contributions of two-point closure to turbulence, in *Frontiers in Fluid Mechanics* (Springer, Berlin, Heidelberg) p. 68
Hunt, J. C. R. (1978): A review of the theory of rapidly distorded turbulent flows and its applications. Fluid Dyn. Trans. **9**, 121–152
Jeandel, D. (1976): Une approche phénoménologique des écoulements turbulents inhomogènes. Thèse Univ. Cl. Bernard, Lyon I, Juillet 1976
Jeandel, D., Brison, J. F., Mathieu, J. (1978): Modeling methods in physical and spectral space. Phys. Fluids **21**, 169
Klebanoff, P. S. (1955): Characteristics of turbulence in a boundary layer with zero pressure gradient. NACA Report 1247
Kraichnan, R. H. (1972): Test-field model for inhomogeneous turbulence. J. Fluid Mech. **56**, 287–304
Lumley, J. L. (1983): Turbulence modeling, J. Appl. Mech. **50**, 1097
Maxey, M. R. (1982): Distortion of turbulence in flows with parallel streamlines. J. Fluid Mech. **124**, 261–282
Ménoret, L. (1982): Contribution à l'étude spectrale des écoulements turbulents faiblement inhomogènes et anisotropes. Thèse Univ. Cl. Bernard, Lyon I, Juin 1982
Ng, K. H., Spalding, D. B. (1972): Turbulence model for boundary layers near walls. Phys Fluids **15**, 20
Ng, K. H., Spalding, D. B. (1976): Predictions of two-dimensional boundary layers on smooth walls with a two-equation model of turbulence. Int. J. Heat Mass Transf. **19**, 1161–1172
Orszag, S. A. (1970): Analytical theories of turbulence. J. Fluid Mech. **41**, 363–386
Patankar, S. V., Spalding, D. B. (1967): *Heat and Mass Transfer in Boundary Layers* (Morgan-Grampian, London)
Rodi, W., Spalding, D. B. (1970): Wärme Stoffübertrag. **3**, 85
Weinstock, J. (1981): Theory of pressure-strain-rate correlation for Reynolds-stress turbulence closures, Part 1. Off-diagonal element. J. Fluid. Mech. **105**, 369–396
Weinstock, J. (1982): Theory of the pressure-strain-rate. Part 2. Diagonal elements. J. Fluid. Mech. **116**, 1–29
Wu, C. T., Ferziger, J. H., Chapman, D. R. (1985): Simulations and modeling of homogeneous, compressed turbulence, Fifth Symposium on Turbulent Shear Flows

The Interaction of Two Distinct Turbulent Velocity Scales in the Absence of Mean Shear

S. Veeravalli and Z. Warhaft

Sibley School of Mechanical and Aerospace Engineering, Cornell University, Ithaca, NY 14853, USA

Abstract

We report experimental results of a mixing layer formed in decaying grid tubulence in which the mean fluid velocity is constant throughout but two distinct scales of turbulence are formed on either side of the stream. This is achieved by using a parallel bar grid with the same solidity over its entire cross section but with a larger mesh length on one side of the grid than the other. Two experiments are being carried out, one with a mesh length ratio of three and another with a mesh length ratio of seven although here we will deal only with the 3:1 grid. A clear distinction between the intermittent penetration of one turbulent field into the other and the mixing (or diffusion) of the two fields is made. The experiment is along the lines of the previous work of Gilbert (1980), however here the scale ratio of the two turbulent fields is greater, revealing vital effects unobserved by Gilbert. The experiment is still in progress and we note that since reporting earlier results in the Fifth Turbulent Shear Flows Conference Proceedings we have modified the grid in order to attain better flow homogeneity on either side of the mixing layer. This has resulted in a change in some of the measured values although the new results are in qualitative agreement with the old ones.

Introduction

The expression "turbulent mixing layer" is normally used to describe turbulent mixing between two parallel semi-infinite streams, each moving at a different speed. In such a flow there is a complex interaction between production (of turbulent kinetic energy and Reynolds stress), advection, diffusion, pressure transport and scrambling and dissipation. Over the years this flow has been studied and modelled extensively (see for example Townsend, 1976) yet many of the more subtle effects such as the intermittency and large scale structures are still far from being properly understood. A central and dominating aspect concerning the evolution and structure of this type of mixing layer is the mean shear which is responsible for the turbulence production. Yet the notion of turbulent mixing does not necessarily imply that mean shear must be present. In this paper we describe an experimental study of a mixing layer in which the mean fluid velocity is constant throughout but two distinct scales of turbulence are formed on either side of the stream. As the flow evolves these two different turbulent fields penetrate and diffuse into one another. The experimental apparatus we have used to achieve such a flow is shown in Fig. 1. Laminar flow is passed through a grid of

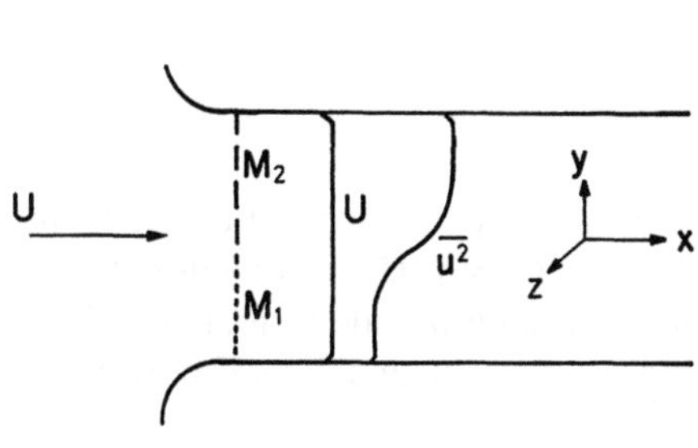

Fig. 1. Sketch of the wind tunnel showing the grid consisting of two different mesh lengths. $M_1 < M_2$. Also shown is a typical mean profile (U) and longitudinal variance profile ($\overline{u^2}$)

Turbulent Shear Flows 5

parallel square cross sectioned bars with one-side of the grid having a mesh length M and bar width d considerably smaller than the other. The ratio M/d is adjusted such that the solidity, σ, and thus the pressure drop and mean velocity is the same on either side of the flow. For the experiments to be described here M_2/M_1 (where M_2 is the large mesh spacing and M_1 is the small mesh spacing) was set at 3. A further experiment with $M_2/M_1 = 7$ is still in progress. (The actual details of the experimental apparatus are given in the next Section.)

The equations for the component velocity variances for this flow (for the high Reynolds number approximation) are

$$U\frac{\partial}{\partial x}(\overline{u^2}) + \frac{\partial}{\partial x}(\overline{u^3}) + \frac{\partial}{\partial y}(\overline{u^2 v}) = \frac{2}{\varrho}\,\overline{p\frac{\partial u}{\partial x}} - \frac{2}{\varrho}\frac{\partial}{\partial x}(\overline{pu}) - 2\nu\left(\overline{\frac{\partial u}{\partial x_j}\frac{\partial u}{\partial x_j}}\right), \tag{1}$$

$$U\frac{\partial}{\partial x}(\overline{v^2}) + \frac{\partial}{\partial x}(\overline{uv^2}) + \frac{\partial}{\partial y}(\overline{v^3}) = \frac{2}{\varrho}\,\overline{p\frac{\partial v}{\partial y}} - \frac{2}{\varrho}\frac{\partial}{\partial y}(\overline{pv}) - 2\nu\left(\overline{\frac{\partial v}{\partial x_j}\frac{\partial v}{\partial x_j}}\right), \tag{2}$$

$$U\frac{\partial}{\partial x}(\overline{w^2}) + \frac{\partial}{\partial x}(\overline{uw^2}) + \frac{\partial}{\partial y}(\overline{vw^2}) = \frac{2}{\varrho}\,\overline{p\frac{\partial w}{\partial z}} - 2\nu\left(\overline{\frac{\partial w}{\partial x_j}\frac{\partial w}{\partial x_j}}\right), \tag{3}$$

where lower case quantities represent fluctuations and upper cases represent means. The coordinate system is shown in Fig. 1. In order to emphasize the form of these relations we have retained the derivatives of fluctuating quantities with respect to x [second terms on the left of (1), (2) and (3) and second term on the right of (1)] although these should be negligible when compared with the other terms. The equations show there is no component energy production; in this "second order" mixing layer the evolution of the flow is determined by third order transverse transport terms and by the pressure transport and scrambling terms. Clearly, proper understanding of this flow can only be achieved by understanding the role of these terms. In some respects such a flow provides greater basic insight into turbulence dynamics than the traditional shear layer since the dominating effect of Reynolds stress production is absent and mixing is achieved solely by means of turbulence-turbulence interactions. The Reynolds stress itself should also be small but not entirely absent as an examination of the mean momentum equations shows:

$$\frac{\partial}{\partial y}\frac{\partial}{\partial x}(\overline{u^2} - \overline{v^2}) = \frac{\partial^2}{\partial x^2}(\overline{uv}) - \frac{\partial^2}{\partial y^2}(\overline{uv}). \tag{4}$$

This equation is derived by taking the difference of the y derivative of the u equation and the x derivative of the v equation. Since the turbulence is never isotropic (even on either side of the mixing layer $\overline{u^2} \neq \overline{v^2}$), and since the turbulence is decaying and the mixing layer is thickening the left hand side of this equation should be not equal to zero. Thus the existence of a (probably small) Reynolds stress is implied. The budget for this Reynolds stress is approximately

$$\frac{\partial \overline{uv^2}}{\partial y} = -\frac{1}{\varrho}\left(\overline{v\frac{\partial p}{\partial x}} + \overline{u\frac{\partial p}{\partial y}}\right). \tag{5}$$

The idea of mixing two distinct scales of turbulence in the way described above is not new; there is at least one other experiment due to Gilbert (1980). However Gilbert only investigated the case $M_2/M_1 = 2$. This produced a turbulent length scale ratio, l_2/l_1, of 1.4, a ratio too small to observe some of the major effects that occur in such a flow. For example, as M_2/M_1 (or l_2/l_1) is increased the profile of the velocity variance departs significantly from a symmetrical error function, the curve Gilbert fits to his data, due to the two distinct eddy viscosities of the flow. The asymmetry of the rms profile is apparent from our measurements (Fig. 8 and 9 below). Gilbert also reports that the u and v fluctuations are normally distributed (kurtosis = 3). Our results show large intermittency and skewness within the mixing layer. These higher order moments are vital to an understanding of the penetration and mixing processes.

Our objective, then, is to examine the shearless second order mixing layer in some detail. Setting up such an experiment poses difficulties. One problem is due to the growth of the mixing layer and its ultimate interaction with the tunnel walls at a downstream location not quite far enough to provide a detailed estimate of whether the mixing layer is self similar or not. Another problem concerns the nature of turbulence downstream of a mono-planar grid. As will be shown, the turbulence generated by such a grid (in the region on either side of the mixing layer) is much less homogeneous than turbulence generated by a biplanar grid. This is due to the production of large coherent eddies behind the mono-planar grid and will be discussed in detail below. This problem had not been properly appreciated in our earlier report (5th Turbulent Shear Flows Conference Proceedings) and correcting it has resulted in a modified grid which suppresses the large eddies. Thus, although the results presented here are in qualitative agreement with those reported earlier, their magnitudes have been affected considerably.

Apparatus

The experiment was carried out in a vertically oriented wind tunnel (Warhaft, 1980) with a test section of 4.25 m and a cross section of $0.406 \times 0.406\ m^2$. As explained above two grids were used ($M_2/M_1 = 3$ and 7) although the results reported here will be for the 3:1 grid. The actual values of M were $M_2 = 30.9$ mm and $M_1 = 10.2$ mm for the case $M_2/M_1 = 3$ and $M_2 = 33.3$ mm and $M_1 = 4.8$ mm for the case $M_2/M_1 = 7$. The grid solidity was 0.3 for all cases. The mean wind speed was 6.0 m/s ($M_2/M_1 = 3$) and 9 m/s ($M_2/M_1 = 7$).

Conventional X-hot wire anemometry was used to measure the U, u, v and w velocity fields and the results were analyzed on a PDP 11/34 computer.

Results and Discussion

Turbulence generated by a mono-planar grid

As mentioned above (see also Fig. 1) the turbulence was generated by a mono-planar grid rather than by the more traditional biplanar grid. Great difficulties would have been encountered in tuning and matching the large and small grids had two biplanar grids been used to generate the small and large scales. The mono-planar grid has its own problems, however.

While the experiment was under way we observed large scale inhomogeneity in the turbulence field produced by a homogeneous mono-planar grid, i.e. for turbulence generated by a grid of constant bar spacing and constant bar thickness over the full cross section of the tunnel. Such grids were constructed in order to determine $\overline{u^2}$, $\overline{v^2}$ and $\overline{w^2}$ decay laws on either side of the mixing layer accurately (see below). The large scale inhomogeneity suggested the presence of organized structures at a scale larger than the mesh spacing. Measured two point correlations bear this out. Figure 2 shows the two point longitudinal correlation $R_{11}(0, y, 0)$ (or $R_{11}(0, 0, z)$ which is identical) for a one inch biplanar grid as well as

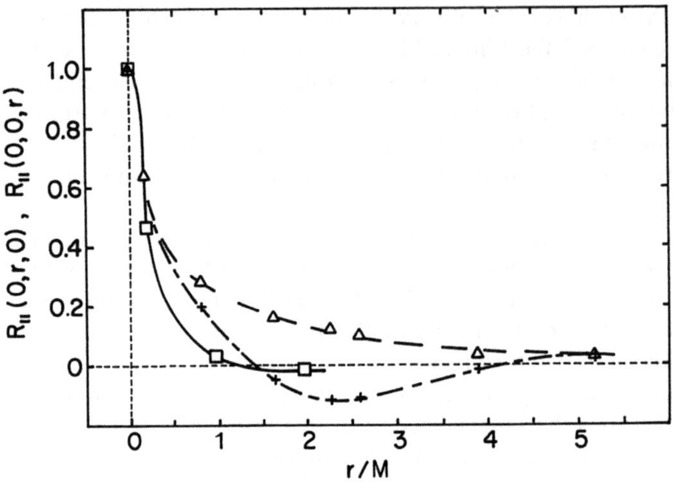

Fig. 2. Transverse correlations for the unmodified mono-planar large scale homogeneous grid (M = 3 cm over the whole cross section) compared with that of a one inch biplanar grid shown as a function of the separation distance normalized by the respective mesh lengths; $x = 64$ cm.
(Note: In the text $R_{11}(0, r, 0)$ and $R_{11}(0, 0, r)$ have been denoted by $R_{11}(0, y, 0)$ and $R_{11}(0, 0, z)$ respectively, for greater clarity.)
△ — — —, $R_{11}(0, r, 0)$ mono-planar grid;
+ — · —, $R_{11}(0, 0, r)$ mono-planar grid;
□ ———, $R_{11}(0, r, 0)$ (same as $R_{11}(0, 0, r)$) biplanar grid

$R_{11}(0, y, 0)$ and $R_{11}(0, 0, z)$ for the mono-planar grid with $M = 3$ cm (i.e. the 'large scale' side of the 3:1 grid). The correlation function goes to zero at $M \sim 1$ for the biplanar grid, as it should. However for the monoplanar grid the long tails suggest the existence of a second, larger scale (Townsend, 1976). The opposite signs of $R_{11}(0, y, 0)$ and $R_{11}(0, 0, z)$ for large separation allow us to infer that large eddies are generated with their axes orthogonal to the grid bars. In a biplanar grid these eddies are chopped up by the second orthogonal set of grid bars. At the suggestion of Professor Abernathy (Harvard University) we wound fine wires (1.41 mm diameter) in a helical fashion around the square grid bars with a lead of 1 mesh length. This has the effect of breaking up the large eddies. Figure 3 shows $R_{11}(0, y, 0)$ and $R_{11}(0, 0, z)$ for this modified mono-planar grid compared to the two point correlation for the biplanar grid. The correlation functions are now the same. Of course some trial and error tests using wires of different thickness and windings of different leads were needed to achieve the results shown in Fig. 3.

The effect on the homogeneity of the velocity variance using the modified and original grids are shown in Figs. 4 and 5. Figure 4 shows the $\overline{u^2}$, $\overline{v^2}$ and $\overline{w^2}$ transverse profiles taken along the bars (z direction) for the monoplanar grid. The fine helical winding reduces the inhomogeneity from over 30% to about 10%. The velocity field also becomes more isotropic. Notice also that for the mono-planar grid $\overline{u^2}$ is less than $\overline{v^2}$ by about 16% (or 8% in the rms). For biplanar grids (Comte-Bellot and Corrsin, 1967; Warhaft, 1980), $\overline{u^2}$ is greater than $\overline{v^2}$ and $\overline{w^2}$ by about this amount. The reason for this difference is unclear but is not of great concern since the degree of anisotropy of the (modified) mono-planar grid (although different in sign) is not worse than that of the biplanar grid. Figure 5 shows the $\overline{u^2}$, $\overline{v^2}$ and $\overline{w^2}$ transverse profiles in the y direction for the modified and original grid. The original profiles were not as inhomogeneous as those in the z direction but they were strongly anisotropic. The fine wire helical windings has reduced this anisotropy to acceptable levels (around 16% in the variance).

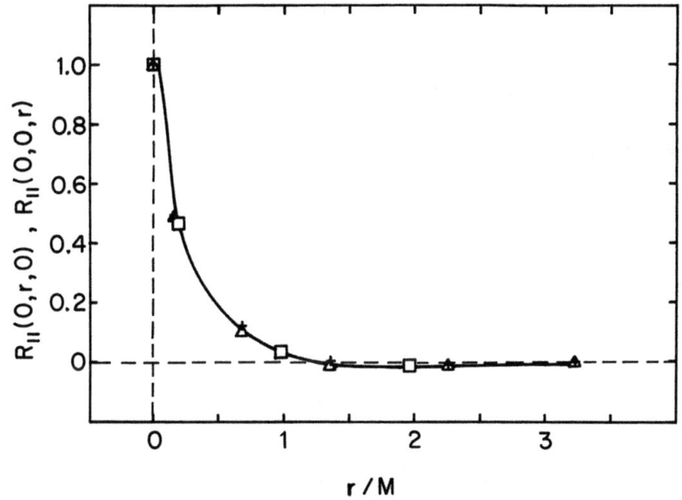

Fig. 3. Transverse correlations for the modified mono-planar large scale homogeneous grid (3:1 case) compared with that of a biplanar one inch grid. The monoplanar grid has been modified by winding fine wire (0.00141 m dia.) helically on the grid bars with a lead of one mesh length. Symbols and normalization, same as in Fig. 2; $x = 64$ cm

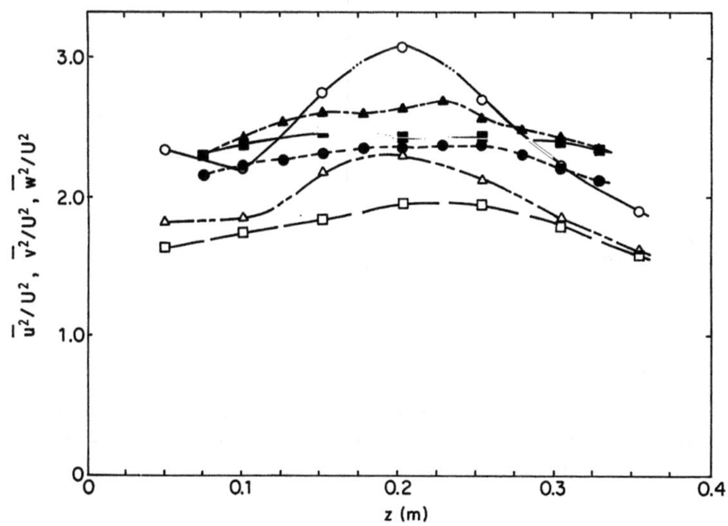

Fig. 4. Variation with the cross stream distance z (along the grid bars) of normalized $\overline{u^2}$, $\overline{v^2}$ and $\overline{w^2}$ for the mono-planar homogeneous large scale grid (3:1 case), before and after modification. The variances are normalized with the mean velocity squared. ($x = 66$ cm). The vertical axis is $\times 10^3$.

○ — ··· —, $\overline{u^2}/U^2$ unmodified grid;
△ — -- —, $\overline{v^2}/U^2$ unmodified grid;
□ —— ——, $\overline{w^2}/U^2$ unmodified grid;
●-------, $\overline{u^2}/U^2$ modified grid;
▲ — - —, $\overline{v^2}/U^2$ modified grid;
■———, $\overline{w^2}/U^2$ modified grid

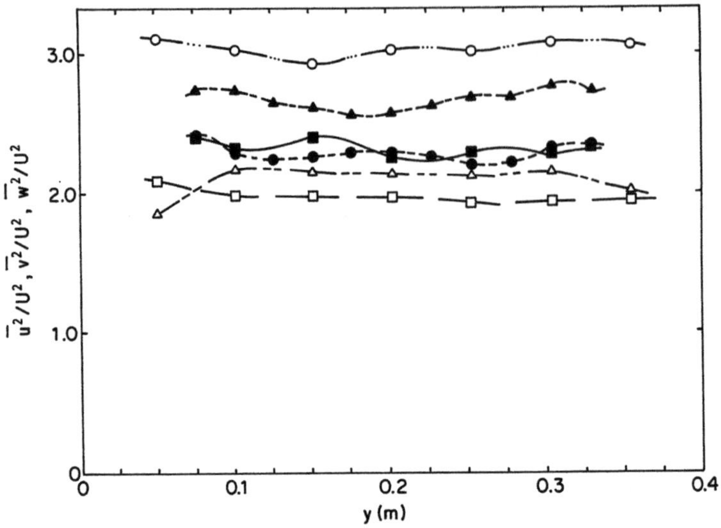

Fig. 5. Variation with the cross stream distance y (perpendicular to the grid bars) of normalized $\overline{u^2}$, $\overline{v^2}$ and $\overline{w^2}$ for the monoplanar homogeneous large scale grid (3:1 case), before and after modification. Profiles are normalized with U^2. Symbols same as in Fig. 4. The vertical axis is $\times 10^3$

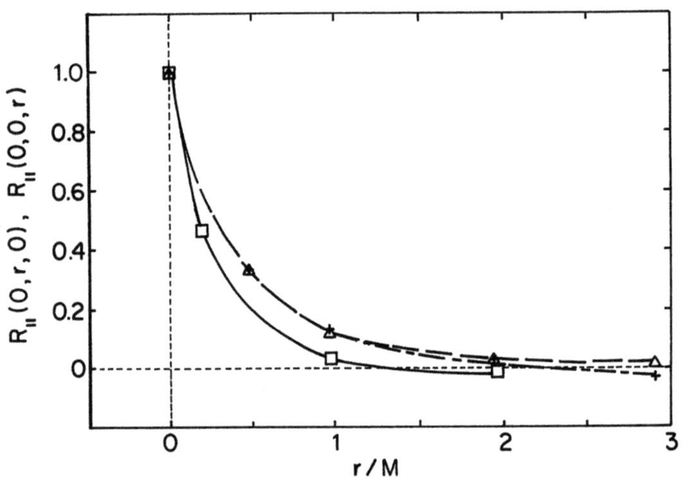

Fig. 6. Transverse correlations for the unmodified small scale homogeneous grid (3:1 case) compared with that of a one inch biplanar grid, shown as a function of the separation normalized by the respective mesh lengths; $x = 66$ cm. Symbols same as in Fig. 2

While the large scale side of the mixing layer grid required the above modifications, experiments on a grid with bar spacing and bar thickness of the same dimensions as the small scale side of the 3:1 mixing layer grid showed that this grid was not as badly affected by large scale structures. Figure 6 shows the two point correlation functions for this smaller mono-planar grid ($M = 1$ cm). While the correlation function does not decrease as rapidly as for the biplanar grid, it is relatively low at large scales; by $M = 1$ it is only 0.1. This should be compared to Fig. 2 for the large scale mono-planar grid. Thus we did not wind the small grid with fine wire. The reason for the contrast between the large degree of inhomogeneity on the (unmodified) large scale grid, and the relatively small amount of inhomogeneity on the small scale grid appears to be related to the aspect ratios of the two grids. The large scale monoplanar grid only has a mesh aspect ratio of 13 (ratio of the width of the tunnel to the spacing between the bars) while that of the small scale side is 39. The length of the large scale structures inferred from the correlation functions (Fig. 2) are of the same order as the tunnel width for the large scale grid; i.e. they are an appreciable size of the total flow and are thus seen like a single large scale inhomogeneous disturbance whereas for the small scale, higher aspect ratio grid, the larger scale disturbances are relatively small in comparison with the tunnel width and thus they have less effect in causing overall inhomogeneity. We note that the aspect ratios of the two sides of Gilbert's grid were 21 and 42, probably high enough so that no severe inhomogeneity would be encountered.

The Mean Velocity Profile

In contrast to the problems associated with the variance profiles, attaining the specified (flat) mean velocity profile was not difficult. Figure 7 shows the cross stream mean velocity profiles measured at 66 cm from the grid. This corresponds to $x/M_1 = 65$ or $x/M_2 = 21$. The deviation of U from the nominal mean of 6.0 m/s is about $\pm 2.5\%$. Using values of the turbulence dissipation rate determined from the decay laws of turbulence kinetic energy on either side of the mixing layer we have determined the ratio of turbulence energy production to turbulence energy dissipation to have a maximum value of approximately 4% at $x = 66$ cm. This is based on maximum local mean velocity gradients. We have not yet determined the dissipation rate within the mixing layer but we do not expect that the production to dissipation ratio to be significantly higher here than the above value. Thus we believe that shear production is a comparatively minor effect compared with dissipation and transport. We note, however, that even small gradients in U produce Reynolds stresses (cross correlations between u and v of as much as 0.2 were observed locally) and while these do not appear to have a dynamical effect they do affect such quantities as skewness coefficients (see below).

The Variance Characteristics

A difficulty in carrying out these experiments in our moderately sized wind tunnel is that in order to achieve modest grid Reynolds numbers (UM/ν) the bar spacing is such that the mixing layer grows to almost the width of the tunnel by about $x/M_1 = 200$. Thus even at about $x/M_1 = 150$ the homogeneous turbulence on either side of the mixing layer is limited in extent. Hence, in order to get an accurate measurement of the $\overline{u^2}$ and $\overline{v^2}$ decay laws on

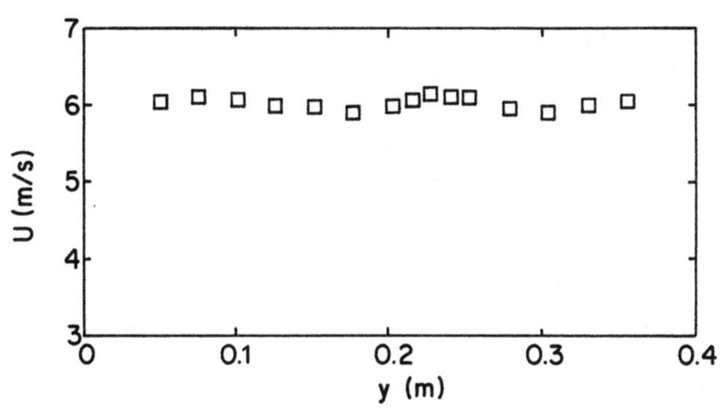

Fig. 7. The mean velocity profile as a function of cross stream distance, y at $x = 66$ cm, 3:1 grid ($x/M_1 = 64.7$)

Table 1. Flow parameters for the 3:1 grid. All fluctuation parameters are evaluated at $x = 66$ cm ($x/M_1 = 64.7$)

$R_{M_2} \equiv UM_2/\nu$	11,588
$R_{M_1} \equiv UM_1/\nu$	3,863
$\overline{u_2^2}(m^2/s^2)$	0.0826
$\overline{v_2^2}(m^2/s^2)$	0.0965
$\overline{w_2^2}(m^2/s^2)$	0.0829
$\overline{u_1^2}(m^2/s^2)$	0.0187
$\overline{v_1^2}(m^2/s^2)$	0.0202
$\overline{w_1^2}(m^2/s^2)$	0.0164
$k_2 \equiv (\overline{v_2^2} + 2\overline{u_2^2})/2(m^2/s^2)$	0.131
$\varepsilon_2 = -dk_2/dt(m^2/s^3)$	2.22
$k_1 \equiv (\overline{v_1^2} + 2\overline{u_1^2})/2\ (m^2/s^2)$	0.0288
$\varepsilon_1 = -dk_1/dt(m^2/s^3)$	0.352
$l_2 = k_2^{3/2}/\varepsilon_2(m)$	0.0214
$l_1 = k_1^{3/2}/\varepsilon_1(m)$	0.0139

either side of the mixing layer we constructed two completely homogeneous grids of parallel bars with $M_1 = 10.2$ mm and $M_2 = 30.9$ mm. The decay laws for the 3:1 data are

$$\overline{u_2^2}/U^2 = 0.027\,(x/M_2 - 10.3)^{-1.011},$$

$$\overline{v_2^2}/U^2 = 0.073\,(x/M_2 - 5.64)^{-1.203},$$

$$k_2/U^2 = 0.056\,(x/M_2 - 8.83)^{-1.074},$$

$$\overline{u_1^2}/U^2 = 0.031\,(x/M_1 - 12.3)^{-1.04},$$

$$\overline{v_1^2}/U^2 = 0.055\,(x/M_1 - 13.1)^{-1.15},$$

where k_2 is defined in Table 1. Here subscripts 1 and 2 refer to the low and high turbulence sides of the mixing layer respectively. The results are in reasonable agreement with the decay laws taken from the extreme edges of the mixing layer grid although far downstream these latter decay laws are influenced by the interaction of the mixing layer and tunnel wall boundary layers. The decay rates for these parallel bar grids are in broad conformity with those measured by other workers for biplanar grids (Comte-Bellot and Corrsin, 1966). The only significant difference (see Sect. 3.1 above) is that the magnitude of $\overline{u^2}$ is slightly less than that of $\overline{v^2}$ while for a biplanar grid $\overline{u^2}$ is slightly greater than $\overline{v^2}$ or $\overline{w^2}$. The main characteristics of the flow field on either side of the mixing layer are listed in Table 1.

Figure 8 shows the cross stream $\overline{u^2}$, $\overline{v^2}$ and $\overline{w^2}$ profiles (at $x = 66$ cm), normalized by $\overline{u_1^2}$ where subscript 1 refers to the small scale side of the mixing layer. From the figure we note that most of the transition from large scale to small scale occurs to the left of the physical center (shown by arrow) of the grid (i.e. on the small bar side). At the right end of the figure are shown the levels attained in the large scale homogeneous mono-planar grid, which these curves must asymptotically approach. The data show some scatter at the large scale end, which we should expect as we were only 21 mesh length downstream with respect to M_2. The deviations from the curves are less than 5% in the rms and are hence well within acceptable limits. It is quite apparent that the curves are not symmetric. The lack of symmetry reflects the difference in eddy diffusivities on either side of the flow. The asymmetry

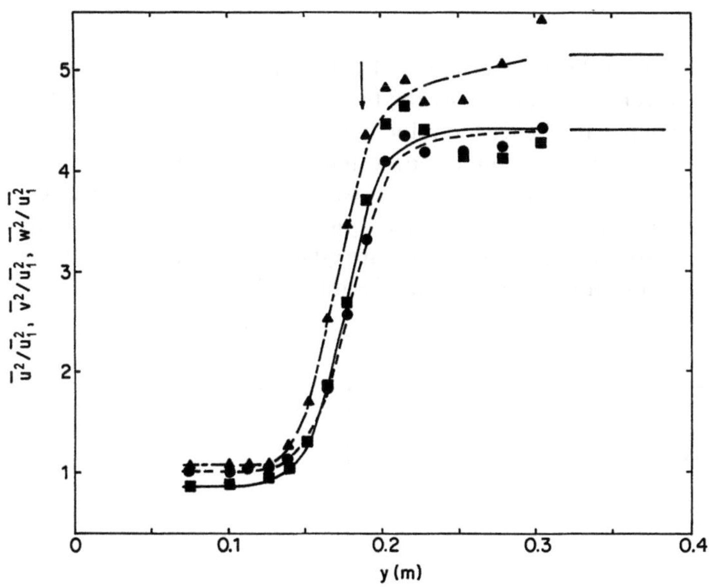

Fig. 8. Normalized $\overline{u^2}$, $\overline{v^2}$ and $\overline{w^2}$ profiles for the 3:1 grid at $x = 66$ cm. The data has been normalized by the local mean velocity prior to rescaling by $\overline{u_1^2}/U^2$. The arrow indicates the physical location of the change from the grid of M_1 to that of M_2. The solid lines at the top right-hand side indicate the values of $\overline{v_2^2}/\overline{u_1^2}$ and $\overline{u_2^2}/\overline{u_1^2}$ $(= \overline{w_2^2}/\overline{u_1^2})$ obtained from the monoplanar large scale homogeneous grid. ● ----, $\overline{u^2}/\overline{u_1^2}$; ▲ – - –, $\overline{v^2}/\overline{u_1^2}$; ■ ——, $\overline{w^2}/\overline{u_1^2}$

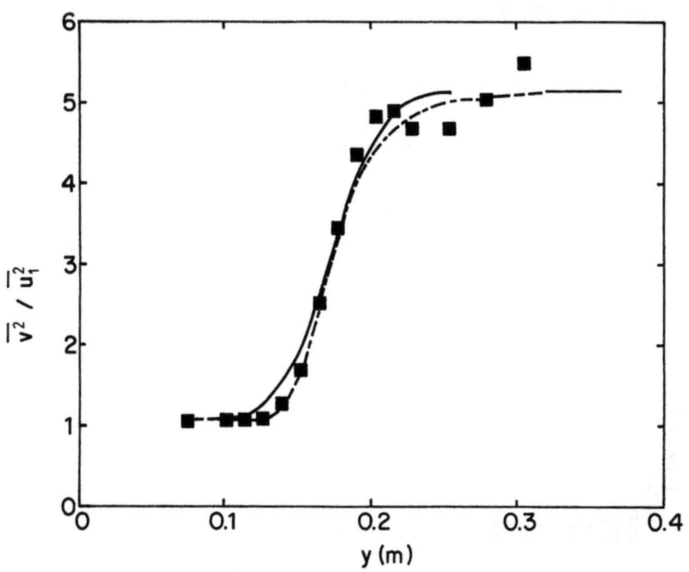

Fig. 9. Comparison of $\overline{v^2}/\overline{u_1^2}$ (same data as Fig. 8) with an error function (solid line)

is most prominent in the $\overline{v^2}$ profile and in Fig. 9 this has been compared with a (symmetric) error function. As expected the comparison is quite poor.

The profiles discussed above were at the closest downstream location and we are in the process of studying the evolution of the mixing layer in order to determine the spreading rate and whether the flow is self similar.

The Kurtosis and Skewness

Figure 10a shows the variation of the kurtosis of u, v and w fluctuations, K_u, K_v and K_w respectively, across the mixing layer at $x = 66$ cm. At the two ends of the mixing layer, the flow is homogeneous in the (y, z) plane and hence the velocity *pdfs* are nearly Gaussian with a kurtosis of 3. However, in the mixing layer all three depart significantly from 3. From these graphs we note that the peaks lie towards the small turbulence side of the mixing layer. Though the peak magnitudes of K_u and K_v have dropped significantly from those reported earlier (Turbulent Shear Flows 5 Proceedings), the profiles are qualitatively quite similar. As before we expect the peaks to diminish in magnitude as the mixing layer evolves.

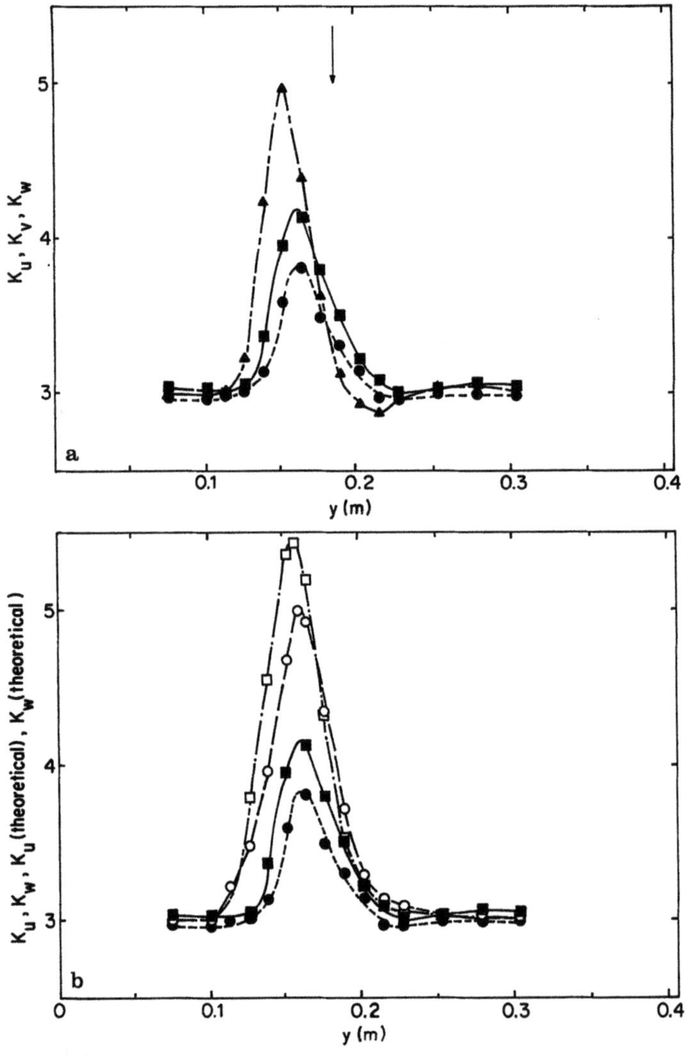

Fig. 10a. Kurtosis (normalized fourth moment) profiles for u, v and w; 3:1 grid $x = 66$ cm. Arrow has the same significance as in Fig. 8. ●----, K_u; ▲ — - -, K_v; ■ ——; K_w

Fig. 10b. K_u and K_w (same data as Fig. 10a) compared with their theoretical estimates based on $\overline{u^2}$ profiles (Eq. 7). ●----, K_u; ○ — — —, K_u (theoretical); ■ ——, K_w; □ —·—, K_w (theoretical)

The behavior of K_u and K_w may be understood as follows: if it is assumed that mixing takes place purely by penetration (rather than by penetration and diffusion together), the *pdf* in the mixing layer is given by a linear combination of the *pdfs* at the two sides. If the intermittency factor is r then

$$\sqrt{\overline{u^2}}(r) = \sqrt{\overline{u_1^2}\, r + (1 - r)\, \overline{u_1^2}}, \tag{6}$$

where $\sqrt{\overline{u_1^2}}$ and $\sqrt{\overline{u_2^2}}$ are the rms values of the (normally distributed) turbulence on either side of the mixing region and $\sqrt{\overline{u^2}}(r)$ is an intermediate rms velocity. It follows that

$$K(r) \equiv \overline{u^4}/(\overline{u^2})^2 = \frac{3((R^4 - 1)\, r + 1)}{((R^2 - 1)\, r + 1)^2}, \tag{7}$$

where $R \equiv \sqrt{\overline{u_2^2}}/\sqrt{\overline{u_1^2}}$. From the measured value of $\overline{u^2}$ (Fig. 8) r can be calculated from

$$r = (\overline{u^2} - \overline{u_1^2})/(\overline{u_2^2} - \overline{u_1^2}). \tag{8}$$

Further, K attains its maximum at $r_m = 1/(R^2 + 1)$. Its value is $K_{u_m} = (3/4)\,(R^2 + 1)^2/R^2$.

Figure 10b shows a comparison between the experimental curves and the theoretical curves estimated from (7). The peaks of the curves show excellent agreement in terms of their transverse position within the mixing layer and their forms are similar. However the maxima of the experimentally observed Kurtosis are much lower than those deduced from (7). This difference is due to the fact that the two turbulent fields do not only penetrate each other (the assumption for the theoretical estimate) but they also diffuse and mix together. If only diffusion were present the *pdf* would be Gaussian and the Kurtosis would be 3 throughout the mixing layer. A similar comparison for the 7:1 grid (not shown here) shows the discrep-

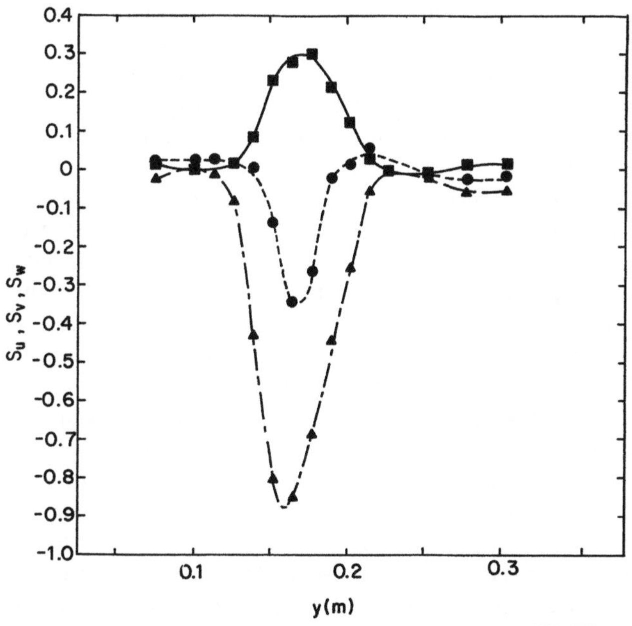

Fig. 11. Skewness profiles (normalized third moments $\overline{u^3}/(\overline{u^2})^{3/2}$, $\overline{v^3}/(\overline{v^2})^{3/2}$ and $\overline{w^3}/(\overline{w^2})^{3/2}$) at $x = 66$ cm for the 3:1 grid. ● ----, S_u; ▲ – – –, S_v; ■ ——, S_w

ancy in the magnitudes of the theoretical and experimental peaks of K_u is not so pronounced. Hence diffusion effects become less dominant in comparison to penetration as the scale ratio of the mixing layer increases. Probably the nearly Gaussian values for the *pdfs* measured by Gilbert (1976) are due to his very low ratio of l_2/l_1 (he used a 2:1 grid ratio).

The analysis for K_v is more difficult than that of K_u since v fluctuations are along the direction of the gradient and only part of the *pdf* of v is transported across the mixing layer. Thus even an estimate based on pure penetration as was done for K_u and K_w is more difficult. Its higher value than K_u is consistent with observations in other flows such as the edge of a boundary layer. Here too $K_v > K_u$.

Figure 11 shows the behavior of the skewness of u, v and w fluctuations, S_u, S_v and S_w respectively for the same downstream locations as K_u (Fig. 10a). At the two sides of the mixing layer the skewness is essentially zero indicating that the turbulence is homogeneous there. The large negative S_v is due to the fact that in the low turbulence side of the mixing layer, large deviations from zero are more likely to have originated from the large scale side and hence are more likely to be negative. The presence of nonzero S_u and S_w is unexpected. The inhomogeneities in the mean velocity profile could account for the presence of S_u, but as yet we do not understand the reason for the nonzero value of S_w.

The results of the 7:1 grid will not be described here since the measurements are still under way. All indications suggest, however, that the 7:1 grid amplifies the effects found for the 3:1 grid; i.e. 3rd and 4th moments are larger and there is greater asymmetry in the variance profiles.

Concluding Comments

The shearless "second order" mixing layer focuses particular attention on third and fourth order moments and has a distinctly different character to the traditional "first order" mixing layer in which turbulent production of kinetic energy and shear stress play dominant roles in the flow dynamics. The results presented here show that the spreading of the shearless mixing layer is due to a combination of intermittent penetration of one scale of turbulence into the other and of turbulent diffusion of the two fields into each other. It should be noted that simple gradient diffusion models will not provide insight into this flow.

The experiment is incomplete, but in view of the significance of this type of flow in aiding in our understanding of transport processes in turbulence we felt this communication was appropriate. The reader interested in attempting to quantitatively model this flow should note that the results presented here are quantitatively different from those reported in the 5th Turbulent Shear Flows Conference Proceedings due to the discovery (and removal) of large scale unwanted inhomogeneous effects caused by the mono-planar grid. However, the results are in qualitative agreement with those reported previously.

We are in the process of determining the kinetic energy budget for this flow. From this we will be able to determine the magnitude and significance of the fluctuating pressure terms. We are also examining the form of the fluctuating velocity derivatives in order to find whether the large kurtosis and skewness observed at the integral scales, extends to the micro-scales. The spectral characteristics of the fluctuating properties are also being studied. In future work we intend to add a passive scalar to this flow. Apart from a basic interest in understanding how this mixing layer smears and mixes a scalar, by tagging one side of the flow with the passive trace we will be able to examine in greater detail the penetration of the one turbulence scale into the other.

We thank Professor F. H. Abernathy for his suggestion concerning the grid modification (3.1). This work has been funded from grants from the Engineering section and the Atmospheric Science section of the U.S. National Science Foundation.

References

Comte-Bellot, G., Corrsin, S. (1966): The use of a contraction to improve the isotropy of grid generated turbulence. J. Fluid Mech. **25**, 657–682

Gilbert, B., (1980): Diffusion mixing in grid turbulence without mean shear. J. Fluid Mech. **100**, 349–365

Gilbert, B., (1976): "An Experimental Investigation of Turbulent Mixing of Fluids with Different Dynamically Significant Scales;" Ph.D. Thesis, University of Illinois at Chicago Circle

Townsend, A. A. (1976): *The Structure of Turbulent Shear Flows* (Cambridge University Press)

Warhaft, Z. (1980): An experimental study of the effect of uniform strain on thermal fluctuations in grid generated turbulence. J. Fluid Mech. **99**, 545–573

The Mixing Layer Between Turbulent Fields of Different Scales

S. B. Pope and D. C. Haworth

Sibley School of Mechanical and Aerospace Engineering, Cornell University, Ithaca, New York 14853, USA

Abstract

A modelling study is reported of the interaction and mixing between two semi-infinite homogeneous turbulent flow fields of different scales. This turbulence mixing layer, in which there are no mean velocity gradients, is ideal for studying turbulent energy transport. The velocity joint *pdf* equation is closed by using the Langevin model and is solved by a Monte Carlo method. A novel – and essential – component in the present model is a Lagrangian equation for the dissipation. The model calculations are compared to the experimental data of Veeravalli and Warhaft. There is good agreement between the calculated and measured velocity skewness and excesses, which are the most revealing statistics.

Introduction

A modelling study is reported of the interaction and mixing between two semi-infinite homogeneous turbulent flow fields of different scales. This flow can be (approximately) realized in wind-tunnel experiments by using a turbulence-generating grid with two different bar and mesh sizes. Downstream of the grid, on each side of the mixing layer, there are regions of nearly homogeneous turbulence. The turbulence energies and length scales are different on the two sides, but by careful adjustment of the grid, the mean velocity profile can be made uniform. The subject of the study is the evolution of the turbulence mixing layer that forms between the two fields of homogeneous turbulence. Experiments have been performed by Gilbert [1, 2] and by Veeravalli and Warhaft [3] whose data are compared to the model calculations [1].

The modelling study of this turbulence mixing layer is based on a transport equation for the joint probability density function (*pdf*) of the three components of velocity [4]. The joint *pdf* equation is closed by using the Langevin equation to model the effects of viscosity and the fluctuating pressure gradient, and the resulting modelled *pdf* equation is solved by a Monte Carlo method [5]. A novel feature of the present model is the use of a Lagrangian evolution equation for the dissipation.

The motivation for the study is that in many practical situations a turbulent shear flow is embedded in a surrounding field of turbulence of a different intensity. Examples are: the turbulent buoyant jet from a smoke stack issuing into the atmospheric boundary layer; the dilution jets in a gas turbine combustor issuing into the turbulent flow from the primary zone; and, the boundary layer on a turbine or compressor blade with a turbulent free stream. The flow studied is simpler than these examples and allows the fundamental turbulence phenomena to be examined in a simple geometry and in the absence of shear.

---

[1] It should be noted that the conditions of the experiments reported by Veeravalli and Warhaft in the Symposium Proceedings [3] differ from those reported in this volume. In particular the energy ratios are different. The comparisons made here are with the data reported in the Symposium Proceedings.

In contrast to the flows mentioned above, most laboratory shear flows are surrounded by non-turbulent, irrotational fluid. Then the phenomenon of intermittency is known to be important. Several researchers have argued that turbulence models can be improved by taking excplicit account of intermittency, and have accordingly constructed "conditional models" [6–8]. These conditional models depend upon the fact that the freestream is in a singular state (i.e. zero vorticity) and hence they cannot simply be generalized to treat shear flows in turbulent surroundings. The present study is a first step towards understanding the modelling questions posed by the interaction of turbulence of different scales.

Turbulence Mixing Layer

In a coordinate system moving with the uniform mean flow, the mixing layer in the wind-tunnel experiment is (to a good approximation) statistically one-dimensional and evolves in time. Ideally, this temporally growing mixing layer is as depicted in Fig. 1. In the regions of homogeneity, remote from the mixing layer, the turbulence is isotropic with kinetic energy $k_L(t)$ on the large-scale side ($y > 0$) and $k_S(t)$ on the small-scale side ($y < 0$). The virtual origin of the turbulence is at $t = 0$, and k_L is taken to be unity at $t = 1$. Then with n being the decay exponent, we have

$$k_L(t) = t^{-n}. \tag{1}$$

We take $n = 1.29$, a typical value for grid turbulence. It is assumed that the decay exponent is the same on both sides and so the energy ratio r is constant:

$$r \equiv k_L(t)/k_S(t). \tag{2}$$

In order to make comparisons with Veeravalli and Warhaft's data, we choose the two values of the energy ratio, $r = 7$ and 25.

From the decay equation (1), the dissipation rate is deduced to be

$$\varepsilon_L(t) = -\frac{dk_L}{dt} = n t^{-(n+1)}, \tag{3}$$

and hence the length and time scales are

$$l_L \equiv k_L^{3/2}/\varepsilon_L = t^{(1-n/2)}/n, \tag{4}$$

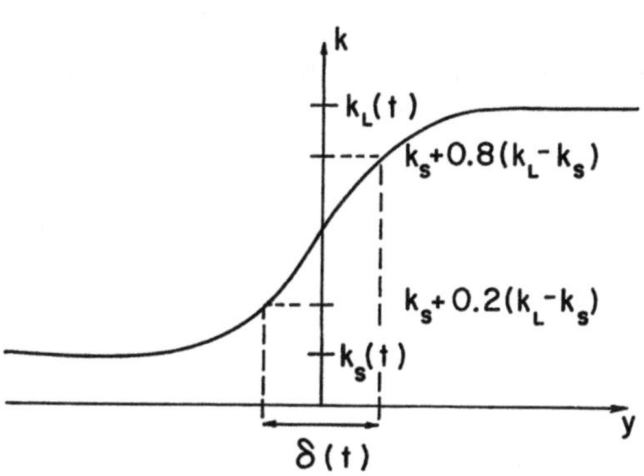

Fig. 1. Sketch of the turbulence mixing layer showing the definition of the thickness δ

and

$$\tau_L \equiv k_L/\varepsilon_L = t/n. \tag{5}$$

On the small-scale side, the dissipation ε_S is smaller by a factor of the energy ratio r, the length scale l_S is smaller by a factor of $\sqrt{r}$, while the time scale is the same.

The virtual origin of the mixing layer is taken to be $t = 1$, at which time the kinetic energy has a step change at $y = 0$. As time evolves, the thickness of the layer $\delta(t)$ (defined on Fig. 1) increases, and statistical quantities develop smooth profiles.

All components of the mean velocity are zero. The mean momentum equation in the y-direction reveals that there is a mean pressure gradient

$$\frac{\partial\langle p\rangle}{\partial y} = -\varrho\frac{\partial\langle v^2\rangle}{\partial y}, \tag{6}$$

where ϱ is the constant density.

Of primary interest are the profiles of the Reynolds stresses. The turbulence is axisymmetric in the $x - z$ plane (normal to the y-direction): hence the normal stresses $\langle u^2\rangle$ and $\langle w^2\rangle$ are equal, and all the shear stresses are zero. But the profile of $\langle v^2\rangle$ may differ from those of $\langle u^2\rangle$ and $\langle w^2\rangle$.

It is possible that the turbulence mixing layer becomes self-similar. Then, all statistical quantities, when normalized by l_L and u_L

$$u_L \equiv (\tfrac{2}{3}k_L)^{1/2}, \tag{7}$$

vary only with the normalized lateral distance η

$$\eta \equiv y/l_L, \tag{8}$$

independent of time. In the experiments [3], the initial conditions for the mixing layer are not easily defined. But it is plausible that the self-similar state has been reached by the last measurement station. Consequently our attention is focussed on this self-similar region, even though the transient model equations are solved, and comparison is made with the state at earlier measurement stations.

Joint *pdf* Model

A Lagrangian viewpoint is useful when modelling, interpreting, and solving *pdf* evolution equations [5]: the behavior of fluid particles in a turbulent flow provides a complete description of the turbulence. At time t, the position and velocity of a fluid particle are denoted by $\hat{\boldsymbol{x}}(t)$ and $\hat{\boldsymbol{u}}(t)$. According to the Navier-Stokes equations, in an infinitesimal time interval dt, $\hat{\boldsymbol{x}}(t)$ and $\hat{\boldsymbol{u}}(t)$ change

$$d\hat{x}_i = \hat{u}_i(t)\,dt, \tag{9}$$

and

$$d\hat{u}_i = -\frac{1}{\varrho}\frac{\partial\langle p\rangle}{\partial x_i}dt + \left(\nu\nabla^2 u_i - \frac{1}{\varrho}\frac{\partial p'}{\partial x_i}\right)dt, \tag{10}$$

where ν is the kinematic viscosity. The Eulerian pressure $p(\boldsymbol{x}, t)$ has been decomposed into its mean $\langle p \rangle$ and fluctuation $p'(\boldsymbol{x}, t)$. (Angled brackets denote means.) The terms on the right-hand side of (10) are evaluated at the particle location $\boldsymbol{x} = \hat{\boldsymbol{x}}(t)$.

In place of the exact expression (10), we model the Lagrangian velocity increment by the Langevin equation [5, 9, 10]:

$$d\hat{u}_i = -\frac{1}{\varrho}\frac{\partial\langle p\rangle}{\partial x_i}\,dt - \left(\frac{1}{2} + \frac{3}{4}C_0\right)\hat{u}_i\frac{dt}{\tau} + (C_0\,\hat{\varepsilon})^{1/2}\,dW_i. \tag{11}$$

Here $\hat{\varepsilon}$ is the dissipation associated with the fluid particle (defined below), $\boldsymbol{W}(t)$ is an isotropic Weiner process, and C_0 is a positive constant. The time scale $\tau(\boldsymbol{x}, t)$ is defined by

$$\tau \equiv k/\varepsilon, \tag{12}$$

where ε is the mean dissipation rate

$$\varepsilon(\boldsymbol{x}, t) = \langle \hat{\varepsilon}(t) \,|\, \hat{\boldsymbol{x}}(t) = \boldsymbol{x} \rangle. \tag{13}$$

(For homogeneous turbulence, $\hat{\varepsilon}$ is equal to ε, and the Langevin equation (11) reverts to the standard form used previously [10, 11].)

By comparing (10) and (11) it may be seen that the effects of viscosity and the fluctuating pressure gradient are modelled by the last two terms in (11). The final term represents a random walk in velocity space: $d\boldsymbol{W}$ is a joint normal random vector with zero mean and covariance

$$\langle dW_i dW_j \rangle = dt\,\delta_{ij}. \tag{14}$$

The form of this term is consistent with Kolmogorov inertial-range scaling [12] and the Kolmogorov constant C_0 has been determined to be 2.1 [11].

The penultimate term in (11) represents a linear return to zero velocity fluctuations. For isotropic turbulence, the precise form of the term is completely determined by the condition that the expected energy decays at the rate $\hat{\varepsilon}$. Haworth and Pope [10, 13] have developed and used a generalization of (11) that is applicable to anisotropic turbulence and to inhomogeneous flows with mean velocity gradients. In the generalized model the linear term is more involved although it reduces to the isotropic form in the appropriate limit. The isotropic form is deemed adequate for the turbulence mixing layer since there are no mean velocity gradients, and departures from isotropy are not large.

Boundary conditions for $\hat{\boldsymbol{u}}$ are simply specified: on each side of the mixing layer $\hat{u}_i$ is an isotropic normal random variable with mean zero and variance $2/3\,k_L$ on the large-scale side and $2/3\,k_S$ on the small-scale side.

In the context of the turbulence mixing layer, one aspect of the modelling should be emphasized. Pressure fluctuations have two effects on the Reynolds stresses: they redistribute the energy among the components; and, they can cause transport of energy in physical space. The Langevin equation models the first of these processes, but the second is neglected. Whether the (neglected) energy flux $\langle \boldsymbol{u}p'/\varrho \rangle$ due to pressure fluctuations is significant compared to the convective flux $\langle \boldsymbol{u}(1/2\,\boldsymbol{u}\cdot\boldsymbol{u})\rangle$ is not known with certainty. One promise of experiments on the turbulence mixing layer is to answer this question.

Dissipation Model

In contrast to previous studies using the Langevin equation, here we use a particle-dependent dissipation rate $\hat{\varepsilon}$ rather than the local mean dissipation ε. This is motivated by the observations that the mean dissipation rate changes by a factor of the energy ratio r (7, say) across the layer, and that (at least at early times) a fluid particle can traverse the layer in a time interval that is short compared to the turbulent time scale τ. Thus a packet of turbulent fluid (of size l_L) can travel from the large-scale side (where the dissipation is ε_L) to the small scale side (where the dissipation is $\varepsilon_S = \varepsilon_L/r$) more rapidly than the turbulence within the packet can respond to its changed surroundings. For a fluid particle within this packet, the appropriate dissipation rate for use in the Langevin model is closer to ε_L than to the local mean on the small-scale side, ε_S.

In order to account for these phenomena, we use a simple relaxation equation for $\hat{\varepsilon}$:

$$\frac{d\hat{\varepsilon}}{dt} = -C_{\varepsilon 3}(\hat{\varepsilon} - \varepsilon)/\tau - C_{\varepsilon 2}\hat{\varepsilon}/\tau, \tag{15}$$

with boundary conditions $\hat{\varepsilon}(t) = \varepsilon_L(t)$ and $\hat{\varepsilon}(t) = \varepsilon_S(t)$ on the large-scale and small-scale sides. Away from the mixing layer, in the regions of uniformity, $\hat{\varepsilon}$ is equal to ε and hence the first term on the right-hand side of (15) is zero. Then the equation reduces to the standard model equation for ε [14] which, with $C_{\varepsilon 2} = 1 + 1/n$, is consistent with the decay law, (1). Within the turbulence mixing layer, the first term on the right-hand side of (15) causes $\hat{\varepsilon}$ to relax to the local mean value ε at a rate determined by the local time scale τ and the constant $C_{\varepsilon 3}$.

In the next section we report results of the calculations with three different choices of $C_{\varepsilon 3}$: 0, 1/4 and ∞. With the extreme value $C_{\varepsilon 3} = \infty$, $\hat{\varepsilon}$ adopts its local mean value; whereas with the other extreme value $C_{\varepsilon 3} = 0$, $\hat{\varepsilon}$ is equal to ε_L or ε_S depending upon whether the fluid particle originated from the large-scale or small-scale side of the layer. The comparison of calculations with Veeravalli and Warhaft's data suggests that the appropriate value of $C_{\varepsilon 3}$ is about 1/4.

For each fluid particle, $\hat{\varepsilon}(t)$ can be interpreted as the mean dissipation, conditional upon the particle's history $\hat{x}(t)$.

We introduce the Eulerian joint *pdf* of velocity and dissipation, $f(V, \tilde{\varepsilon}; x, t)$. In terms of particle properties, this is the joint probability density of the events $\hat{u}(t) = V$ and $\hat{\varepsilon}(t) = \tilde{\varepsilon}$, conditional upon $\hat{x}(t) = x$. From (11) and (15), we deduce the evolution equation for $f(V, \tilde{\varepsilon}; x, t)$ to be

$$\frac{\partial f}{\partial t} + V_i \frac{\partial f}{\partial x_i} = \frac{\partial}{\partial V_i}\left\{\frac{f}{\varrho}\frac{\partial\langle p\rangle}{\partial x_i} + \left(\frac{1}{2} + \frac{3}{4}C_0\right) f \frac{V_i}{\tau}\right\} + \frac{1}{2} C_0 \tilde{\varepsilon} \frac{\partial^2 f}{\partial V_i \partial V_i}$$
$$+ \frac{1}{\tau}\frac{\partial}{\partial \tilde{\varepsilon}}\{f[C_{\varepsilon 3}(\tilde{\varepsilon} - \varepsilon) + C_{\varepsilon 2}\tilde{\varepsilon}]\}. \tag{16}$$

This joint *pdf* equation was solved by a Monte Carlo method [5]. For the flow considered, the method amounts to solving (11) and (15) for a large number ($\approx$ 20,000) of notional particles.

Results

We report, first, results for the energy ratio of 7 (corresponding approximately to Veeravalli and Warhaft's 3:1 grid) with the model constant $C_{\varepsilon 3} = 1/4$. Then the influence of the constant $C_{\varepsilon 3}$ is demonstrated by results obtained with the extreme values $C_{\varepsilon 3} = 0$ and $C_{\varepsilon 3} = \infty$. Finally some results are reported for the energy ratio 25, corresponding approximately to the 7:1 grid.

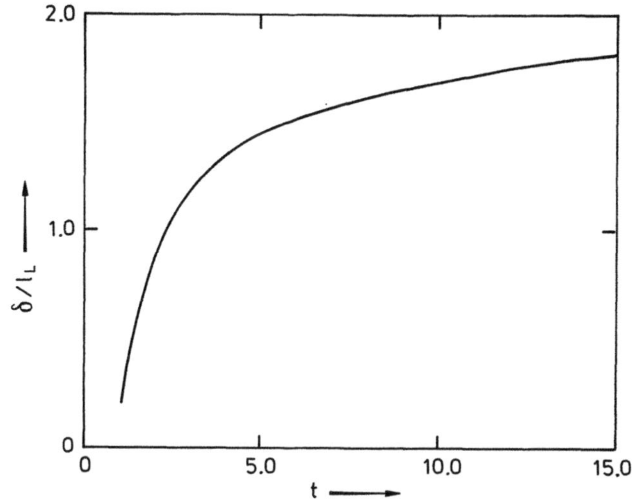

Fig. 2. Calculated evolution of normalized thickness ($r = 7, C_{\varepsilon 3} = 1/4$)

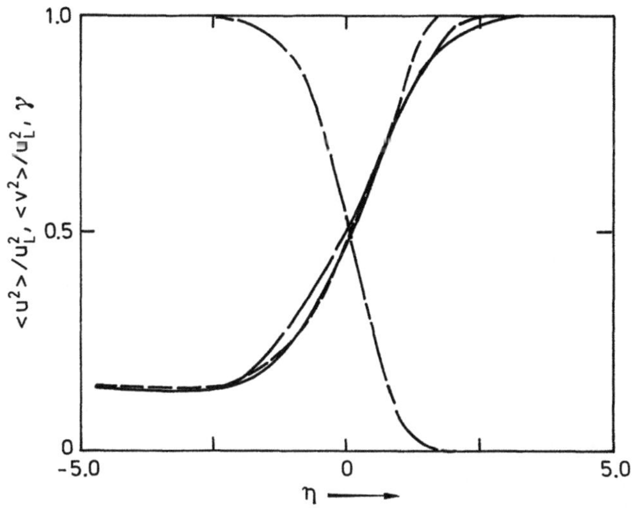

Fig. 3. Calculated normalized variances and intermittency profiles ($r = 7$, $C_{\varepsilon 3} = 1/4$, $t = 15$). ---- $\langle u^2 \rangle$; — — — $\langle v^2 \rangle$; —— error function; — - — γ

For the energy ratio $r = 7$ and $C_{\varepsilon 3} = 1/4$, Fig. 2 shows the temporal growth of the mixing layer: the thickness $\delta(t)$ (defined on Fig. 1) is normalized by $l_L(t)$. It may be seen that an asymptotic value of $\delta/l_L \approx 1.9$ is being approached, but the asymptote is not quite attained by the end of the calculations ($t = 15$).

In Fig. 3, the variances $\langle u^2 \rangle$ and $\langle v^2 \rangle$ at $t = 15$ (normalized by u_L) are plotted against the normalized cross-stream coordinate η (8). An error function curve (with the same maximum slope as $\langle u^2 \rangle / u_L^2$) is shown for comparison. It may be seen that the $\langle u^2 \rangle$ profile differs significantly from the error function on the large-scale side. The calculated spread of the $\langle v^2 \rangle$ profile is slightly greater than that of $\langle u^2 \rangle$.

Also shown in Fig. 3 is the "intermittency factor" γ, defined to be the probability of the fluid having come from the small-scale side.

In performing calculations with different values of $C_{\varepsilon 3}$ and different energy ratios r, we found that the spreading rate and the variance profiles changed little. The most sensi-

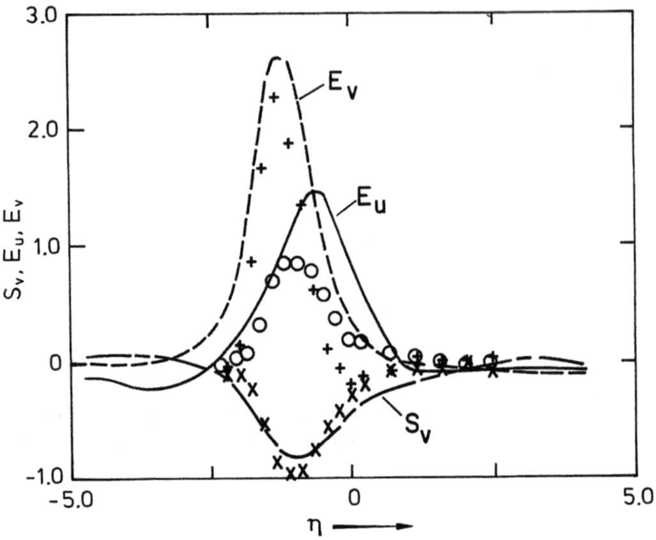

Fig. 4. Excesses of u and v and skewness of v. Lines are calculations ($r = 7$, $C_{\varepsilon 3} = 1/4$, $t = 15$), symbols are experimental data of Veeravalli and Warhaft (3:1 grid, $x = 150$ cm)

tive quantities were the excesses E_u and E_v ($E_u \equiv \langle u^4 \rangle / \langle u^2 \rangle^2 - 3$) and the skewness ($S_v \equiv \langle v^3 \rangle / \langle v^2 \rangle^{3/2}$). (The skewness of u is zero by symmetry.) Figure 4 shows the calculated values of E_u, E_v and S_v at $t = 15$ compared to the measurements at the furthest measurement station ($x = 150$ cm). It may be seen that both in shape and magnitude the calculations are in good agreement with the measurements. The peaks in all the profiles are significantly displaced towards the small-scale side, with the peak of E_v being further out than that of E_u. (It should be noted that the cross stream coordinate has not been normalized by the width of the layer, nor has the origin been shifted.) The measured peak value of E_v is 2.3 and that calculated is 2.5. On the other hand, with $C_{\varepsilon 3} = 0$ and $C_{\varepsilon 3} = \infty$, the calculated values are 3.2 and 0.8, respectively. The intermediate value $C_{\varepsilon 3} = 1/4$ was chosen to give approximately correct peak values of E_u, E_v and S_v at this location.

The measurements at the upstream station ($x = 45$) clearly show that self-similarity has not been attained at that location (see Fig. 6a of [3]): the magnitudes of the excesses and skewness are higher than at the downstream station. In the calculations too these magnitudes decrease with time. But there is no unambiguous means of determining the appropriate time t at which to compare calculations with the measurements at $x = 45$ cm. In Fig. 5 the calculated profiles of E_u, E_v and S_v at $t = 6$ are compared to those measured at $x = 45$ cm. This time ($t = 6$) was chosen for the comparison because the peak values were in general agreement. Again, in shape and magnitude there is good agreement between calculations and measurements. The agreement would be improved if the width of the calculated layer were slightly greater, and hence the calculated profiles were shifted slightly to the left.

The fundamental effect of different values of $C_{\varepsilon 3}$ is most evident in the conditional variances. We define $\langle u^2 \rangle_s$ and $\langle u^2 \rangle_l$ to be the variances of u conditional upon the fluid having originated on the small-scale and large-scale sides respectively. Figure 6 shows $\langle u^2 \rangle_s$ plotted against the unconditional variance $\langle u^2 \rangle$ for the three values of $C_{\varepsilon 3}$: 0, 1/4 and ∞. With $C_{\varepsilon 3} = 0$, the dissipation $\hat{\varepsilon}(t)$ of all fluid particles originating from the small-scale side is $\varepsilon_s(t)$, irrespective of the particle location $\hat{x}(t)$. Consequently the conditional variance $\langle u^2 \rangle_s$ is almost independent of position (and hence of $\langle u^2 \rangle$). (There is a small dependence on

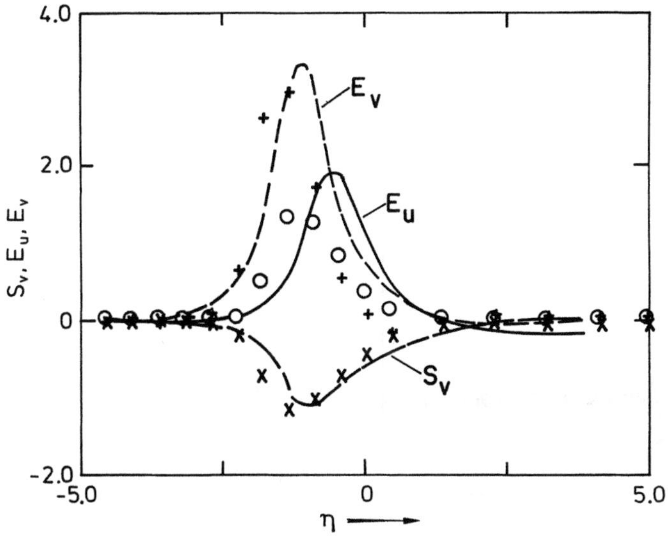

Fig. 5. Excesses of u and v and skewness of v. Lines are calculations ($r = 7$, $C_{\varepsilon 3} = 1/4$, $t = 6$), symbols are experimental data of Veeravalli and Warhaft (3:1 grid, $x = 45$ cm)

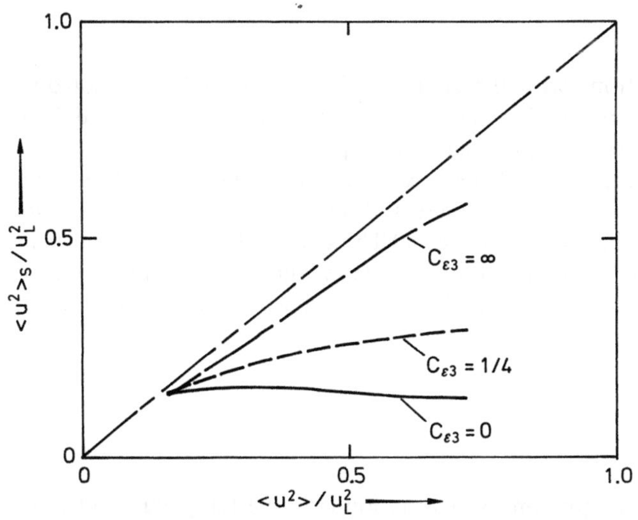

Fig. 6. Calculated variances of u, conditional on the fluid having originated on the small-scale side, against the unconditional variance ($r = 7$, $t = 15$, $C_{\varepsilon 3} = 0, 1/4, \infty$)

position because the time scale τ varies across the mixing layer, even though it has the same value in both uniform streams.) At the opposite extreme, with $C_{\varepsilon 3} = \infty$, the particle dissipation $\hat{\varepsilon}(t)$ is always equal to the local mean ε. These fluid particles rapidly lose the influence of their origin and the conditional variance $\langle u^2 \rangle_s$ is close to the unconditional variance $\langle u^2 \rangle$. With the intermediate value $C_{\varepsilon 2} = 1/4$, this conditional variance $\langle u^2 \rangle_s$ increases above its boundary value, but is significantly below the value with $C_{\varepsilon 3} = \infty$.

We now consider Veeravalli and Warhaft's data for the 7:1 grid. In the experiment, because the uniform streams are not completely isotropic, the variance ratios $\langle u^2 \rangle_L / \langle u^2 \rangle_s = 30$ and $\langle v^2 \rangle_L / \langle v^2 \rangle_S = 22$ are not the same. In order to make comparisons

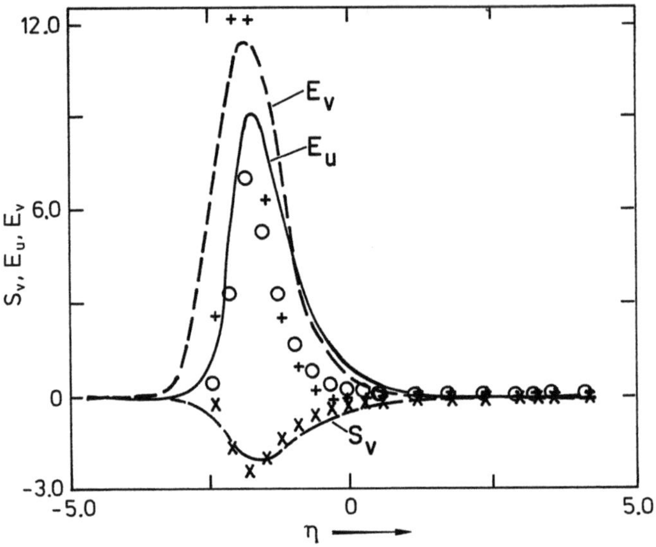

Fig. 7. Excesses of u and v and skewness of v. Lines are calculations ($r = 25$, $C_{\varepsilon 3} = 1/4$, $t = 15$), symbols are experimental data of Veeravalli and Warhaft (7:1 grid, $x = 62$ cm)

with these data, we made calculations with the energy ratio $r = 25$ (and with isotropic streams). Since the one new model constant $C_{\varepsilon 3}$ was chosen by reference to the 3:1 grid data, the following calculations can, to an extent, be regarded as predictions.

Figure 7 shows the calculated excesses and skewness at $t = 15$ compared to Veeravalli and Warhaft's 7:1 grid data at the single measurement station $x = 62$ cm. (It is not known whether self-similarity has been attained at this location.) The shapes of the curves and their relative positions are similar to the 3:1 grid results, but the magnitudes are much greater. The calculated peak excess of v is $E_v = 11.5$ which agrees well with the measured value of 12.1.

Discussion and Conclusion

With the use of a novel Lagrangian equation for dissipation (15), the Langevin model has been successfully applied to the turbulence mixing layer. The most revealing statistics in the flow are the velocity skewness and excesses. These quantities are calculated in good agreement with the data of Veeravalli and Warhaft [3] for both 3:1 and 7:1 grids. The calculated $\langle v^2 \rangle$ profile is slightly wider than the $\langle u^2 \rangle$ profile, but the distinctive measured profile shapes [3] were not observed.

The standard Langevin model, in which the local mean dissipation rate ε is used, is recovered from the present model with the model constant value $C_{\varepsilon 3} = \infty$. Then the particle dissipation $\hat{\varepsilon}$ relaxes instantaneously to the local mean ε. But it is significant that the appropriate value of $C_{\varepsilon 3}$ is quite small ($C_{\varepsilon 3} = 1/4$). This indicates that the relaxation is quite slow and hence the particle's history significantly influences its evolution.

The Lagrangian dissipation equation (15) is stated here for the case of isotropic turbulence with no mean velocity gradients. But the equation is readily generalized and its use in *pdf* calculations of inhomogeneous flow (e.g. [13]) will be investigated.

Further insight into the statistical nature of the dissipation can be found in Kolmogorov's similarity hypotheses of 1962 and their consequences [15]. A refined particle dissipation model based on these considerations is in its early stages of development [16]. The new model replaces (15) with a stochastic diffusion process such that $\hat{\varepsilon}(t)$ has a log-normal distribution.

Until recently, the turbulence mixing layer had not been studied experimentally, and the flow has been ignored by turbulence modellers. It is an extremely interesting flow from which much can be learned. In future experiments it would be useful to include a small temperature difference between the streams. Then, not only could scalar mixing be studied, but temperature could serve as an (imperfect) indicator of the origin of fluid particles. Consequently it would be possible to obtain conditional statistics which reveal a great deal about the processes involved (see Fig. 6, for example).

Acknowledgement. We are grateful to S. Veeravalli and Z. Warhaft for making their preliminary results available to us.

References

1. Gilbert, B. (1980): Diffusion mixing in grid turbulence without mean shear. J. Fluid Mech. **100**, 349–365
2. Gilbert, B. (1976): "An Experimental Investigation of Turbulent Mixing of Fluids with Different Dynamically Significant Scales," Ph.D. Thesis, University of Illinois at Chicago Circle
3. Veeravalli, S., Warhaft, Z. (1986): "The Interaction of Two Distinct Turbulent Velocity Scales in the Absence of Mean Shear," this volume, p. 21
4. Pope, S. B. (1981): Transport equation for the joint PDF of velocity and scalars in turbulent flow. Phys. Fluids **24**, 588–596
5. Pope, S. B. (1985): PDF methods for turbulent reactive flows. Prog. Energy Combust. Sci. **11**, 119–192
6. Libby, P. A. (1975): On the predicition of intermittent turbulent flows. J. Fluid Mech. **68**, 273–295
7. Byggstoyl, S., Kollmann, W. (1981): Closure model for intermittent turbulent flows. Int. J. Heat Mass Transfer **24**, 1811–1822
8. Pope, S. B. (1984): Calculations of a plane turbulent jet. AIAA J. **22**, 896–904
9. Pope, S. B. (1983): A Lagrangian two-time PDF equation for turbulent flows. Phys. Fluids **26**, 3448–3450
10. Haworth, D. C., Pope, S. B. (1985): A generalized Langevin model for turbulent flows. Phys. Fluids **29**, 387–405
11. Anand, M. S., Pope, S. B. (1984): "Diffusion Behind a Line Source in Grid Turbulence," in *Turbulent Shear Flows 4*, ed. by L. J. S. Bradbury et al. (Springer, Berlin, Heidelberg), pp. 46–61
12. Obukhov, A. M. (1959): Description of turbulence in terms of Lagrangian variables. Adv. Geophys. **6**, 113–116
13. Haworth, D. C., Pope, S. B. (1985): "Application of a Generalized Langevin Model to the Two-Dimensional Mixing Layer," Fifth Symposium on Turbulent Shear Flows, Cornell University
14. Launder, B. E., Spalding, D. B. (1972): *Mathematical Models of Turbulence*, (Academic, New York)
15. Monin, A. S., Yaglom, A. M. (1975): *Statistical Fluid Mechanics*, Vol. 2, (MIT Press, Cambridge)
16. Haworth, D. C., Pope, S. B. (1985): "Langrangian dissipation models for turbulent flows," Bull. Am. Phys. Soc. **30**, 1694

On the Structure of Homogeneous Turbulence

Moon J. Lee[1] and William C. Reynolds

Department of Mechanical Engineering, Stanford University, Stanford, California 94305, U.S.A.

Abstract

Full turbulence simulation has been conducted on homogeneous turbulence in irrotational strain and relaxation processes. Effects of the mode and strength of the imposed mean strain rate on the Reynolds stress, dissipation and vorticity fields are analyzed, yielding new insight. The Reynolds stress anisotropy depends on both the strain and the strain rate. The vorticity anisotropy is always bigger than the Reynolds stress anisotropy at least in the flows we computed. During relaxation, the trajectories of the large-scale (Reynolds stress) anisotropy invariants relax in a nonunique manner, while those of the small-scale turbulence (dissipation and vorticity fields) relax uniformly. Non-unique relaxation trajectories on the Reynolds stress invariant map indicate that the evolution of the large-scale turbulence cannot be modeled solely in terms of the large scale itself, contrary to the hypothesis that current second-moment models make. On the other hand, the tensorial evolution of the small-scale turbulence can be modeled in terms of the small scales.

1. Introduction

We consider homogeneous turbulence subject to irrotational 'simple' strains (plane strain, axisymmetric contraction and expansion) and the relaxation of these flows from strain. In a distorting turbulent flow, dynamics and structure of turbulence are affected by the mean field deformation. Relaxing turbulent flow is simpler to model, because there is no mean velocity gradient imposed on the turbulence field. In homogeneous relaxing turbulence, the terms explicitly involving the mean velocity gradient vanish from the Reynolds stress transport equation, leaving the pressure-strain-rate and dissipation rate terms only. However, the instantaneous absence of the mean deformation does not necessarily imply that turbulence behavior is independent of the previous strain history.

Experimental study of homogeneous turbulent flow subject to irrotational strain has been conducted in a wind-tunnel (e.g., Townsend, 1954; Uberoi, 1956, 1957; Tucker and Reynolds, 1968; Gence and Mathieu, 1979, 1980; Choi, 1983). Effects of the mean field deformation on turbulence have been studied in a distorting section, and experiments on relaxing flows have been performed by extending straight duct to the end of the distorting section. These experiments provide precious information for enhancement of our understanding of turbulence behavior. However, due to inherent difficulties in measuring techniques, quantities involving pressure fluctuations and higher-order statistics were often unavailable. Full turbulence simulation (FTS) is best suited to complement laboratory experiments, providing *data* useful for studying and modeling turbulence (Rogallo, 1981).

This paper provides a preliminary report on our FTS work. In Sect. 2, the mean field deformations are specified, and the anisotropy tensors and their invariants are briefly intro-

[1] Present address: MS 202A-1, NASA-Ames Research Center, Moffett Field, Calif.

duced. Section 3 describes the effects of the mean field deformation on turbulence. Discussion on relaxation of turbulence from strain and modeling procedure is presented in Sect. 4.

2. Formulation

The dynamical equations are taken as

$$\nabla \cdot \boldsymbol{u} = 0, \tag{1}$$

$$\frac{\bar{D}\boldsymbol{u}}{Dt} = -\boldsymbol{u} \cdot \nabla \boldsymbol{U} - \boldsymbol{u} \cdot \nabla \boldsymbol{u} - \frac{1}{\varrho} \nabla p + \nu \nabla^2 \boldsymbol{u}, \tag{2}$$

where $\boldsymbol{u}$ and p are the fluctuating vector velocity and pressure, $\boldsymbol{U}$ the specified mean velocity and $\bar{D}/Dt = \partial/\partial t + \boldsymbol{U} \cdot \nabla$.

The mean fields were all of the form

$$U_i(\boldsymbol{x}, t) = U_{i,j}(t)\, x_j. \tag{3}$$

We write the mean strain rate tensors as

(i) Plane strain (PS):

$$U_{i,j} = S \begin{pmatrix} 0 & 0 & 0 \\ 0 & -1 & 0 \\ 0 & 0 & 1 \end{pmatrix}; \tag{4}$$

(ii) Axisymmetric contraction (AC):

$$U_{i,j} = \frac{2}{\sqrt{3}} S \begin{pmatrix} 1 & 0 & 0 \\ 0 & -\frac{1}{2} & 0 \\ 0 & 0 & -\frac{1}{2} \end{pmatrix}; \tag{5}$$

(iii) Axisymmetric expansion (AE):

$$U_{i,j} = \frac{2}{\sqrt{3}} S \begin{pmatrix} -1 & 0 & 0 \\ 0 & \frac{1}{2} & 0 \\ 0 & 0 & \frac{1}{2} \end{pmatrix}; \tag{6}$$

(iv) Relaxation:

$$U_{i,j} = 0. \tag{7}$$

Here, the *equivalent mean strain rate*

$$S = (S_{ij} S_{ij}/2)^{1/2} \tag{8}$$

denotes the strength (or intensity) of the mean strain rate tensor S_{ij} in a way consistent with the invariant modeling procedure. In a distorting flow, the *mean strain rate parameter*

$$S^* = \frac{S q^2}{\varepsilon} \tag{9}$$

is an important quantity, since it represents the time-scale ratio of the turbulence field to the mean deformation.

The Navier-Stokes equations were transformed to a coordinate system moving with the mean flow, and solved numerically. The basic numerical procedure was that of Rogallo (1981). The computations were carried out using a modified version of Rogallo's computer code for homogeneous turbulence with a 128 × 128 × 128 mesh on the Cray X-MP computer at the NASA-Ames Research Center. See Lee (1985) for more details of the FTS work.

In order to understand the structural behavior of turbulence, evolution of the anisotropy tensors and their invariants was investigated. The anisotropy tensor b_{ij} and the anisotropy invariants II_b and III_b of the Reynolds stress tensor $R_{ij} = \overline{u_i u_j}$ are given by

$$b_{ij} = \frac{R_{ij}}{R_{kk}} - \frac{1}{3}\,\delta_{ij}, \qquad II_b = -\frac{1}{2}\,b_{ij}b_{ji}, \qquad III_b = \frac{1}{3}\,b_{ij}b_{jk}b_{ki}. \tag{10}$$

By the exactly same procedure, we can define the anisotropy tensors and their invariants of the dissipation rate tensor $D_{ij} = 2\nu\overline{u_{i,k}u_{j,k}}$ and the vorticity correlation tensor $V_{ij} = \overline{\omega_i \omega_j}$, respectively:

$$d_{ij} = \frac{D_{ij}}{D_{kk}} - \frac{1}{3}\,\delta_{ij}, \qquad II_d = -\frac{1}{2}\,d_{ij}d_{ji}, \qquad III_d = \frac{1}{3}\,d_{ij}d_{jk}d_{ki}, \tag{11}$$

$$v_{ij} = \frac{V_{ij}}{V_{kk}} - \frac{1}{3}\,\delta_{ij}, \qquad II_v = -\frac{1}{2}\,v_{ij}v_{ji}, \qquad III_v = \frac{1}{3}\,v_{ij}v_{jk}v_{ki}. \tag{12}$$

Notice that anisotropy tensors are dimensionless and trace-free by definition, and that each anisotropy invariant map (AIM in short) has the boundaries first identified by Lumley and Newman (1977) for the Reynolds stress AIM. It should be emphasized that these AIM's are useful tools in the study of anisotropy evolutions and turbulence structures.

3. Distortion of Turbulence by Strain

Run specifications for the strained flows are given in Table 1. Within each class of flows all runs had the same initial conditions, but were strained at rates differing by two orders of magnitude ($S^* = 0.7 \sim 154.0$). Due to limitations of mesh size and requirements on the

Table 1. Specifications for strain runs

Run	Strain mode	Initial field	Final field	Re_l	S^*
PXB	PS	HIA9	PXB4	41.8 ~ 70.1	2.0 ~ 3.7
PXD		HIA9	PXD4	43.8 ~ 57.9	6.6 ~ 8.0
PXF		HIA9	PXF4	39.0 ~ 46.7	63.6 ~ 154.0
AXK	AC	HIC6	AXK6	48.1 ~ 61.2	1.0 ~ 2.1
AXL		HIC6	AXL6	54.4 ~ 61.2	5.7 ~ 9.7
AXM		HIC6	AXM5	49.2 ~ 61.2	38.2 ~ 96.5
EXO	AE	HID6	EXO6	44.7 ~ 69.7	0.7 ~ 3.0
EXP		HID6	EXP6	55.0 ~ 84.3	7.0 ~ 8.5
EXQ		HID6	EXQ6	55.0 ~ 58.2	47.6 ~ 70.7

PS, plane strain; AC, axisymmetric contraction; AE, axisymmetric expansion.
$Re_l = q^4/(\nu\varepsilon)$; $S^* = Sq^2/\varepsilon$ where $S = (S_{ij}S_{ij}/2)^{1/2}$.

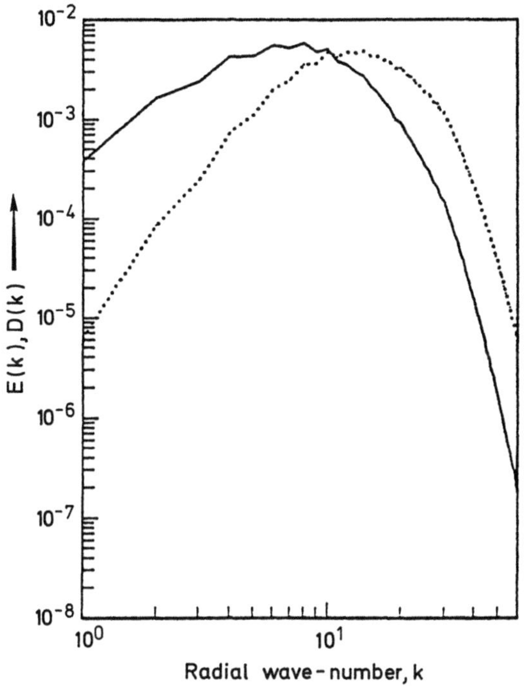

Fig. 1. Initial energy and dissipation rate spectrum functions of isotropic field HIA9 (initial conditions for plane strain runs): ——, $E(k)$; ·····, $D(k)$

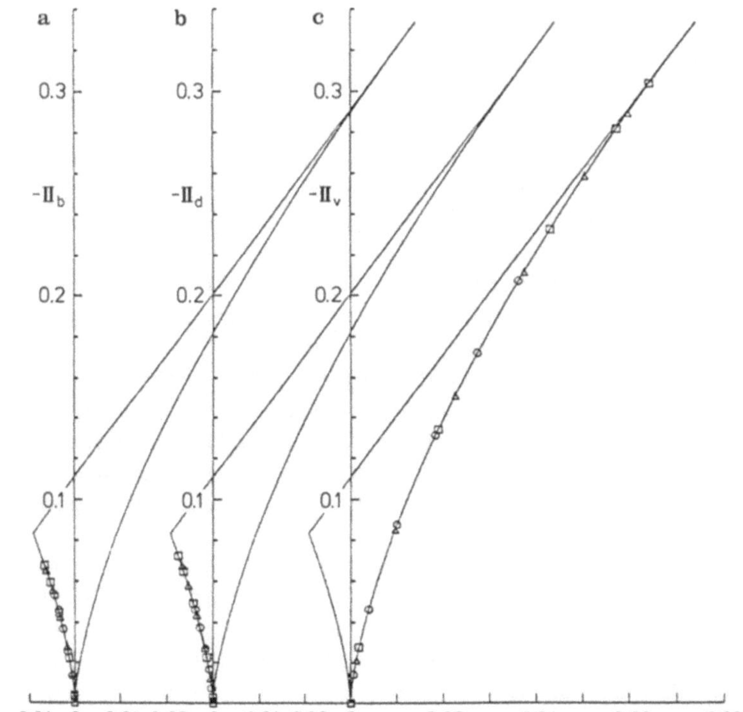

Fig. 2. Trajectories for axisymmetric contraction flows on the AIM's: (a) velocity; (b) dissipation rate; (c) vorticity. ⊙, AXK; △, AXL; □, AXM (Trajectories follow symbols moving up, away from isotropy)

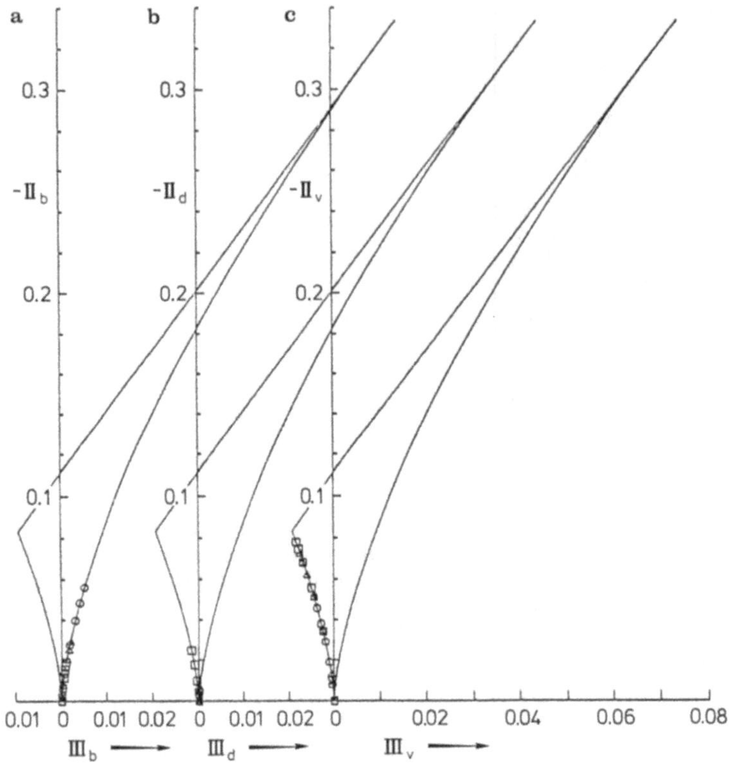

Fig. 3. Trajectories for axisymmetric expansion flows on the AIM's: (a) velocity; (b) dissipation rate; (c) vorticity. ○, EXO; △, EXP; □, EXQ (Trajectories follow symbols moving up, away from isotropy)

small-scale resolution, the turbulent Reynolds numbers $q^4/(\nu\,\varepsilon)$ were limited to less than 100.

Typical initial energy and dissipation rate spectra (before strain) are shown in Fig. 1. Note that the small-scale contributions are captured. (The ratio of the Kolmogorov micro-scale $(\nu^3/\varepsilon)^{1/4}$ to the nominal grid spacing was $O(1)$ for all the cases listed in Table 1.)

Figure 2 shows the AIM's for axisymmetric contraction runs. The (0, 0) point on each plot corresponds to isotropic turbulence (the initial conditions), the upper line corresponds to a two-dimensional anisotropy in the variable, and the upper-most point to a one-dimensional field in the variable. In axisymmetric contraction runs, the flow state moves up the left boundary in the Reynolds stress and dissipation rate AIM's, and up the right boundary in the vorticity AIM. Contraction tends to produce a nearly one-dimensional vorticity field (two-dimensional velocity field). The final points shown are for the same total strain, and it is seen that the highest strain-rate case (run AXM) produces the greatest anisotropy in all variables.

Figure 3 shows the AIM's for axisymmetric expansion, for which the trajectories on the Reynolds stress AIM follow the right boundary and those on the vorticity AIM (and apparently those on the dissipation rate AIM) follow the left boundary. Again, the final points all have the same total strain. Here the case with the greatest strain rate (run EXQ) produces the least the Reynolds stress anisotropy, but the largest vorticity and dissipation rate anisotropy.

Figure 4 shows the AIM's for plane strain, which can be viewed as a linear combination of axisymmetric contraction and expansion. Trajectories show that a stronger mean strain rate creates more anisotropy in the small scales, i.e., the dissipation rate and vorticity fields.

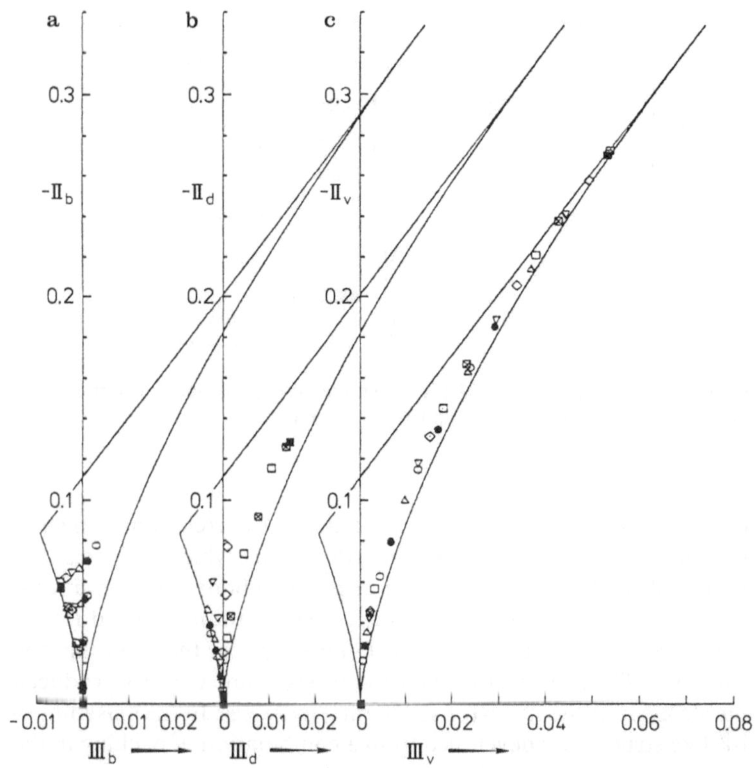

Fig. 4. Trajectories for plane strain flows on the AIM's: (a) velocity; (b) dissipation rate; (c) vorticity. △, PXB; ◇, PXD; □, PXF. For other symbols, see Lee (1985) (Trajectories follow symbols moving up, away from isotropy)

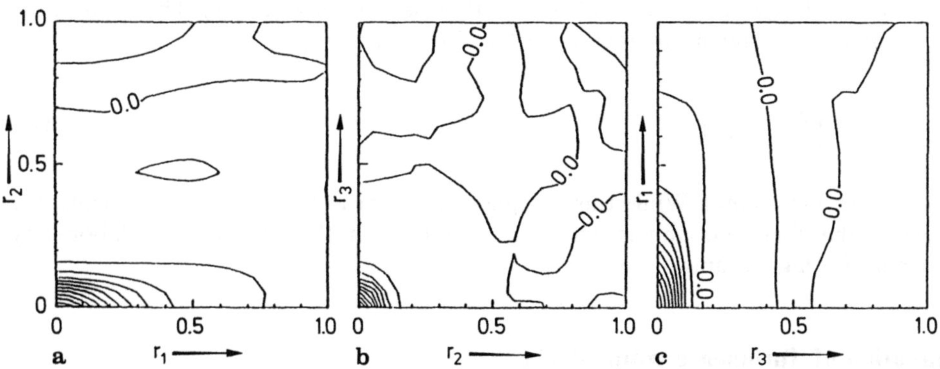

Fig. 5. Isocorrelation contour plots of field AXM2: $S = 86.60$, axisymmetric contraction, half-domain size = (2.903, 3.268, 3.268). (a) $W_{ii}^*(r_1, r_2, 0)$; (b) $W_{ii}^*(0, r_2, r_3)$; (c) $W_{ii}^*(r_1, 0, r_3)$

However, the Reynolds stress anisotropy is almost independent of the mean strain rate imposed during distortion, and depends more on the total strain. Trajectories on the vorticity AIM exhibit the similarity of the structure in the vorticity field that exists in this flow.

In Fig. 5 and 6, the isocorrelation contour plots of the two-point vorticity correlation functions

$$W_{ii}^*(\boldsymbol{r}) = \frac{\overline{\omega_i(\boldsymbol{x})\,\omega_i(\boldsymbol{x}+\boldsymbol{r})}}{\overline{\omega_k(\boldsymbol{x})\,\omega_k(\boldsymbol{x})}} \tag{13}$$

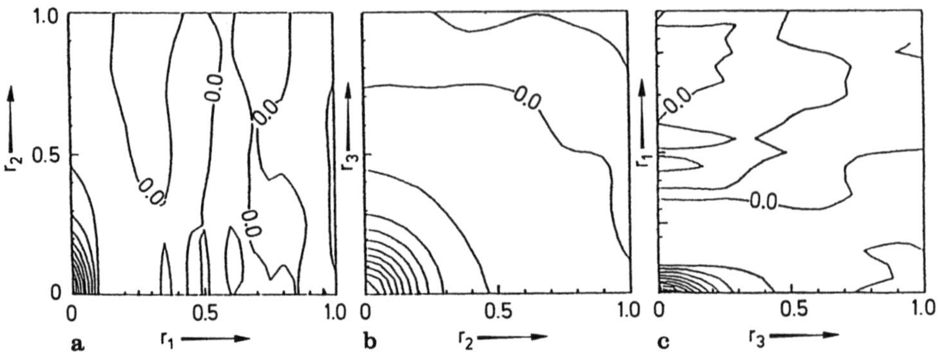

Fig. 6. Isocorrelation contour plots of field EXQ3: $S = 62.13$, axisymmetric expansion, half-domain size $= (3.138, 3.143, 3.143)$. (a) $W_{ii}^*(r_1, r_2, 0)$; (b) $W_{ii}^*(0, r_2, r_3)$; (c) $W_{ii}^*(r_1, 0, r_3)$

are presented for fields AXM2 and EXQ3, respectively. Note that the correlations vanish at distances much shorter than the half-domain size in all three directions, which indicates that the computational domain is sufficiently large for the use of periodic boundary conditions. When the axisymmetric contraction is imposed, the principal correlation ellipsoid (PCE) deduced from Fig. 5 has a *rod-like* shape, while it has a *disk-like* shape for the axisymmetric expansion case as shown in Fig. 6. The rod-like turbulence consists of line vortices produced by stretching; these become stronger as the strain (or strain rate) increases. In the axisymmetric expansion flow, the disk-like structure is developed from a combination of packing in the direction of negative strain rate and stretching in the plane of the disk.

Surprisingly enough, comparison of all the trajectories on the Reynolds stress and vorticity AIM's during distortion indicates that the vorticity field is always *more* anisotropic than the velocity field, contrary to the widely held belief. Indeed, anisotropy in vorticity is essential for anisotropy in the Reynolds stress (Biot-Savart consequence). The ratio of the time scale for vortex filament interaction to that for the strain

$$\frac{S}{\omega} = \frac{S^*}{\sqrt{\mathrm{Re}_l}} \quad \left(\mathrm{Re}_l = \frac{q^4}{\nu \varepsilon}\right) \tag{14}$$

is $0.1 \sim 20.0$ for these flows. While this is higher than typical laboratory experiments, the FTS suggests that there probably is significant anisotropy in the small scales in laboratory experiments with high strains.

4. Relaxation of Turbulence from Strain

Table 2 shows the run specifications for relaxing flows. Relaxation runs were performed starting with the fields developed under strain, i.e., using the fields from runs described in Table 1 as initial conditions. There was not much variation in the Reynolds numbers throughout each run.

In Fig. 7–9, it is shown that the anisotropies in the dissipation rate and vorticity fields 'return' toward isotropy. It is also seen that the dissipation rate anisotropy decreases faster than the Reynolds stress and vorticity anisotropy invariants, which confirms the traditional notion that the turbulence becomes isotropic faster at the small scales than at the large scales. Note that the return trajectories are uniform (well-behaved) on the dissipation rate and

Table 2. Specifications for relaxation runs

Run	Previous strain	Initial field	Final field	Re_l
B4R	PS	PXB4	B4R6	65.7 ~ 70.1
D4R		PXD4	D4R6	57.9 ~ 68.9
F4R		PXF4	F4R6	39.0 ~ 64.1
K6R	AC	AXK6	K6R6	39.5 ~ 52.3
L6R		AXL6	L6R6	55.5 ~ 58.4
M5R		AXM5	M5R6	49.2 ~ 61.0
O6R	AE	EXO6	O6R6	74.5 ~ 88.5
P6R		EXP6	P6R6	84.3 ~ 93.7
Q6R		EXQ6	Q6R6	56.7 ~ 83.6

PS, plane strain; AC, axisymmetric contraction; AE, axisymmetric expansion. $Re_l = q^4/(\nu\varepsilon)$.

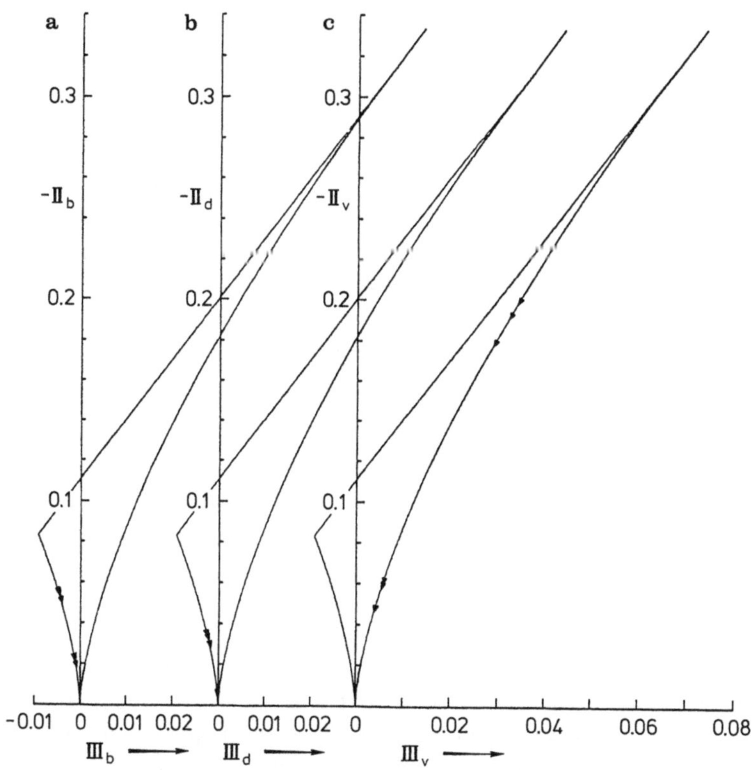

Fig. 7. Trajectories for relaxation from axisymmetric contraction on the AIM's: (a) velocity; (b) dissipation rate; (c) vorticity

vorticity AIM's. Comparison is made between values of II_d and II_v in Fig. 10 to show existence of definite correlations between them:

$$|II_d| \simeq |II_v|^2. \tag{15}$$

(However, the components d_{ij} are not correlated well with v_{ij}.)

Assuming that the tensorial evolution of the vorticity anisotropy tensor v_{ij} during relaxation is influenced by v itself only, the most general expression for the vorticity anisotropy

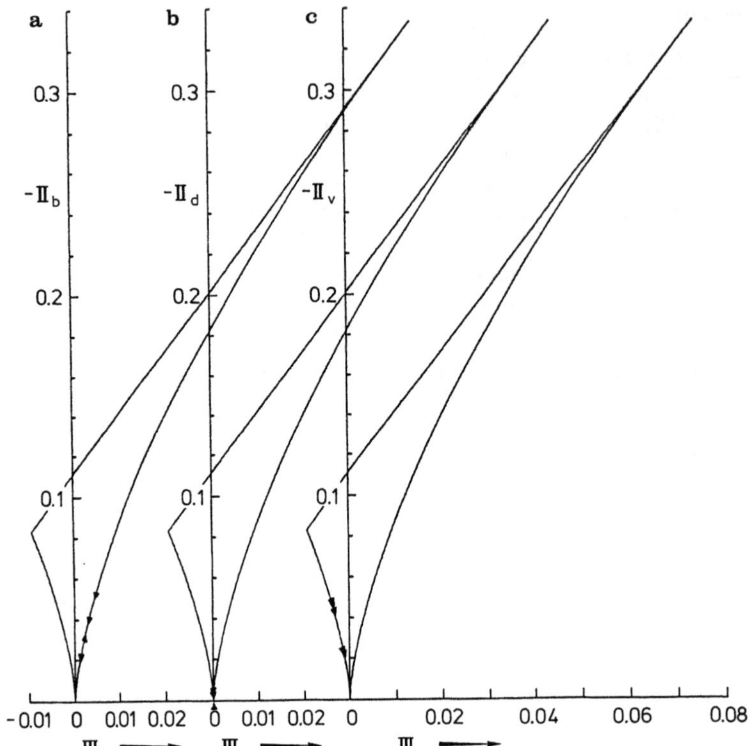

Fig. 8. Trajectories for relaxation from axisymmetric expansion on the AIM's: (a) velocity; (b) dissipation rate; (c) vorticity

during relaxation is given by

$$\frac{dv_{ij}}{dt} = -\frac{\varepsilon}{q^2}\left[\alpha_v v_{ij} + \beta_v\left(v_{ik}v_{kj} + \frac{2}{3}II_v\delta_{ij}\right)\right], \tag{16}$$

where α_v and β_v can be functions of the invariants. Note that α_v is the fractional rate of the vorticity anisotropy relaxation (or the vorticity relaxation rate). From Fig. 7–9 and others for cases not reported here, it is observed that the relaxation trajectories of the vorticity anisotropy invariants are *symmetric* about the axis $III_v = 0$ regardless of prior strain history. In order to have this feature, the coefficient β_v must be an odd function of III_v. The simplest choice of the coefficient β_v is to set

$$\beta_v = 0. \tag{17}$$

Thus, the relaxation trajectories of the dissipation anisotropy invariants are given by a family of hyperbolae

$$\frac{dIII_v}{dII_v} = \frac{3}{2}\frac{III_v}{II_v}, \tag{18}$$

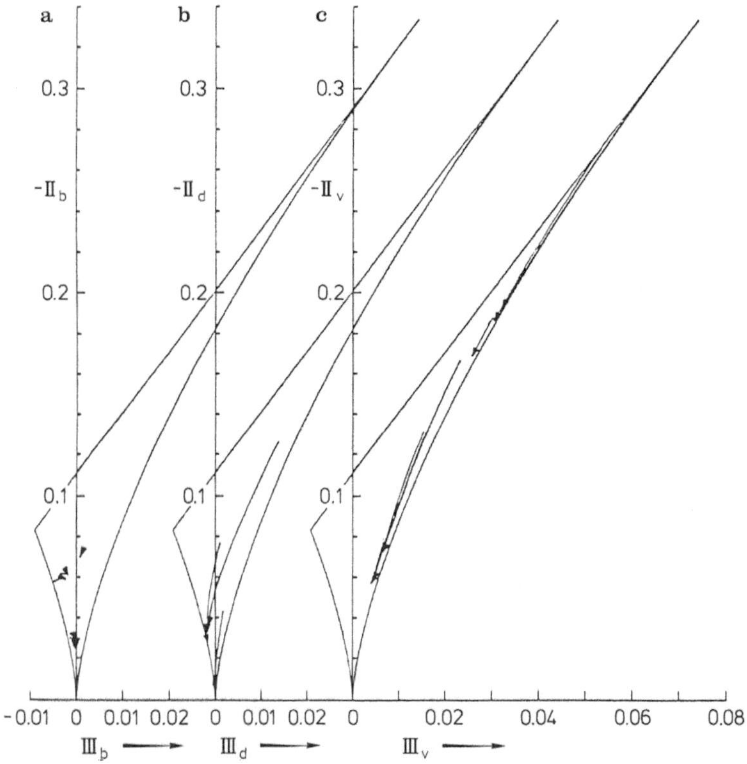

Fig. 9. Typical trajectories for relaxation from plane strain on the AIM's: (a) velocity; (b) dissipation rate; (c) vorticity

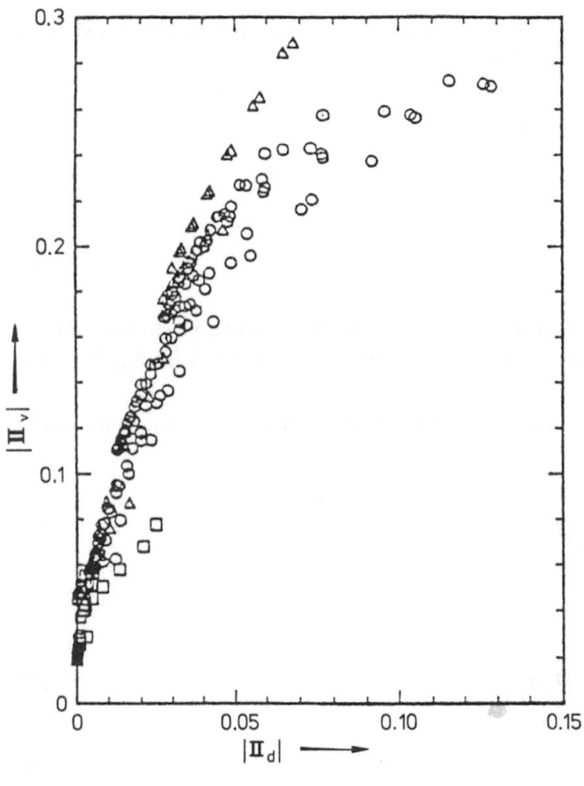

Fig. 10. Comparison of $|II_v|$ *vs.* $|II_d|$ during relaxation from simple strains: ⊙, relaxation from plane strain; △, relaxation from axisymmetric contraction; □, relaxation from axisymmetric expansion

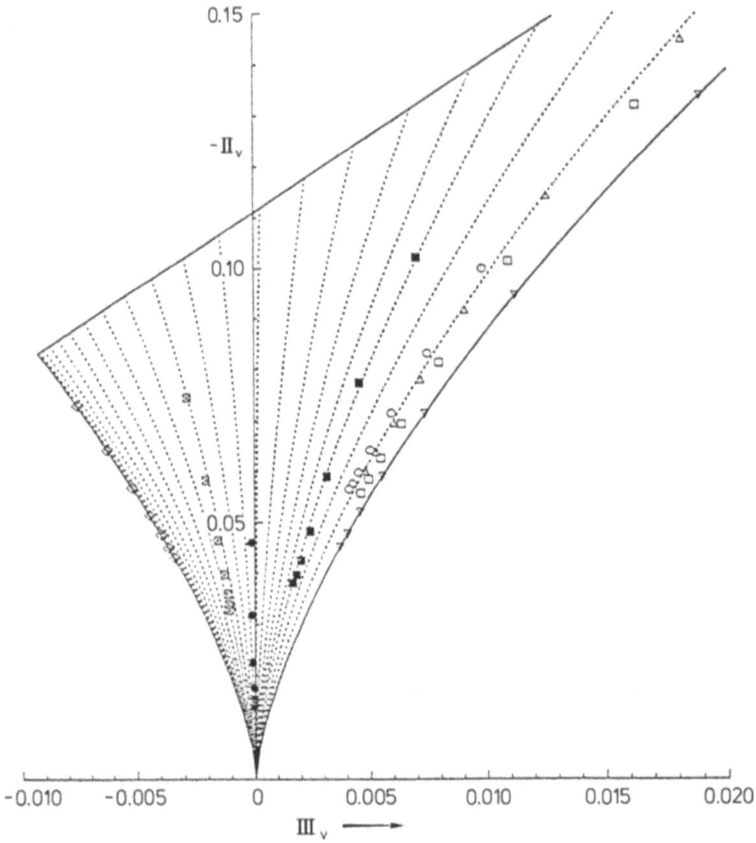

Fig. 11. Comparison of relaxation trajectories on the vorticity AIM: ······, trajectories predicted by the present model; symbols are the data taken from Lee (1985)

or, after integration,

$$III_v = \pm\, C_v\, |II_v|^{3/2}, \tag{19}$$

where the numerical factor C_v can be determined by initial values. Figure 11 shows that this model is in good agreement with data from the FTS on the vorticity AIM (including cases not covered by Table 2).

The dissipation anisotropy transport during relaxation also can be modeled in an analogous form:

$$\frac{dd_{ij}}{dt} = -\frac{\varepsilon}{q^2}\left[\alpha_d d_{ij} + \beta_d\left(d_{ik}d_{kj} + \frac{2}{3}\, II_d \delta_{ij}\right)\right], \tag{20}$$

or, equivalently,

$$\frac{dd_{ij}}{dt} = -\frac{\varepsilon}{q^2}\,\alpha_d\left[d_{ij} + \gamma_d\left(d_{ik}d_{kj} + \frac{2}{3}\, II_d \delta_{ij}\right)\right] \tag{21}$$

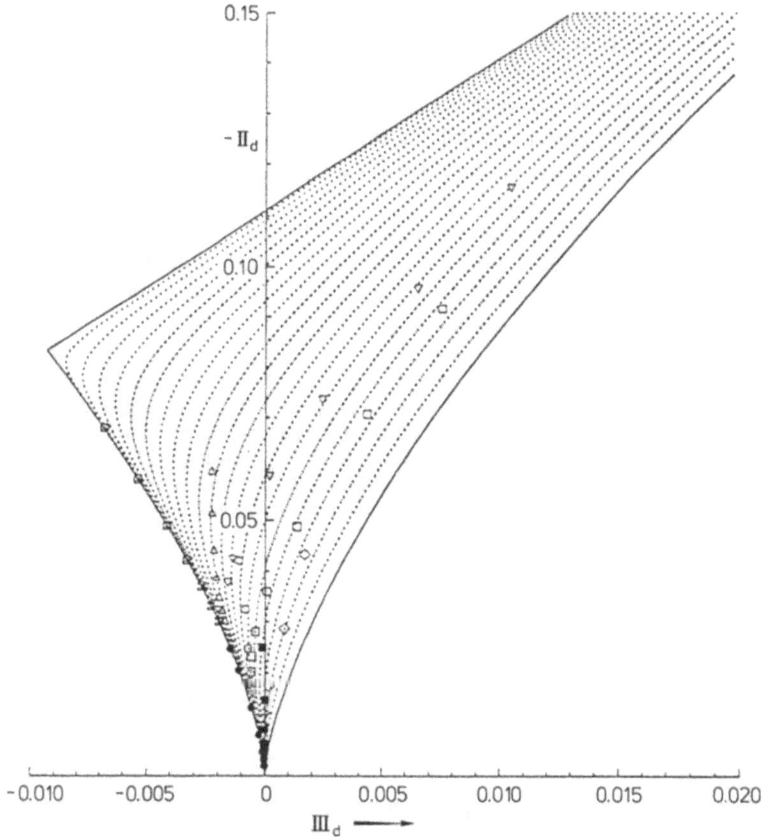

Fig. 12. Comparison of relaxation trajectories on the dissipation rate AIM: ······, trajectories predicted by the present model; symbols are the data taken from Lee (1985)

where

$$\gamma_d = \beta_d/\alpha_d. \tag{22}$$

Requiring that this model be consistent with the realizability condition (Schumann, 1977; Lumley, 1978) applied for the two-dimensional turbulence (where one component of the fluctuating velocity remains zero identically everywhere), one finds

$$\gamma_d = \frac{3}{1 + 6\,II_d}. \tag{23}$$

Of course, this must be true only for the two-dimensional turbulence. If we assume that this relationship holds for general case, then the relaxation trajectories on the dissipation rate AIM are given by

$$\frac{dIII_d}{dII_d} = \frac{3\,III_d + 2\,II_d^2 + 18\,II_d III_d}{2\,II_d - 9\,III_d + 12\,II_d^2}. \tag{24}$$

Figure 12 shows that the relaxation trajectories of this model agree well with the data from the FTS. Therefore, it is concluded that this model is quite good for general case.

In order to complete the models, we need to determine the relaxation rates α_v and α_d. Preliminary study indicated that the small-scale relaxation rates are strongly dependent upon the large-scale anisotropy (II_b). This work will be reported elsewhere.

In contrast with the small-scale relaxation, the relaxation trajectories on the Reynolds stress AIM are not unique as shown in Fig. 7–9. Thus, the tensorial evolution of the Reynolds stress anisotropy may not be modeled solely in terms of the Reynolds stress anisotropy itself, which most current turbulence models assume (e.g., Launder et al, 1975; Lumley and Newman, 1977; Lumley, 1978). It is possible that the relaxation of the large-scale anisotropy is also coupled with the small-scale turbulence.

Acknowledgements. We are grateful to Dr. R. S. Rogallo at the NASA-Ames Research Center for sharing his expertise and cooperation in numerical work. Prof. J. H. Ferziger at Stanford University and Dr. A. Leonard at the NASA-Ames Research Center (now at Caltech, Pasadena) deserve special thanks for their suggestions and interests. The computer time made available through Grant NCC-2-15 by the NASA-Ames Research Center is gratefully acknowledged. One of us (M.J.L.) was supported in part through the Korean Government Scholarship by the Ministry of Education, Korea and in part through Grant NCC-2-15 by the NASA-Ames Research Center during the course of this study.

References

Choi, K. S. (1983): "A Study of the Return to Isotropy of Homogeneous Turbulence;" Ph.D. Thesis, Cornell University, Ithaca, New York

Gence, J. N., Mathieu, J. (1979): On the application of successive plane strains to grid-generated turbulence. J. Fluid Mech. **93**, 501–513

Gence, J. N., Mathieu, J. (1980): The return to isotropy of an homogeneous turbulence having submitted to two successive plane strains. J. Fluid Mech. **101**, 555–566

Launder, B. E., Reece, G. J., Rodi, W. (1975): Progress in the development of a Reynolds stress turbulence closure. J. Fluid Mech. **68**, 537–566

Lee, M. J. (1985): "Numerical Experiments on the Structure of Homogeneous Turbulence;" Ph.D. Thesis, Stanford University, Stanford, California

Lumley, J. L. (1978): Computational modeling of turbulent flows. Adv. Appl. Mech. **18**, 123–176

Lumley, J. L., Newman, G. R. (1977): The return to isotropy of homogeneous turbulence. J. Fluid Mech. **82**, 161–178

Rogallo, R. S. (1981): Numerical experiments in homogeneous turbulence. NASA Tech. Memo. 81 315

Schumann, U. (1977): Realizability of Reynolds stress turbulence models,. Phys. Fluids **20**, 721–725

Townsend, A. A. (1954): The uniform distortion of homogeneous turbulence. Q. J. Mech. Appl. Math. **7**, 104–127

Tucker, H. J., Reynolds, A. J. (1968): The distortion of turbulence by irrotational plane strain. J. Fluid Mech. **32**, 657–673

Uberoi, M. S. (1956): Effect of wind-tunnel contraction on free-stream turbulence. J. Aero. Sci. **23**, 754–764

Uberoi, M. S. (1957): Equipartition of energy and local isotropy in turbulent flows. J. Appl. Phys. **28**, 1165–1170

Turbulence in a Stably Stratified Shear Flow: A Progress Report

J. J. Rohr[1], K. N. Helland[2], E. C. Itsweire[3] and C. W. Van Atta[4]

Institute for Pure and Applied Physical Sciences and Department of Applied Mechanics and Engineering Sciences, University of California, San Diego, La Jolla, Ca 92093, USA

Abstract

A preliminary investigation of an evolving stably stratified uniform gradient turbulent shear flow has been conducted in a ten-layer, salt-stratified water channel. Simultaneous single-point measurements of the mean and fluctuating density and the horizontal and vertical velocities were made over a wide range of downstream positions. This study is a first attempt to extend the stratified, unsheared grid-generated turbulence study reported in Stillinger, Helland and Van Atta (1983) to a stratified, uniform shear flow. A primary objective of the present study is to generalize the dynamics described by the length scale models developed in SHV by the addition of a uniform shear in order to obtain criteria which can be compared with more general atmospheric and oceanic measurements. Other objectives are the examination of the dependence of turbulent energy and length scales on the gradient Richardson number, refining and extending the earlier work of Webster (1964).

Nomenclature

g	Earth gravity
L_k	Kolmogorov scale
L_0	Ozmidov scale
L_t	Turbulent length scale
L_T	Thorpe scale
M	Grid mesh size
N	Brünt-Väisälä frequency
$q^2/2$	Total turbulent kinetic energy per unit mass
Ri	Gradient Richardson number
$\bar{U}$	Mean velocity
$\bar{U}_{℄}$	Centerline velocity
u	Downstream velocity fluctuation
w	Vertical velocity Fluctuation
x	Downstream coordinate axis
z	Vertical coordinate axis

Greek Symbols

ε	Turbulent kinetic energy dissipation rate
ε_{tr}	Transition dissipation rate
ν	Kinematic viscosity
$\bar{\varrho}$	Mean density
ϱ	Density fluctuation
τ	Dimensionless time

Suffixes

$'$	Denotes root mean square value
$\overline{(\ )}$	Denotes time average value

Introduction

In the initial studies of unstratified uniform mean shear flow, Rose (1966, 1970), Champagne, Harris and Corrsin (1970) and Mulhearn and Luxton (1975) did not observe growth in the turbulent kinetic energy because their wind tunnels were not sufficiently long enough for their chosen values of applied shear. In terms of a dimensionless development time τ ($\tau = (x/\bar{U})\,[(\partial\bar{U}/\partial z)]$, none of the aforementioned experiments exceeded a τ of 3.6. The growth of turbulent kinetic energy in a homogeneous shear flow was first documented

[1] Present address: Naval Ocean Systems Center, San Diego, CA
[2] Also Data Ready, P.O. Box 19549, Birmingham, AL
[3] Present address: Chesapeake Bay Institute, Johns Hopkins University
[4] Also Scripps Institution of Oceanography

Turbulent Shear Flows 5

by Harris, Graham and Corrsin (1977) for τ's exceeding 4. Tavoularis and Corrsin (1981) extended these measurements by superimposing a passive scalar (temperature) on the same velocity field.

Even before the earliest results for unstratified uniform mean shear appeared, Webster (1964) succeeded in generating a stratified shear flow, using temperature in air to produce buoyancy effects. At both 1.73 and 4.57 m downstream of the shear generating grid, he measured turbulent diffusivities, kinetic energies and length scales as a function of the gradient Richardson number [$\mathrm{Ri} = N^2/(\partial\bar{U}/\partial z)^2$]. Unfortunately, Webster (1964) found a different dependence at each position, which he attributed to the flow not reaching a 'steady state'. A rough estimate of the τ associated with Webster's grid can be achieved from the formula of Owen and Zienkiewicz (1957):

$$\bar{U}(z) = \bar{U}_{\mathrm{C\!L}}[1 + 0.85(z/h - 1/2)],$$

where $\bar{U}(z)$ is the velocity at the heigth z measured from the tunnel floor, $\bar{U}_{\mathrm{C\!L}}$ is the centerline velocity and h the height of the tunnel. This formula predicts $\tau \simeq 3$ and 7 at the respective measuring positions of 1.73 and 4.57 meters downstream of the grid. Apparently the measurements at the first position ($\tau < 4$) were appreciably affected by the decaying turbulence of the shear generating grid.

For each of our experiments τ ranged from 0 to 9, and as expected for the unstratified shear, a distinct growth in the turbulence was observed between $\tau = 5$ and 9. By repeating these measurements with different density gradients the effect of buoyancy on the evolution of turbulence in a uniform mean shear could be investigated.

Experimental Facility

The UCSD ten-layer water channel was designed to produce a stably stratified, uniform gradient shear for sufficiently long time to permit detailed investigation of turbulent statistical quantities. The facility has been described in Stillinger, Head, Helland and Van Atta (1983). The velocity profile was independent of time and varied only slightly with downstream position x. The density profile decayed in time, but the rate of decay was slow enough that statistics of velocity and density at a point were not significantly contaminated. A biplane grid of mesh length $M = 1.905$ cm was used at the inlet to the test section to remove the effects of the discrete layers required to produce the shear flow. Measurements of detailed turbulence statistics and profiles were made at selected downstream locations x, where x was measured from the turbulence producing grid. The water channel had the same configuration used for the Stillinger, Helland and Van Atta (1983) (hereinafter referred to as SHV) measurements thereby providing the opportunity for direct comparison with their unsheared case.

A commercial TSI X-film probe was used to measure downstream ($\bar{U}$, u) and vertical (w) velocity components, while a four-wire conductivity probe developed by Head (1983) measured the local instantaneous conductivity of the salt solution. The center-to-center separation between the conductivity and X-film probes was small enough (< 1 mm) to measure the buoyancy flux $(g/\bar{\varrho})\,\overline{\varrho w}$ without correcting for sensor separation.

Measurements

Measurements were first made without density stratification in order to compare the results with other uniform mean shear flow facilities and to establish a baseline for comparison with

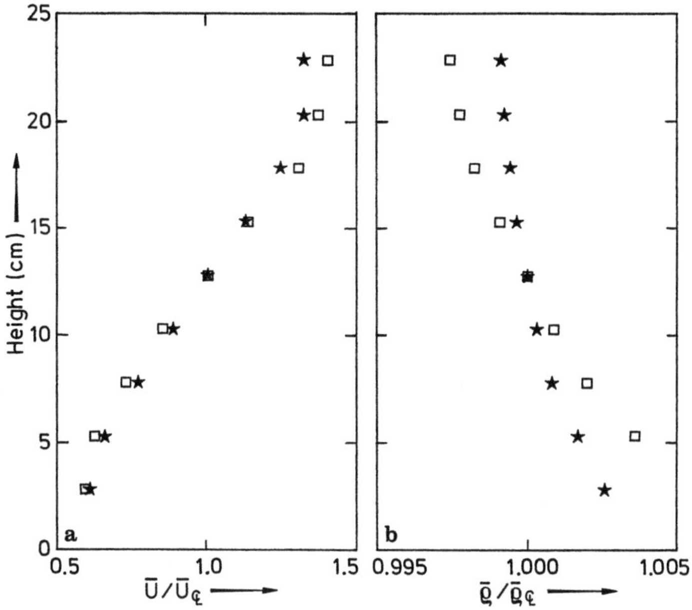

Fig. 1. Vertical velocity (a) and density (b) profiles. □, $x/m = 20$; ★, $x/M = 70$ (30 minutes later)

stratified shear experiments in the present facility. The development of the turbulent component energies $u'^2/\bar{U}^2_{\mathbb{C}}$ and $w'^2/\bar{U}^2_{\mathbb{C}}$ and shear-stress correlation coefficient $-\overline{uw}/u'w'$ compare very well with both the grid decay and shear dominated regions of Champagne, Harris and Corrsin (1970) and Harris, Graham and Corrsin (1977). The Taylor microscale was estimated using isotropic assumptions and the resulting variation with downstream position x/M exhibits a linear growth region followed by an asymptotic approach to a constant value, also in agreement with their data. The measured one-dimensional energy spectra display large relative increases in the low wavenumber regions consistent with the expected downstream increases in integral length scales. Inferring length scales from these spectra, we find that these scales are smaller than the vertical region over which the velocity and density gradients are constant.

Measurements were then made with a stratification added to the shear. The velocity profiles in Fig. 1a are shown at two downstream locations, x/M of 20 and 70. As the density profile decays in time, sets of downstream measurements characterized by different stabilities can be compiled. The two density profiles shown in Fig. 1b were from two such sets taken thirty minutes a part and characterized by Brünt-Väisälä frequencies N $[N^2 = (-g/\bar{\varrho})(\partial\bar{\varrho}/\partial z)]$ of 0.60 and 0.37 rad/sec. Both the velocity and density profiles remain nearly constant over the central half of the test section height. All detailed turbulence measurements were taken at the midpoint of the channel.

Results

The downstream development of the vertical turbulent kinetic energy, $w'^2/\bar{U}^2_{\mathbb{C}}$, is shown in Fig. 2a for five different stratifications. The longitudinal component $u'^2/\bar{U}^2_{\mathbb{C}}$ behaves in much the same way, but evolves faster. The reason for this difference is that the turbulent production term, $\overline{uw}(\partial\bar{U}/\partial z)$, appears only in the u' component of the turbulent kinetic energy

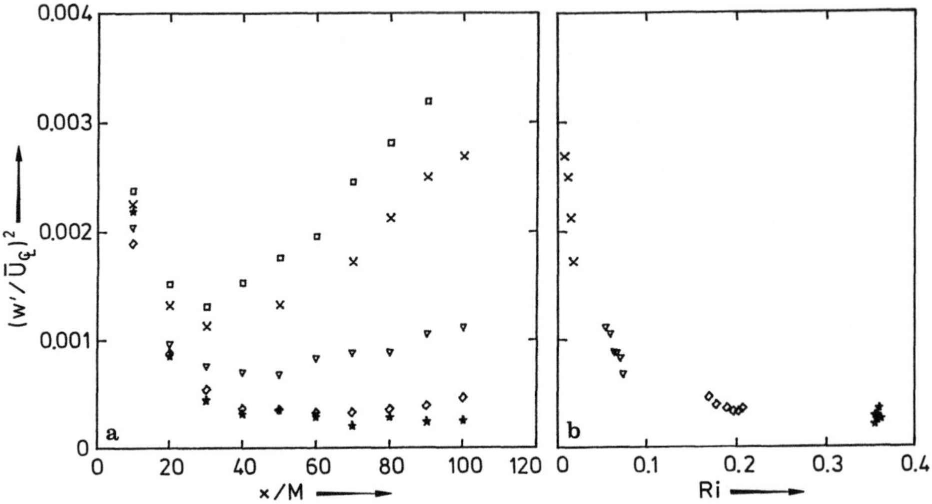

Fig. 2. Downstream development of vertical turbulent energy with x/M in (a) and stability Ri in (b). ★, $N = 0.65$ rad/sec; ◇, $N = 0.49$ rad/sec; ▽, $N = 0.29$ rad/sec; ×, $N = 0.16$ rad/sec; □, $N = 0$ rad/sec

equation. The growth of w'^2 is a consequence of the pressure-velocity correlation which redistributes the production $\overline{uw}(\partial\bar{U}/\partial z)$ among the remaining components. In all cases, at small x/M the turbulence is grid dominated regardless of the imposed shear and stratification, and the turbulent kinetic energy is observed to decay. However, at large x/M the shear and buoyancy forces determine the characteristics of the turbulence. The growth rates of u'^2 and w'^2 with stratification can be dramatically less than for the shear flow without stratification. As the stability of the flow decreases, the growth rates approach the passively stratified case and the limiting behavior of the correlation coefficients $\overline{\varrho w}/\varrho' w'$, $\overline{uw}/u'w'$ and $\overline{\varrho u}/\varrho' u'$ are found consistent with similar measurements reported by Tavoularis and Corrsin (1981).

The normalized vertical turbulent kinetic energy $w'^2/\bar{U}_{\mathbb{C}}^2$ for all stratifications is shown in Fig. 2b as a function of the local gradient Richardson number. Only data points outside the influence of the decaying grid turbulence region are included in Fig. 2b. The overall decrease of $w'^2/\bar{U}_{\mathbb{C}}^2$ with Ri is similar to that found by Webster (1964). Fig. 2a and 2b suggest that if these measurements were to be extended downstream, increasing τ while keeping Ri nearly constant, a dependence of $w'^2/\bar{U}_{\mathbb{C}}^2$ on τ would be apparent. This has been recently verified by Rohr (1985). For Webster (1964), at his only location not influenced by the grid, τ remained nearly constant because the shear generating grid produced nearly constant $(1/\bar{U})$ $(\partial\bar{U}/\partial z)$. Karnik and Tavoularis (1983) found a similar result when they used a shear generator consisting of 12 separate channels, each filled with a series of interchangeable screens to produce the shear. Only when each layer of the flow is driven separately, which the present facility allows, can $\bar{U}_{\mathbb{C}}$ and $\partial\bar{U}/\partial z$ be varied independently.

The gradient Richardson number of the largest stratification varied monotonically between 0.39–0.36 with increasing x/M and nowhere is the turbulent kinetic energy observed to grow. The remaining stratifications were all at gradient Richardson numbers less than 0.2, and downstream of the region influenced by the grid the turbulent intensities grow at a rate which depends on the level of stratification. The Miles (1961) criterion for an inviscid continuously stratified flow to be stable to infinitesimal disturbances requires a gradient Richardson number greater than 0.25. Although this value is consistent with the present

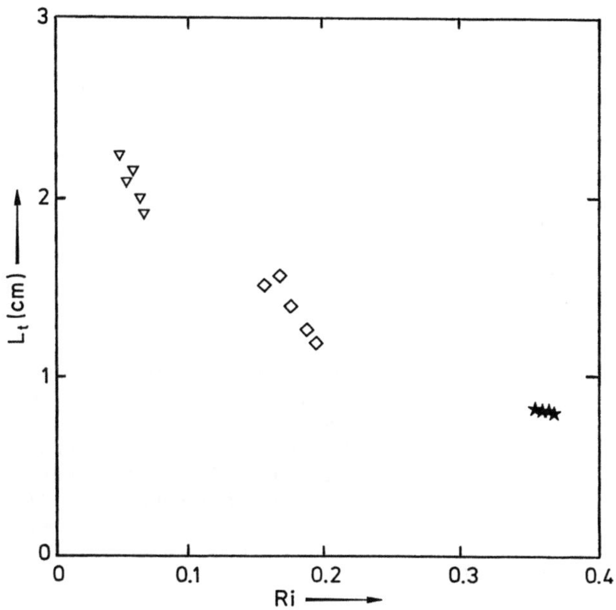

Fig. 3. Dependence of the turbulent length scale L_t on stability Ri. Symbols as defined in Fig. 2

data, the Miles criterion cannot be strictly applied since the grid at the inlet creates finite disturbances not amenable to a linear stability analysis.

The dependence on the gradient Richardson number of the rms density fluctuation ϱ' is shown in Fig. 3 in terms of the tubulent scale $L_t = \varrho'/(\partial\bar{\varrho}/\partial z)$ introduced by Ellison (1957) as "a length typical of the vertical distance travelled by particles before either returning towards their equilibrium levels or mixing." This definition of L_t differs from that of SHV by a factor of two. As seen in this figure, like Webster (1964), we also find a reduction in the vertical scale of the turbulence as the stability of the flow increases. The data of Fig. 3 were chosen from measurement positions beyond the region characterized by decaying grid turbulence. Recent experiments by Rohr (1985) have also shown a τ dependence of L_t.

The ratio K_z/K_m, determined at positions not influenced by the grid, was found to be about 1.34 ± 0.3 and not noticeably dependent on Ri over the range 0.01 to 0.10. As reported by Webster (1964), the majority of values at low Reynolds numbers seem to cluster around 1.35 for small gradient Richardson numbers Ri. His data, however, do exhibit a weak Ri dependence with K_z/K_m decreasing with increasing Ri. A roughly similar treatment has been made for the SHV unsheared data by Rohr, Itsweire and Van Atta (1984). Here too it is found that the ratio K_z/K_m is nearly constant at low stabilities.

The scale of motion at which buoyancy forces become of the same order of magnitude as the inertial forces is known as the Ozmidov scale $L_0 = (\varepsilon/N^3)^{1/2}$ and defines the upper limit permissible for the size of tubulent motions. The smallest scales are characterized by the Kolmogorov scale $L_k = (\nu^3/\varepsilon)^{1/4}$ where the viscous and inertial forces are of the same order. SHV obtained quantitative estimates for dynamically significant relationships between the Ozmidov and Kolmogorov scales by using density and turbulent mass flux measurements. Their results suggest that active (overturning) turbulence exists at scales λ in a decaying stratified flow for which $1.4\,L_0 > \lambda > 15.4\,L_k$. Gibson's (1980) theoretical estimate of $1.2\,L_0 > \lambda > 15\,L_k$ is in fair agreement with these results. These results were based on an overturning scale defined as $2\varrho'/(\partial\bar{\varrho}/\partial z)$. In keeping with the present definition, $L_t = \varrho'/(\partial\bar{\varrho}/\partial z)$, SHV's results can be rewritten as $0.7\,L_0 > \lambda > 7.7\,L_k$.

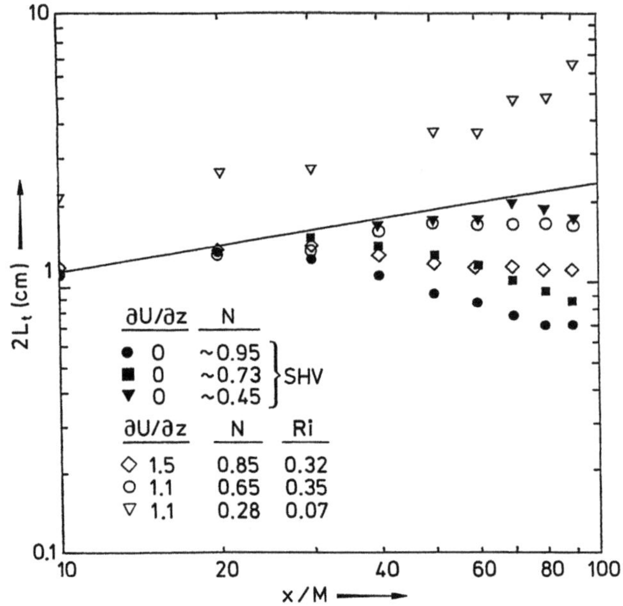

Fig. 4. Overturning turbulent length scale vs. downstream distance from the grid x/M ($M = 1.905$ cm). Solid symbols $= 2L_t$ of SHV. Open symbols $= 2L_t$ of present data. —— $= 2L_t$ of Montgomery (1974) with $\partial \bar{U}/\partial z = 0$ and $N \approx 0$

Our experiments can be used to determine whether these proportionality constants are applicable to the decaying uniform shear cases using the same procedures as SHV in their study of decaying grid turbulence. The coefficients are determined from: measuring the ratio L_t/L_0 at the estimated point where the growth of L_t first falls from the curve describing its passive growth, and measuring the ratio L_0/L_k at the estimated point where $\overline{\varrho w}$ first goes to zero. We find $1.1\ L_0 > \lambda > 8.8\ L_k$ for the two decaying shear cases studied, where Ri ~ 0.36. It appears then that for high enough Ri number (greater than 0.3) the shear may only marginally influence the developing flow. It is important to emphasize the fact that the same grid was used in both the uniform and shear flows. Measurements taken close to the grid are in both cases characterized by decaying grid turbulence. Only farther from the grid, and when the gradient Richardson number is small enough to allow turbulent growth, do the features of the flow distinctively depart from decaying grid turbulence.

In Fig. 4 the downstream development of L_t is shown for relatively large and small Richardson numbers. Also included in Fig. 4, for the sake of comparison, are the uniform mean shear experiments of SHV and Montgomery (1974) which were actively and passively stratified, respectively. Montgomery's data were normalized by the ratio of his grid mesh length to that used in the present facility, since the grid mesh length induces a given initial length scale. For the larger Richardson numbers, where the turbulence did not grow downstream, the departures of L_t from the passive growth law are reminiscent of those found by SHV for their unsheared study. However, for smaller Ri, where the turbulence grows, the development of L_t is no longer amenable to comparison with Montgomery's passively stratified uniform mean flow measurements.

SHV found that when buoyancy forces are important in an unsheared decaying turbulent flow, L_t decreased with L_0 and the ratio of L_t/L_0 approached a constant, as shown by the solid line in Fig. 5. The unsheared data of SHV suggests that the ratio L_t/L_0 approaches the same constant at large x/M independent of stratification. When a shear is present and the Ri > 0.25, the ratio L_t/L_0 for large x/M approaches the same ratio observed for grid turbulence without shear as shown in Fig. 5 (★ symbols). When the Ri $\leqq 0.2$ the shear provides a sufficiently large source of turbulent energy that both L_t and L_0 are observed to

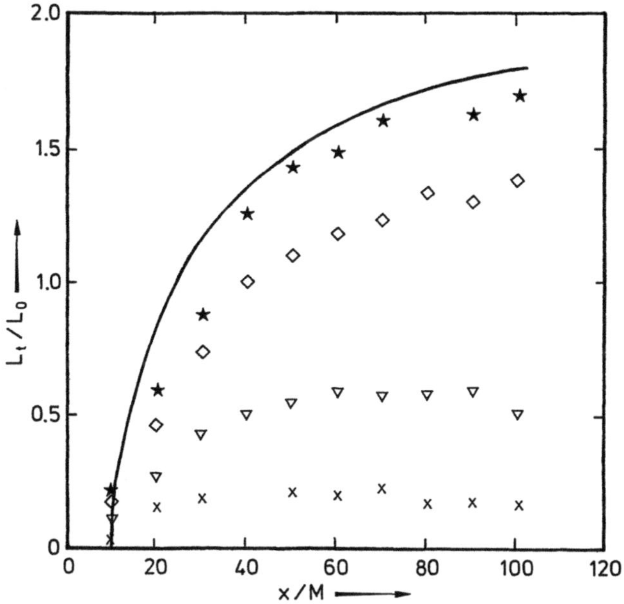

Fig. 5. Downstream development of the ratio of observed to limiting overturning scales. ——, $N = 0.96$ (unsheared). Symbols (shear data) as in Fig. 2

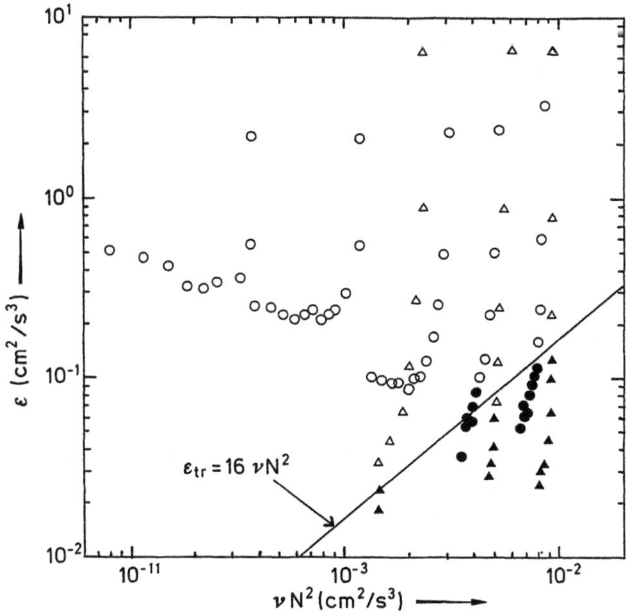

Fig. 6. Transition boundary between turbulent and nonturbulent domains. ——, $\varepsilon_{tr} = 16\,\nu N^2$; △, SHV unsheared data, $\overline{\varrho w} > 0$; ▲, SHV unsheared data, $\overline{\varrho w} \sim 0$; ○, present sheared data, $\overline{\varrho w} > 0$; ●, present sheared data, $\overline{\varrho w} \sim 0$

grow while their ratio remains nearly constant. As illustrated in Fig. 5 this ratio strongly depends on stability. For weak stability, the ratios are small because the limiting buoyancy length scale L_0 increases faster than the overturning scale L_t.

For the decaying shear flow data, when $L_0 = 8\,L_k$ all overturning scales have been effectively suppressed ($\overline{\varrho w} = 0$) due to the combined effect of buoyancy and viscosity. From this criterion and the definition of the Ozmidov and Kolmogorov length scales, the value of the transition rate ε_{tr} is calculated to be $16\,\nu N^2$, rather close to the unsheared case value of $15\,\nu N^2$ (Itsweire, Helland and Van Atta, 1985). In the region beyond the influence of the grid,

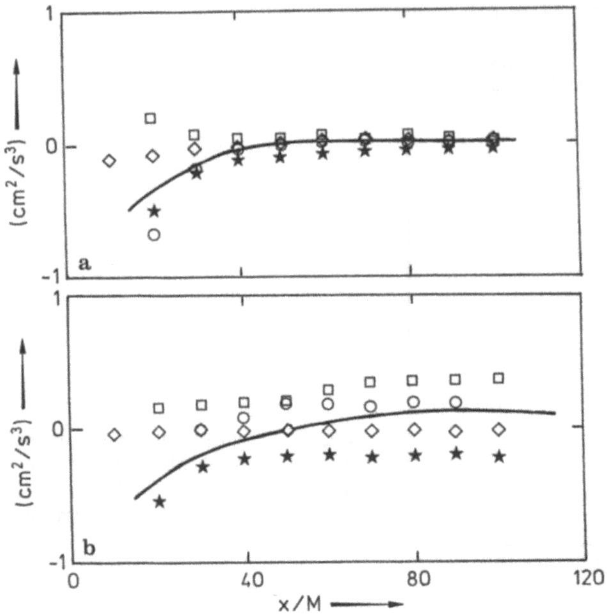

Fig. 7. Downstream development of the principal terms in the turbulent kinetic energy equation for two stabilities: (a) Ri = 0.36; (b) Ri = 0.06. ○, $\bar{U}(\partial/\partial x)\,(\overline{q^2}/2)$; □, $-\overline{uw}(\partial\bar{U}/\partial z)$; ◇, $(-g/\bar{\varrho})\,\overline{\varrho w}$; ★, $-\varepsilon$. The solid line represents the sum of the last three terms (to be compared with the first term)

it is observed that for Ri $\geqq 0.3$, $\varepsilon < \varepsilon_{tr}$ and the turbulence decays, while for Ri $\leqq 0.2$, $\varepsilon > \varepsilon_{tr}$ and the turbulence grows. Therefore, the criterion for the minimum dissipation rate required for the existence of tubulence applies more generally than to the specialized case of unsheared decaying turbulence. Figure 6 shows ε vs νN^2 for our data as well as the SHV data. As shown for either data set, $\varepsilon_{tr} = 16\,\nu N^2$ serves well in separating the turbulent and non-turbulent domains. For smaller Ri, as the shear becomes a more effective source for turbulence, ε is observed to be much larger than ε_{tr}.

Figure 7a, b contrasts the development of the dominating turbulent kinetic energy terms for decaying and growing turbulent uniform shear flows. The buoyancy flux $g\overline{\varrho w}/\bar{\varrho}$ acts as a relatively weak sink for turbulent kinetic energy in either case. The dissipation (ε) always decreases immediately behind the grid, but further downstream if Ri is low enough it is observed to grow along with the turbulence. The production term $(\overline{uw}(\partial\bar{U}/\partial z))$ shown in Fig. 7b is observed to be more sensitive to the mean shear.

Comparison with Ocean Microstructure Measurements

These studies of stratified turbulence have direct application to natural flows such as those occurring in the ocean. For decaying grid turbulence the laboratory overturning scale L_t has been found by Itsweire (1984) to be roughly equal to the commonly defined Thorpe scale L_T. Length scale observations from the oceanographic literature may thus be compared with the laboratory measurements. Figure 8 compares the overturning scale observed in the flow with the largest possible scale allowed by buoyancy for the present and UCSD laboratory data and the oceanographic data of Dillon (1984). It is observed that the unsheared decaying grid data exhibit a positive slope similar to the oceanographic data only when buoyancy domi-

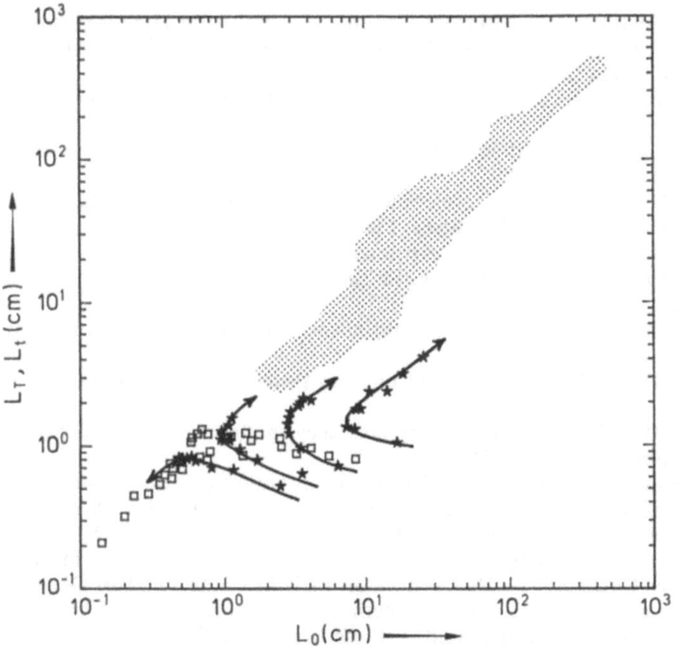

Fig. 8. Comparison of laboratory and oceanic length scale measurements. ★, present sheared measurements of L_t, arrows indicate increasing distance from the grid; □, unsheared measurements of L_T from Itsweire (1984), increasing distance from the grid corresponds to decreasing L_0; ⁘, oceanic measurements of L_T from Dillon (1984)

nates during the turbulence decay, but not near the grid where buoyancy is relatively unimportant. The laboratory sheared turbulence data exhibit trends similar to the ocean data while the turbulence is growing under the combined influences of shear and buoyancy, but not in the region of initial decay where buoyancy is again relatively unimportant. This similarity of the behavior suggests that the laboratory stratified shear flow data will provide flow fields of direct relevance to the interpretation of oceanic microstructure data obtained under a great variety of decaying, growing, and steady-state conditions.

Acknowledgements. Financial support from the National Science Foundation under Grants OCE82-05946, OCE85-11290 and MEA81-00431 is gratefully acknowledged.

References

Champagne, F. H., Harris, V. C, Corrsin, S. (1970): Experiments on nearly homogeneous turbulent shear flow. J. Fluid Mech. **41**, 81

Dillon, T. M. (1982): Vertical overturns: a comparison of Thorpe and Ozmidov length scales. J. Geophys. Res. **87**, 9601

Ellison, T. H. (1957): Turbulent transport of heat and momentum from an infinite rough plane. J. Fluid Mech. **2**, 456

Gibson, C. H. (1980): "Fossil Temperature, Salinity and Vorticity in the Ocean," in *Marine Turbulence*, ed. by J. C. T. Nihoul, (Elsevier, Amsterdam) p. 221

Harris, V. C., Graham, A. A., Corrsin, S. (1977): Further experiments in nearly homogeneous turbulent shear flow. J. Fluid Mech. **81**, 657

Head, M. J. (1983): "The Use of Miniature Four-Electrode Conductivity Probes for High Resolution Measurements of Turbulent Density or Temperature Variations in Salt-Stratified Water Flows;" Ph.D. Thesis, University of California, San Diego

Itsweire, E. C. (1984): Measurements of vertical overturns in a stably stratified turbulent flow. Phys. Fluids **27**, 764

Itweire, E. C., Helland, K. N., Van Atta, C. W. (1986): The evolution of grid generated turbulence in a stably stratified fluid. J. Fluid Mech. **162**, 299

Karnik, V, Tavoularis, S. (1983): "The Asymptotic Development of Nearly Homogeneous Turbulent Shear Flow," in Fourth Symposium on Turbulent Shear Flows, University of Karlsruhe

Miles, J. W. (1961): On the stability of heterogeneous shear flows. J. Fluid Mech. **10**, 496

Montgomery, R. D. (1974): "An Experimental Study of Grid Turbulence in a Thermally-Stratified Flow;" Ph.D. Thesis, University of Michigan

Mulhearn, P. J., Luxton, R. E. (1975): The development of turbulence structure in a uniform shear flow. J. Fluid Mech. **68**, 577

Owen, D. R., Zienkiewicz, M. K. (1957): The production of uniform shear flow in a wind tunnel. J. Fluid Mech. **2**, 521

Rohr, J. J., Itsweire, E. C., Van Atta, C. W. (1984): Mixing efficiency in stably stratified decaying turbulence. J. Geophys. Astrophys. Fluid Dyn. **29**, 221

Rohr, J. J. (1985): "An Experimental Study of Evolving Turbulence in Uniform Mean Shear Flows With and Without Stable Stratification;" Ph.D. Thesis, University of California San Diego

Rose, W. G. (1970): Interaction of grid turbulence with a uniform mean shear. J. Fluid Mech. **44**, 767

Stillinger, D. C. (1981): "An Experimental Study of the Transition of Grid Turbulence to Internal Waves in a Salt-Stratified Water Channel;" Ph.D. Thesis, Universtiy of California, San Diego

Stillinger, D. C., Helland, K. N., Van Atta, C. W. (1983): Experiments on the transition of homogeneous turbulence to internal waves in a stratified fluid. J. Fluid Mech. **131**, 91

Stillinger, D. C., Head, M. J., Helland, K. N., Van Atta, C. W. (1983): A closed loop gravity-driven water channel for density-stratified shear flows. J. Fluid Mech. **131**, 73

Tavoularis, S., Corrsin, S., (1981): Experiments in nearly homogeneous turbulent shear flows with a uniform mean temperature gradient. Part 1. J. Fluid Mech. **104**, 311

Webster, C. G. A. (1964): An experimental study of turbulence in a density-stratified shear flow. J. Fluid Mech. **19**, 221

Part II

Free Shear Flows

Introduction

R. J. Adrian

Department of Theoretical and Applied Mechanics, University of Illinois, Urbana, IL 61801, USA

The three classical free turbulent shear flows are jets, wakes and free shear layers, and true to this tradition, the papers in this section deal with each of these flows. There has been a trend in research, however, to emphasize the study of the free shear layer, partly because its structure is simpler than that of a jet or a wake, but largely because it is a "building block" flow that occurs in many other types of turbulent motion. Thus, the axisymmetric shear layer surrounding the potential cone of a submerged jet dominates the properties of the early jet, and the separated boundary layer in the wake of a cylinder behaves much like a simple free shear layer.

The current issues in free turbulent shear flow research are concerned mainly with the properties of the large scale, "coherent structures" that characterize such flows. More than any other flow, the free shear layer exhibits striking and irrefutable evidence of the existence of coherent structures, as seen so clearly in the work of Brown and Roshko (1974). Early questions about coherent structures centered on the significance of their contribution to the Reynolds stresses, their existence at large Reynolds numbers, and their significance in the production of energy. Bradshaw (1980) gives an excellent discussion of the evolution of coherent structure concepts in his introduction to an earlier chapter in the Turbulent Shear Flow Series.

It is now generally agreed that coherent structures are indeed significant phenomena worthy of careful study, and while many of the original questions have not gone away, they have become sharper. For example, early speculation which supposed that large scale structures might vanish in the limit of large Reynolds number, have been replaced by the realization that the "age" of a shear layer structure, as measured by the time since its initial instability, is much more important than the Reynolds number, per se. Thus, two-dimensional vortices with spanwise vorticity exist in the early stages of shear layers despite very high Reynolds numbers. They may, and most certainly do, evolve substantially as they interact with other vortices through pairing and tearing motions, and undergo three-dimensional instabilities.

The issues raised in this section on free shear flows revolve around two broad questions. What is the relationship between coherent structures and the instabilities and transitional flows that preceded them? And, what is the relationship between coherent structures and the structures that exist in fully turbulent flow, long after the coherent patterns cease to be readily observable?

The papers by Kourta, Boisson, Braza, Chassaing and Minh and Louchez, Kawal and Keffer offer an interesting contrast in that the former authors study the earliest stages of wake formation, while the latter study the wake in its late, self-preserving stage. Kourta et al. show that the separated laminar boundary layer in the wake of a circular cylinder undergoes free shear layer instabilities whose frequencies, f_t, interact with the fundamental von Karman vortex frequency f_s to produce beat-tones in the turbulence spectrum. The ratio f_t/f_s is proportional to $\mathrm{Re}_D^{1/2}$ so that f_t and f_s become "disconnected" at high Reynolds number, but

Turbulent Shear Flows 5

interact strongly at lower values of Re_D. This sort of study offers insights into the mechanisms by which relatively simple shear flows evolve into chaotic turbulence.

Louchez et al. ask two questions. Firstly, do wakes from different bodies differ in their macroscopic development? Secondly, is there a universal, self-preserving state attained by wakes created from different generators? On the basis of self-preserving characteristics exhibited by second order moments they conclude that sufficiently far downstream the plane wake attains a universal, self-preserving structure, independent of the initial body. This suggests that macroscopic structure of the self-preserving wake is unaffected by initial intricat of the type observed by Kourta et al. Perhaps differences would be observed on a more detailed level of comparison.

Lasheras and Maxworthy study the instabilities of the young shear layer, with particular interest in the mechanisms leading from large scale two-dimensional vortices to three-dimensional vortices with streamwise vorticity. They show that three-dimensional disturbances to the upstream flow can produce streamwise vorticity in the form of counter-rotating axial vortices that exist in both the braids between the spanwise vortices and in the spanwise vortices themselves. This mechanism is similar to the vortical motion studied by Corcos (1979) and Corcos and Lin (1984) due to straining motion in the braids.

Proceeding yet further into the turbulent regime, Metcalfe, Hussain, Menon and Heyakawa compare measurements in a high quality shear layer with three dimensional Navier-Stokes computations on a 64^3 grid. The comparison allows them to assess the significance of several parameters of the experiment and of the computation, including the Reynolds number, differences in the age, as measured by the number of pairings, and the difference between computational shear layers evolving in time and real shear layers evolving in space. The generally good qualitative agreement between educed coherent structures indicates that none of these factors have a strong influence.

In fully turbulent flow the identification and measurement of large scale structure is made difficult by the presence of small scale energy which obscures the structure in noise, even when probed by conditional sampling methods. Lumley (1970, 1981) proposed his now famous "characteristic eddy" method as a means of extracting the large scale structure from such flows, but efforts to use this method have been hampered by the task of measuring the co-spectra on a large grid of points. Glauser, Leib and George present a very interesting set of results computed from a small 7×7 grid of co-spectral data in the axisymmetric shear layer. One of the most interesting points in this paper is the rate at which the terms in the orthogonal decomposition converge. Over 40% of the mixing layer energy is contained in the first eigenmode and the next two eigenmodes contain another 40%. Thus, Lumley's orthogonal decomposition is a very efficient representation of the random process.

The last paper in this section, by Saripalli, is an excellent example of applied turbulence research in a complicated flow with twin jets impinging on a surface. The study was performed to develop basic understanding of the upwash region between the jets in order to assess its effect on the lift experienced by VTOL aircraft.

Research on free shear flows is likely to concentrate on coherent structures for many years. Work on initial instabilities and transition is needed, and if it can be shown that the coherent structures retain some elements of their initial forms, this work will lead the way in determining the anatomy of these flows. The problem of linking fully turbulent structure to early structure will be a great challenge.

References

Brown, G. L., Roshko, A. (1974): J. Fluid Mech. **64**, 775

Bradshaw, P. (1980): In *Turbulent Shear Flows 2*, ed. by L. J. S. Bradbury et al. (Springer Berlin, Heidelberg) p. 259

Corcos, G. M. (1979): The mixing layer: Deterministic models of a turbulent flow. U. C., Berkeley, College Engr. Rept. FM-79-2

Corcos, G. M., Lin, S. J. (1984): J Fluid Mech. **139**, 67

Lumley, J. L. (1970): *Stochastic Tools in Turbulence* (Academic, New York)

Lumley, J. L. (1981): In *Transition and Turbulence*, ed. by R. Meyer (Academic, New York) p. 215

Wake-Shear Layer Interaction and the Onset of Turbulence Behind a Circular Cylinder

A. Kourta, H. C. Boisson, M. Braza, P. Chassaing and H. Ha Minh

Institut de Mecanique des Fluides de l'Institut National Polytechnique de Toulouse,
Departement de Recherche de l'Enseeiht, Laboratoire Associe au CNRS LA-04005,
2 rue Charles Camichel, 31071 Toulouse, France

Abstract

The formation zone of the large regular vortices of the wake of a circular cylinder ($0 < X/D < 3$) in the subcritical regime ($2{,}400 < \mathrm{Re}_D < 60{,}000$) is studied with and without the introduction of a splitter plate. Spectral properties are used to describe the different stages of the interaction between shear layer vortices and alternating ones. Some physical properties of this interaction, emerging from a numerical simulation using a pressure-velocity formulation, are examined separately. Both unexcited and excited 2D-plane mixing layers are studied using streakline maps and time traces of the dynamical properties.

Nomenclature

b	Shear layer width (distance $0.95\,U_1$ and $1.05\,U_2$)
b_0	Numerical reference length
D	Diameter of the cylinder
f	Frequency
f_s	Vortex shedding frequency
f_t	Shear layer frequency
f_f	Forcing frequency
d_f	Frequency resolution (sampling frequency/number of points)
$P_{uu}(f)$	Discrete power spectrum of velocity
P_0	Total power of the signal $= \int_{-\infty}^{+\infty} P_{uu}(f)\,df$
R	$= (U_1 - U_2)/(U_1 + U_2)$: velocity ratio
Re_D	$= (U_0 * D)/\nu$: wake Reynolds number
Re_b	$= U_1 * b_0/\nu$: shear layer Reynolds number
St_D	$= f_s * D/U_0$: wake Strouhal number
T	Numerical time value
U_0	External uniform velocity
U_m	$= (U_1 + U_2)/2$: shear layer reference velocity
U_1	(resp. U_2) External higher (lower) velocity
X, Y	Geometrical position
α	Non integer coefficient
ν	Kinematic viscosity

Introduction

The transition to turbulence in the wake of a circular cylinder has been studied under various aspects. The onset of turbulence seems to happen at Reynolds numbers around 150 as pointed out by Roshko [1]. Up to Reynolds numbers around 1000 the situation is quite complex. In a recent paper Sreenivasan [2] has denumbered at least four successive sequences of emergence of ordered and chaotic behaviour for the Reynolds number range of 100 to 200. In numerical studies by Braza et al. [3–5], the simultaneous development of secondary and separated mixing layer eddies around the cylinder is analyzed for Reynolds numbers varying fro 200 to 10,000 and different types of vortex interactions are also found.

In the present paper a restricted part of the subcritical regime is addressed namely $2{,}400 < \mathrm{Re}_D < 60{,}000$. The situation, in this case, is quite well defined since the transition occurs in the shear layer developing from the separation point in the laminar boundary layer before the rolling down of the first vortex shed by the wake.

The previous experimental studies of Bloor [6] or Bloor and Gerrard [7] and some visual evidence provided by Crausse [8] or Domptail [9], among others, has given a first description of the flow structure. However some fundamental questions are still unresolved concerning

Turbulent Shear Flows 5

the mechanism of vortex development and their participation within the framework of the establishment of the chaos.

The aim of this paper is to analyze the vortex interactions in the formation zone of the large scale regular vortices of the wake.

In the first part, a study using hot wire anemometry is performed to detect the vortical structures and their spectral properties in the subcritical regime where the periodic character of the wake is well established (Sect. 3.1).

The second part is devoted to the intrinsic description of the structures in the shear layer as an independent instability. This purpose is reached by comparing the previous results with the wake of the same cylinder fitted with a splitter plate damping the direct vortex shedding contribution (Sect. 3.2).

In order to analyze some of the observed features of the actual complex situation, a numerical simulation of the flow in a mixing layer is carried out. The phenomena associated to instability developments and interactions are studied both in the non excited and excited $2D$-plane mixing layer (Sect. 4).

Experimental Set Up

The experimental set up is the same as the one used by Boisson et al. [10] and Kourta et al. [11]. The measurements are conducted in a closed return wind tunnel, the initial turbulence level of which is less than 0.1%. The cylinder of diameter $D = 42$ mm spans the test section vertically. End plates are added in order to minimize the interaction between the cylinder and the tunnel boundary layer and to obtain a two-dimensional mean flow as shown by Stansby [12]. The plates are desinged in agreement with his prescriptions (Fig. 1) and have the same dimensions as the one used by West and Apelt [13]. These authors have also found that the corresponding aspect ratio of 13 and blockage coefficient of 7% adopted here are sufficiently low and do not introduce significant distorsions compared to infinite external flow conditions.

The signals from single hot wire probes providing the time variation of the modulus of velocity are digitized and processed on a specialized analyser PLURIMAT-S INTERTECHNIQUE. Power spectra are calculated using a classical FFT algorithm. The probe displacments are also computer controlled (step 5 µm), in order to ensure an adequate accuracy of the positioning specially in the thin zones near the wall and in the shear layer.

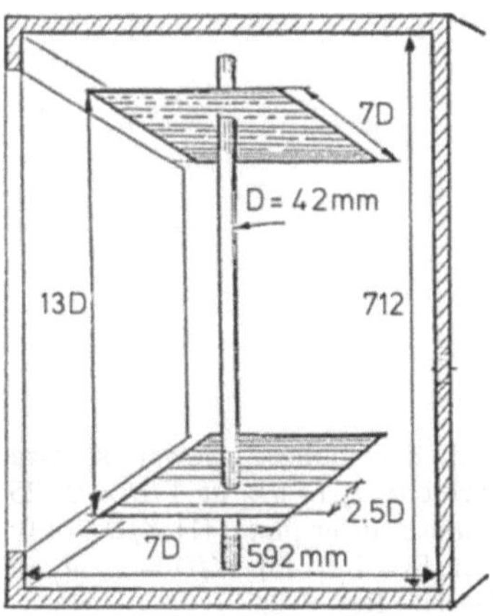

Fig. 1. The cylinder with end plates

Flow visualizations are also made in an open wind tunnel of circular section (diameter = 2,400 mm) at Reynolds numbers ranging from 2,400 to 6,800. The flow is seeded at the surface of the cylinder with traces of carbon dioxide and a Schlieren technique is used to make slides and high speed films of the flow configuration in the near wake.

Results

Cylinder Without Splitter Plate

Two different flow regimes are first identified from the properties of the probability density functions and the power spectra of the anemometer signals, depending upon the Reynolds number. The corresponding flow configuration is schematically sketched on Fig. 2.

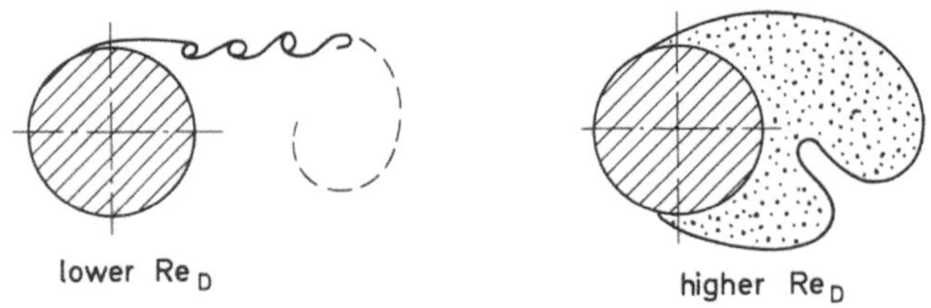

Fig. 2. Schematic diagram of the flow configuration

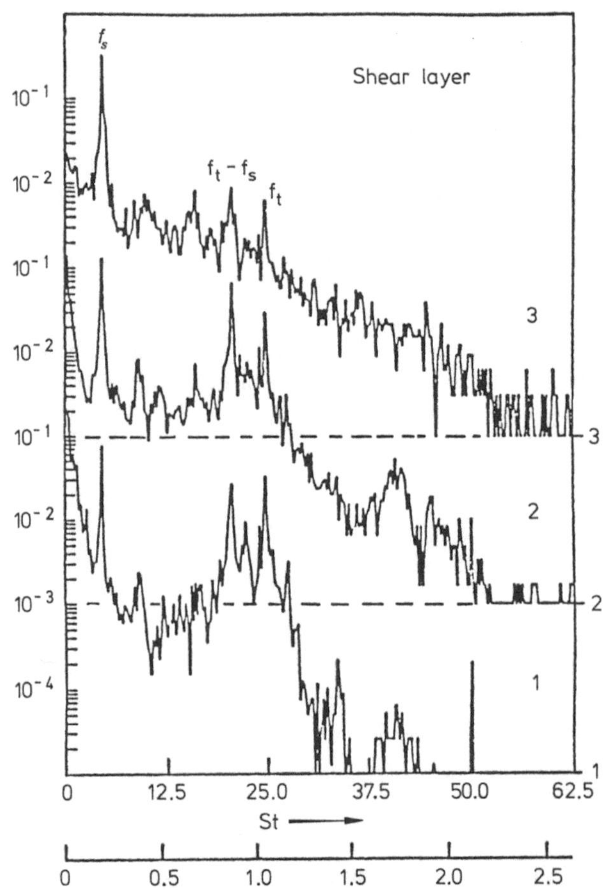

Fig. 3. Normalized power spectra of velocity $P_{uu}(f)/P_0$ at $Y/D = 0.625$ (*without splitter plate*) (1) $X/D = 1.0$; (2) $X/D = 1.4$; (3) $X/D = 2$; $d_f = 0.244$ Hz; $\mathrm{Re}_D = 2{,}677$

Table 1. Characteristic frequencies in the wake *without Splitter Plate*

Re_D	f_s (Hz)	f_t (Hz)	f_t/f_s
2667	5.13	24.66	4.81
4800	9.76	60.50	6.20

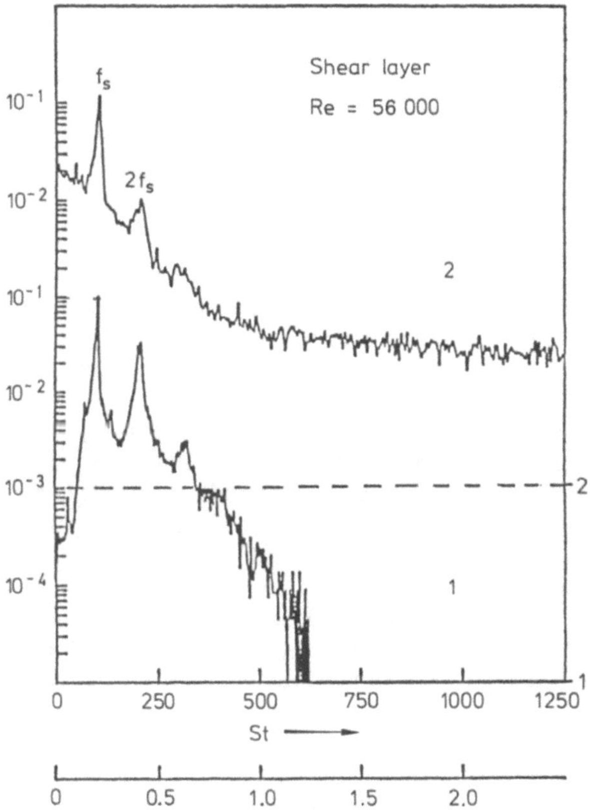

Fig. 4. Normalized power spectra of velocity $P_{uu}(f)/P_0$ at $Y/D = 0.575$ (*without splitter plate*) (1) $X/D = 0$; (2) $X/D = 0.25$; $d_f = 4.88$ Hz; $Re_D = 56{,}000$

a) At low values of Re (2,667 and 4,800), a dead fluid zone is observed at the back of the cylinder. It is bounded by two shear layers nearly parallel to the uniform external flow in the first stage. The regular vortex shedding formation takes place downstream of this zone at a frequency f_s. Another discrete frequency f_t is observed on the power spectrum (Fig. 3). It characterizes the initial instability and the developing vortices of the shear layer. The values of the observed frequency f_t and f_s are given on Table 1. The frequency ratio f_t/f_s is found to vary as the square root of the Reynolds number as found previously by Bloor [6].

At the beginning of the shear layer, the power corresponding to the frequency f_t is very low. Moving downstream on a line parallel to the uniform flow direction, this power is considerably increased and discrete interaction frequencies ($f_t \pm nf_s$) arise from the combination between f_t and the general oscillation due to the main vortex shedding f_s. The peak at ($f_t - f_s$) rapidly becomes very strong initiating a system driven by three frequencies. This type of interactions is described by Kim et al. [14], Miksad et al. [15] or Motohashi [16]. It is the signature of non linear phase and amplitude modulations gradually leading to a chaotic

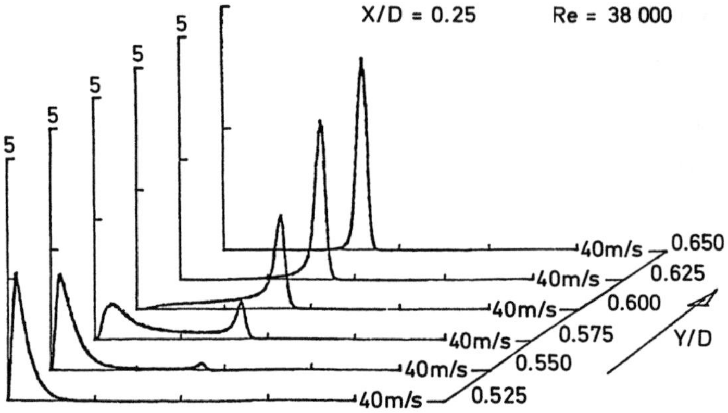

Fig. 5. Probability density function of the velocity signal for the higher Reynolds numbers

behaviour. It appears that the power rise is mainly localized in the interaction band between $(f_t - f_s)$ and $(f_t + f_s)$. This band plays an essential role in the structuration of the spectrum of the random fluctuation. Downstream of the formation zone the only significant resolute peaks are found for f_s and its harmonics. The flow is therefore considered as the organized motion of a turbulent fluid medium.

b) At high values of Re (from 16,000 to 60,000), the frequency ranges of the two types of vortical structures are disconnected. The spectra exhibit a discrete periodic set of peaks based on f_s and a rather smooth random distribution (Fig. 4). The intermittent nature of the flow is evident from the bimodal probability density functions (Fig. 5) at a given space station resulting from successive crossings of turbulent and nonturbulent flow according to the periodic formation of large scale structures.

For this last regime the transition of the shear layer occurs close to the separation point and the whole flow can be treated as if the background turbulence was distributed as a continuous viscous medium rather than associated with a localized interactive process.

Cylinder With Splitter Plate

A wake splitter plate parallel to the external flow (Fig. 6) is extended to 4.12 diameters downstream from the base of the cylinder. Such a device deeply modifies the flow pattern in the wake as studied by Roshko [17], Bearman [18] or Apelt et al. [19]. It produces a considerable decrease in the absolute value of the base pressure coefficient and in the mean drag coefficient (from 25% for Re = 2,400 to 39% for Re = 56,000). The plate stabilizes the flow behind the cylinder and avoids direct interaction between both sides modifying the

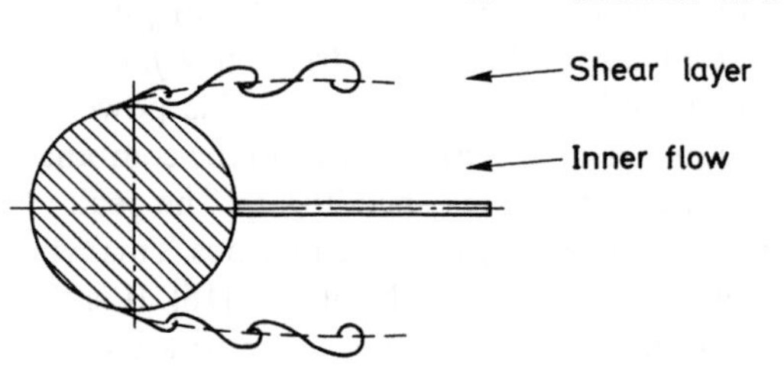

Fig. 6. Schematic diagram of the flow with splitter plate

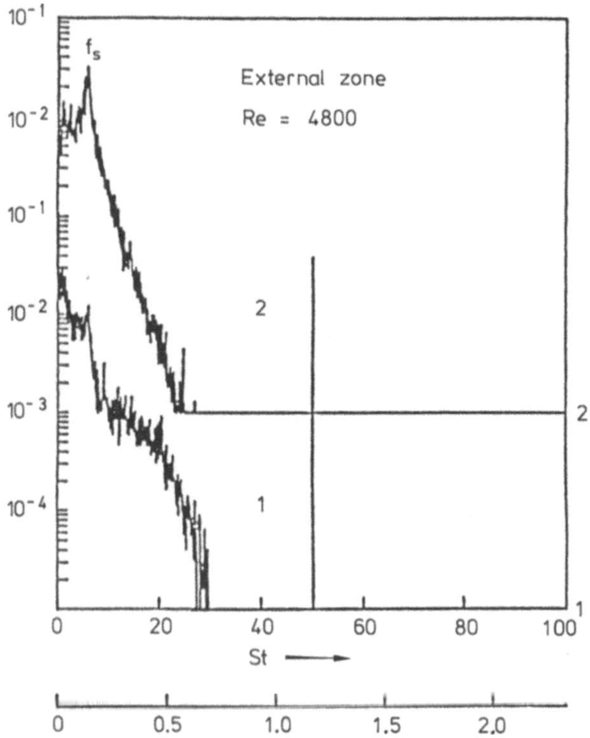

Fig. 7. Normalized power spectra of velocity $P_{uu}(f)/P_0$ at $Y/D = 2$ (*with splitter plate*) (1) $X/D = 3$; (2) $X/D = 5$; $d_f = 0.0976$ Hz; $\mathrm{Re}_D = 4{,}800$

vortex shedding process. The position of the separation point is also different and is observed to be located at an angle of approximately 70 degrees from the front stagnation point.

The impact of vortex shedding on the oscillating flow near the cylinder wall is much weaker than without splitter plate. The root mean square (rms) of the pressure fluctuations is reduced about 75% but a discrete peak at f_s is still found in the front part of the cylinder, upstream of the separation point (West et al., [20]). The same peak is observed in the potential flow. It is stronger near the end of the plate where the alternating vortices are formed. The corresponding spectra are given in Fig. 7.

The Strouhal number St_D based upon this frequency can be considered as a constant in the analyzed Reynolds number range and is found to be:

$$\mathrm{St}_D = f_s * D/U_0 = 0.130 \pm 0.004 .$$

This value agrees with the results of Roshko [17] for a short plate at the corresponding distance from the cylinder.

In the shear layer developing at the edges of the cylinder and being nearly parallel to the uniform flow direction, the intensity of the fluctuations rises considerably. A maximum is found near the line $Y/D = 0.79$ for which the typical spectra are represented on Fig. 8. The vortex shedding frequency is blurried in the ambient random spectrum and a corresponding peak cannot be distinguished. Conversely for the small scale vortices, a peak or at least a bump is clearly detected in the first stages of development of the shear layer ($X/D > 1$). It can be noticed that the frequency f_t is not significantly different from the one obtained

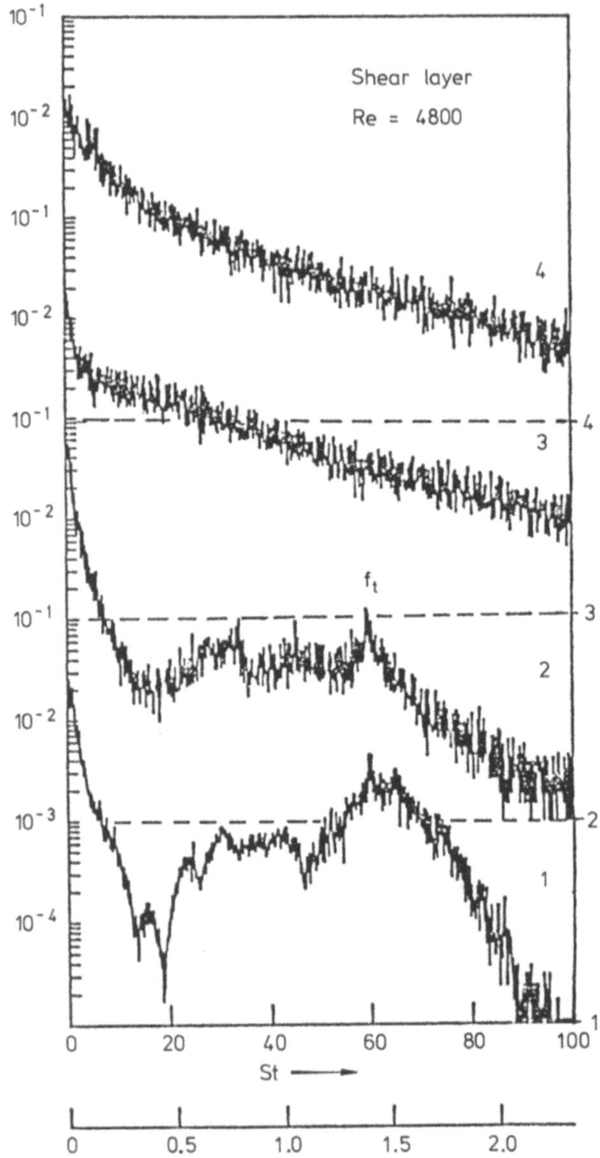

Fig. 8. Normalized power spectra of velocity $P_{uu}(f)/P_0$ at $Y/D = 0.7857$ (*with splitter plate*) (1) $X/D = 0.59$; (2) $X/D = 1$; (3) $X/D = 3$; (4) $X/D = 5$; $d_f = 0.0976$ Hz; $\mathrm{Re}_D = 4{,}800$

without the splitter plate as shown in Fig. 9. In both cases the frequency follows the non dimensional relationship

$$f_t * D^2/\nu = 0.0236 * \mathrm{Re}_D^{3/2}.$$

It is deduced from these results that the frequency f_t of the shear layer vortices is independent of the vortex shedding one (Table 2). This seems to exclude a feed back mechanism in the establishment of this frequency. However this frequency probably affects the size of the large scale wake vortices, mainly in the case without splitter plate.

In the development of the shear layer a subharmonic at $f_t/2$ is observed while moving downstream. This is typically associated with the pairing process as described by Ho and Huerre [21] among others for the $2D$ plane mixing layer.

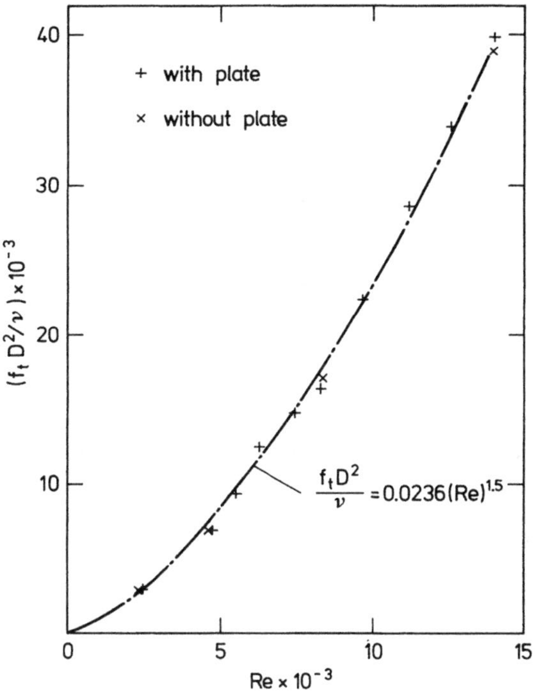

Fig. 9. Non dimensional variations of the frequency f_t with respect to the Reynolds number

Table 2. Characteristic frequencies in the wake *with Splitter Plate*

Re_D	f_s (Hz)	f_t (Hz)	f_t/f_s
2667	3.1	24.66	7.95
4800	5.57	60.5	10.86

However the process here is more complicated. In fact the interaction with the vortex shedding, though weaker than in the absence of the plate, still exists and is observed by the $(f_t - f_s)$ frequency. Also observable are higher order subharmonics $f_t/3$ and $f_t/4$ or $f_t(1 - 1/n)$. These last interactions cannot be attributed to the pure regular pairing process of the classical plane mixing layer.

The numerical simulation of a natural and excited plane mixing layer is developed in the following Section to provide information on these mechanisms.

Study Governing Mechanism

In order to analyze separately the dynamics of the mixing layer eddies and the interaction mechanism with the alternating ones, a $2D$ unsteady mixing layer flow is computed by direct solution of the time dependent Navier-Stokes equations.

The Theoretical Numerical Approach

The governing equations of an incompressible homogeneous viscous fluid are the continuity and the momentum equations written in conservative form in a $2D$ cartesian coordinate system.

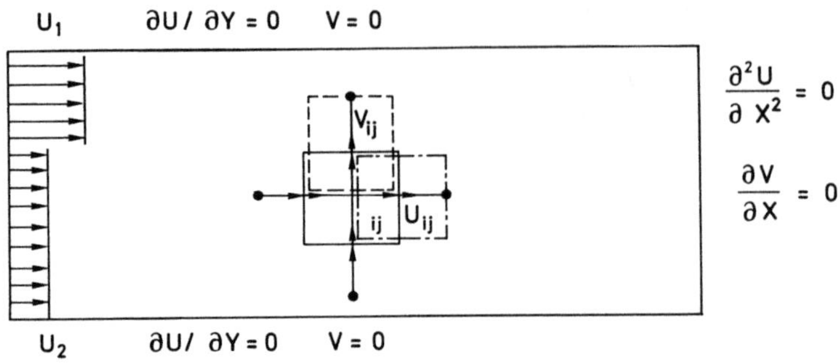

Fig. 10. The geometrical domain and the boundary conditions

The numerical code is the one developed by Braza [3]. It is a pressure-velocity, finite volume, semi-implicit second order accurate numerical method, based on a predictor-corrector pressure scheme proposed by Chorin [22]. The momentum equations are solved by the Alternating Direction Implicit (ADI) method.

The computation domain is rectangular (Fig. 10). The boundary conditions are sketched on this figure and detailed in [23]. The inlet boundary conditions are those of a shear flow created by two different velocities and having an inflexion point. The outlet boundary conditions are those of a free boundary layer, in order to minimize confining effects. The conditions for the upper and the lower boundaries are ($\partial U/\partial y = 0$) and ($V = 0$) by assuming that these frontiers are streamlines.

The Results

The dynamics of the shear layer are related to the following parameters: U_m, R, b_0, Re_b (see Nomenclature).

Owing to the existence of an inflexion point in the initial velocity profile, the free shear layer is unconditionally unstable. The instability consists of an acoustic-like wave, propagating downstream. Streaklines formed by marker particules issued from the inlet section are used to visualize the generation of the above instability.

The results are discussed in three different situations: the natural instability occurring at the frequency f_t and the shear layer excited either at $f_t/2$ or at $f_t/(\alpha)$ (α being a non integer equal here to 6.94).

In the first case the calculations are conducted without introducing any external perturbation. It is found that the flow is parallel and rather steady near the inlet section. Further downstream the instability wave grows and rolls up in convected vortices. The development of this natural instability is obtained spontaneously owing to the important shear stress rate applicated in the inlet section ($R = 33\,\%$) for the explored Reynolds number range $200 < R_b < 1{,}000$.

The spread of the mixing layer, which is defined by the streamlines corresponding to $U = 0.95 * U1$ and $1.05 * U2$ is plotted on Fig. 11 for a Reynolds number $\mathrm{Re}_b = 200$. It can be seen that the spreading rate fits to the analytical law established by Liepman and Laufer [24] in the laminar case for $0 < X/b_0 < 20$:

$$b = k * \sqrt{X/b_0}.$$

Beyond $X/b_0 = 20$ the expansion rate is found to be constant, a result that corresponds to the beginning of the transition to turbulence.

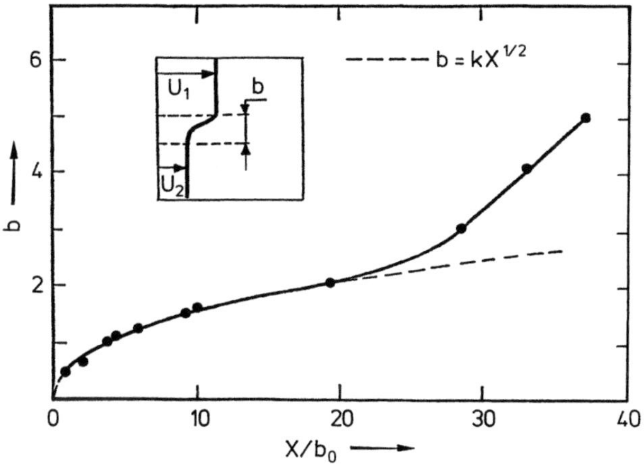

Fig. 11. Spread of the non excited layer at $Re_b = 200$

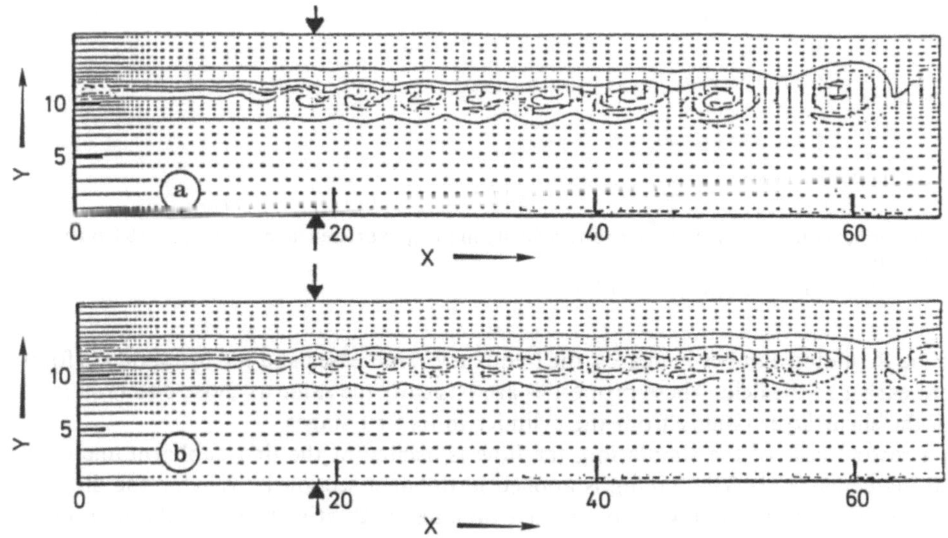

Fig. 12. Streaklines pattern of the non excited shear layer ($R = 0.33$; $Re_b = 500$; (a) $T = 90$; (b) $T = 100$)

The streakline pattern of the convected vortices is represented on Fig. 12 for a Reynolds number $Re_b = 500$ for which regular downstream vortices are observed initiating a process similar to the one described by Brown and Roshko [25].

The time dependent velocity components and pressure fluctuations (Fig. 13) show the periodicity in the mixing layer at a natural frequency f_t, which appears more obviously on the V-component. The maximum amplitude is observed on the central axis ($Y/b_0 = 10.2$). On both sides of the shear layer ($Y/b_0 = 7.9$ and 12.1) weaker induced oscillations are observed. The U-Component is more marked by the downstream convection process and the oscillations look less regular.

For the pressure fluctuations, the subharmonic $f_t/2$ is observed after a dimensionless time value of 100 at the point $X/b_0 = 18.8$ for which these signals are displayed. This indicates the possibility of a future pairing of vortices (not observed in these calculations up to a time of 160).

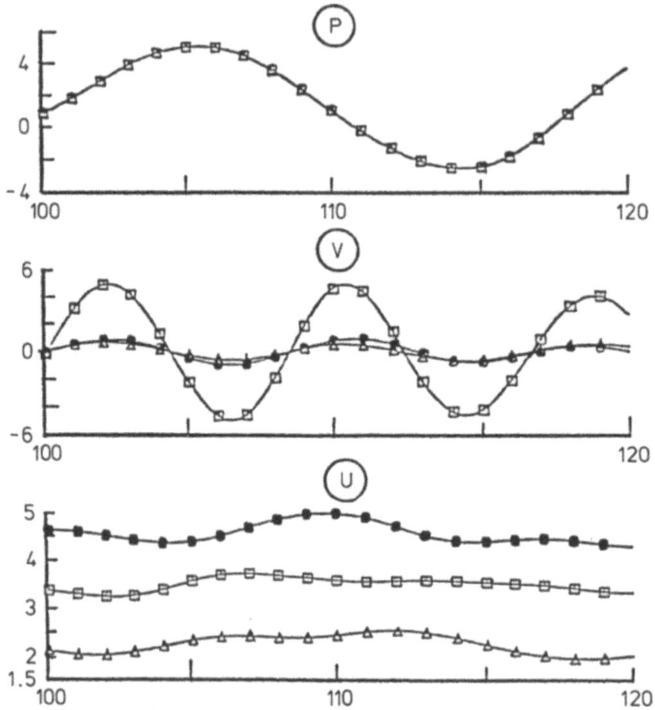

Fig. 13. Numerical time traces of pressure and velocity in the non excited layer at $X/b_0 = 18.8$ (arbitrary scales) ($R = 0.33$; $Re_b = 500$).
(△ $Y/D = 7.92$; □ $Y/D = 10.2$; ● $Y/D = 12.14$)

As the Reynolds number increases, the frequency f_t increases and the instability is found to occur earlier in the field. These properties are basically the same as those observed in the case of the shear layer issuing from the cylinder in its first stages.

In a second case of the simulation, the flow is excited at the first subharmonic of the natural frequency ($f_t/2$). The forcing is applied in the inlet section by varying the direction of the velocity V at the inflexion point. It simulates qualitatively the oscillations of the separation point and the boundary layer in the case of the cylinder. The forcing at this frequency was chosen because of its importance in the pairing phenomenon as observed experimentally by Freymuth [26] or Ho and Huerre [21].

When the forcing is applied, two neighbouring vortices are laterally displaced, due to different convective velocities. Hence they wrap around each other to form a single structure in the classical pairing process (Fig. 14). The shear layer spreading is therefore increased by the occurrence of the pairing. This interaction is observed in a transient phase after which the forcing frequency controls the shedding of vortices. These are then emitted at a frequency of $f_t/2$ and have a wavelength which is twice the natural one.

The time traces of pressure and velocity are displayed on Fig. 15 on the line $Y/b_0 = 10.2$. The forced oscillations at $f_t/2$ is found in all signals near the inlet section ($X/b_0 = 0.1$) and also in the pressure fluctuations at all downstream positions ($X/b_0 = 18.8$ and 37.8). The V-component at these same points oscillates at a frequency which is due to the simultaneous effect of the natural frequency and of the subharmonic $f_t/2$.

The pairing process is observed in several flow regimes in the shear layer issued from the cylinder and specially when splitter plates are added. In this case it becomes the predominant process in the coupling mechanism.

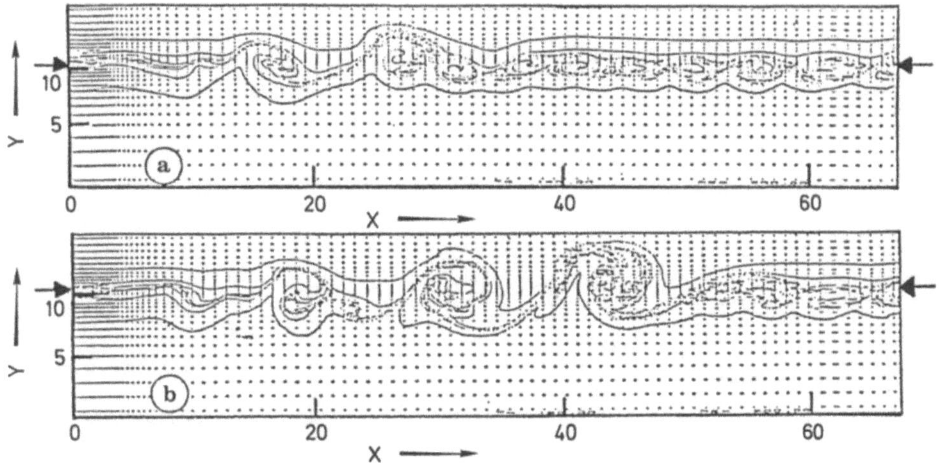

Fig. 14. Streaklines pattern of the shear layer excited at $f_t/2$ ($R = 0.33$; $\mathrm{Re}_b = 500$; (a) $T = 140$; (b) $T = 160$)

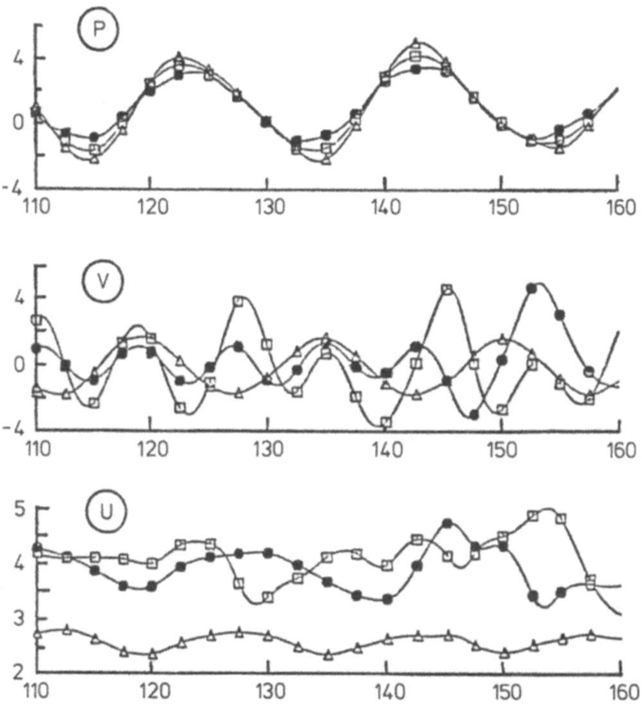

Fig. 15. Numerical time traces of pressure and velocity in the shear layer excited at $f_t/2$ (arbitrary scales) ($R = 0.33$; $\mathrm{Re}_b = 500$; $Y/b_0 = 10.25$).
(△ $X/b_0 = 0.1$; □ $X/b_0 = 18.8$; ● $X/b_0 = 37.8$)

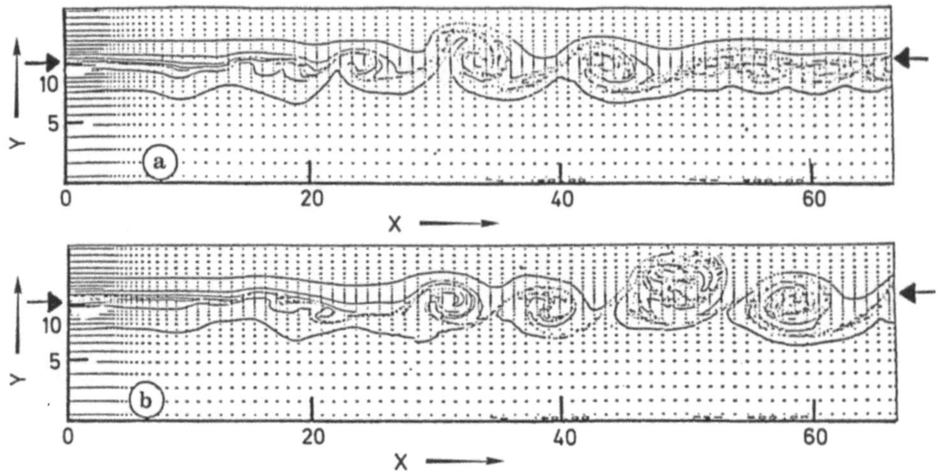

Fig. 16. Streaklines pattern of the shear layer excited at $f_t/6.94$ ($R = 0.33$; $\mathrm{Re}_b = 500$; (a) $T = 160$; (b) $T = 180$)

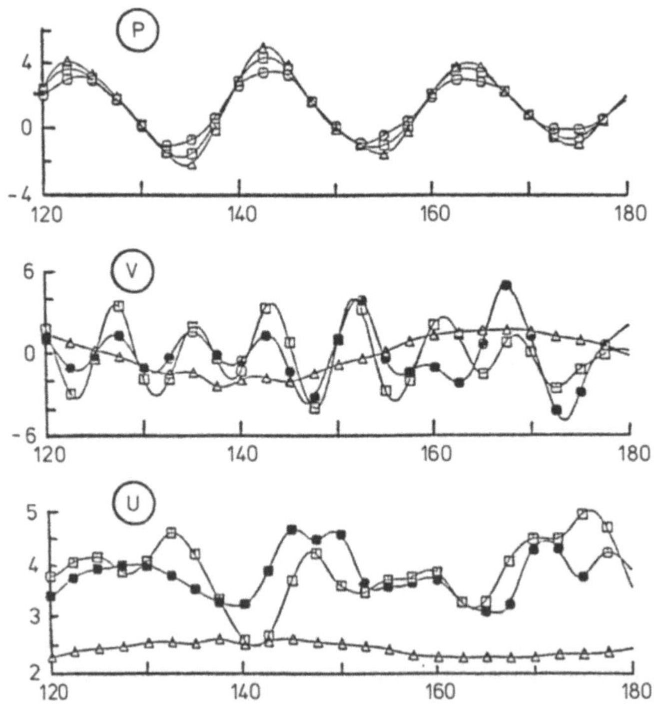

Fig. 17. Numerical time traces of pressure and velocity in the shear layer excited at $f_t/6.94$ (arbitrary scales) ($R = 0.33$; $\mathrm{Re}_b = 500$; $Y/b_0 = 10.25$).
(△ $X/b_0 = 0.1$; □ $X/b_0 = 18.8$; ● $X/b_0 = 37.8$)

The last case simulated is the forcing at a frequency (f_f) incommensurate with the natural one f_t. This excitation is superimposed to simulate the chopping effect of the alternating eddies frequency on the separated layer. In the sample configuration, the non linear interaction between the above frequencies is observed. The forcing frequency f_f is chosen to be equal to $f_t/6.94$ which corresponds in the circular cylinder case to the vortex shedding frequency f_s at a Reynolds number Re_D of 5,000 for which a strong coupling exists between both vortical motions.

The effect of this forcing is shown on Fig. 16. Pairings are also observed but in contrast with the previous case, this mechanism is irregular and does not appear automatically. Single vortices at the natural frequency can be found among those forming pairs and the spectrum is likely to contain a complex set of interaction frequencies as it is observed in the case of the formation zone of the wake vortices of the cylinder.

The traces on the line $Y/b_0 = 10.2$ (Fig. 17) show a non linear interaction between both incommensurate frequencies. The pressure fluctuation is rather regular and a frequency of $(f_t/2 - f_f)$ corresponds to the time interval between two maxima. The V-component oscillates at a frequency of $(f_t - f_f)$, which was found to be the main interaction frequency in the cylinder case (see Sect. 3.1).

Conclusion

The transition to turbulence in the shear layer behind a circular cylinder has been studied pointing out the dynamical properties of the interaction between structures of different sizes and frequencies.

The different flow configurations can be distinguished depending upon the Reynolds number and the frequency ratio between the large scale wake vortices and the smaller ones from the shear layer.

At low Reynolds number $2{,}400 < \mathrm{Re}_D < 15{,}000$, a strong coupling is observed and the spectra exhibit discrete peaks at the natural frequency f_t.

The frequency ratio f_t/f_s being proportional to $\mathrm{Re}_D^{1/2}$, with increasing Reynolds numbers ($16{,}000 < \mathrm{Re}_D < 60{,}000$), the frequency ranges become disconnected. The transition takes place on a very short length scale compared to the vortex shedding mechanism. Therefore peaks at the frequency f_t are not observed, even in sections directly over the wall of the cylinder (for instance at $X/D = 0$). Apart from the discrete set of harmonics based upon f_s, the major part of the energy in the spectra is contained in a continuous part corresponding to the random motion of the small scale disorganized vortices. Thus this motion acts as an eddy viscosity with respect to the regular wake vortex shedding.

When the ratio f_t/f_s is small enough, in the lower Reynolds number range, the non-linear coupling which controls the transition process is studied in details both experimentally and numerically.

Moving downstream in the shear layer, a set of non linear frequency interactions $(f_t \pm nf_s)$ is progressively generated. Nevertheless the continuous spectrum of the random-like motion which appears subsequently grows preferably in the interaction band, showing the predominant effect of the coupling mechanism on the establishment of turbulence.

Both sources of discrete frequency f_t and f_s are shown to be independent mechanisms by the results of the experiments with the splitter plate. Indeed, at the same Reynolds number, the vortex shedding frequency is lower than in the case without plate while the shear layer frequency is unchanged. However, the interaction mechanism, owing to the weak coupling due to the effect of the plate, is quite similar to the one of a classical plane mixing layer.

The basic phenomenon is identified as the development of the plane mixing layer vortices submitted to the excitation of the large scale motion. It is simulated numerically by solving the unsteady Navier-Stokes equations. The main features observed for the cylinder are predicted by the code. The natural instability frequency appears spontaneously for the velocity ratio $R = 0.33$ and the pairing is stabilized when exciting at $f_t/2$. For an incommensurate frequency $f_t/6.94$, corresponding to a Reynolds number of 5,000 in the case of the cylinder, irregular pairings are observed and the signals are modulated as a consequence of the interaction between f_t and f_f.

In fact it turns out that the main mechanism for the establishment of the chaos is a two-dimensional process in a first stage and the numerical approach used for obtaining circumstancial information is justified, even though a more complete simulation is still necessary.

Acknowledgements. The authors wish to thank Professor G. S. West for his participation to this work during a long time stay in their laboratory. They are also indebted to Dr. A. Sevrain for valuable discussions, to Mr J. C. Pons and to Mr J. C. Bonnefont for their technical assistance and to Mrs Nguyen-Tristani for the presentation of this paper.

This work was supported by the French Army Department under Contract number DRET.84/1412/DS/SR.

References

1. Roshko, A. (1954): On the development of turbulent wakes from vortex streets. N.A.C.A. Report 1191
2. Sreenivasan, K. R. (1985): "Transition to Turbulence in Fluid Flows and Low-Dimensional Chaos," in *Frontiers in Fluid Mechanis*, ed. by S. H. Davis, J. L. Lumley, (Springer, Berlin, Heidelberg) pp. 41–67
3. Braza, M. (1981): "Simulation Numérique du décollement Instationnaire Externe par une Formulation Vitesse-Pression, Application à l'écoulement autour du cylindre;" Thèse de Docteur-Ingénieur, I.N.P., Toulouse, No. 182
4. Braza, M., Ha Minh, H., Chassaing, P. (1984): "Numerical Simulation of the Vortex Shedding Past a Circular Cylinder Using a Pressure-Velocity Formulation," in *Numerical Methods for Transient and Complex Problems* (Pineridge Press)
5. Braza, M., Chassaing, P., Ha Minh, H. (1985): "Numerical Study of Different Scale Structures in the Near Wake of a Circular Cylinder in Laminar to Turbulent Transition," in *Numerical Methods in Laminar and Turbulent Flow* (Pineridge Press)
6. Bloor, M. S. (1964): The transition to turbulence in the wake of a circular cylinder. J. Fluid Mech. Vol. **19**, Part 2, 290–304
7. Bloor, M. S., Gerrard, J. H. (1966): "Measurements on Turbulent Vortices in a Cylinder Wake." Proceedings of Royal Society, Serie A, pp. 319–342
8. Crausse, E. (1936): "Contribution Expérimentale à l'étude des Phénomènes Transitoires et Périodiques se produisant dans les Lignes en Mouvement;" Thèse de Doctorat ès Sciences, Université de Toulouse
9. Domptail, C. (1979): "Sillage Turbulent en aval de deux Barreaux parallèles en Tunnel Hydrodynamique: Visualisation et Vélocimétrie Laser;" Thèse de 3e Cycle, Université d'AIX-Marseille II
10. Boisson, H. C., Chassaing, P., Ha Minh, H. (1983): Conditional analysis of intermittency in the near wake of a circular cylinder. Phys. Fluids **26**, 653–658
11. Kourta, A., Boisson, H. C., Chassaing, P., Ha Minh, H. (1984): "Transition to Turbulence in the Wake of a Circular Cylinder by Interaction Between Shear Layers and a Vortex Street," Fifth Physico-Chemical Hydrodynamics Conference, Tel-Aviv
12. Stansby, P. K. (1974): The effect of end plates on the base pressure coefficient of a circular cylinder. Aeronaut. J. 36–37
13. West, G. S., Apelt, C. J. (1982): The effects of tunnel blockage and aspect ratio on the mean flow past a circular cylinder in the range $10^4 < \mathrm{Re} < 10^5$. J. Fluid Mech. **114**, 361–371
14. Kim, Y. C., Khadra, L., Powers, E. J. (1980): Wave modulation in a non linear dispersive medium. Phys. Fluids **28**, 2250–2257

15. Miksad, R. W., Jones, F. L., Powers, E. J., Kim, Y. C., Khadra, L. (1982): Experiments on the role of amplitude and phase modulations during the transition to turbulence. J. Fluid Mech. **123**, 1–29
16. Motohashi, T. (1979): A Higher-order non-linear interaction among spectral components. Phys. Fluids **22**, 1212–1213
17. Roshko, A. (1955): On the wake and drag of bluff bodies. J. Aeronaut. Sci. **22**, 124–132
18. Bearman P. W. (1965): Investigation of the flow behind a two-dimensional model with blunt trailing edge and fitted with splitters plates. J. Fluid Mech. **21**, 241–255
19. Apelt, J., West, G. S., Szewczyk, A. A. (1973): The effects of wake splitter plate on the flow past a circular cylinder in the range $10^4 < Re < 5.10^5$. J. Fluid Mech. **61**, 187–198
20. West, G. S., Boisson, H. C., Kourta, A. (1985): Influence d'une plaque de recollement sur les propriétés spectrales de l'écoulement autour d'un cylindre circulaire en régime subcritique. IMF-Toulouse, Report No. 48
21. Ho, C. M., Huerre, P. (1984): Perturbed free shear layers. Ann. Rev. Fluid Mech. **16**, 365–424
22. Chorin, A. J. (1967): A numerical method for solving incompressible viscous flow problems. J. Comput. Phys. **2**, 12–26
23. Kourta, A. (1984): Analyse Physique et simulation numérique des structures tourbillonaires du sillage proche d'un cylindre circulaire. Thèse I.N.P. Toulouse, No. 346
24. Liepman, H. W., Laufer, J. (1947): Investigation of free turbulent mixing. N.A.C.A. Tech. Note, No 1257
25. Brown, G. L., Roshko, A. (1971): "The Effects of Density Difference on the Turbulent Mixing Layer," AGARD Meeting on Turbulent Shear Flow, London
26. Freymuth, P. (1966): On transition in a separated boundary layer. J. Fluid Mech. **25**, 983–704

Investigation of the Detailed Spread Characteristics of Plane Turbulent Wakes

P. R. Louchez, J. G. Kawall and J. F. Keffer

Department of Mechanical Engineering, University of Toronto, USA

Abstract

An experimental investigation of plane turbulent wakes generated by solid bodies of various cross-sectional shapes and by strips of screens of various solidities has been carried out by means of hot-wire anemometry for the purpose of establishing the detailed spread characteristics of these flows. Each wake was found to attain a state in which the lateral distribution of its principal statistical properties, viz., U_d, u', v' and $-\overline{uv}$, were self-similar and the characteristic intensity and length scales associated with these distributions varied in accordance with predicted self-preserving laws; moreover, all wakes except those produced by screen-strips having characteristic Reynolds numbers based on screen-wire diameter less than a critical value of about 100, were found to approach a universal self-preserving state.

Nomenclature

a_i	Virtual origin for a characteristic intensity scale ϕ_i, $i = 1, 2, 3, 4$
b_i	Virtual origin for a characteristic length scale λ_i, $i = 1, 2, 3, 4$
C_D	Drag coefficient
d	Screen-wire diameter, mm
F_D	Drag force per unit length, N/m
$g(\ )$	Universal function
L	Wake generator length, mm
l_d	Characteristic length scale for a mean velocity defect profile, mm
l_u	Characteristic length scale for an RMS streamwise velocity profile, mm
l_v	Characteristic length scale for an RMS lateral velocity profile, mm
l_{uv}	Characteristic length scale for a Reynolds stress profile, mm
R	Reynolds number
r_i	Intensity scale ratio, equal to $(\phi_{i+1}/\phi_1)^2$, $i = 1, 2, 3$
s_i	Length scale ratio, equal to λ_{i+1}/λ_1, $i = 1, 2, 3$
U	Instantaneous streamwise velocity, ms^{-1}
$\bar{U}$	Mean streamwise velocity, ms^{-1}
U_d	Mean velocity defect, ms^{-1}
U_{dc}	Characteristic intensity scale for a mean velocity defect profile, ms^{-1}
U_0	Free-stream velocity, ms^{-1}
u	Fluctuating component of the streamwise velocity, ms^{-1}
u'	RMS streamwise velocity, equal to $(\overline{u^2})^{1/2}$, ms^{-1}
u'_C	Characteristic intensity scale for an RMS streamwise velocity profile, ms^{-1}
V	Instantaneous lateral velocity, ms^{-1}
$\bar{V}$	Mean lateral velocity, ms^{-1}
v	Fluctuating component of the lateral velocity, ms^{-1}
v'	RMS lateral velocity, equal to $(\overline{v^2})^{1/2}$, ms^{-1}
v'_c	Characteristic intensity scale for an RMS lateral velocity profile, ms^{-1}
$-\overline{uv}$	Reynolds stress, $m^2 s^{-2}$
$\overline{uv}_c$	Characteristic intensity scale for a Reynolds stress profile, $m^2 s^{-2}$
W	Wake-generator width, mm
W_a	Wake number, equal to θ^*/θ
x, y, z	Streamwise, lateral and spanwise Cartesian coordinates, mm
α_i	Growth rate for $1/\phi_i^2$, $i = 1, 2, 3, 4$
β_i	Growth rate for λ_i^2, $i = 1, 2, 3, 4$
θ	Mean momentum thickness, mm
θ^*	Fluctuation momentum thickness, mm
λ_i	Generalized characteristic length scale, representing $l_d (i = 1)$, $l_u (i = 2)$, $l_v (i = 3)$ and l_{uv} $(i = 4)$, mm
ϱ	Density, kg/m^3
ϕ_i	Generalized characteristic intensity scale, representing $U_{dc} (i = 1)$, $u'_c (i = 2)$, $v'_c (i = 3)$ and $\overline{uv}_c (i = 4)$, mm

Turbulent Shear Flows 5

Introduction

On the basis of numerous experimental findings, e.g., [1–4], it is now generally accepted that coherent structures exist within fully developed turbulent shear flows. In the case of the plane wake generated by a circular cylinder, the results of Kawall and Keffer [2] have shown that such structures are, in effect, organized motions possessing definite mean circulation, and that they are characterized by a frequency which is significantly lower than the Strouhal (or vortex shedding) frequency. This is consistent with the work of Cimbala et al. [1], who have found that large-scale coherent structures with a characteristic wavenumber which is lower than that associated with the Kármán vortices evolve in the fully developed wake, and that these structures appear to result from instability of the mean velocity profiles of the developing flow.

While there is no doubt about the existence of these structures, the questions of how they originate, how they evolve and what role they play in turbulent transport processes remain unresolved. In an effort to obtain an explanation of their behaviour, we have carried out an experimental investigation of plane turbulent wakes produced by solid bodies of various cross-sectional shapes and by strips of screens of various solidities, using hot-wire anemometry and flow visualization. Detection of the coherent structures within the flows was accomplished with the aid of (a) coherence functions of pairs of velocity signals from the flows – the presence of broadband peaks (frequency-centred-activity) in these functions signifying the occurrence of the structures and (b) digital image processing of flow visualization pictures.

To enable us to develop a comprehensive analysis of the behaviour of the coherent structures in the aforementioned wakes, it was essential that we first carry out a systematic assessment of the basic spread characteristics of these wakes, since such an assessment did not exist. The present work focuses upon this assessment and provides answers to the following two questions: (i) Do these wakes differ in their overall development? (ii) Is there a universal self-preserving state attained by plane wakes produced by all types of wake generators?

Analytical Details

A Cartesian (x, y, z) reference frame is used with its centre, 0, located at the centre of gravity of each wake generator, $0x$ in the streamwise (or potential flow) direction, $0z$ in the spanwise direction (aligned with the main axis of the wake generator) and $0y$ in the lateral (or vertical) direction. Since no mean spanwise velocity and no z gradients exist, each wake can be treated as a two-dimensional flow in the xy plane.

The relevant flow variables are the streamwise velocity, U, and the lateral velocity, V. These can be decomposed as follows:

$$U = \bar{U} + u = U_0 - U_d + u$$

and

$$V = \bar{V} + v,$$

where $(\overline{\ \ })$ represents a time-averaged quantity, U_0 is the free-stream (or potential-flow) velocity, U_d is the velocity defect, and $\bar{u} = \bar{v} = 0$.

We introduce intensity scales, U_{dc}, u'_c, v'_c, which are the centre-line values of U_d, $u' = (\overline{u^2})^{1/2}$, $v' = (\overline{v^2})^{1/2}$ respectively, and $\overline{uv}_c$, which is the maximum value of the Reynolds shear stress, $-\overline{uv}$; and length scales, l_d, l_u, l_v, which are the lateral locations of the "half-centre-line" values of U_d, u', v' respectively, and l_{uv}, which is the lateral location of uv_c.

Now for any fully developed "small-deficit" wake, the equation governing the mean velocity field is given by

$$U_0 \frac{\partial U_d}{\partial x} = \frac{\partial \overline{uv}}{\partial y}. \tag{1}$$

This is subject to the boundary condition,

$$U_d = 0 \quad \text{at} \quad y = \pm\infty,$$

and the integral condition,

$$\bar{F}_d = \varrho U_0 \int_{-\infty}^{\infty} U_d dy, \tag{2}$$

where $\bar{F}_d$ is the average drag force per unit (spanwise) length due to the wake generator, and ϱ is the fluid density.

On the assumption that the spread of the flow depends only on (1) – the effect of the equations governing the fluctuating velocity field being neglected – the spread characteristics, i.e., the intensity and length scales, can be uniquely scaled by means of U_0 and the momentum thickness given by

$$\theta = \bar{F}_D/(\varrho U_0^2) = C_D W/2, \tag{3}$$

where W is the width (measured in the y-direction) of the wake generator. Moreover, on the assumption that the statistical properties $U_d(x, y)$, $u'(x, y)$, $v'(x, y)$ and $-\overline{uv}(x, y)$ are self-similar, their normalized lateral distributions must be independent of x, i.e.,

$$\begin{aligned} U_d/U_{dc} &= g_d(y/l_d), \\ u'/u'_c &= g_u(y/l_u), \\ v'/v'_c &= g_v(y/l_v), \end{aligned} \tag{4}$$

and

$$-\overline{uv}/\overline{uv_c} = g_{uv}(y/l_{uv}).$$

It then follows from (1–4) that

$$U_0^2/\phi_i^2 = \alpha_i(x/\theta + a_i)$$

and (5)

$$\lambda_i^2/\theta^2 = \beta_i(x/\theta + b_i),$$

where $i = 1, 2, 3, 4$; $\phi_1 = U_{dc}$, $\phi_2 = u'_c$, $\phi_3 = v'_c$, $\phi_4^2 = \overline{uv_c}$; $\lambda_1 = l_d$, $\lambda_2 = l_u$, $\lambda_3 = l_v$, $\lambda_4 = l_{uv}$; a_i and b_i are referred to as virtual origins; α_i and β_i are referred to as growth rates. From

(5), we obtain the following ratios:

$$
\begin{aligned}
r_1 &= (u'_c/U_{dc})^2 = (\alpha_1/\alpha_2)\,(x + a_1\theta)/(x + a_2\theta),\\
r_2 &= (v'_c/U_{dc})^2 = (\alpha_1/\alpha_3)\,(x + a_1\theta)/(x + a_3\theta),\\
r_3 &= \overline{uv}_c/U_{dc}^2 = (\alpha_1/\alpha_4)\,(x + a_1\theta)/(x + a_4\theta),\\
s_1 &= l_u/l_d = [(\beta_2/\beta_1)\,(x + b_2\theta)/(x + b_1\theta)]^{1/2},\\
s_2 &= l_v/l_d = [(\beta_3/\beta_1)\,(x + b_3\theta)/(x + b_1\theta)]^{1/2},\\
s_3 &= l_{uv}/l_d = [(\beta_4/\beta_1)\,(x + b_4\theta)/(x + b_1\theta)]^{1/2}.
\end{aligned}
\tag{6}
$$

In the case of a specific wake, given that either $a_1 = a_2 = a_3 = a_4$ and $b_1 = b_2 = b_3 = b_4$ or x/θ is sufficiently large that the "unmatched" virtual origins have a negligible effect on the r_i and s_i, then that wake will be in a self-preserving state. Furthermore, given that θ and U_0 are the only relevant scaling parameters, the asymptotic values of the r_i and s_i will be the same for all wakes and hence, a *universal* self-preserving state will be possible. It should be noted that if the spread of a wake depends on the fluctuating velocity field as well as the mean velocity field, then another length-scaling parameter (in addition to θ) must be considered. This parameter is given by

$$\theta^{*2} = (1/U_0^2) \int_{\omega}^{\infty} \int_{\omega}^{\infty} \overline{u(y)\,u(y')}\,dy\,dy'. \tag{7}$$

Since θ and θ^* are independent variables for wakes produced by different wake generators, the implication is that, *in principle*, a universal self-preserving state is not possible: the r_i and s_i will be functions of the "wake number" given by

$$W_a = \theta^*/\theta.$$

If, however, W_a is sufficiently small, then the self-preserving state that is attained will effectively be the universal state.

Experimental Details

The wakes were produced in a variable-speed, low-turbulence recirculating wind tunnel with a 5 m × 1.2 m × 1 m test section. Some of the wake generators that were used are listed along with their pertinent characteristics in Table 1. It should be noted here that the detailed behaviour of a wake produced by a screen-strip depends upon the characterstic Reynolds number for the screen-strip based on screen-wire diameter, R_d, varying according to whether R_d is above or below a critical value of about 100 (see Louchez [5]). The length of each wake generator, L, was approximately equal to the spanwise extent of the test section, viz., 1.2 m, so that the aspect ratio, L/W, was in excess of 100 for all the wakes, which ensured that they were two-dimensional flows. The Reynolds numbers based on wake-generator width (W) were between about 2,000 and about 13,000. Basic wake characterstics are listed in Table 1. Velocity measurements were carried out by means of DISA 55 M anemometer systems in conjunction with normal hot-wire and X-wire probes, and data were obtained at stations located between a few wake widths and about 1000 widths from the wake generators.

Table 1. Wake-generator and wake characteristics

Code	Type	Width W (mm)	Thick-ness (mm)	Solidity (%)	Wire dia. d (mm)	Reynolds number R_d	Reynolds number R_w	Drag coeff. C_D	Momentum thick. θ (mm)
CC9	circular cylinder	9.53	9.53	100.0	–	–	6300	1.03	4.89
SA9	screen-strip	9.00	0.90	52.0	0.43	290	6000	0.79	3.57
SB10	screen strip	10.00	0.80	59.0	0.38	250	6600	0.91	4.54
SC10	screen strip	10.00	0.20	70.0	0.11	70	6600	1.06	5.30
SC5	screen-strip	5.00	0.20	70.0	0.11	70	3300	1.04	2.60
VP10	vertical plate	10.00	0.50	100.0	–	–	6600	1.96	9.82

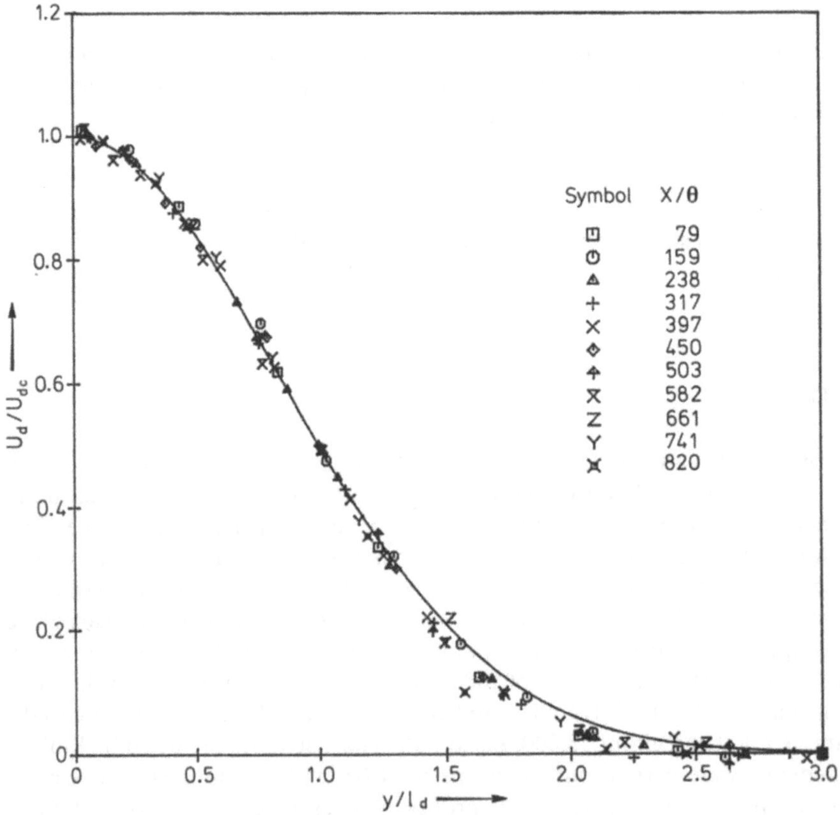

Fig. 1. Mean velocity defect profiles for SA9

Results and Discussion

Self-similarity

Representative normalized U_d, u', v' and $-\overline{uv}$ profiles are presented in Figs. 1–4 respectively. Figure 1 pertains to the 52% solidity screen-strip SA9, Fig. 2 to the circular cylinder CC9, Fig. 3 to the vertical plate VP10 and Fig. 4 to the 59% solidity screen-strip SB10 (see Table 1). The solid line in each of these figures represents an average curve based on "far-wake" results for all wake generators except those for the 70% solidity screen-stips SC5 and SC10 (see Table 1), which will be referred to as sub-critical wake generators, their characteristic Reynolds numbers being substantially less than the critical value of about 100. These results, in conjunction with similar results pertaining to the other wake generators, establish that each wake produced was self-similar beyond some location downstream from its generator. Moreover, these results serve to show that all the wakes *except the subcritical ones*, i.e., those produced by the sub-critical screen-strips, attained identical self-similar states, in the sense that the functions $g_d(y/l_d)$, $g_u(y/l_u)$, $g_v(y/l_v)$ and $g_{uv}(y/l_{uv})$ describing the normalized U_d, y', v' and $-\overline{uv}$ profiles [see (4)] were in fact universal, i.e., the same $g_d(y/l_d)$ described all U_d profiles, etc. These wakes will henceforth be referred to as regular wakes.

The downstream location beyond which self-similarity prevailed was found to be the same for all profiles (U_d, u', v' and $-\overline{uv}$) associated with a given wake. This location was, however, dependent upon the type of wake generator involved (see Louchez [5]).

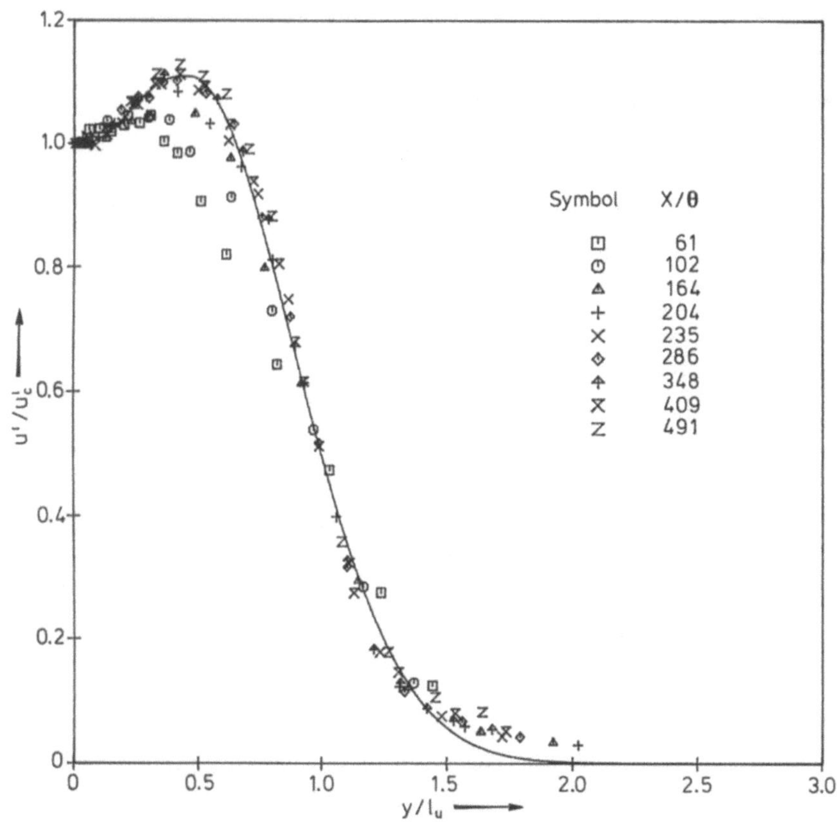

Fig. 2. RMS streamwise velocity profiles for CC9

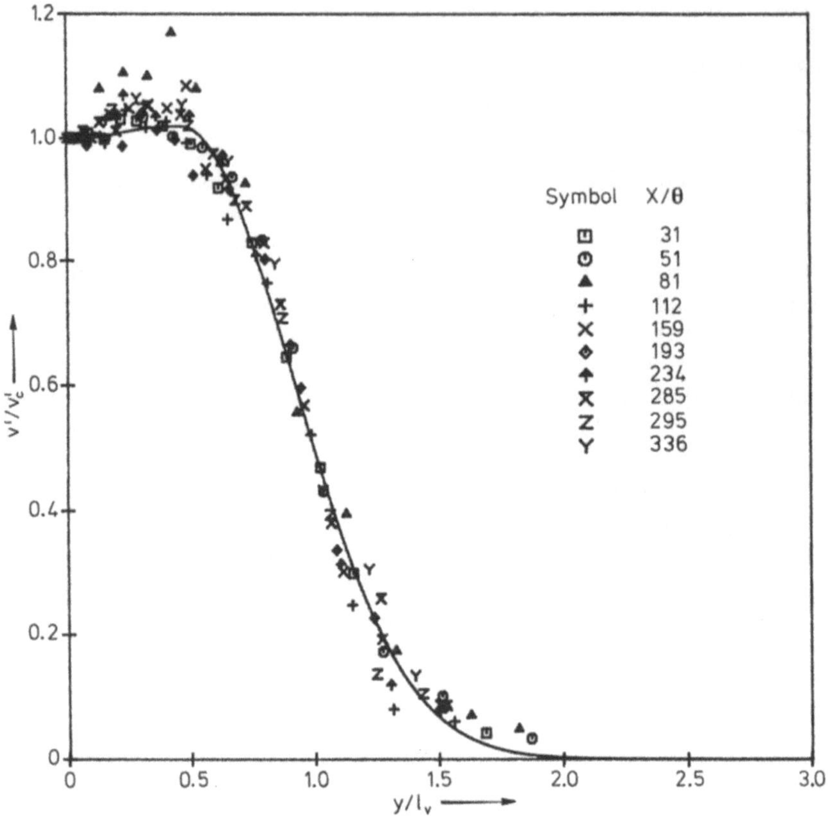

Fig. 3. RMS lateral velocity profiles for VP10

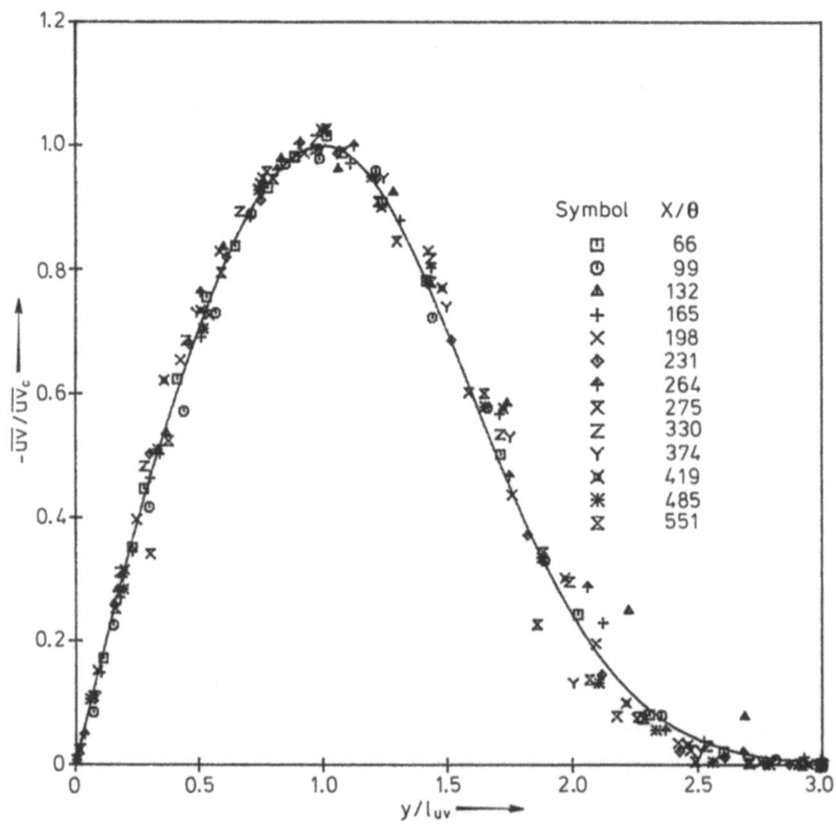

Fig. 4. Reynolds stress profiles for SB10

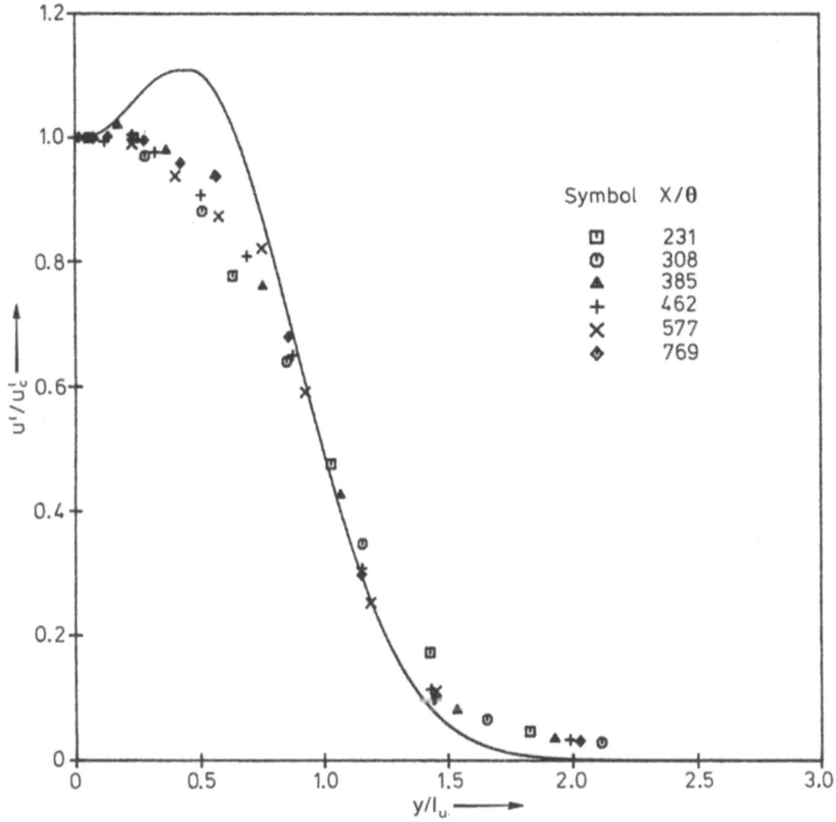

Fig. 5. RMS streamwise velocity profiles for SC5

Figure 5 depicts the normalized u' profiles for a sub-critical (or irregular) wake. It is evident from the figure that this wake did not reach the same self-similar state as did the regular wakes. We have concluded that this anomaly in wake behaviour results from a fundamental difference between the large eddies present in irregular wakes and those present in regular wakes. It seems probable that the large eddies of all regular wakes are coherent structures (Kawall and Keffer [2], Louchez [5]). Accordingly, there would be, for these flows, significant bulk convection from the coherent structures, in addition to the usual gradient diffusion which occurs in any turbulent shear flow. It may be speculated that the large eddies of the subcritical wakes are not really coherent, in which case, bulk convection would be relatively insignificant with respect to these flows. It then follows that their self-similar state, as manifested by u' and v' profiles, would be distinctly different from that of the regular wakes – as demonstrated by the present results.

Quasi Self-preservation

In order to determine whether or not the various wakes attained quasi self-preserving states, the actual streamwise variations of the characteristic length and intensity scales were determined for each wake (via plots of the measured scale values versus x/θ) and compared with predicted variations, viz., those dictated by the self-preserving "growth" laws given by (5). In every case, the agreement between the actual variation and the predicted variation was at least reasonably good for x/θ in excess of a certain value (see Louchez [5]). In the case of the

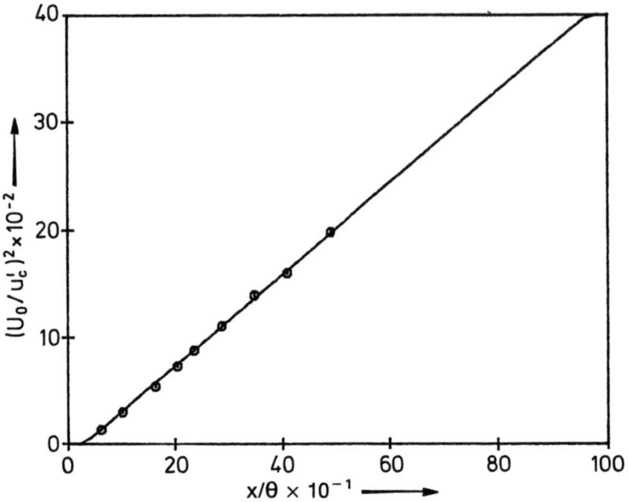

Fig. 6. RMS streamwise velocity intensity scales for CC9

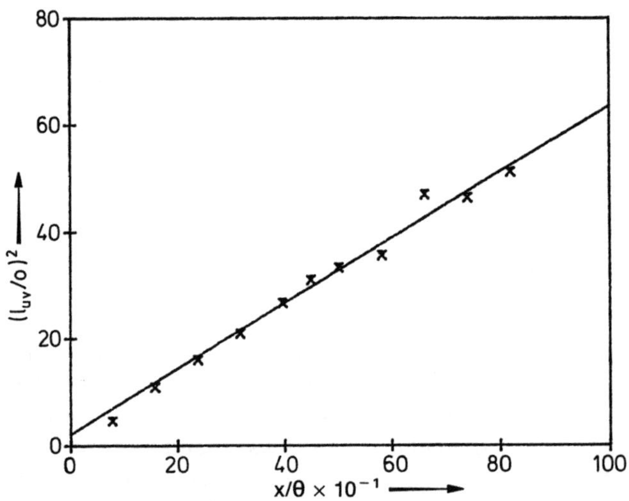

Fig. 7. Reynolds stress length scales for SA9

length scales, the latter value corresponded roughly to the downstream location beyond which self-similarity existed and depended upon the wake generator involved. Clearly, each wake attained at least a quasi self-preserving state. Typical results are presented in Figs. 6 and 7. Figure 6 depicts the behaviour of the characteristic RMS streamwise velocity intensity scale pertaining to the circular cylinder CC9, and Fig. 7 depicts the behaviour of the characteristic Reynolds stress length scale pertaining to the 52% solidity screen-strip SA9. The solid lines in these figures represent the predicted variations. As can be seen, the actual variations are concordant with the self-preserving "growth" laws.

Universal Self-preservation

As pointed out earlier, a universal self-preserving state for all wakes is possible if the wake number, W_a, is small enough. Whether or not such a state exists or is approached in practice,

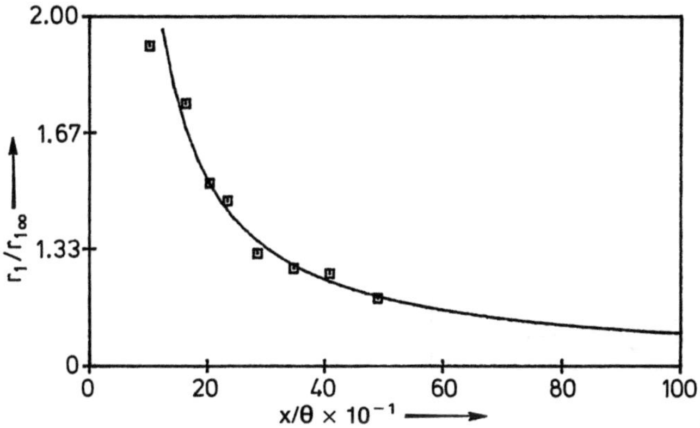

Fig. 8. RMS streamwise velocity intensity scale ratios for CC9

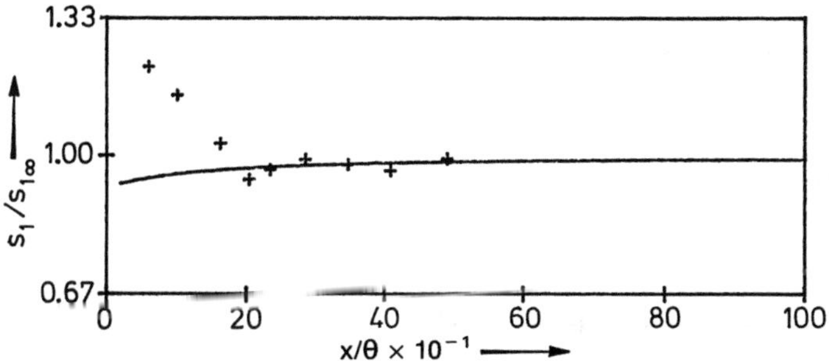

Fig. 9. RMS streamwise velocity length scale ratios for CC9

with respect to a specific set of wakes, can be ascertained from an examination of the streamwise evolutions of the intensity scale ratios and the length scale ratios, r_i and s_i, $i = 1$, 2, 3, for the wakes in question. If each of these ratios reaches or approaches a constant value at some x/θ, then a universal self-preserving state exists; otherwise, such a state does not exist. By plotting the RMS streamwise velocity (intensity and length) scale ratios, r_1 and s_1, the RMS lateral velocity scale ratios, r_2 and s_2, and the Reynolds stress scale ratios, r_3 and s_3, versus x/θ for the various wakes considered, we obtained the requisite evolution data. These data establish that all the wakes, with the exception of the irregular wakes (produced by SC10 and SC5), approached the same self-preserving state. In other words, a universal self-preserving state existed for the regular wakes. It should be noted that, for the most part, the x/θ values at which the intensity scale ratios, r_1, r_2 and r_3, reached their asymptotic values, $r_{1_\infty} = \alpha_1/\alpha_2$, $r_{2_\infty} = \alpha_1/\alpha_3$ and $r_{3_\infty} = \alpha_1/\alpha_4$ [see (6)], were found to be significantly larger than the x/θ values at which the length scale ratios, s_1, s_2 and s_3, reached their asymptotic values, $s_{1_\infty} = (\beta_4/\beta_1)^{1/2}$ $s_{2_\infty} = (\beta_3/\beta_1)^{1/2}$, and $s_{3_\infty} = (\beta_4/\beta_1)^{1/2}$ [see (6)]. This difference between the intensity scale ratios and the length scale ratios is a consequence of the fact that the virtual origins associated with the various scales are not in general the same (see Louchez [5]).

Figures 8 and 9 depict the evolution of the RMS streamwise velocity scale ratios, r_1 and s_1, pertaining to the circular cylinder CC9, and Figs. 10 and 11 depict the evolution of the RMS lateral velocity scale ratios, r_2 and s_2, pertaining to the 52% solidity screen-strip SA9. The solid lines in these figures are based on the predicted self-preserving "growth" laws given by (5). One notes that the intensity ratios r_1 (Fig. 8) and r_2 (Fig. 10) reach their asymptotic

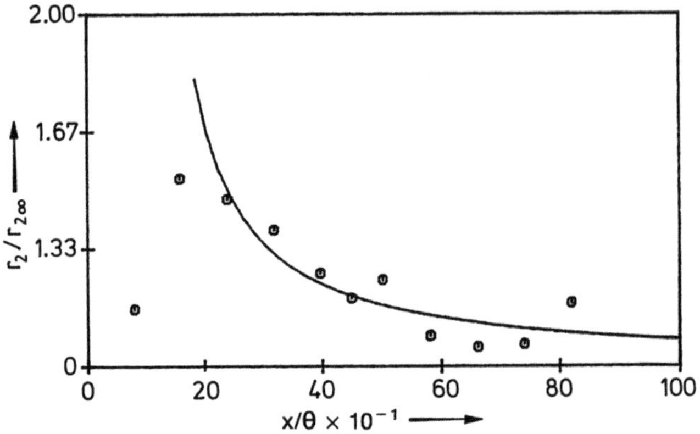

Fig. 10. RMS lateral velocity intensity scale ratios for SA9

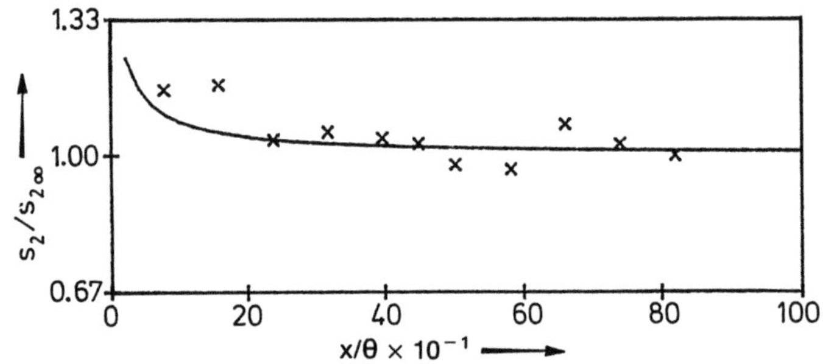

Fig. 11. RMS lateral velocity length scale ratios for SA9

values at fairly large x/θ values; in contrast the length scale ratios s_1 (Fig. 9) and s_2 (Fig. 11) reach their asymptotic values at relatively small x/θ values.

Conclusions

On the basis of the present findings, the following two conclusions can be drawn.

1. Plane wakes produced by various wake generators (e.g., circular cylinders and screen-strips) attain quasi self-preserving states in which the lateral distributions of their statistical properties are self-similar and their characteristic intensity and length scales vary in accordance with the predicted self-preserving "growth" laws. Thus, the answer to the first question posed earlier is an unequivocal 'no' – plane wakes produced by different wake generators do not differ in their gross development.
2. Plane wakes produced by solid bodes and by super-critical porous bodies can achieve a *universal* self-preserving state. Thus, the answer to the second question posed earlier is a qualified 'yes' – there is, in practice, an attainable universal self-preserving state for wakes produced by most, but not all, wake generators.

Acknowledgement. This research was supported by the Natural Sciences and Engineering Research Council of Canada through Grant A-2746.

References

1. Cimbala, J., Nagib, H., Roshko, A., (1981): Wake instability leading to new large scale structures downstream of bluff bodies. Bull. Am. Phys. Soc. Ser. II, **26**, 1256
2. Kawall, J. G., Keffer, J. F. (1982): "The Role of Coherent Structures in the Development of a Uniformly Strained Turbulent Wake," in *Turbulent Shear Flows 3*, by L. J. S. Bradbury, F. Durst, B. E. Launder, F. W. Schmidt, J. H. Whitelaw (Springer, Berlin, Heidelberg) pp. 132–145
3. Moum, J. N., Kawall, J. G., Keffer, J. F. (1983): Coherent structures within the plane turbulent jet. Phys. Fluids **26**, 2939–2945
4. Mumford, J. C. (1983): The structure of the large eddies in fully developed turbulent shear flow. Part 2: The plane wake. J. Fluid Mech. **137**, 447–456
5. Louchez, P. R. (1985): "An Investigation of Plane Turbulent Wakes Generated by Solid and Porous Bodies;" Ph.D. Thesis, University of Toronto

Coherent Structures in a Turbulent Mixing Layer: A Comparison Between Direct Numerical Simulations and Experiments

Ralph W. Metcalfe, A. K. M. F. Hussain*, Suresh Menon, and M. Hayakawa*

Flow Research Company, 21414 68th Avenue South, Kent, Washington 98032, USA

Abstract

An eduction scheme has been developed in an attempt to determine the characteristics of large-scale vortical structures in a turbulent mixing layer. This analysis scheme has been applied to a set of experimental data taken in a new, large mixing layer facility designed to minimize boundary and resonance effects. The scheme is based on detection of large-scale vorticity concentrations from smoothed vorticity maps, accepting structures of certain size and strength and aligning the realizations through correlation of vorticity. A similar scheme has been developed to apply to the results of a direct numerical simulation of a temporally growing mixing layer. A comparison of the two approaches shows important similarities in details of the coherent structures educed both ways. The numerical simulations indicate that low levels of coherent forcing can dramatically change the evolution of the mixing layer. In the absence of such forcing, the numerical simulations and experiments show a lack of regularity in the transverse position, spacing, amplitude, shape and spanwise coherence of the large-scale vortical structures.

Nomenclature

Re_λ	Reynolds number based on Taylor microscale
RE_x	Reynolds number $= x\,U_e/\nu$
$\bar{S}_M$	Peak mean shear rate $= \left.\dfrac{\partial U}{\partial y}\right\|_{max}$
T	$= t\,U_c/\theta$
$\bar{T}$	$= t\,\bar{S}_{Mo}$
U	Mean velocity difference (simulations)
U_c	Streamwise advection velocity of coherent structures (experiments) $= 0.5\,U_e$
U_e	Mean exit velocity (experiment)
$\bar{U}(y)$	Mean velocity
u', v', w'	rms perturbation velocities
u_c, v_c	Coherent velocity fields
u_r, v_r	Incoherent velocity fields
x, y, z	Streamwise, transverse and spanwise coordinates
Y	$= y/\theta$
δ_ω	Mean vorticity thickness $= U_e/\bar{S}_M$
θ	Local momentum thickness
θ_e	Exit boundary layer momentum thickness (experiment)
λ	Wavelength of fundamental (most unstable) mode
ω	Vorticity

Introduction

The mixing layer is presently one of the most active areas of research in fluid mechanics. Mixing layer evolution embodies some topics of great current interest in the fluid dynamics community: primary and secondary instabilities, mechanisms of entrainment and mixing, aerodynamic noise generation, transition, turbulence production, and turbulence modification by artificial excitation, to name a few. One important aspect of this process is the existence of large-scale, coherent, two-dimensional structures at very high Reynolds numbers, which were identified in a mixing layer by Brown and Roshko (1974). The extensive recent literature on mixing layer research is well reviewed by Ho and Huerre (1984). We have performed direct numerical simulations of a temporally growing mixing layer and compared

* Department of Mechanical Engineering, University of Houston, USA

the results of these simulations with new experimental data. These results provide some insight into the topological details of the large-scale vortical structures characteristic of mixing layer evolution.

Riley and Metcalfe (1980) have previously successfully applied the direct numerical simulation technique to the study of mixing layers. This technique was instrumental in examining anomalous behavior, such as countergradient momentum fluxes and negative turbulent energy production, observed in laboratory experiments with a forced mixing layer (Zaman and Hussain, 1977, 1980; Wygnanski et al., 1979). They used direct numerical simulations to predict the effects of the phase relationship between the fundamental and subharmonic modes during vortex pairing. They also showed that large-scale structures could evolve from a random, high-intensity initial field, but that the lateral coherence was not as strong as indicated in some laboratory experiments. In addition, these simulations showed that many critical turbulence correlations depend very strongly on the behavior of the large-scale structures.

Some of these earlier three-dimensional mixing layer simulations were performed on $32 \times 32 \times 32$-point domains or on $64 \times 64 \times 64$-point domains, but for modest integration times that permitted at most one vortex pairing. While these runs provided useful information about the dynamics of the mixing layer evolution, they have not been sufficient to reach a totally self-similar state. Our recent simulations, however, have involved much longer time integration runs on high-resolution ($64 \times 64 \times 64$-point) grids, so that two complete vortex pairings can take place. These simulations now approach a statistically self-similar state for the important velocity correlation quantities and are consistent with the experimental data of Wygnanski and Fiedler (1970) and Kleis and Hussain (1979). In addition, the higher resolution simulations also exhibit some of the secondary instability features seen in laboratory experiments.

We have performed three series of simulations. The first was initialized with a low-amplitude, uncorrelated, perturbation velocity field (Run R, random initial field); the second with low-amplitude, coherent, unstable two-dimensional pairing modes added in the initial state (Run WF, weak forcing); and the third the same as Run WF but with high-amplitude, two-dimensional modes present (Run SF, strong forcing). The data from these simulations are analyzed using techniques originally developed for analysis of coherent structure properties in experimental data (Hussain and Zaman, 1981 a). A comparison with the experimental data is then made.

Methodology

The laboratory experiments have been performed in a new, larger, plane mixing layer facility. This apparatus was designed to reduce the effects of wall interactions and possible feedback that downstream structures could exert on the initial conditions near the separation lip. The experiments were performed in a room 9 m long, 4 m wide and 3.5 m high from which air was sucked out by a blower. The air circulates back into the room from an adjacent large hall via honeycombs and screens and a contraction terminating into a 48 cm × 96 cm rectangular hole. A single-stream plane mixing layer with a spanwise extent of 96 cm is produced. Entrainment air is fed into the room through a porous wall that is lined with thick filter cloth and is positioned opposite to the mixing layer. This porous wall runs the entire length of the room so vortical mixing layer air is not reentrained.

At 10 cm upstream of the flow separation lip, the boundary layer was tripped by a serrated edge, resulting in a fully turbulent boundary layer at the lip. The three-dimensionality of the trip was found to be important in enhancing the development of the turbulent boundary layer. At the separation lip, detailed hot-wire measurements confirmed that the mean velocity, turbulence intensity, spectra and other characteristics were consistent with those for equilibrium flat plate turbulent boundary layers. The mean velocity profile was characterized by adequate widths of the logarithmic and wake regions, with the wake strength agreeing with that expected for a flat plate boundary layer at the corresponding value of Re_θ. It was felt that by starting with a fully turbulent boundary layer, a minimal downstream distance would be required for the development of an "equilibrium" coherent structure that was independent of upstream conditions. The free-stream velocity U_e was 12 ms^{-1}. The mean velocity distribution across the entrance was uniform to within 2%. The free-stream turbulence intensity was less than 0.2%. The exit mean and fluctuating velocity profile details were found to be uniform in the spanwise direction for 95% of the span.

Measurements up to a distance of 4 m downstream of the separation lip were made, so that the effective Reynolds number was about $\mathrm{Re}_x = 3 \times 10^6$. The value of Re_λ exceeded 1000 everywhere within the measurement range. Data acquisition and probe (standard linearized hot-wire anemometers) traverses were performed using automated computer control. At various downstream locations, data were taken with a rake of eight X-wire probes. These were recorded on magnetic tape, and subsequently linearized, processed, and analyzed to obtain instantaneous vorticity maps in the plane of the rake. Additional details regarding the experimental apparatus are given in Hussain and Zaman (1985).

A detailed eduction scheme was developed to recognize the passage of coherent structures, and the topological details of these structures were computed through phase-locked ensemble averaging. Basically, the data were first smoothed to eliminate localized high-frequency peaks, and the time evolution of the smoothed, spanwise vorticity maps in the (y, t) plane was computed. A structure was accepted for inclusion in the analysis if the strength of the structure denoted by the peak value of the smoothed vorticity was above a selected level, i.e., $\omega_p > 2.5\,\bar{S}_M$, and if the streamwise and transverse dimensions of the vortex, Δx and Δy (defined as the sizes of domains of contour levels having a value of 0.2 ω_p), satisfied Δx, $\Delta y > \delta_\omega/4$. The experiment was run for a sufficiently long time so that data on about 2000 selected structures could be collected. Pairing modes were excluded by accepting only those structures having a peak at a given transverse location and having well-defined, isolated peaks. Note that the transverse rake of X-wires was placed such that it was centered at the transverse distance corresponding to the most probable centers of passing structures, and only structures centered at the middle of the rake were accepted.

The next step was to compute the *first-estimate ensemble average* $\langle\omega\rangle_0$ of the spanwise vorticity by aligning the accepted smoothed vorticity maps on their peaks. Each realization ω_i was then correlated with $\langle\omega\rangle_0$ and the phase shifts δx determined so that the cross-correlation was maximized. A new ensemble average $\langle\omega\rangle_1$ was then determined and this process repeated until no further significant improvement in the correlation values could be achieved. Final ensemble averages of velocities and spanwise vorticity were then computed using cumulative phase shifts for each realization, but employing the original *raw* (rather than smoothed) data. Further enhancement was achieved by discarding realizations requiring excessive shifts or producing low peak values of cross correlation. The departure of each individual (unsmoothed) realization from the corresponding ensemble average produced incoherent turbulence. Further details of the eduction scheme have been discussed by Hussain (1985).

Our direct numerical simulations of a turbulent mixing layer were performed by solving the (time-dependent, nonlinear) Navier-Stokes equations in three spatial dimensions plus time. Pseudospectral numerical methods were used, with periodic boundary conditions in the streamwise and spanwise directions (x, z) and free-slip conditions in the transverse (y) direction (Gottlieb and Orszag, 1977). No subgrid-scale modeling was used. The Reynolds

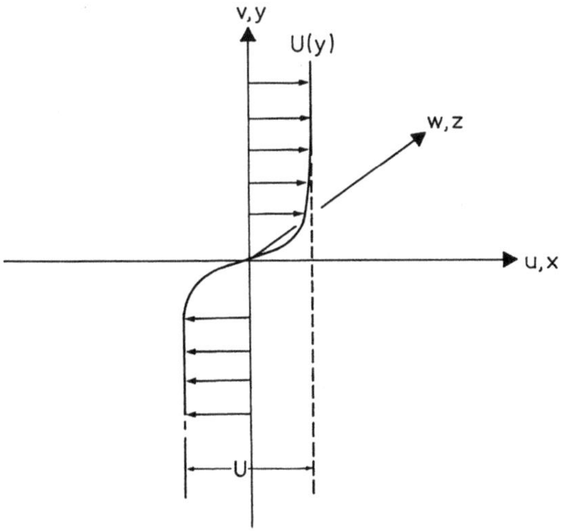

Fig. 1. Problem geometry

number of the flow was kept sufficiently small so that all critical scales of motion were accurately resolved (Reynolds number based on Taylor microscale was typically in the range $\mathrm{Re}_\lambda \cong 50-80$). As described above, three different types of initial conditions were defined. In all three cases, a background, quasi-random, three-dimensional perturbation velocity field was superimposed onto a hyperbolic tangent mean velocity profile (Fig. 1). Details of this rather complex initialization process can be found in Riley and Metcalfe (1979).

The main difference between the previous simulations and those presented in this paper is that the initial peak intensities are much smaller in the present runs. In Run R, with no coherent mode, the initial peak intensities at $y = 0$ were $u' = 0.012\ U$, $v' = 0.011\ U$, and $w' = 0.013\ U$, where U corresponds to U_e in the experiments. At these amplitudes, the initial stages of the calculation were dominated by the mostly linear growth of the unstable modes. In the second simulation, Run WF, the linearly most unstable two-dimensional mode (Michalke, 1964) and its first and second subharmonics were added to the velocity field used in Run R. Thus, two complete vortex pairings were possible, and the phase relation among these modes was such as to maximize the vortex pairing process (Riley and Metcalfe, 1980). The amplitude of the two-dimensional modes was of the same order as the three-dimensional random field so that the peak intensities were $u' = 0.028\ U$, $v' = 0.017\ U$ and $w' = 0.013\ U$. The third case, Run SF, was the same as Run WF but the amplitude of the most unstable two-dimensional mode was increased so that the corresponding peak intensities became $u' = 0.152\ U$, $v' = 0.091\ U$, and $w' = 0.013\ U$.

The procedure employed for the eduction of the coherent structures in the numerical simulations was similar to that described above for the laboratory experiments. At a predetermined time in the calculation, a nine-point averaging formula was used to compute smoothed u and v velocity fields on each spanwise plane. From these, a smoothed, spanwise vorticity field ω_s was computed on each spanwise (x, y) plane using finite differences. Peak vorticity values were then determined, and those planes having peak values less than $\bar{S}_M$ values were discarded. An ensemble average $\langle\omega\rangle_0$ was then formed from the remaining planes, which were phase shifted so that the vortex peaks were all aligned at the center of the computational domain. A cross-correlation between ω_s at each spanwise plane and $\langle\omega\rangle_0$ was computed, and the streamwise shift x_s and transverse shift y_s required to maximize this cross-correlation were determined. Once the optimal shift for each spanwise data plane was known, the final ensemble-averaged coherent velocities u_c and v_c are computed from the original *raw* data. The coherent vorticity ω_c was then computed from second-order central differences of these fields. Subtracting the coherent velocities from the raw velocities $(u - u_c, v - v_c)$ gives the incoherent velocities (u_r, v_r) for each spanwise plane. Once the coherent and incoherent velocities are known, all correlations of interest involving these quantities, such as the coherent and incoherent intensities, strains and stresses, can easily be computed.

This eduction process can be further refined by using $\langle\omega\rangle_1$ (instead of $\langle\omega\rangle_0$) to maximize the cross-correlations by determining the appropriate time shifts for phase alignment. The above process can then be repeated further until convergence is achieved. In the simulations described here, three iterations have been used.

There is a major difference between the analysis performed on the experimental and numerical data sets. The experimental eduction process is performed on a set of about 2000 *independent* accepted structures as they pass the X-wire rake. For the numerical simulation data, the analysis is performed on a subset of the 64 spanwise planes of data at a given instant of time. Thus, the numerical data are *related*, being a part of the same three-dimensional flow field. This difference in analysis is a necessary compromise reflecting the present impossibility of performing large numbers of independent flow realizations on the computer due to memory and time limitations. A logical follow-on to this first intercomparison effort is to compute a larger number of flow realizations and extend the averaging process to include them. The inclusion of additional realizations will probably have the least effect in Runs SF and WF, where the flow evolution is dominated by the coherent, two-dimensional modes.

Results

One of the most important aspects of the direct numerical simulation of turbulent flows is a careful comparison with experimental data. This serves to determine the limits of validity of the simulation, which in turn provides confidence in making use of the simulation to provide details of the flow field evolution not available from experimental data. In the work described in this paper, we are attempting to extend these comparisons to an examination of the coherent structures that evolve in this flow field.

It should be recalled that there are some very significant differences between the experiments and the numerical simulations. The numerical simulations are of a temporally growing flow rather than a spatially growing one, as in the experiments. The Reynolds numbers of the experiments based on mean velocity and mean vorticity thickness are about 10^5, while those of the simulations are about 10^3. The simulations are of a relatively immature structure after one vortex pairing, while the experimental data are of a mature structure downstream from a fully turbulent initial state and after many pairings. Nonetheless, previous simulations have shown that there are very important aspects of the dynamical, nonlinear evolution of the temporally growing mixing layer that are similar to the spatially growing case. Ultimately, the question of the significance of the differences between the two flows will probably be addressed by larger-scale, more sophisticated simulations of the spatially growing layer itself. While the Reynolds number of the simulations is much smaller than that for the experiments, it is still large enough that the flow does become turbulent, with the microscale Reynolds number at the center of the mixing layer in the range of 50–80. Since there is significant experimental evidence of the insensitivity of the characteristics of large-scale structures to Reynolds number (Zaman and Hussain, 1984), useful information about their evolution can be obtained even from these relatively low Reynolds number simulations.

Figure 2 shows contour plots of the instantaneous spanwise vorticity normalized by the peak mean shear, $\bar{S}_M$, from the laboratory experiment at a downstream distance of 2000 θ_e.

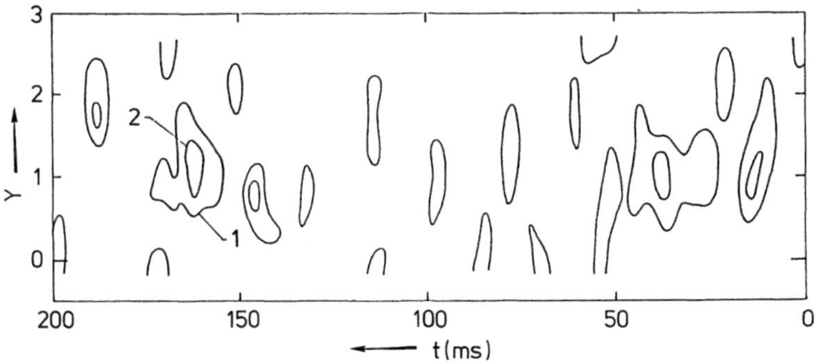

Fig. 2. Experimental instantaneous vorticity contours $\omega_z/\bar{S}_M$ at $x/\theta_e = 2000$

The abscissa shows the time in milliseconds at which the data passed the X-wire probe rake, and the ordinate shows the transverse distance from the separation lip normalized by the local shear layer momentum thickness, θ. For this unforced mixing layer evolving from a tripped, turbulent boundary layer on the splitter plate, there is an apparent lack of regularity or coherence in the transverse position, spacing, amplitude and shape of the vorticity contours. Although there is evidence from some experiments of strong, spanwise coherent features in dye visualizations or reactant products (e.g., Brown and Roshko, 1974; Breidenthal, 1981), the instantaneous vorticity fields may be less regular. Simultaneous measurements from two different probe rakes at different spanwise positions showed a correlation that fell off very rapidly with increasing spanwise separation. As suggested by this typical experimental data record, a very large amount of data had to be collected in order to acquire a sufficiently large sample of coherent structures ($\sim$ 2000) centered at the middle of the rake and satisfying other constraints.

Figure 3a is a contour plot of the spanwise vorticity normalized by peak mean shear from the numerical simulation with weak forcing, Run WF at $\bar{T} = 60$, which is between the first and second vorticity pairings. All numerical data presented in this paper correspond to this time, and the computational domain for all figures is a cube with sides of length 4λ. The vortex cores of the subharmonic mode are very sharply defined (the cores corresponding to the fundamental mode have already been assimilated into them), and the second pairing (at the second subharmonic) is just beginning. These structures have a very strong spanwise coherence, and plots at other spanwise locations are nearly identical except for small details.

Figures 3b and 3c are corresponding plots for the random forcing case, Run R, at two different spanwise planes separated by the wavelength of a streamwise subharmonic mode. These vortical structures are clearly not as regular as those from Run WF, having multiple vorticity peaks and lower peak values. The presence of irregular small-scale features indicates more three-dimensionality than in Run WF. In addition, the streamwise positions of the cores are shifted at the two spanwise locations, suggesting that these structures lack spanwise coherence. Profiles of $\bar{U}(y)$ and $u'(y)$ are given in Fig. 4.

The lack of spanwise coherence is demonstrated more graphically in Figs. 5a and 5b. These are contour plots of spanwise vorticity on the plane $y = 0$, at the middle of the mixing layer. Run WF (Fig. 5a) shows very strong spanwise coherence, with small modulations in ω_z caused by secondary spanwise instabilities that eventually evolve into streamwise, counterrotating vortices or "ribs" (cf. Lin and Corcos, 1984; Pierrehumbert and Widnall, 1982; Bernal, 1981; Hussain, 1983). In the case of random forcing alone (Run R, Fig. 5b), the spanwise vorticity shows very little spanwise coherence, and there are no structures that span the entire domain. A comparison with Figs. 3b and 3c shows that large structures which

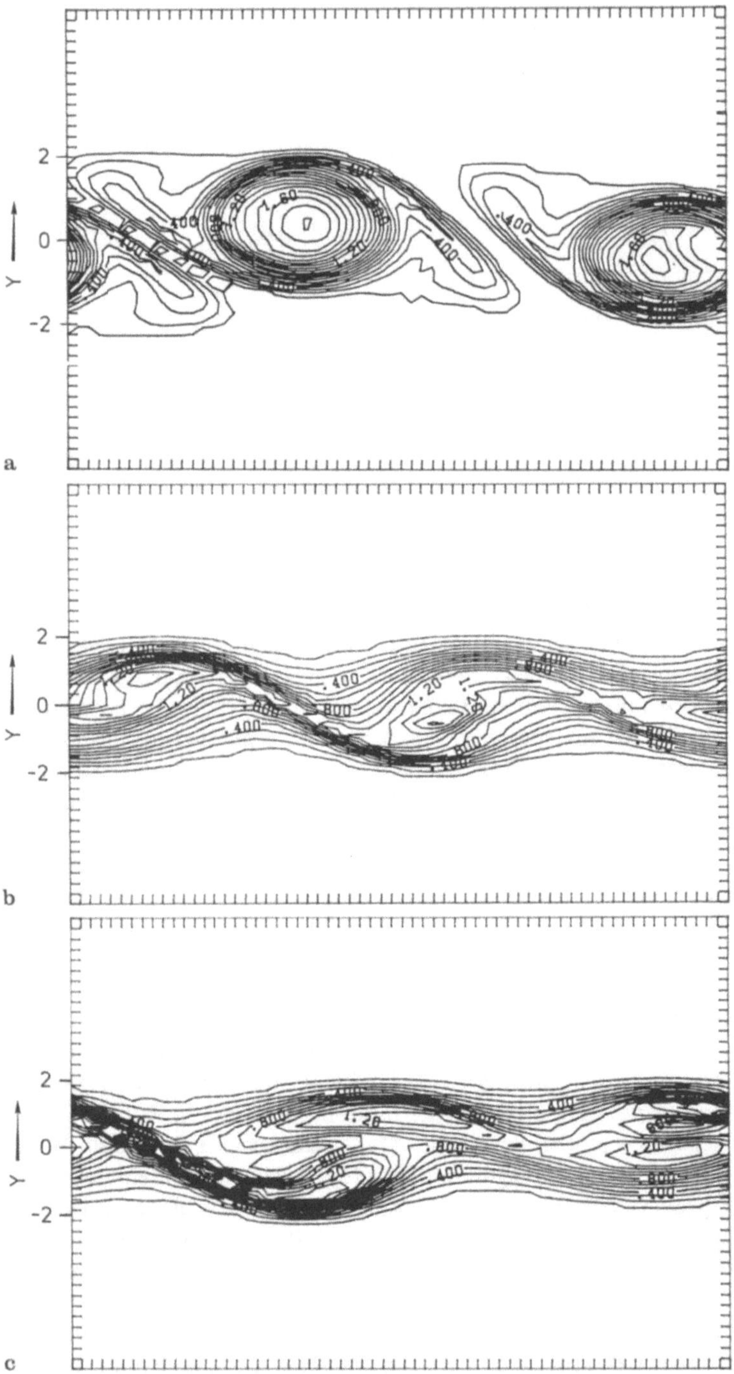

Fig. 3 a. Calculated instantaneous vorticity contours $\omega_z/\bar{S}_M$ in the $(x-y)$ plane at $z=0$ for Run *WF* at $\bar{T}=60$. The range of contour values is from 0 to 1.9 in increments of 0.1; **b** Calculated instantaneous vorticity contours $\omega_z/\bar{S}_M$ in the $(x-y)$ plane at $z=0$ for Run *R*. The range of contour values is from 0 to 1.8 in increments of 0.1; **c** Calculated instantaneous vorticity contours $\omega_z/\bar{S}_M$ in the $(x-y)$ plane at $z=2\lambda$ for Run *R*. The range of contour values is from 0 to 1.9 in increments of 0.1

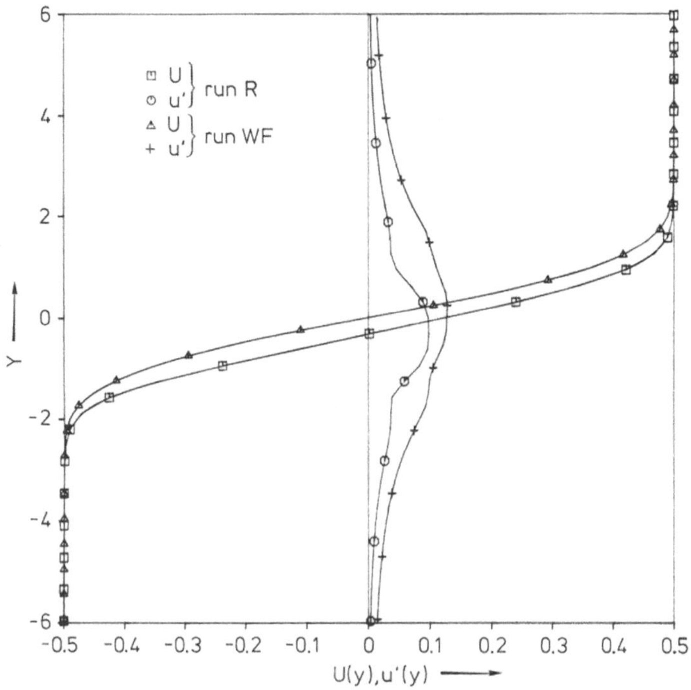

Fig. 4. The mean and rms velocity profiles for Run *WF* and Run *R* at $\bar{T} = 60$ as a function of *Y*

may appear from their spanwise cross sections to have the shape of coherent, two-dimensional pairing modes are in fact strongly three-dimensional. In terms of its irregularity and the lack of spanwise coherence of the large structures, Run *R* more closely resembles the experimental data than Runs *WF* and *SF*. However, the dramatic change in the flow caused by the presence of low-amplitude, coherent two-dimensional modes in the initial conditions shows that the dynamical evolution of this flow is very sensitive to the presence of low-amplitude coherent forcing.

Figure 6a is a plot of the coherent vorticity of the laboratory experiment computed by the eduction procedure described in the previous section (Methodology). The abscissa is normalized by U_c/θ and the ordinate as in Fig. 2. Note that if $U_c t$ is equivalent to x (according to the Taylor hypothesis), then $T = x/\theta$. Thus, the experimental plots will have the right aspect ratio. In addition, although the mean velocity is subtracted out of the u_c fields plotted in Fig. 7, the vorticity fields plotted have the mean component. A similar analysis of the data taken at a distance half as far downstream from the separation lip gives essentially identical results, which indicates that this is a self-similar or equilibrium structure. The corresponding plot for the weakly forced numerical simulation (Run *WF*) is shown in Fig. 6b. Note the strong similarity of even the structure shapes and inclinations between vorticity contours educed experimentally and numerically (Figs. 6a and 6b). Due to the strong spanwise coherence, this plot compares quite well with the vorticity on any particular spanwise location, such as that in Fig. 3a. The coherent vorticity for Run *R* is shown in Fig. 6c. Although this structure is not as large or as intense as that for Run *WF*, the eduction procedure does clearly extract a vortical structure. The differences are due partly to the fact that Run *R* has not evolved as far as Run *WF* since the vortex pairing modes have lower initial amplitudes in Run *R* and, hence, take longer to reach comparable amplitudes to those in Run *WF*. Also, in the early stages of rollup, vorticity levels are higher in the braids.

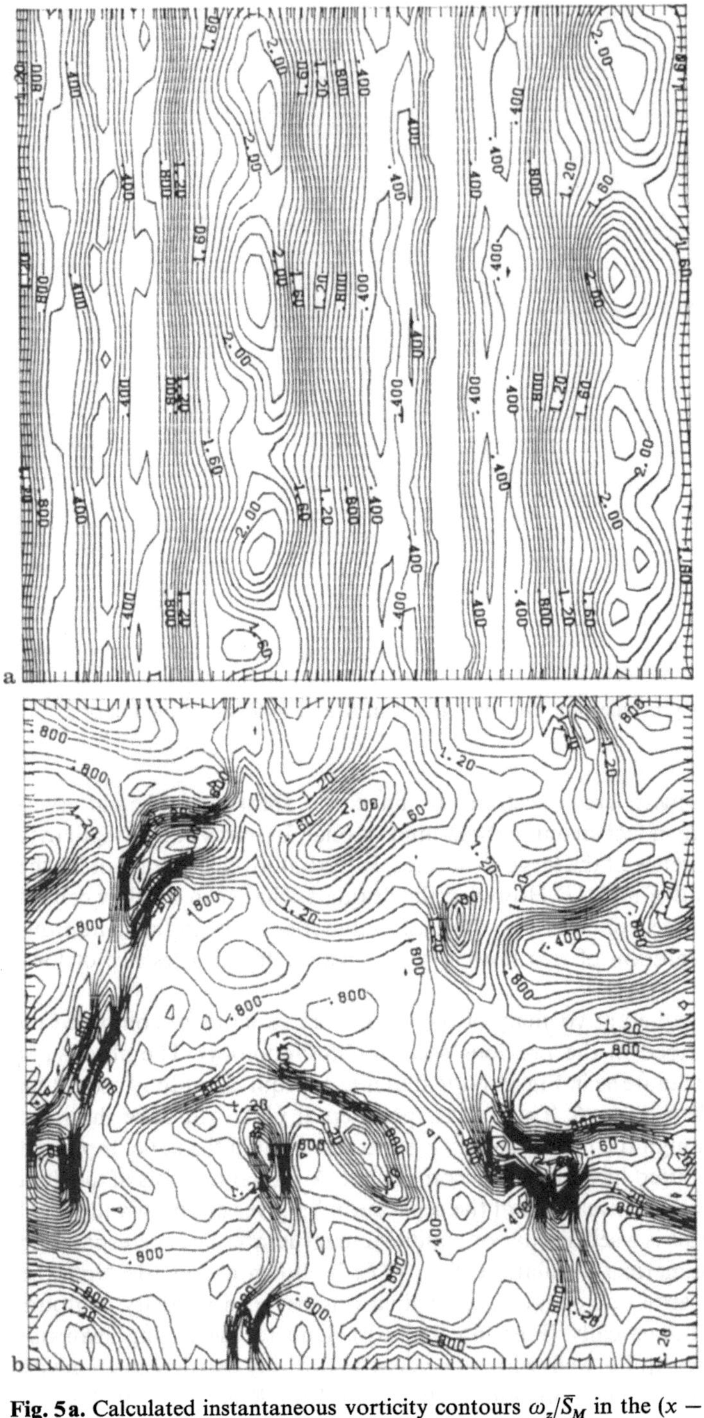

Fig. 5a. Calculated instantaneous vorticity contours $\omega_z/\overline{S}_M$ in the $(x - z)$ plane at $y = 0$ for Run *WF*. The range of contour values is from 0 to 2.4 in increments of 0.1; **b** Calculated instantaneous vorticity contours $\omega_z/\overline{S}_M$ in the $(x - z)$ plane at $y = 0$ for Run *R*. The range of contour values is from 0 to 2.5 in increments of 0.1

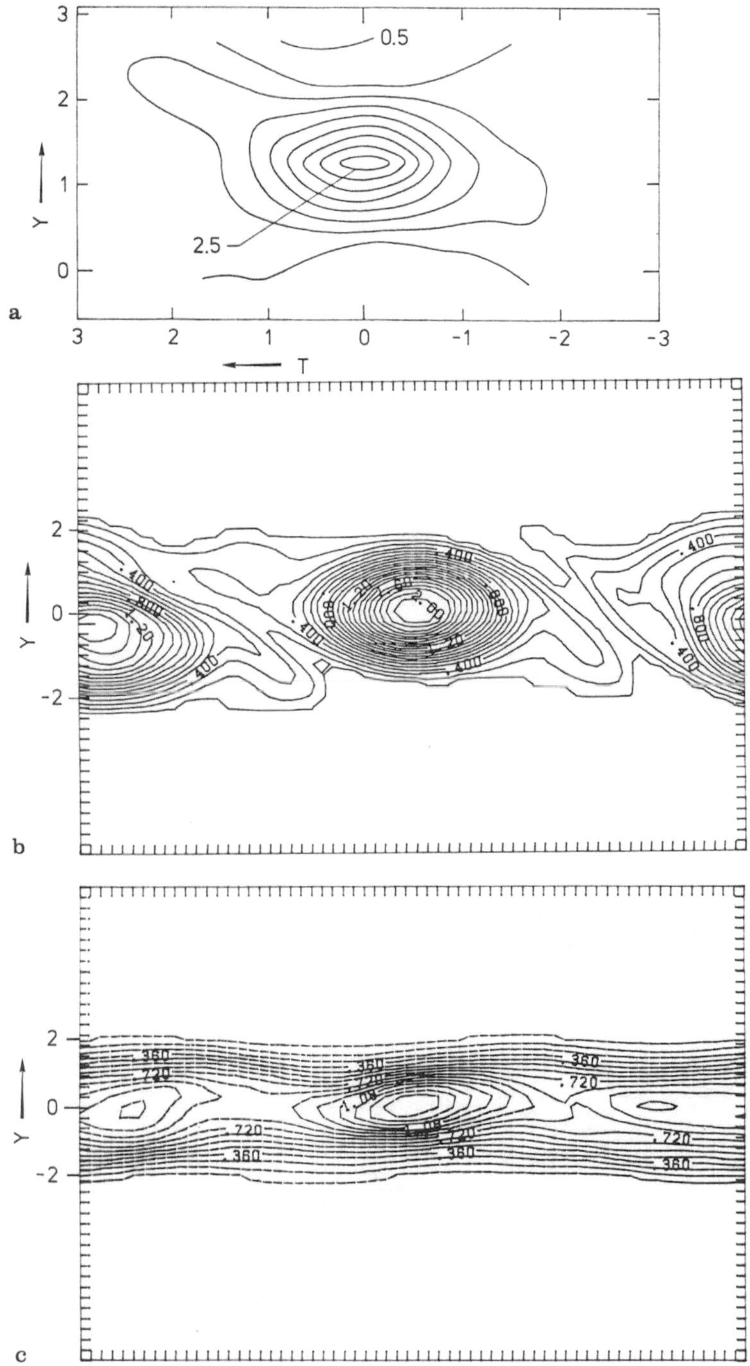

Fig. 6a. Experimental coherent vorticity field $\langle\omega_z\rangle/\bar{S}_M$ at $x/\theta_e = 2000$. Contour interval is 0.25; **b** Calculated coherent vorticity field $\langle\omega_z\rangle/\bar{S}_M$ for Run *WF*. The range of contour values is from 0 to 2.1 in increments of 0.1; **c** Calculated coherent vorticity field $\langle\omega_z\rangle/\bar{S}_M$ for Run *R*. The range of contour values is from 0 to 1.35 in increments of 0.09

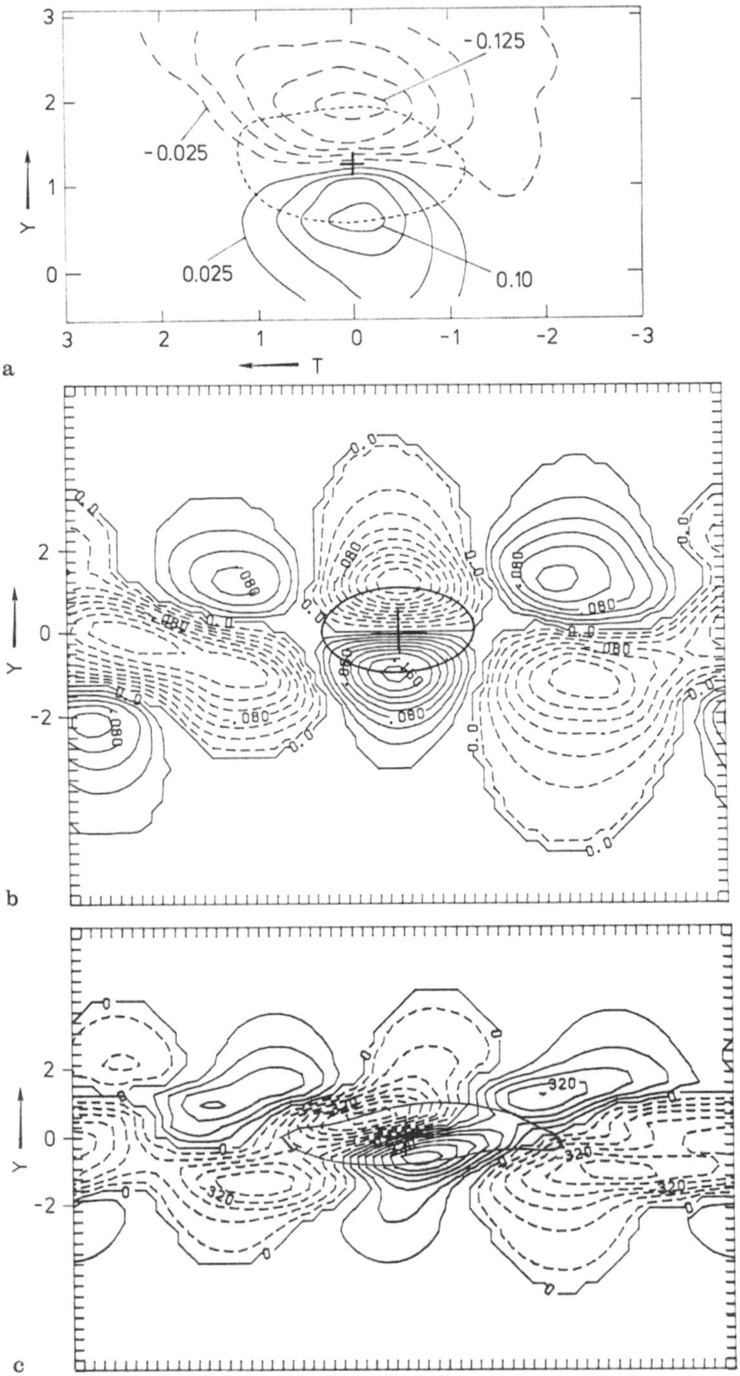

Fig. 7 a. Experimental coherent streamwise velocity field u_c/U at $x/\theta_e = 2000$. Contour interval is 0.025. Coherent vorticity field at 50% peak value has been overlaid in the following figures; **b** Calculated coherent streamwise velocity field u_c/U for Run *WF*. The range of contour values is from -0.18 to 0.18 in increments of 0.02; **c** Calculated coherent streamwise velocity field u_c/U for Run *R*. The range of contour values is from -0.056 to 0.056 in increments of 0.008

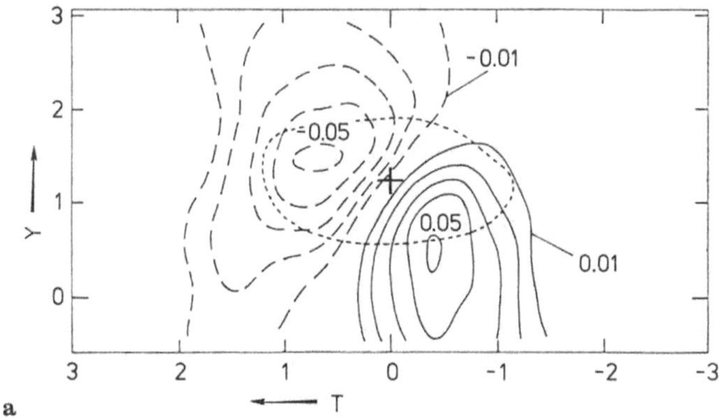

a

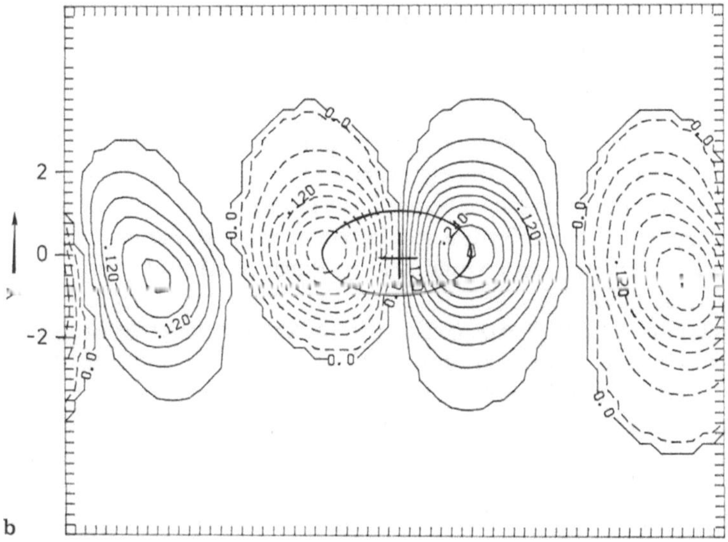

b

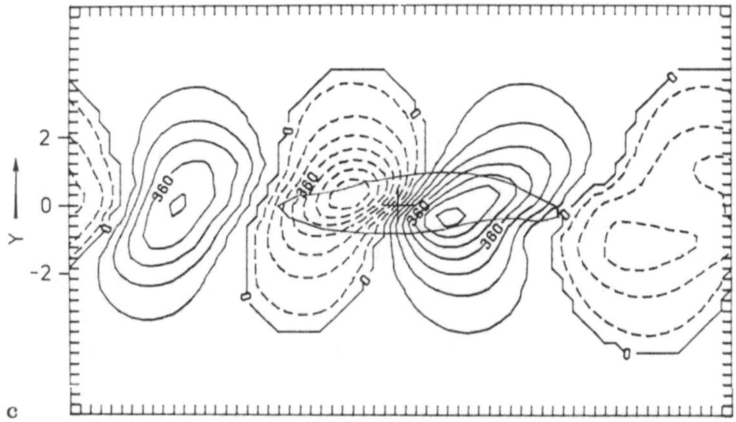

c

Fig. 8a. Experimental coherent transverse velocity field v_c/U at $x/\theta_e = 2000$. Contour interval is 0.01; **b** Calculated coherent transverse velocity field v_c/U for Run *WF*. The range of contour values is from -0.24 to 0.30 in increments of 0.03; **c** Calculated coherent transverse velocity field v_c/U for Run *R*. The range of contour values is from -0.063 to 0.063 in increments of 0.009

Nonetheless, at later stages of this simulation, the amplitude of the coherent vorticity does reach a value of 3.1 and the core develops a shape more like that of Run *WF* in Fig. 6b.

In Figures 7a and 8a are plots of the coherent perturbation velocity fields u_c and v_c from the experiment normalized by the peak mean velocity. For reference, superimposed on these plots are dotted lines showing the corresponding coherent vorticity contours at 50% of their peak values. The + sign marks the vortex peak (i.e., the structure center) identified by the peak of the vorticity contour. Comparable plots for Runs *WF* and *R* are shown in Figs. 7b, 7c, 8b and 8c. The coherent vorticity contour superimposed on Figs. 7b and 8b is at 50% of the peak value, while for Figs. 7c and 8c, it is at 70% due to the relatively lower peak value found in Run *R*. The structures of the u_c fields are similar. The amplitude of the velocity peaks within the vortex core is nearly identical in Fig. 7b as would be expected from symmetry considerations for a temporally growing flow. For the spatially growing structure in Fig. 7a, the retardation of the flow at $Y = 2$ is slightly greater than the corresponding forward motion at $Y = 0.5$. Similarly, the asymmetry in Fig. 8a may also be due to spatial growth.

The presence of a strong, coherent two-dimensional mode can significantly increase the amplitude of the perturbation velocity fields near the vortex core. For example, the peak value of u_c in Fig. 7a is about 0.13 U, while that in 7b is 0.18 U. Likewise, v_c in Fig. 8a has a peak of about 0.05 U, while 8b peaks at 0.30 U. The corresponding amplitudes in Run *R* are 0.06 U for u_c and 0.06 U for v_c, which are much closer to the experimental values. A similar enhancement in the amplitudes of these velocities has been observed experimentally in work on the "preferred mode" of the axisymmetric jet (Hussain and Zaman, 1981b). Smaller peak values for the experimental data would also be expected since the resolution is coarser than in the numerical simulations.

Run *SF* was quite similar to Run *WF* except that the spanwise coherence was even stronger. The initial stages of rollup proceeded more rapidly since the amplitude of the fundamental mode was much larger, but the later stages were quite similar to Run *WF* since the amplitudes of the subharmonics were the same as in Run *WF*. Another interesting feature of Run *SF* was that stronger secondary instabilities developed. These counterrotating streamwise vortices, or ribs, were more intense than in Run *WF* and had a wavelength of about one half that of the most unstable two-dimensional streamwise mode.

Concluding Remarks

We have performed a series of direct numerical simulations of a turbulent, temporally growing mixing layer with and without forcing by coherent two-dimensional modes. The results of these simulations have been compared with new experiments performed in a large facility designed to minimize wall interference and feedback effects. A data analysis scheme has been developed to map out the structure of coherent vortical modes that develop during mixing layer evolution. A similar analysis scheme has been applied to the numerical simulations. Since many studies of coherent structures have focused on flow visualization, this work is an attempt toward a more quantifiable analysis.

The experimental analysis scheme has been designed to elicit the details of the coherent structures. Although it is possible to average over thousands of samples, there are significant limitations in resolution. Conversely, the numerical simulations, while being limited by computer resources in the number of independent realizations that can be calculated, can resolve the evolution of the coherent structures in much greater detail than is possible experimentally. Thus, the two approaches are complementary if care is taken to account for the inherent differences in geometry and Reynolds number.

Acknowledgements. This work is supported by ONR Contracts N00014-82-C-0600 and N00014-85-K-0126 and by NASA Contract NAS3-23531.

References

Bernal, L. P. (1981): "The Coherent Structure of Turbulent Mixing Layers. I. Similarity of the Primary Vortex Structure. II. Secondary Streamwise Vortex Structure;" Ph. D. Thesis, California Institute of Technology

Breidenthal, R. (1981): Structure in turbulent mixing layers and wakes using a chemical reaction. J. Fluid Mech. **109**, 1–24

Brown, G. L., Roshko, A. (1974): On density effects and large structure in turbulent mixing layers. J. Fluid Mech. **64**, 775–816

Gottlieb, D., Orszag, S. (1977): Numerical Analysis of Spectral Methods: Theory and Applications, CBMS-NSF Regional Conference Series in Applied Mathematics, Society for Industrial and Applied Mathematics

Ho, C. M., Huerre, P. (1984): Perturbed free shear layers. Ann. Rev. Fluid Mech. **16**, 365–424

Hussain, A. K. M. F. (1983): "Coherent Structures and Incoherent Turbulence," in *Turbulence and Chaotic Phenomena in Fluids,* ed. by T. Tatsumi (North Holland, Amsterdam) pp. 245–249

Hussain, A. K. M. F. (1985): "Measurements of Large Scale Organized Motions in Turbulent Flows," in *Forum on Unsteady Flows in Biological Systems,* ed. by M. H. Friedman, D. C. Wiggert, ASME, 8–12

Hussain, A. K. M. F., Zaman, K. B. M. Q. (1981 a): The 'Preferred Mode' of the axisymmetric jet. J. Fluid Mech. **110**, 39–71

Hussain, A. K. M. F., Zaman, K. B. M. Q. (1981 b): The natural large-scale coherent structure in an initially turbulent mixing layer. Bull. Am. Phys. Soc. **26**, 1273

Hussain, A. K. M. F., Zaman, K. B. M. Q. (1985): An experimental study of organized motions in the turbulent plane mixing layer. J. Fluid Mech. **159**, 85–104

Kleis, S., Hussain, A. K. M. F. (1979): The asymptotic state of the plane mixing layer. Bull. Am. Phys. Soc. **24**, 1132

Lin, S. J., Corcos, G. M. (1984): The mixing layer: deterministic models of a turbulent flow. Part 3. The effect of plane strain on the dynamics of streamwise vortices. J. Fluid Mech. **141**, 139–178

Michalke, A. (1964): On the inviscid instability of the hyperbolic-tangent velocity profile. J. Fluid Mech. **19**, 543–556

Pierrehumbert, R. T., Widnall, S. E. (1982): The two- and three-dimensional instabilities of a spatially periodic shear layer. J. Fluid Mech. **114**, 59–82

Riley, J. J., Metcalfe, R. W. (1979): "Direct Numerical Simulations of the Turbulent Wake of an Axisymmetric Body," in *Turbulent Shear Flows 2,* ed. by L. J. S. Bradbury et al. (Springer, Berlin, Heidelberg) pp. 78–93

Riley, J. J., Metcalfe, R. W. (1980): Direct numerical simulation of a perturbed turbulent mixing layer. AIAA Paper 80-0274

Wygnanski, I., Fiedler, H. E. (1970): The two-dimensional mixing region. J. Fluid Mech. **41**, 327–361

Wygnanski, I., Oster, D., Fiedler, H. (1979): "A Forced, Plane, Turbulent Mixing-Layer; A Challenge for the Predictor," in *Turbulent Shear Flows 2,* ed. by L. J. S. Bradbury et al. (Springer, Berlin, Heidelberg) pp. 314–326

Zaman, K. B. M. Q., Hussain, A. K. M. F. (1977): "Vortex Pairing and Organized Structures in Circular Jets Under Controlled Excitation," *Turbulent Shear Flows I,* Penn. State Univ., pp. 11.23–11.31

Zaman, K. B. M. Q., Hussain, A. K. M. F. (1980): Vortex pairing and organized structures in circular jets under controlled excitation. J. Fluid Mech. **101**, 449–492

Zaman, K. B. M. Q., Hussain, A. K. M. F. (1984): "Natural large-scale structures in the axisymmetric mixing layer. J. Fluid Mech. **138**, 325–351

Structure of the Vorticity Field in a Plane Free Shear-Layer

J. C. Lasheras and T. Maxworthy

Department of Mechanical Engineering, University of Southern California,
Los Angeles, CA, 90089-1453, USA

Abstract

A plane isothermal free shear layer between two streams of reactive solutions has been investigated to moderate Reynolds numbers (10^2). Both Laser Induced Fluorescence and a Direct Interface Visualization were used to monitor the interface between the two fluids. The structure of the vorticity field was then investigated through an analysis of the evolution of the position of the interface over time.

The response of the plane, shear layer to localized three dimensional disturbances of varying magnitude in the upstream conditions was investigated. The results show that the streamwise vortical structure is sensitive to localized perturbations while the large scale spanwise structure is relatively insensitive to this type of forcing.

In addition, the interaction of a plane, free shear layer with vertical solid boundaries was also studied. It was found that this interaction with the side walls leads to the generation of streamwise vortices which spread laterally by self-induction.

Introduction

In a plane free shear layer, large scale spanwise vortical structures have been observed to exist in a very wide range of Reynolds numbers [1, 2]. Many researchers have documented the formation and persistence of this highly organized, two-dimensional vortical structure [3–7, among many others]. In addition, it has been found that the plane shear layer also contains well organized, streamwise vortices which superimpose onto the spanwise structure. The occurrence of this secondary vortical structure, which is composed of pairs of counter rotating axial vortices, has been experimentally documented in recent studies [8–10, among others].

Corcos [12] and Corcos and Lin [13] studied the volution of a row of weak, alternating vortices whose axes were parallel to the direction of a uniform straining flow. Their numerical analysis shows that for certain values of the strain and distance between axial vortices, this streamwise vorticity may evolve into concentrated, round vortices. Pierrehumbert and Widnall [14], through stability analysis of a family of coherent, Stuart vortices, suggest that the streamwise vortices could be the result of a linear instability "the translative instability". Although these theoretical studies provide possible mechanisms for the generation of the streamwise vortical structure, the origin and the evolution of this three dimensional vorticity has not yet been determined experimentally.

Recently, Lasheras et al [10] have shown that small random upstream disturbances lead to the formation of streamwise vorticity which spreads laterally by strong self-induction. This paper further analyzes the response of a free shear layer to localized three dimensional perturbations introduced in the initial conditions and focuses on the response of both vortical structures (span and streamwise) to this type of forcing.

A large portion of the experimentally investigated shear layers are bounded flows in contact with solid walls where vorticity is always present. Our observations show that

Turbulent Shear Flows 5

through a complex interaction of the plane, shear layer with the wall's vorticity, a transfer of the latter away from the wall may occur. Axial vortices are then generated near the wall region, which further interact with the streamwise vorticity existing in the layer. In this study, the interaction of a plane, shear layer with vertical, lateral boundaries is also investigated.

Experimental Apparatus

The experiments were conducted in a blow-down water channel facility schematically shown in Fig. 1. A turbulent reacting mixing layer was generated between two streams of aqueous reactants which were supplied separately and allowed to meet and mix in a 100 cm long test section. The two reactants were fed from two 120 l capacity tanks through control valves and flow meters into a settling chamber consisting of a distributor, a layer of honey comb, two layers of 60 mesh-size and two of 40 mesh-size screens. The settling chamber was followed by a fifth order polynomial nozzle with an inlet to outlet ratio of 15.4 : 1. The settling chamber and nozzle were divided symmetrically by a 1/16″ stainless steel splitter plate whose thickness was gradually reduced at the trailing edge. A 7.8 cm high × 7.4 cm wide × 100 cm long test section followed the nozzle. A dumping section was divided into four parts, each of which were independently monitored through flow meters and control valves to minimize any feedback effect of the dump control on the upstream flow pattern.

The velocities of the high and low speed streams selected for these experiments were 6.5 cm sec^{-1} and 3.5 cm sec^{-1} respectively. The boundary layers of both flows on the splitter plate were laminar with momentum thicknesses of 0.58 mm and 0.62 mm for the fast and slow flows respectively, (both measured at the geometrical origin of the shear layer). The value of the Reynolds number based on the velocity difference and the momentum thickness of the layer varied from 36 at 25 mm from the trailing edge of the splitter plate to 210 at 150 mm downstream. A detailed description of the mean axial velocity profiles, momentum thickness, and vorticity thickness corresponding to these flow conditions obtained using hot film annemometry can be found in Lasheras et al. [10].

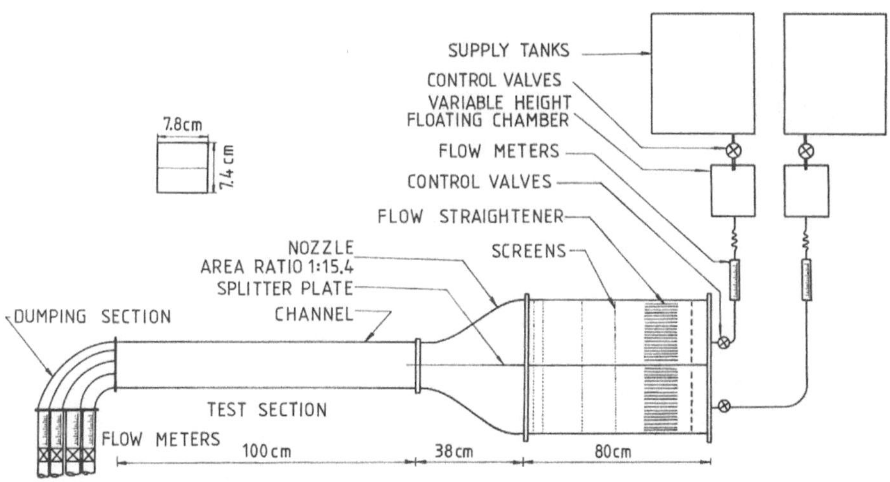

Fig. 1. Experimental apparatus

Flow Visualization

A diffusion controlled reaction between two dilute hydrochloric acid and sodium hydroxide water solutions (PH of 6.0 and 11.5, respectively) was used to obtain a "Direct Interface Visualization" (DIV) of the interface. Water resolved Cresol Red was added to the acid solution at a rate of 0.04 grm/l as a PH indicator. At the interface, the reaction products became highly visible as the Cresol Red responded to changes in the PH level. This acid-base technique, which has been used previously by Breidenthal [11], among others, is a remarkably accurate and ideal one for visualizing the interface between the two fluids in shear layers at low and moderate Reynolds numbers. Figure 2 shows the visualization of the top and side view of the reacting, plane shear layer using the method described above. It shows that the reaction is confined to a very thin, diffusion controlled region at the interface. It can be observed that for our moderate Reynolds number cases, due to the high local strain rate and large Schmidt number, very little reaction occurs at the interface. This remains to be the case as the interfacial surface is streteched and augmented by the roll-ups. Well organized, three dimensional streamwise vortices can also be seen coexisting as identifiable coherent vortical structures superimposed onto the spanwise ones (top view, Fig. 2). In addition to this DIV technique, visualization of cross sectional cuts of the layer were obtained using Laser Induced Fluorescence (LIF), a technique used in the past by Dewey [15] and Bernal [8] among others. Water resolved fluorescent dye (fluorescein) was added to the sodium chloride water solution and its fluorescence was excited by a plane sheet of light of approximately 0.5 mm thickness. The orientation of the plane of illumination was varied according to the experimental needs. To obtain optimum information from this second visualization technique, both vertical (spanwise) cuts and horizontal (longitudinal) cuts were obtained for each case studied.

A Perspective Three Dimensional Reconstruction

A three dimensional reconstruction was obtained using a method similar to the one recently applied by Jimenez et al. [9]. This method consisted of analyzing consecutive frames of 16 mm high speed movie film of vertical and horizontal cross cuts of the plane mixing layer

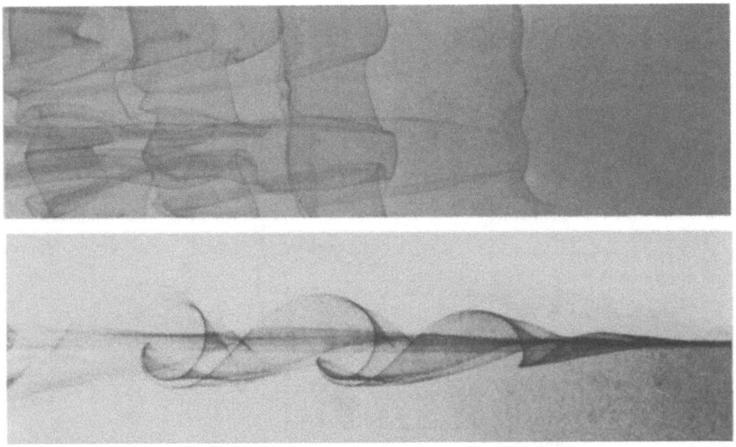

Fig. 2. Top and side views of the chemically reacting shear layer. ($U = 6.5$ cm sec^{-1} and $U =$ 3.5 cm sec^{-1}.) The flow is from left to right

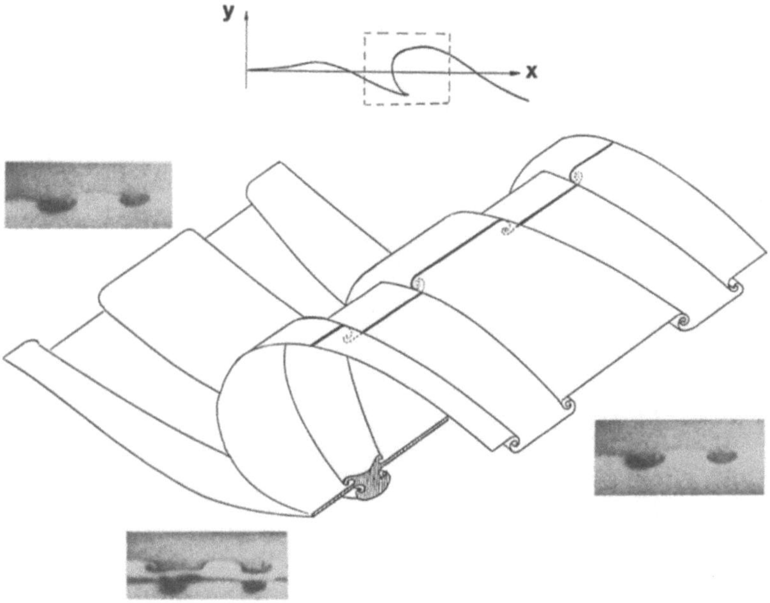

Fig. 3. A three dimensional reconstruction of the plane shear layer

using LIF. A perspective view of that reconstruction is presented in Fig. 3. It shows the same structure of the vorticity field already presented in Fig. 2. Observe how early in its evolution the shear layer is composed of a secondary streamwise vortical structure superimposed onto the primary spanwise one. The schematic picture of the braids is essentially similar to the one described by Bernal [8] and further refined by Jimenez et al [9], i.e. pairs of counter-rotating axial vortices of approximately the same strength. However, the present results also show that the well organized streamwise vortices can be noted, not only on the braids, but also inside the cores of the spanwise vortices. Furthermore, these vortices and their interactions inside the cores of the primary roll-ups were found to increase the mixing. This may be one of the mechanisms by which smaller scale structures inside the large vortex core are produced, thus leading to a substantial increase in mixing.

Response of a Plane Shear Layer to a Three Dimensional Perturbation

Early results recently obtained by Lasheras, et al. [10], have indicated that the plane free shear layer is unstable to localized, three dimensional disturbances in the upstream conditions. Coherent, streamwise vortices were observed to form as a result of this instability. Figure 4 shows the response of the mixing layer to localized three dimensional disturbances introduced in the boundary layer of the fast flow (bottom) 1 cm upstream from the tip of the splitter plate. The three dimensional disturbances consisted of a single cylinder of 6 mm height placed vertically on the splitter plate in the boundary layer of the fast side. The selected cylinder diameters were 1.5 mm, 2.8 mm, and 3.2 mm.

The figure presents top views of the plane shear layer showing this response to the localized perturbations. The unperturbed flow condition is shown in the top picture of the figure. Observe how, with the absence of any upstream disturbances, the layer remains "almost" two dimensional at a distance of even more than five wavelengths downstream. The resulting response of the layer to each of the perturbations is presented in the order of

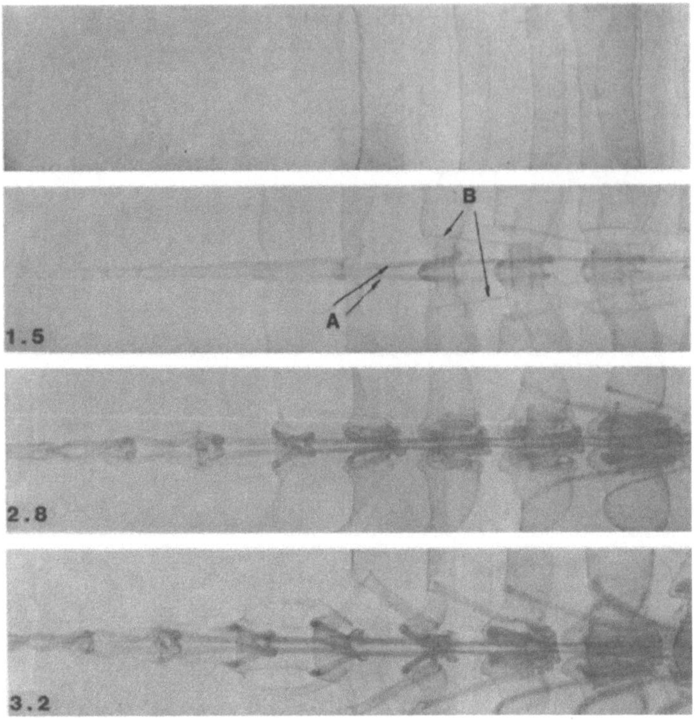

Fig. 4. Plan views of the shear layer showing the response to localized perturbations. Top picture corresponds to the unperturbed flow condition. The response of the layer to localized perturbations is presented in order of increasing magnitude from top to bottom

increasing cylinder diameter from top to bottom. From these plan views, it can be noted that under the effect of the positive strain rate created by the large spanwise vortices, the three dimensional perturbations concentrate to form a pair of counter rotating streamwise vortices (A in the second plan view, 1.5 mm case) in a fashion similar to the one predicted by Corcos' theory [12], i.e. the three dimensional vorticity shed by the cylinder evolves into a horseshoe pair of axial vortices which are stretched under the strain created by the large spanwise vortices. However, as this perturbation is convected downstream, it spreads laterally inducing the formation of concentrated streamwise vorticity on either side (vortex tubes indicated by B in the second plan view, 1.5 mm case) suggesting a "wave-like" type of instability. The persistence of this effect remains highly noticeable as the size of the perturbation is increased. It is also important to notice that, although the streamwise structure is thus sensitive to small scale perturbations, the large scale spanwise vortices are relatively insensitive to this type of forcing. The figure shows how the frequency and wavelength of the spanwise vortices are practically unaffected by the localized perturbation.

A proposed mechanism for this axial vorticity generation [10] is presented in Fig. 5. A small three dimensional disturbance, with either or both axial and vertical vorticity affecting the interface, will produce a small kink of the vortex filaments. The strain field created by the spanwise vortex cores causes stretching of the perturbed vortex filament in the axial direction resulting in a pair of counter rotating streamwise vortices. Self-induced motion raises the base of the filament near the base of the loop; these portions are convected in opposite directions. This process continues propagating laterally by induction.

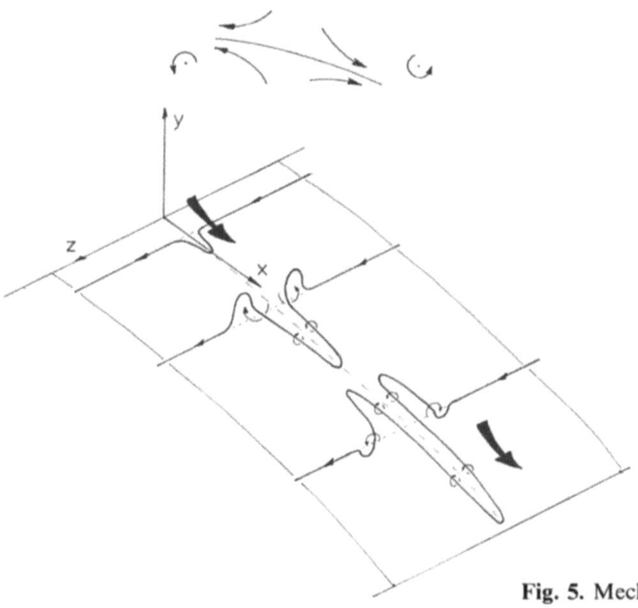

Fig. 5. Mechanism for axial vorticity generation

Interaction with the Vertical Side Wall

To investigate the complex interactions of a plane, shear layer with a vertical boundary both Direct Interface Visualization and Laser Induced Fluorescence techniques were used.

Figure 6 shows a plan view of the interaction of the mixing layer and the vertical side wall. The side wall is clearly visible (top of the figure) and the flow is from left to right. Notice how, through the interaction with the boundary layers on the side wall, the axes of the large, spanwise vortices bend backwards upon approaching the walls. The location where the axis of the first spanwise vortex was observed bending was at a distance of 8 mm from the wall

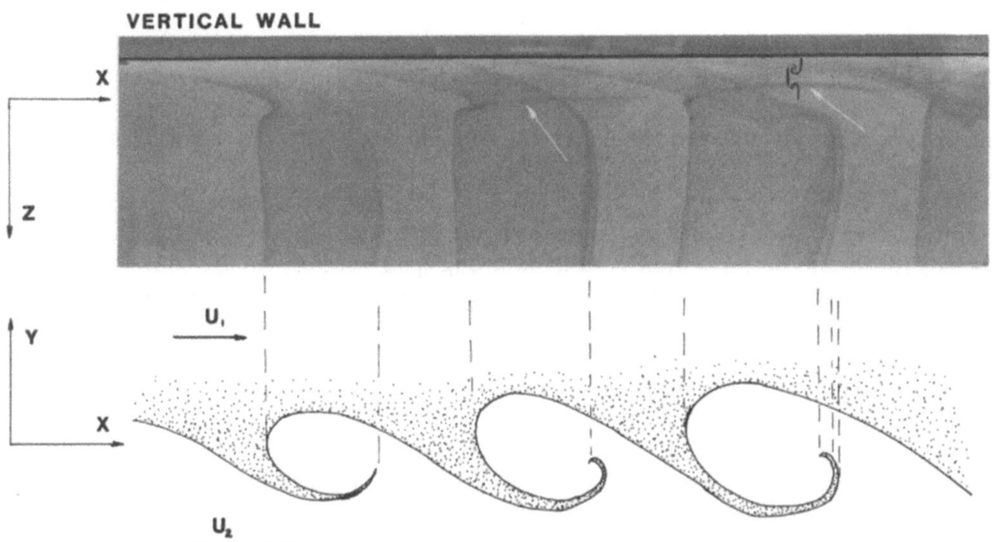

Fig. 6. Direct interface visualization of the Wall's interaction

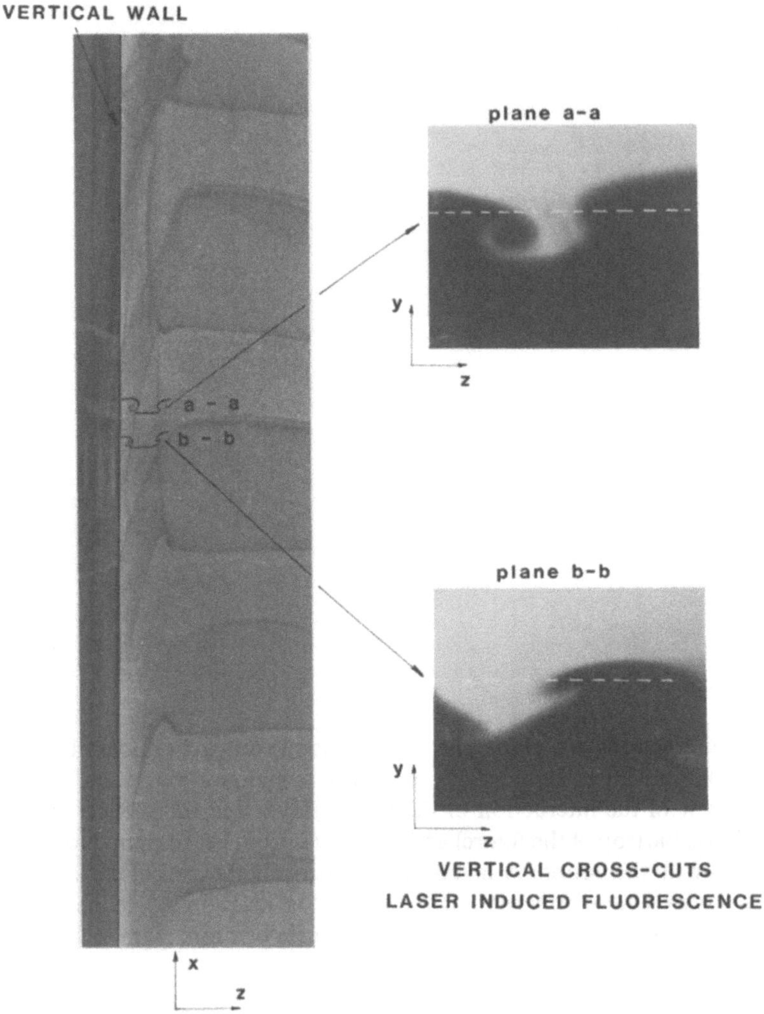

Fig. 7. Combined Laser induced fluorescence cross-cuts and plan view of the Wall's interaction

(this is approximately equal to the thickness of the boundary layer at the wall of the slow flow).

After the Kelvin Helmholtz instability has caused the redistribution of the layer's vorticity into regions of strong vorticity (the cores of the spanwise vortices) and regions of weak vorticity (the braids between these cores), the strong spanwise vortices are convected downstream with a velocity approximately equal to the average of both free stream velocities. However, due to the presence of the boundary layers on the wall, the convection velocity of the spanwise vortices decreases as they approach the side boundary. This decrease in the convection velocity is the cause of the observed backward bending of the axis of the spanwise vortex upon approaching the wall.

Figure 6 also shows that the large spanwise vortices weaken upon approaching the wall, while a kink in the axial direction $(+X)$ forms at the bend of their axis. Finally, observe how additional streamwise vortices result from the interaction of the large spanwise eddies as they are retarded and weakened upon approaching the wall. This new concentrated region of axial vorticity is shown by the arrows in the figure.

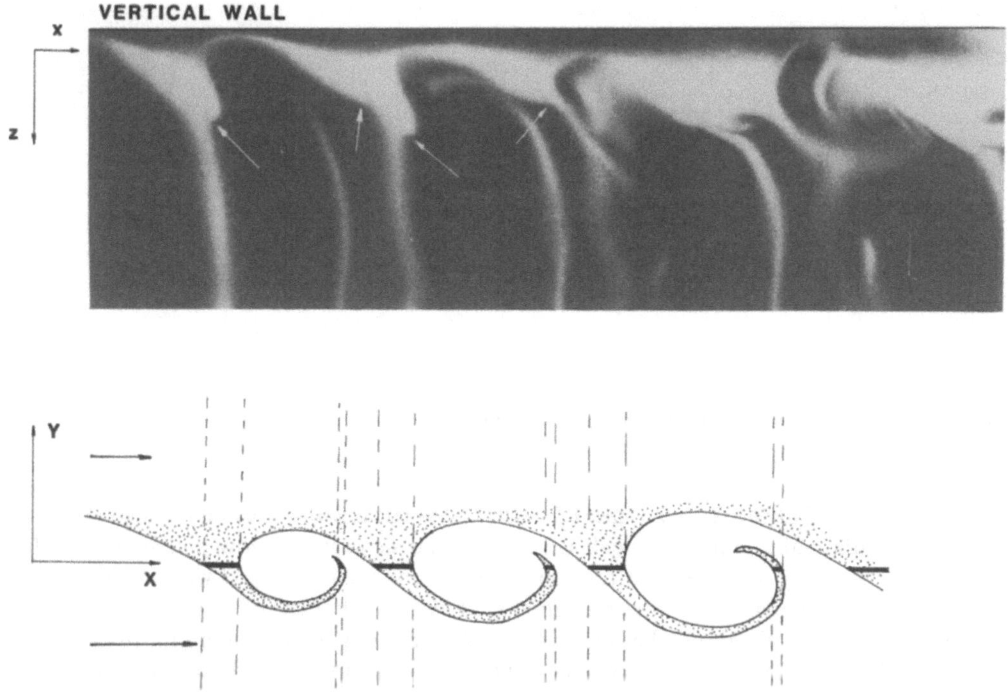

Fig. 8. Laser induced fluorescence horizontal cross-cuts near the Wall's region

Figure 7 is a composition of a detail of the direct flow visualization previously described and two vertical cross cuts (planes perpendicular to the X axis). These were obtained unsing Laser Induced Fluorescence of fluorescine particles added to the slow (upper) flow. These vertical cross cuts again clearly show the presence of both the primary vortex being retarded toward the wall and the induced streamwise vortex formed by this interaction. Notice how the strength of the spanwise vortex decreases as it bends in approaching the side wall (the vortex located in the lefthand side of the cross cuts). Also see how the axis of this vortex heads downward toward the negative Y direction as it interacts with the vertical vorticity existing at the wall (in the $+\ Y$ direction). The vertical vorticity along the wall is much stronger in the bottom, faster flow than in the upper, slower flow. Thus, as the spanwise vortex is retarded and weakened upon approaching the wall, it beocomes tangled with the strong, vertical wall vorticity of the fast flow.

Figure 8 shows a horizontal cross cut ($X - Z$ plane) of the layer on a plane located along the center line of the shear layer using Laser Induced Fluorescence. The bright portions of the figure correspond to the regions occupied by the upper (slow) flow where the fluorescine particles were added. The streamwise vortices existing on the braids which have resulted from the wall interaction, are again indicated by arrows. The figure also shows how the axis of the spanwise primary vortices is retarded backward toward the wall and then tilts downward as it becomes tangled with the vertical wall's vorticity.

Mechanism for the Interaction with the Wall

A perspective view of the interaction with the wall is schematically represented in Fig. 9 # 1, # 2, and # 3 represent three large spanwise vortices with axes initially along Z direction.

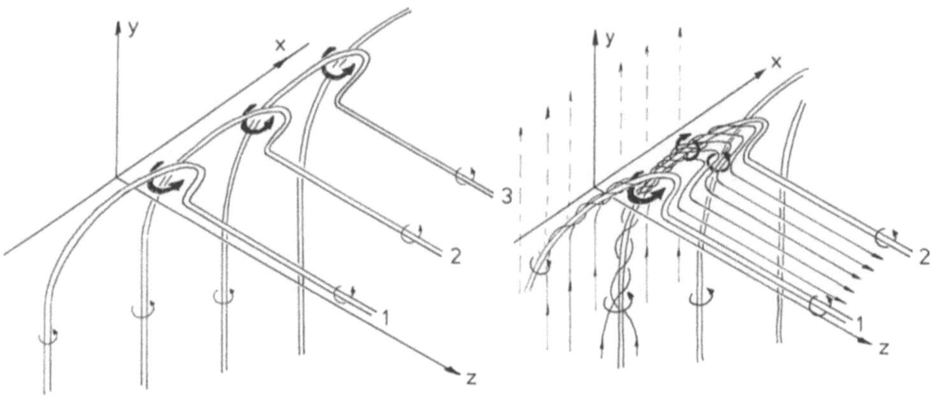

Fig. 9. A schematic of the Wall's interaction mechanisms

The wall is located at the $X - Y$ plane ($Z = 0$). The vertical wall's vorticity is also sketched in the figure (dashed line for the weak vorticity of the slow, upper flow and solid lines for the stronger vorticity of the fast, bottom flow). As Vortex # 2 bends backward, approaching the wall, it tilts downward as it becomes entangled with the vertical vorticity at the wall and in turn induces a kink along the streamwise (X) direction in the incoming Vortex # 1. Similarly, Vortex # 3 has bent backwards and tilted downwards, thus inducing a kink along the X direction in the incoming Vortex # 2. The strain field between the large vortex cores (# 1 and # 2) produces a stretching of the perturbed vortex filaments existing on the braids, concentrating the vorticity along the streamwise direction as indicated in Fig. 9. This concentrated axial vorticity corresponds to the vortex lines which were shown in Figs. 6–8 (indicated by arrows).

Once that this streamwise vorticity is found in a concentrated form (vortex tube), it spreads laterally by self-induction in a fashion very similar to the one described for the streamwise vortices created by the effect of the localized perturbation (shown in Fig. 5). Figure 9 schematically represents this mechanism of lateral propagation.

Conclusions

An isothermal plane free shear layer has been visually investigated with Laser Induced Fluorescent techniques and a Direct Interface Visualization obtained by the reactive nature of the flow. The structure of the vorticity field was then investigated by monitoring the interface separating the two fluids and its evolution over time. The results show that in the absence of any upstream disturbances, the plane, shear layer remains two-dimensional even five to six spanwise vortex wavelengths downstream of its origin. It was also found that streamwise vortical structures are formed due to the effect of localized three dimensional disturbances in the upstream conditions. The streamwise vorticity generated through this effect was found to spread laterally by strong self-induction under the effect of the strain field created by the spanwise vortices.

In addition, the interaction of a plane shear layer with lateral, vertical walls was also investigated. Axial streamwise vortices which were found to spread laterally by self-induction were observed forming through this interaction.

Although yet inconclusive, the results seem to indicate that the lateral spreading rate of both effects is comparable.

Acknowledgements. This work has been funded by a University of Southern California Research and Innovation Fund given to Prof. J. C. Lasheras. The help received from Dr. J. S. Cho in the acquisition of the photographic data is greatly appreciated. Discussions with Prof. F. K. Browand were very useful in clarifying some of the ideas presented in this paper.

References

1. Brown, G. L., Roshko, A. (1974): J. Fluid Mech. **64**, 775
2. Dimotakis, P., Brown, G. L. (1976): J. Fluid Mech. **78**, 535
3. Winant, C. D., Browand, F. K. (1974): J. Fluid Mech. **63**, 237
4. Konrad, J. H. (1976): Ph.D. Thesis, California Institute of Technology
5. Rebollo, M. (1973): Ph.D. Thesis, California Institute of Technology
6. Wygnanski, I., Fiedler, H. E. (1970): J. Fluid Mech. **41**, 327
7. Jimenez, J. (1980): J. Fluid Mech. **96**, 447
8. Bernal, L. P. (1981): Ph.D. Thesis, California Institute of Technology
9. Jimenez, J., Cogollos, M., Bernal, L. P. (1985): J. Fluid Mech. **152**, 125
10. Lasheras, J. C., Cho, J. S., Maxworthy, T. (1986): On the origin and scale of streamwise vortical structure in a plane, free shear layer. J. Fluid Mech. **172**, 231
11. Breidenthal, R., (1978): Ph.D. Thesis, California Institute of Technology
12. Corcos, G. M. (1979): The mixing layer: deterministic models of a turbulent flow. U. C. Berkeley, Coll. Engng. Rep. FM-79-2
13. Corcos, G. M., Lin, S. J. (1984): J. Fluid Mech. **139**, 67
14. Pierrehumbert, R. T., Widnall, S. E. (1982): J. Fluid Mech. **114**, 59
15. Dewey, C. F., Jr. (1976): AGARD-CP-193, 17-1 to 17-7

Coherent Structures in the Axisymmetric Turbulent Jet Mixing Layer

Mark N. Glauser, Stewart J. Leib, William K. George

Department of Mechanical and Aerospace Engineering, State University of New York at Buffalo, Buffalo, New York 14260, USA

Abstract

In 1967 Lumley proposed two different, but complimentary approaches to the objective determination of coherent structures. The first uses an orthogonal decomposition to extract eigenvectors from two point velocity measurements, the lowest order eigenvector representing the largest structure. Where there are partial homogeneities, or when the flow is stationary, the eigenfunctions are the harmonic ones and the coherent features are impossible to identify. To organize these fluctuating Fourier modes into coherent features, a second decomposition is used, the shot-noise decomposition.

The initial experiment (on which this paper is based) has generated cross-spectral data at seven radial positions across the jet mixing layer. The 49 cross-spectra have then been decomposed to obtain the eigenvectors and the time development of the streamwise velocity component of the large eddies in the mixing layer. The results to date show clearly the existence of a large scale structure in the mixing layer containing 40% of the turbulent energy. The second and third order structures contain another 40% of the energy. Thus nearly all the energy is contained in the first three modes.

Nomenclature

$a_{(n)}$	Random coefficients
f	Characteristic eddy
g	Random distribution function
R_{ij}	Velocity correlation tensor
S	Streamwise velocity spectrum
t	Time
$\boldsymbol{u}$	Velocity vector
u	Streamwise velocity
$\hat{u}$	Fourier transform of velocity
W	Weighting function
$\boldsymbol{x}$	Special vector
y	Radial distance across jet mixing layer
α	Inner product of velocity vector and candidate eddy
δ_{mn}	Kroneker delta function
$\lambda^{(n)}$	Eigenvalues
Φ_{ij}	Cross spectral tensor
ϕ	Eigenvectors
ψ	Eigenvectors in transformed domain
ω	Frequency

Introduction

The earliest investigations of the statistical characteristics of the turbulence in an axisymmetric jet mixing layer were carried out by Laurence (1956), Davies et al. (1963), and Bradshaw et al. (1964). Qualitative models for the large scale structures in a number of turbulent flows were formulated by Grant (1958) and Townsend (1956, 1976). Crow and Champagne (1971) found identifiable structures in a jet which resembled large scale vortical puffs.

Recently it has been found that these eddies may be more orderly and energetic than was first imagined. Also experiments indicate that these eddies are coherent in some sense. That is, they travel downstream as an identifiable entity for some distance before mean shear distortion and energy exchange with smaller scales destroy them.

While the precise role of the eddies in the dynamics of the flow is not yet established, they are suspected of being responsible for the entrainment process in addition to acting as

Turbulent Shear Flows 5

sources of noise. For a review of some of the more recent work on coherent structures see Cantwell (1981).

Lumley (1967) has proposed two different, but complimentary approaches to the objective determination of coherent structures, the orthogonal decomposition for inhomogeneous directions and the shot noise decomposition for homogeneous directions and for time if the process is stationary. These two approaches are implemented in this paper.

In the current work, experiments were carried out in the mixing region of an axisymmetric turbulent jet to acquire the necessary data to perform a one-dimensional shot noise decomposition and a scalar two-dimensional version of the full orthogonal decomposition. Numerical computations were carried out and the eigenfunction and characteristic eddies were extracted from these measurements. The power spectra as well as the instantaneous velocity signals were reconstructed from these eigenfunctions.

Experiment and Apparatus

Briefly, the experiment was performed in a circular air jet facility. The jet nozzle was of fifth order polynomial design and has an exit diameter of 0.098 m with a contraction ratio of 10:1. The exit conditions of the jet are as follows: The turbulence intensity was 0.35%, the Reynolds number was 110,000 and the boundary layer thickness at the exit ($u = 0.99\, u_e$) was 0.0012 m. A specially-designed probe containing seven single hot wire sensors positioned at $x/d = 3$ was used to carry out the experiment. The individual sensors were spaced 0.0127 m apart with the center sensor in the middle of the mixing layer. Leib et al. (1984) provide a more complete description of the experiment.

The Orthogonal Decomposition

In 1967 Lumley proposed that the coherent structure should be that structure which has the largest mean square projection on the veloctiy field. If $\boldsymbol{\phi}(\boldsymbol{x}, t)$ is our candidate structure, then we should choose $\boldsymbol{\phi}$ to maximize

$$\overline{|\boldsymbol{u} \cdot \boldsymbol{\phi}|^2} = \overline{|\alpha|^2}, \tag{1}$$

where $\boldsymbol{u} = \boldsymbol{u}(\boldsymbol{x}, t)$ is the instantaneous velocity. $\boldsymbol{\phi}$ is then, in some sense at least, the most likely occurrence of $\boldsymbol{u}$. The quantity in (1) is assumed normalized by the amplitude of the candidate structure so that the projection is only affected by the shape of the candidate and not by its amplitude.

Maximizing $\overline{|\boldsymbol{u} \cdot \boldsymbol{\phi}|^2}$ leads to the following eigenvalue problem:

$$\iiint R_{ij}(\boldsymbol{x}, \boldsymbol{x}', t, t')\, \phi_j(\boldsymbol{x}', t')\, d\boldsymbol{x}'\, dt' = \lambda^{(n)}\, \phi_i(\boldsymbol{x}, t), \tag{2}$$

where $R_{ij}(\boldsymbol{x}, \boldsymbol{x}', t, t') = \overline{u_i(\boldsymbol{x}, t)\, u_j(\boldsymbol{x}', t')}$ and $\lambda = |\alpha|^2$.

Equation (2) has an infinity of orthogonal solutions and the original random field can be reconstructed from them, i.e.,

$$\boldsymbol{u}(\boldsymbol{x}, t) = \sum_{n=1}^{\infty} a_n \boldsymbol{\phi}^{(n)}(\boldsymbol{x}, t). \tag{3}$$

The a's are random and uncorrelated, i.e.,

$$\overline{a_n a_m} = \lambda_n \delta_{mn},$$

where the λ_n are the eigenvalues.

If there are stationarities or homogeneities these must first be removed by Fourier analysis, that is,

$$\boldsymbol{u}(\boldsymbol{x}, t) \rightarrow \hat{\boldsymbol{u}}(\boldsymbol{x}, \omega).$$

Now decompose $\hat{u}$ to get

$$\int \Phi_{ij}(\boldsymbol{x}, \boldsymbol{x}', \omega)\, \psi_j(\boldsymbol{x}_j\omega)\, d\boldsymbol{x}' = \lambda^{(n)}(\omega)\, \psi_i(\boldsymbol{x}, \omega), \tag{4}$$

where the integral is over inhomogeneous directions and $\Phi_{ij}(\boldsymbol{x}, \boldsymbol{x}', \omega)$ is the cross spectral tensor.

The eigenvalues now become eigenspectra

$$\lambda^{(n)} \rightarrow \lambda^{(n)}(\omega)$$

and the eigenvectors are frequency dependent,

$$\phi_i^{(n)}(\boldsymbol{x}, t) \rightarrow \psi_i^{(n)}(\boldsymbol{x}, \omega).$$

The Fourier transform of the velocity can be reconstructed as

$$\hat{\boldsymbol{u}}(\boldsymbol{x}, \omega) = \sum_{n=1}^{\infty} a_n(\omega)\, \Psi^{(n)}(\boldsymbol{x}, \omega), \tag{5}$$

where the a's are random and uncorrelated functions of frequency, i.e., $\overline{|a_n(\omega)|^2} = \lambda^{(n)}(\omega)$, and can be obtained for a single ensemble by

$$a_n(\omega) = \int \hat{\boldsymbol{u}}(\boldsymbol{x}, \omega)\, \boldsymbol{\Psi}^{(n)*}(\boldsymbol{x}, \omega)\, d\boldsymbol{x}.$$

Note that $\hat{\boldsymbol{u}}(\boldsymbol{x}, \omega)$ and $a_n(\omega)$ are random and $\lambda^{(n)}(\omega)$ and $\psi^{(n)}(\boldsymbol{x}, \omega)$ are deterministic.

The spectrum can also be reconstructed at each position by

$$S(\boldsymbol{x}, \omega) = \sum_{n=1}^{\infty} \lambda^{(n)}(\omega)\, |\boldsymbol{\psi}^{(n)}(\boldsymbol{x}, \omega)|^2 \tag{6}$$

as can the cross-spectra

$$\Phi_{ij}(\boldsymbol{x}, \boldsymbol{x}'\omega) = \sum_{n=1}^{\infty} \lambda^{(n)}(\omega)\, \psi_i^{(n)}(\boldsymbol{x}, \omega)\, \psi_j^{(n)*}(\boldsymbol{x}', \omega). \tag{7}$$

A Simpler Version

A scalar, one-dimensional version of the orthogonal decomposition, which utilizes the one-dimensional cross-spectra measured in the jet can be derived and formulated as

$$\int \Phi_{11}(y, y', \omega)\, \psi^{(n)}(y', \omega)\, dy' = \lambda^{(n)}(\omega)\, \psi^{(n)}(y, \omega), \tag{8}$$

where y is the radius in the mixing layer. The Fourier transform of the streamwise velocity can be reconstructed by

$$\hat{u}(y, \omega) = \sum_{n=1}^{\infty} a_n(\omega)\, \psi^{(n)}(\omega, r), \tag{9}$$

where

$$a_n(\omega) = \int \hat{u}(y, \omega)\ \psi^{(n)*}(y, \omega)\ dy.$$

The spectrum is reconstructed from

$$S(y, \omega) = \sum_{n=1}^{\infty} \lambda^{(n)}(\omega)\ |\psi^{(n)}(y, \omega)|^2, \tag{10}$$

where λ_n are the eigenspectra.

Using the measured values of the cross spectrum, (8) can be solved numerically for the eigenvalues and eigenfunctions.

The Numerical Approximation

Briefly, the numerical approximation consists basically of replacing the integral in (8) by a suitably chosen quadrature rule. This can be written in general as

$$\sum_{i=1}^{m} W(y_i)\ \Phi(y_j, y_i, \omega)\ \psi^{(n)}(y_i, \omega) = \lambda^{(n)}(\omega)\ \psi^{(n)}(y_j, \omega), \qquad j = 1, 2, \ldots, m, \tag{11}$$

where m is the maximum number of radial locations and $W(y_i)$ is a weighting function. In the present case $m = 7$. Equation (11) is now an algebraic eigenvalue problem.

Now the matrix formed from the values of Φ is complex Hermitian. That is

$$\Phi_{ij} = \Phi^*_{ji}. \tag{12}$$

However, the weighted matrix formed by $W(y_i)\ \Phi(y_j, y_i)$ no longer has this property. It is desirable, in the numerical solution of eigenvalue problems, to keep the coefficient matrix Hermitian (Baker, 1977).

Operating on both sides of (11) by $W(y_j)^{1/2}$ results in the following eigenvalue problem,

$$\sum_{i=1}^{m} [W^{1/2}(y_j)\ \Phi(y_j, y_i, \omega)\ W^{1/2}(y_i)]\ W^{1/2}(y_i)\ \psi^{(n)}(y_i, \omega) = \lambda^{(n)}(\omega)\ W^{1/2}(y_j)\ \psi^{(n)}(y_j, \omega), \tag{13}$$

where now the matrix $W(y_j)^{1/2}\ \Phi(y_j, y_i, \omega)\ W(y_i)^{1/2}$ is Hermitian. Since this matrix is similar to that in (11), they have the same eigenvalues. The original eigenvectors can be recovered from those obtained from (13) by operating by $W(y_j)^{-1/2}$, i.e.

$$\psi^n(y_j, \omega) = W(y_j)^{1/2}\, \psi^n(y_j, \omega)\ W(y_j)^{-1/2}.$$

Simpson's one-third rule was chosen as the quadrature rule for its accuracy and simplicity. For a more detailed discussion of the numerical approximation see Leib et al. (1984).

Results of the Orthogonal Decomposition Analysis

The results show clearly the existence of a large structure in the mixing layer containing 40% of the turbulent energy. The second and third order structures contain together another 40% of the energy, the remainder being contained in structures impossible to resolve because of statistical and numerical errors.

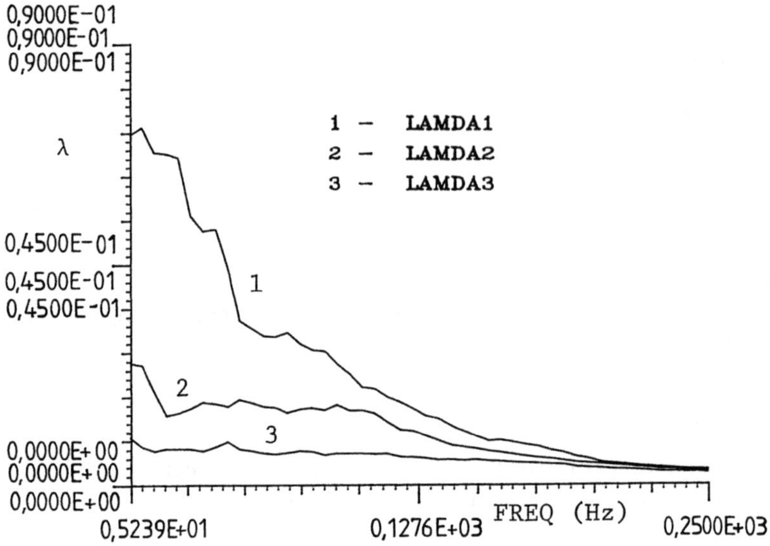

Fig. 1. Eigenspectra showing how energy is distributed among the first three eigenvalues

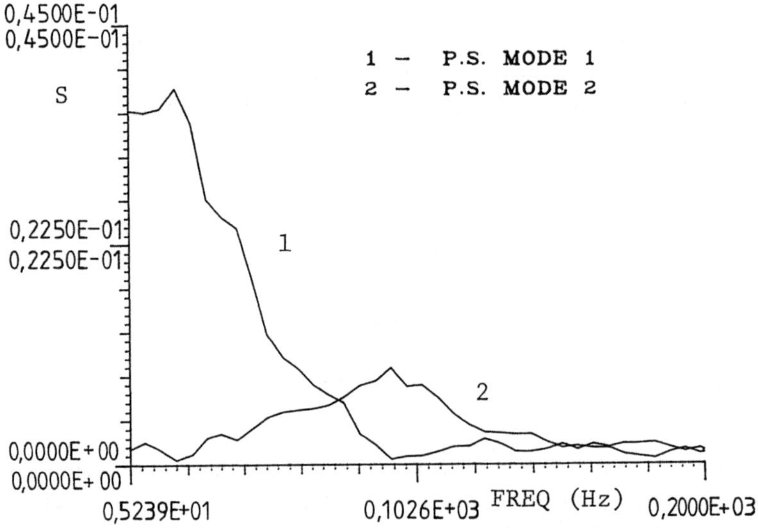

Fig. 2. Individual contributions to the power spectrum from first 2 modes

Figure 1 shows the eigenspectra for the three dominant modes. These eigenvalues represent the contribution to the total energy (integrated across the shear layer) from the various modes. It was found that the first mode contained about 40% of the total energy.

As was shown earlier (10), the power spectra at each radial position can be expanded in a series of the eigenmodes since each mode makes an independent contribution to the spectrum. Figure 2 shows the contributions from the first two modes to the power spectrum at a position on the low speed side of the mixing layer. It can be seen from this figure that the contribution to the power spectrum from the first mode peaks at very low frequency, while that of the second peaks at a higher frequency. This was found to be the case at the

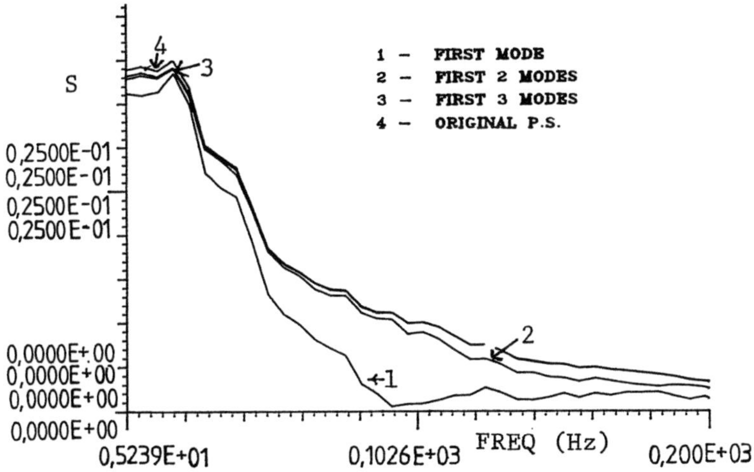

Fig. 3. Convergence of power spectrum

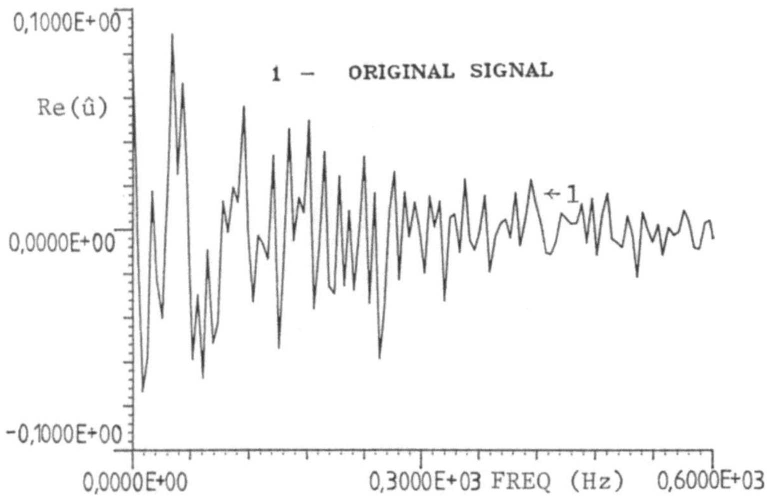

Fig. 4. Real part of F.T. of original signal

other radial positions as well. In this sense then, we see that the dominant mode is indeed representative of the larger scales in the flow.

Figure 3 [using (10) for $n = 1, 2, 3$] shows the convergence of the expansion for this power spectrum. Similar results were obtained at the other locations. These results show that nearly all the energy is contained in the first three modes.

From the spectral data it might be expected that the instantaneous velocity signal could be represented adequately with these three modes. After application of (9) this is indeed seen to be the case (see Figs. 4–7). Figure 4 is one record of the original velocity signal in the center of the mixing layer (real part of its Fourier transform). Figure 5 shows the super-

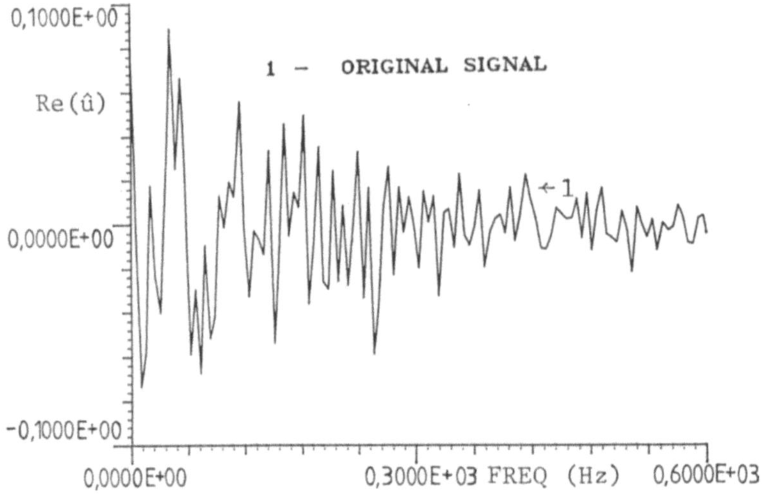

Fig. 5. First mode superimposed on original

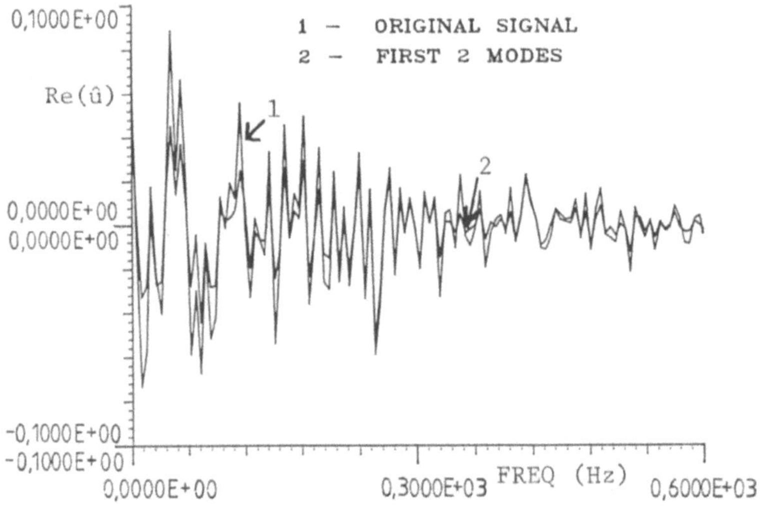

Fig. 6. First 2 modes superimposed on original

position of the 1st mode [$n = 1$ in (9)] on the original signal. Figure 6 shows the first 2 modes. Figure 7 shows the first 3 modes. As can be seen from Fig. 7, almost the entire signal has been reconstructed from the summation of the first 3 modes.

The question may be asked, what do these eigenfunctions that maximize the mean square projection look like? The fact that the eigenvalue problem was solved as a function of radial position for each discrete frequency dictates that there is an individual set of eigenfunctions for each frequency. Since the eigenfunctions are functions of frequency and radial position in the jet mixing layer it is of interest to look at the surfaces generated by them. However these surfaces have been found to be quite complex. Because of this, different cross-sections were examined. The real and imaginary parts of the first three eigenfunctions at a frequency of 5.2 Hz are shown, plotted as a function of radial distance in the jet mixing layer, in Figs. 8

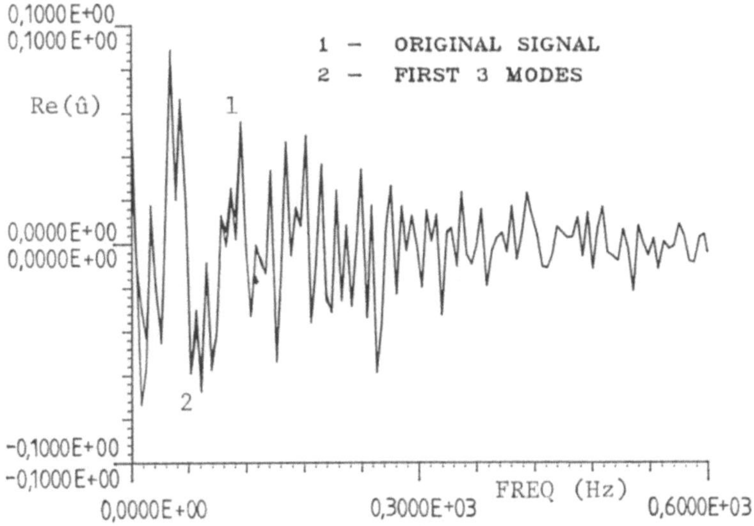

Fig. 7. First 3 modes superimposed on original. Note: Essentially no difference

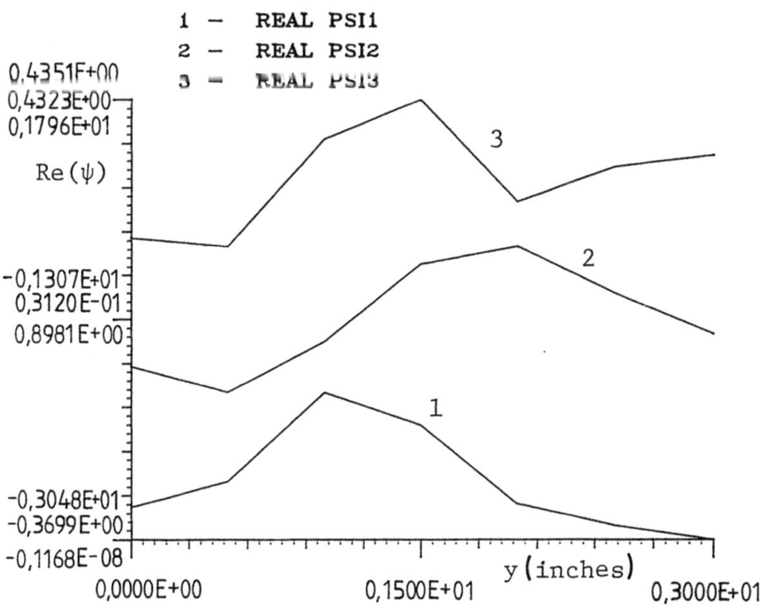

Fig. 8. Real parts of first 3 eigenfunctions at 5.2 Hz plotted as a function of radial distance in jet mixing layer

and 9 respectively. When examining these figures it appears that the first mode is symmetric and bears resemblence to the RMS plots across the jet mixing layer. The second mode is asymmetric and the third mode is even more complicated.

The real and imaginary part of the first eigenfunction for a radial position at the center of the mixing layer are plotted as a function of frequency in Fig. 10. This first eigenfunction eventually decays as the frequency approaches 1000 Hz. This was not the case for the second eigenfunction at the same radial position. Its real and imaginary parts are shown in Fig. 11.

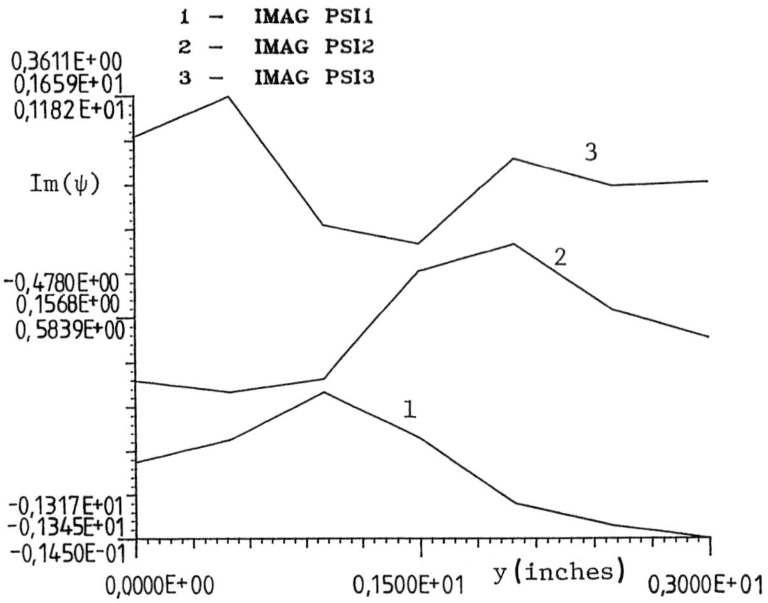

Fig. 9. Imaginary parts of first 3 eigenfunctions at 5.2 Hz plotted as a function of radial distance in the jet mixing layer

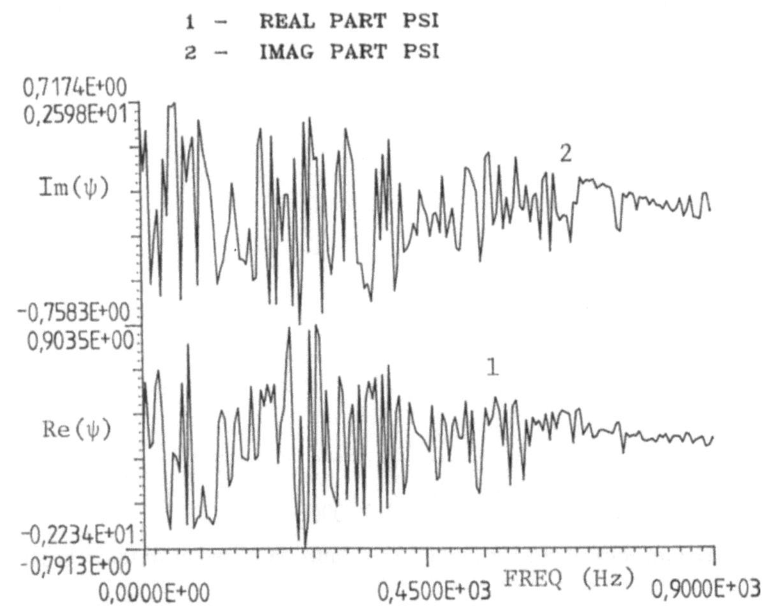

Fig. 10. Real and imaginary parts of first eigenfunction at the center of the mixing layer plotted as a function of frequency

When examining Fig. 11 one observes that the eigenfunction decays with frequency up to approximately 900 Hz but then starts to become large again. A plausible explanation for this observation is as follows. The smallest time scale for a typical eddy that can be resolved is dictated by the probe spacing and the appropriate convection velocity (since eddies smaller than this will not be seen by both probes). If it is assumed that the turbulence field is frozen

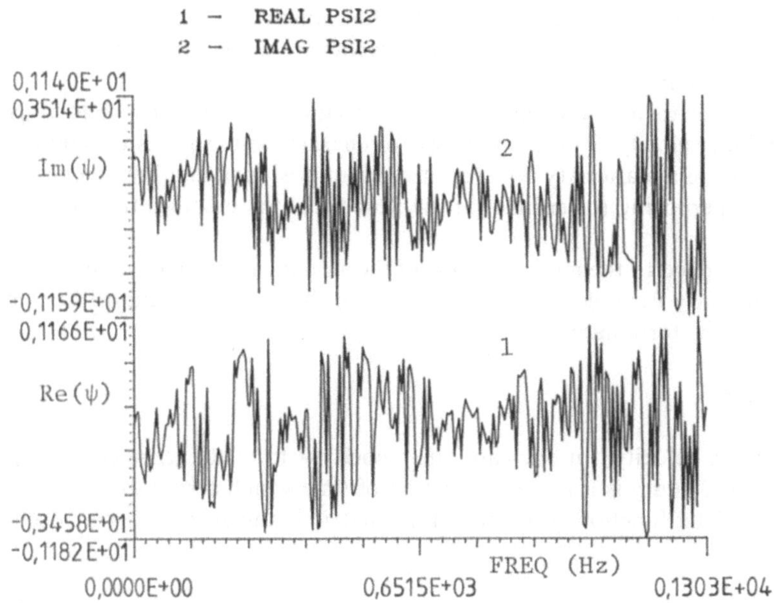

Fig. 11. Real and imaginary parts of 2nd eigenfunction at the center of the mixing layer plotted as a function of frequency

then this time scale is approximately equal to $\Delta y/u_c = 0.00115$ sec $= \Delta t$. Now the resolvable bandwidth in frequency would approximately be equal to $1/\Delta t \cong 870$ Hz. It is felt then that the results in Fig. 11 beyond 900 Hz are just a result of numerical noise.

The Shot Noise Decomposition

How is the coherent structure extracted from the analysis so far and in particular, what does the large eddy look like in physical space? This is difficult to answer. Remember that the coherent structure is defined to be the dominant mode, i.e.

$$\hat{u}^{(1)}(y, \omega) = a_{(1)}(\omega)\, \psi^{(1)}(y, \omega) \tag{14}$$

or $$u^{(1)}(y, t) = \int_{-\infty}^{\infty} e^{i\omega t} a_{(1)}(\omega)\, \psi^{(1)}(y, \omega)\, d\omega, \tag{15}$$

where because $a_1(\omega)$ is random, so is $\hat{u}^{(1)}$.

The visibility of the coherent structure depends on the bandwidth of the random coefficient $a_{(1)}(\omega)$. For low Reynolds number flow there are only a few Fourier coefficients so that the structure is visible, however with high Reynolds flow there are many Fourier coefficients resulting in the structure being invisible. Therefore a method is needed to organize these random pieces into 'groups'. This motivates the use of the shot noise decomposition.

The shot effect can be illustrated by the randomly varying intensity of a flow of electrons from cathode to anode in a vacuum tube. The signal produced by this stream of electrons is random in nature. The arrival of a single electron at the anode would result in a signal characteristic of this event. The signal which results from the arrival of many electrons at random times is then a random superposition of these characteristic signals. It can be seen then that, while the overall signal is indeed random, it is composed of characteristic signals which are deterministic and occur at random times. The shot noise decomposition provides a means of determining the nature of the individual events from measurements of the spectrum of the random signal.

In extending these ideas to a homogeneous or stationary turbulent flow field it is supposed that the random velocity field is composed of characteristic signals or 'eddies', occurring at random times in a stationary flow (or at random positions in a homogeneous flow). Upon randomly superposing (or sprinkling the flow with) these eddies the random velocity field is realized.

These concepts were applied to the measurements of the streamwise velocity component as a function of the stationary variable time. In this case the shot noise decomposition, as suggested by Lumley, can be formulated as

$$u(t) = \int f(t - \tau)\, g(\tau)\, d\tau . \tag{16}$$

In (16), u is the streamwise velocity, (in our case, the velocity field of the large eddy $u^{(1)}$), f is the characteristic eddy, and g is a random function of time which serves to distribute the characteristic eddies. The function g is defined so that it is uncorrelated with itself. That is

$$\overline{g(t)\, g(t')} = \delta(t' - t) . \tag{17}$$

The problem of determining the characteristic eddy then reduces to determining the function f. It is straightforward to show (Lumley, 1967) that this function is given by

$$f(y, t) = (1/2\pi)^{1/2} \int e^{-i\omega t} \sqrt{\frac{\lambda^{(1)}(\omega)}{2\pi}}\, |\psi^{(1)}(y, \omega)|\, d\omega . \tag{18}$$

It is important to note at this time that only the magnitude of the characteristic eddy can be recovered by (18), all phase information is lost. The phase can be recovered by examining the third-order statistics in a manner somewhat like that suggested by Lumley (1981). This will be discussed in a subsequent paper.

At each radial position (18) was used to construct the characteristic eddy using only its magnitudes (ie., ignoring possible phase differences across the mixing layer). The results of these computations are shown in Fig. 12. In this figure the form of the characteristic eddy as a function of time is shown for each radial position. A dominant peak occurred at a

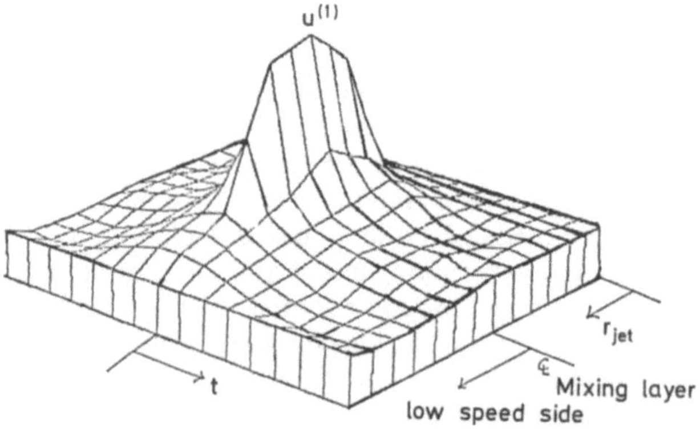

Fig. 12. Coherent structure calculated from Eq. (15)

position in the center of the mixing layer, as might have been expected. It is possible to argue from continuity and axisymmetry that this structure corresponds to a vortex ring-like structure not unlike that educted by Hussain (1981).

Conclusion

The orthogonal decomposition has been shown to be remarkably efficient at organizing data (i.e., 7 × 7 grid appears adequate) and the instantaneous properties of the random signal have not been lost, only organized. So efficient has the scheme been at organizing the energy that only a few terms were needed to completely represent the instantaneous signal.

The implications of these findings on the problem of turbulence modelling could be profound. If, in fact, the large scale features of turbulent flows can be resolved on such a coarse grid, a direct decomposition of the Navier-Stokes equation (Lumley, 1981) might represent a fruitful approach. The largest eddies would be calculated directly and the remainder of the turbulence modelled in more traditional ways.

It is difficult to determine the significance of the form of the coherent structures here due to the fluctuating nature of stationary random signals. Nonetheless an attempt to 'visualize' the large eddy has been made using the shot noise decomposition to obtain the magnitude of the eddy's velocity field, and ignoring for now possible phase differences across the jet. The results look very much like the roller eddies reported by various authors.

Acknowledgements The work described above has been carried out by Mark Glauser, Stewart Leib and Muhammed S. S. Khwaja as a part of their thesis research. This research was initiated under support from the National Science Foundation under Grant No. ENG76-17466 and the Air Force Office of Scientific Research under Contract Nos. F4962078C0047 and F4962080C0053. Work has continued under NSF Grant No. MEA 8316833 and with the support of the NASA trainee program, SJL at NASA/Lewis and MNG at NASA/Ames. These sources of support are gratefully acknowledged.

References

Baker C. T. H. (1977): *The Numerical Treatment of Integral Equation* (Clarendon, Oxford, New York)
Bradshaw, P., Ferriss, D. H., Johnson, R. F. (1964): Turbulence in the noise-producing region of a circular jet. J. Fluid Mech. **19**, 591
Cantwell, B. J. (1981): Organized motion in turbulent flow. Ann. Rev. Fluid Mech. **13,** 457
Crow, J. C., Champagne, F. H. (1971): Orderly structure in jet turbulence. J. Fluid Mech. **48,** 547
Davies, P. O. A. L., Barratt, M. J., Fisher, M. J. (1963): The characteristics of turbulence in the mixing region of a round jet. J. Fluid Mech. **15,** 337
Grant, H. L. (1958): The Large Eddies of Turbulent Motion. J. Fluid Mech. **4,** 149
Hussain, A. K. M. F. (1981): "Coherent Structures and Studies of Perturbed and Unperturbed Jets", Proc. of Int. Conf. on Role of Coherent Structures in Modelling Turbulence and Mixing, U. Politechnic, Madrid
Laurence, J. C. (1956): Intensity, scale and spectra of turbulence in mixing region of free subsonic jet. NACA Report No. 1292
Leib, S. J., Glauser, M. N., George, W. K. (1984): An application of Lumley's orthogonal decomposition to the axisymmetric turbulent jet mixing layer. Presented at 9th Rolla Symposium.
Lumley, J. L. (1967): "The Structure of Inhomogeneous Turbulent Flows," in Atm. Turb. and Radio Wave. Prop., ed by A. M. Yaglom and V. I. Tatarsky, (Nauka, Moscow) pp. 166–178
Lumley, J. L. (1981): "Coherent structures in turbulence". Trans. and Turb., Academic Press, edited by R. E. Meyer NY 215–241
Townsend, A. A. (1956): *The Structure of Turbulent Shear Flow* (Cambridge University Press)
Townsend, A. A. (1976) 2nd ed: The Structure of Turbulent Shear Flow (Cambridge University Press)

Laser Doppler Velocimeter Measurements in 3-D Impinging Twin-Jet Fountain Flows

K. R. Saripalli

McDonnell Douglas Research Laboratories

Abstract

Mean velocity and turbulence measurements were conducted on the three-dimensional fountain flow-field generated by the impingement of two axisymmetric turbulent jets on a ground plane; application was to vertical-take-off and landing (VTOL) aircraft. Three configurations with different jet spacing and height above the ground were considered resulting in different characteristics of the opposing wall jets that form the fountain. Basic instantaneous velocity data were obtained by use of a two-component laser Doppler velocimeter (LDV) in a plane connecting the nozzle centerlines at different heights above the ground. The jet impingement region and the fountain upwash region formed by the collision of the wall jets were emphasized. The distributions of mean velocity components and turbulence quantities, including the turbulence intensity and the Reynolds shear stress, were derived from the basic velocity data. Detailed studies of the characteristics of the fountain revealed self-similarity in the mean velocity and turbulence profiles across the fountain. The spreading of the fountain was found to be linear and the spreading rates equal for all three jet impingement configurations. The mean velocity decay was found to be inversely proportional to the height above the ground; the rate of decay was equal for the three configurations. The spread and mean velocity decay characteristics closely resemble those of radial jets. The fountain was observed to be highly turbulent with maximum turbulence intensities in the range of 50–60%.

Nomenclature

A, B, C	Constants
D	Exit internal diameter of the nozzle or the exit jet diameter
H	Height of the nozzle exit above the ground plate
M	Constant
Re	Reynolds number based on the exit jet diameter and the exit jet velocity
S	Center distance between the jets
U_i	Instantaneous velocity in the streamwise (Z) direction
V_i	Instantaneous velocity in the cross stream (X) direction
W_i	Instantaneous velocity in the Y direction
U	Mean velocity in the streamwise direction
V	Mean velocity in the cross-stream direction
u	Fluctuating component of the velocity in the streamwise direction
v	Fluctuating component of the velocity in the cross-stream direction
U_J	Jet centerline velocity at the exit
U_{max}	The local maximum streamwise velocity in the fountain
$X_{1/2}$	The fountain half-width where $U = \frac{1}{2} U_{max}$
X	Distance parallel to the line connecting the nozzle centerlines
Y	Distance perpendicular to the line connecting the nozzle centerlines and parallel to the ground plane
Z	Perpendicular distance from the ground plane

Introduction

An axisymmetric turbulent free jet impinging normal to a wall generates an axisymmetric ground wall-jet spreading uniformly in all directions. However, when two jets impinge in close proximity (Fig. 1), the individual radial wall-jets collide with each other to form a

Turbulent Shear Flows 5

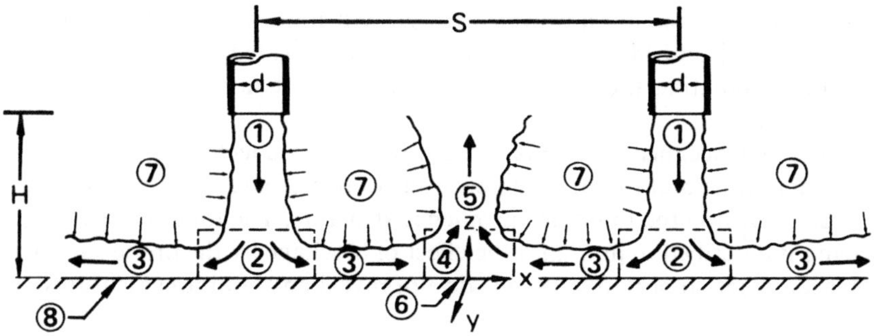

Fig. 1. Schematic illustration of twin-jet impingement flow

1. Lift jet flow;
2. Jet impingement region;
3. Wall jet flow;
4. Fountain formation region;
5. Fountain up-wash flow;
6. Wall jet interaction stagnation line;
7. Entrainment;
8. Ground plane

fountain upwash flow. The fountain (Fig. 1) is observed to be fan-shaped with radially directed upward flow and increasing width away from the ground. The impinging lift jets of a vertical take-off and landing (VTOL) aircraft hovering in ground proximity produce such complicated three-dimensional flowfields. The fountain plays an important role since its impingement on the aircraft undersurface increases lift, elevates skin temperatures and increases the possibility of reingestion into the inlets. Detailed studies of the structure and development of the fountain upwash flow are thus essential to the successful design of efficient VTOL aircraft. The objectives of the current studies are to increase the basic understanding of the turbulent upwash and to provide a reliable data base for use in predictive computational models. Towards these objectives, a unique set of mean velocity and turbulence data on axisymmetric twin-jet fountain flows are presented here.

Background

Some of the earlier measurements of the flow properties of fountain upwash are reported in [1–4]. Kotansky and Glaze [1] reported mean velocity measurements in the wall jet and the fountain for twin axisymmetric jet impingement with $S/D = 12$ and $H/D = 5$ using hot-wire techniques. The fountain formation and development characteristics were determined for a range of the important parameters that influence the flow. The flow angularity in the fountain in the plane connecting the nozzle centerlines was reported to vary randomly. Jenkins and Hill [2] presented pitot pressure profiles in the fountain for twin-axisymmetric jet impingement with $S/D = 6$ and $2 \leqq H/D \leqq 6$. Strong fluctuations were reported in the upwash flow properties, and the magnitude of these fluctuations increased with jet height above the ground. Also, highly fluctuating flow angularity, with frequent complete velocity reversals, was reported in the regions outside of the main upwash fan and the shear layers of the incident jets, and sometimes even in the central upwash region. Therefore, it was recommended that pitot probes be used in the measurement of mean flow properties in the fountain upwash.

Kind and Suthanthiran [3] reported measurements of a two-dimensional fountain upwash and indicated high turbulence levels and spreading rates in the fountain. The first set of careful measurements conducted in a two-dimensional fountain upwash by use of an x-wire hot-film anemometer were presented by Gilbert [4], who reported detailed mean-velocity and turbulence measurements that included decay and spread characteristics of the fountain. The observed levels of turbulent intensities in the fountain were similar to those in an ordinary two-dimensional free jet; however, high growth rates were observed in the fountain. The fountains in [3] and [4] were generated by two isolated, two-dimensional opposing wall jets, thus eliminating the important jet impingement region as part of the fountain formation process.

Mean velocity and turbulence measurements made using hot-wire techniques in a fountain generated by the impingement of two axisymmetric jets were presented in [5]; no definite conclusions were made regarding the fountain turbulence structure. The fountain behavior for close nozzle spacing is described in [6, 7] and the effect of blockage by the probe support is also indicated. Measurements reported in [8–11] do not reveal any additional features of fountain flows.

Although several investigations of the fountain flow have been carried out, interpretations of the measurements vary widely, primarily because of the difficulty in measuring highly unsteady flows using hot-film and pitot-probe techniques. In addition, turbulence measurements on practical configurations have been very limited. Computational codes require a better definition of the fountain flow and its turbulence structure to yield reliable predictions; therefore, there is a need for a reliable, detailed mean-velocity and turbulence data base, obtained using a suitable measurement technique, for a realistic 3-D fountain generated by impinging axisymmetric jets.

Measurement Technique and the Working Medium

The measurement technique for studying the fountain flow was required to meet the following specifications: a) nonintrusive, causing no flow disturbance, especially in the case of closely spaced jets, b) able to sense the direction of flow in the recirculating regions, c) have a linear response, essential for accurate measurements in areas where the turbulence intensities are high, such as in the fountain. A laser Doppler velocimeter (LDV) meets these specifications [12, 13]; in addition, an LDV responds to a specific velocity component in 3-D flows and measures velocity directly without the need to correct for temperature effects.

As a working medium, water offers specific advantages compared to air; the tracers suitable for flow visualization in water are more numerous with better light-reflecting characteristics, and aerodynamic phenomena can be observed at a relatively slow speed for the same Reynolds number and model scale because of the difference in kinematic viscosities of air and water. Moreover, for laser Doppler velocimetry, the seeding of the flow is normally unnecessary in water because the natural suspended particles act as light scatterers. If seeding is necessary, it can be accomplished more easily in water than in air by adding neutral-density plastic particles of the proper size.

Experimental Techniques

Experimental Apparatus

Figure 2 shows the jet-impingement facility used for conducting laser-Doppler-velocimeter (LDV) measurements on a twin-jet impingement flow, with water as the working medium. The primary components are (1) the main Plexiglas tank, (2) the header tank, (3) nozzle units, and (4) the ground plate. The large transparent main tank (1.88 × 1.52 × 1.37 m) has an approximate capacity of 3800 liters. The pressurized stainless steel cylindrical header tank (91.4-cm diam.) supplies water to the nozzle units and acts as a plenum to damp fluctuations produced by the pump.

The function of the Plexiglas nozzle units (Fig. 3) is to reduce turbulence and mean-velocity nonuniformities in the existing jet flow to acceptable levels. Each nozzle unit consists

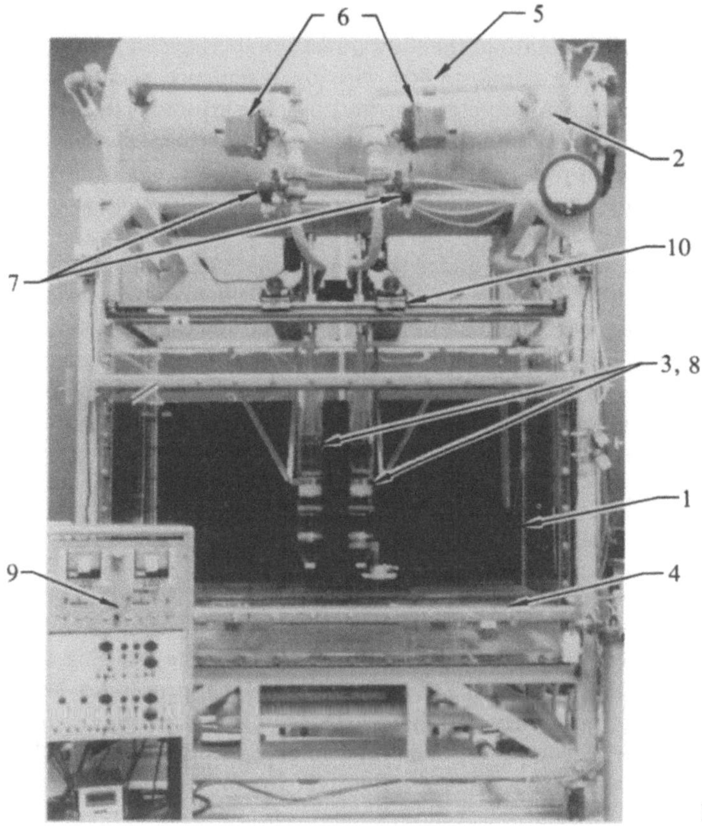

1. Main plexiglas tank
2. Header tank
3. Nozzle units
4. Ground plate
5. Turbine flow meters
6. Motor-operated flow control valves
7. Shut-off valves
8. Settling chamber of 3
9. Control panel

Fig. 2. Jet impingement facility

GP41-1579-1-R

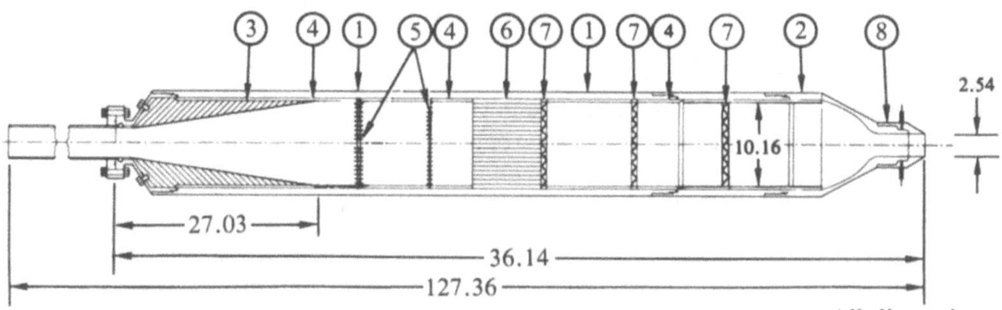

All dimensions are in Cm.

1. Settling chamber
2. Nozzle
3. Flow distributor
4. Spacers
5. Perforated plates
6. Honeycomb
7. Screen rings with screen
8. Adapter for marking the shear layer

Fig. 3. Schematic of the nozzle units

GP41-1579-2-R

of a flow distributor (diffuser) for decelerating the flow, a set of perforated plates, a honeycomb and a set of screens to establish a uniform flow with low turbulence, and a nozzle (2.54-cm exit diam., 16:1 contraction ratio) for accelerating the flow. The ground plate is held at a fixed height above the bottom of the main tank and acts as an impingement surface for the jets with sufficient edge clearance for passage of the flow. The nozzles can be positioned into place by a traversing unit mounted on the main tank accommodating as many as four nozzle units.

The circulation system draws off water from under the ground plate and pumps it into the header tank, thus supplying water to the jets through a series of flow-control devices. With the available pumping capacity, jet Reynolds numbers up to 200 000 can be obtained in a typical twin-jet (2.54-cm-exit diam.) impingement configuration. The three-dimensional fountain flow is visualized through a series of two-dimensional images. Fluorescein-sodium, a fluorescent dye, is injected into the jet flow which then fluoresces bright yellowish-green when the desired cross-section is illuminated with a thin (1-mm) sheet of 488-nm wavelength light from an Argon-ion laser.

Laser Doppler Velocimeter

A two-color (two-component) TSI laser Doppler velocimeter (LDV) system was used in the dual-beam off-axis (about 30°) backward scattering mode. Bragg-cell frequency shifting was used in both channels to detect the flow reversals. TSI counter-type signal processors were used to convert the Doppler signal into a form suitable for recording on a magnetic disk through a dedicated DEC MINC 11/23 Computer. The probe volume was positioned at the required location by use of a remotely driven $X - Y - Z$ traversing unit. A TSI hot-film anemometer system with a TSI 1231 W conical hot-film probe was also used in conjunction with LDV for diagnostic measurements of the jet flow at nozzle exits. A detailed description of the experimental apparatus and the flow visualization and measurement techniques is given in [14–17].

Data Reduction: Six groups of 256 samples each were taken at each datapoint. The data were collected in the coincidence mode with a time window (70 μs) for data acceptance smaller than the estimated smallest time-scale of turbulence expected in the measured flows. The recorded data on the magnetic disk were then processed on a DEC PDP 11/70 minicomputer and plotted on a Benson-Varian electrostatic plotter.

The time-dependent streamwise $U_i(U + u)$ and cross-stream $V_i(V + v)$ velocities in the Z and X directions, respectively, were measured directly by the LDV. The mean velocities (U, V); turbulence intensities $(\sqrt{\overline{u^2}}/U_J, \sqrt{\overline{v^2}}/U_J)$, Reynolds shear stress $(-\overline{uv}/U_J^2)$, and correlation coefficient $(-\overline{uv}/\sqrt{\overline{u^2}}\sqrt{\overline{v^2}})$ were derived from the basic velocity data. All measurements were taken in the plane of symmetry connecting the nozzle centerlines.

Velocity Bias Correction: Velocity bias is one of the possible sources of error in measurements by individual-realization laser Doppler velocimetry. For uniformly seeded, incompressible flows during periods of high velocity, more particles can be measured per unit time than in periods of relatively low velocity. McLaughlin and Tiederman [18] were the first to show that, in such cases, the velocity calculated as the arithmetic mean of all measured particle-velocities will be higher than the true mean velocity and thus biased towards higher velocities. The errors attributable to this form of biasing become important when the turbulence levels are extremely high, as suspected in the present case of fountain flow.

A compensating effect to the above form of velocity bias is discussed in [19]. This compensating effect is based on the argument that the response characteristics of the detector allow slower moving particles to produce signals of higher amplitude than faster moving particles, and hence a higher probability of producing a validated output. Adams et al. [20] and Adams and Eaton [21] give a comprehensive account of the controversial subject of velocity bias and examine several methods of bias elimination, applied to turbulent separated flow behind a backward-facing step. Velocity bias correction methods based on the time-average concept, such as those employing interdetection time were recommended. In regions of high velocity and high turbulence ($\approx 90\%$ based on the local velocity) where the maximum correction was expected, the velocity bias of the mean velocity was found to be on the order of 4% of the reference free-stream velocity.

In the present case, the appropriate averaging times were not recorded; hence, a 3-D version of McLaughlin-Tiederman correction was applied to both mean velocity and turbulence [22] quantities. Each individual sample was weighted inversely to the magnitude of the instantaneous velocity; $\sqrt{(U_i^2 + V_i^2 + W_i^2)}$. An approximation to W_i suggested in [23] was used. The corrected data reveal that a correction is always applied even though [20] and [21] suggest that it may not always be necessary. Also, the data were overcorrected at some points, as indicated by the discontinuities in the profiles. Furthermore, [24] suggests that, in the absence of the ability to apply appropriate time-averaging correction schemes, it is better to leave the data uncorrected than to apply the McLaughlin-Tiederman type of correction. Therefore, the data are reported here without any correction for velocity bias.

Test Conditions

Jet Exit Flow

Measurements were made in the free-jet flow exiting the nozzles to establish uniformity of the mean velocity profile and the core turbulence levels. Both the conical hot-film probe and the LDV were used to cross check the data. The hot-film measurements also facilitated spectral analysis of the velocity signals. Figure 4 shows the mean-velocity and turbulence-intensity profiles across the jet taken 1.5 mm below the nozzle exit. The jet centerline velocity

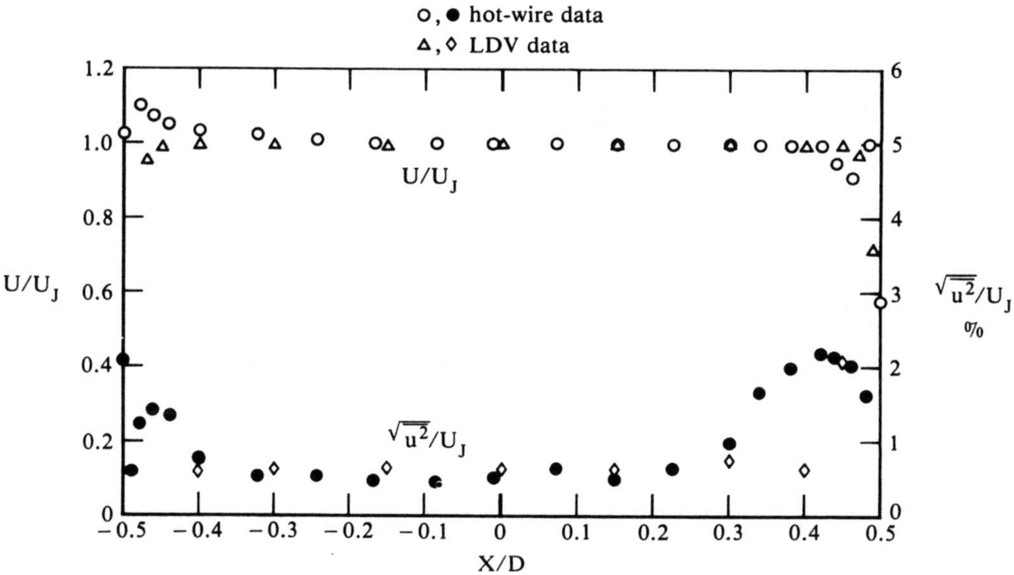

Fig. 4. Mean velocity and turbulence-intensity profiles at the jet exit

at the exit, U_J equaled to 6.71 m/s resulting in a jet Reynolds number of $Re = 1.70 \times 10^5$. The jet flow was uniform with a low level of turbulence in the core region; the shear layers were thin. The agreement between the LDV data and hot-film data was good. Because of the large velocity gradients across the probe volume, which was larger than the shear-layer thickness, the LDV showed higher turbulence levels in the shear layers. The frequency spectrum of the hot-film signal taken at the center of the jet showed no selective or specific peaks indicative of unwanted disturbances.

Test Configurations

Extensive diagnostic flow-visualization studies were conducted for two jets of equal strength impinging normal to the ground at varying nozzle separation distances and heights above the ground. The objective was to select twin-jet impingement configurations having a centrally located, strong isolated fountain rising from the ground plate without interference from the free jets. Based on these flow-visualization studies, a normally impinging, twin-axisymmetric-jet configuration (1) with $S/D = 9.0$ and $H/D = 3.0$ was selected for detailed LDV measurements to provide the basic benchmark data. The fluorescent dye/laser light-sheet technique was used to visualize the corresponding flowfield in a plane connecting the nozzle centerlines as shown in Fig. 5.

Two additional normally impinging, equal strength, twin-jet configurations (2 and 3) with $S/D = 14.0$; $H/D = 3.0$ and $S/D = 9.0$; $H/D = 5.5$ were also selected for LDV measurements to study the influence of geometric parameters on the fountain. Figure 6 shows flowfields visualized in the plane connecting the nozzle centerlines for the configurations 2 and 3. It should be noted here that the straightline distance along the flowpath between the jet exit and the base of the fountain is constant, i.e., $(S/2D + H/D)$ configuration (2) = $(S/2D + H/D)$ configuration (3). In configuration (2), the wall jet develops over a longer distance, and in configuration (3), the free jet develops over a longer distance compared to configuration (1), resulting in different mean velocity and turbulence characteristics for the opposing wall jets that form the fountain. The objective is to study the influence of these configurational changes on the fountain.

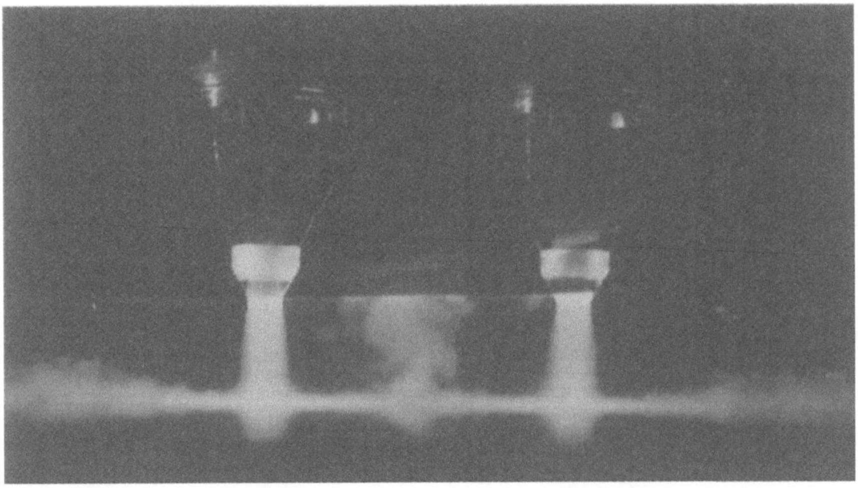

Fig. 5. Visualization of twin-jet impingement flow; configuration (1); $S/D = 9.0$, $H/D = 3.0$

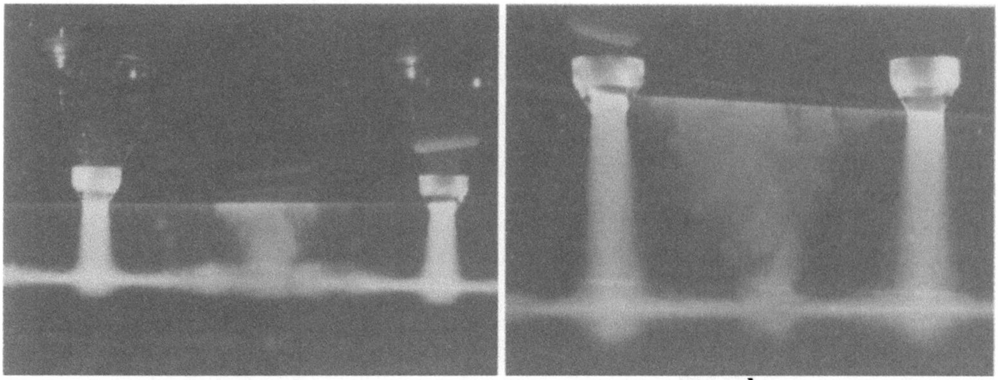

a　　　　　　　　　　　　　　b

S/D = 14.0, H/D = 3.0　　　　　S/D = 9.0, H/D = 5.5

Fig. 6a and b. Visualization of twin-jet impingement flow; **a** configuration (2); $S/D = 14.0$, $H/D = 3.0$; **b** configuration (3); $S/D = 9.0$, $H/D = 5.5$

Experimental Results

The primary experimental data are classified according to the type of twin-jet configuration. All data were taken in the plane of symmetry connecting the nozzle centerlines. Special emphasis was placed on the characterization of the fountain. The spread and decay characteristics of the fountain for the three different configurations are analyzed in a separate Section.

Configuration (1); $S/D = 9.0$, $H/D = 3.0$

Detailed benchmark data on the jet impingement flow and the fountain flow for this configuration are presented. The data were taken at twelve heights ($Z/D = 0.5$ to 2.94) between the nozzle exits and the ground.

Jet Impingement Flow: The mean-velocity and turbulence quantities in the jet were normalized with U_J. Only the data between $Z/D = 1.0$ and the ground are presented to emphasize the jet impingement region. Additional data on the jet are reported in [25]. The region of the jet between the jet exit and $Z/D = 1.0$ is essentially characterized by large-potential core, cross-stream velocities near zero and comparatively slow-developing shear layers. Figure 7 shows the streamwise (U/U_J) and cross-stream (V/U_J) components of the mean velocities across the impinging jet at several streamwise locations, emphasizing the jet impingement region. The streamwise ($\sqrt{\overline{u^2}}/U_J$) and cross-stream ($\sqrt{\overline{v^2}}/U_J$) turbulence intensities and the Reynolds shear stress ($-\overline{uv}/U_J^2$) profiles for the corresponding locations are shown in Figs. 8 and 9, respectively. These data for the jet impingement region are believed to be the first reported; conventional measurement techniques have not been easily applicable to this region.

The measurements in general, and especially the V/U_J data near the ground plate, show a persistent symmetry expected in the flow about the centerline of the jet. The influence of the ground plate extends to a height of $Z/D = 0.75$, where the jet still has a potential core in the streamwise mean-velocity profile and the cross-stream component of the velocity is still close to zero. Below $Z/D = 0.75$, the jet starts deflecting along the ground with a rapid decrease in U and an increase in V. The jet impingement region (Fig. 1) can be observed to extend $0.3\,D$ vertically and $2\,D$ laterally within which bounds both U and V are significant.

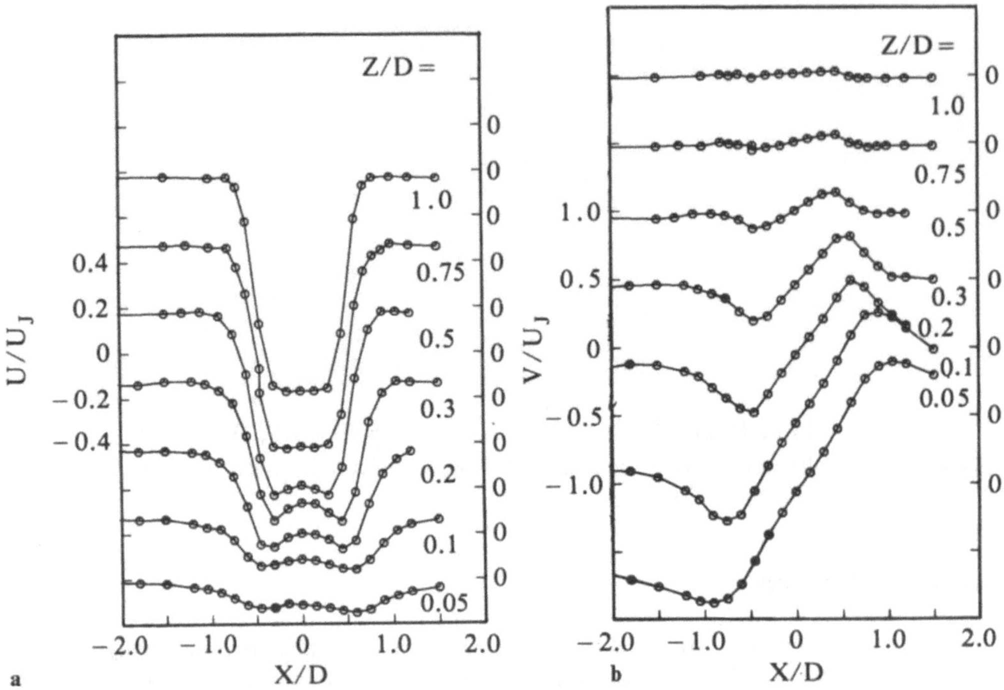

Fig. 7. Variation of the mean velocities (U, V) across the impinging jet in the near ground region for configuration (1); $S/D = 9.0$, $H/D = 3.0$

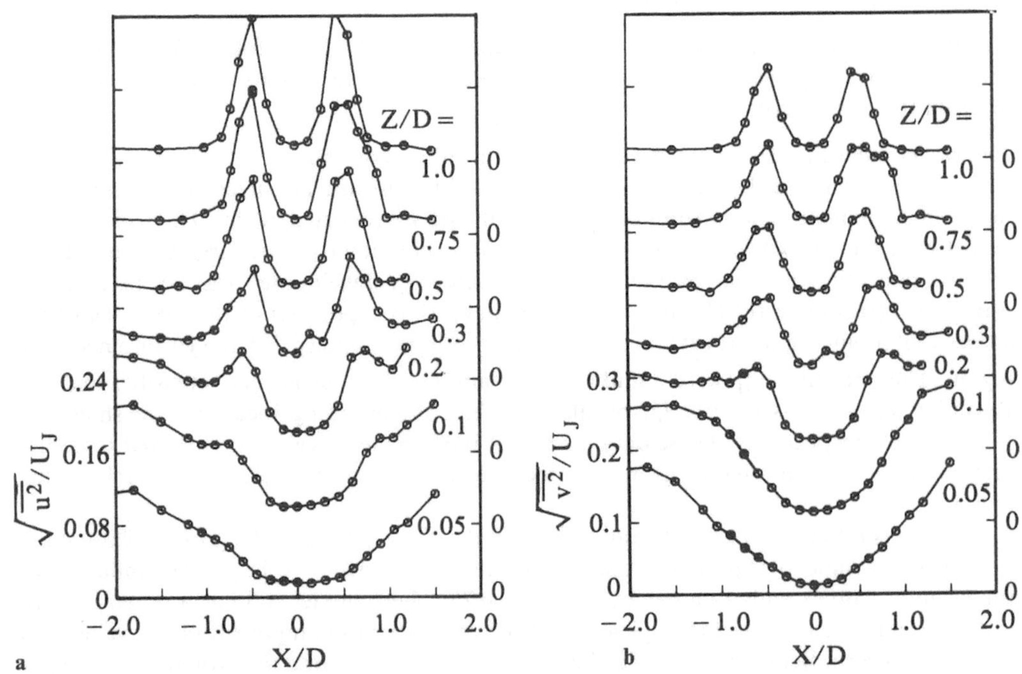

Fig. 8. Turbulence intensities in the near ground region of the impinging jet for configuration (1); $S/D = 9.0$, $H/D = 3.0$

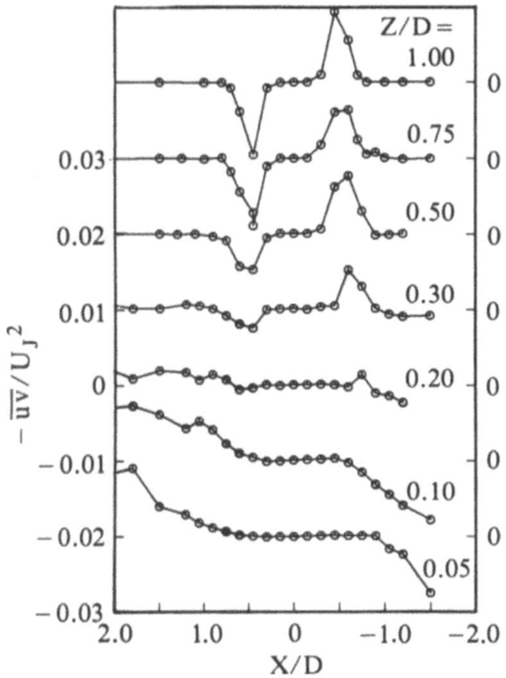

Fig. 9. Distribution of Reynolds shear stress ($-\overline{uv}$) across the impinging jet in the near ground region for configuration (1); $S/D = 9.0$, $H/D = 3.0$

The cross-stream component (V) attains its maximum within the jet impingement region. The variation of V in the central region of the jet can be observed to be linear between the $+ve$ and $-ve$ maxima. The region of constant slope in the distribution of V approximately coincides with the region where U is constant. The slight dip in the U profiles in the central region of the jet near the ground seems typical of jet-impingement flows, as also shown in the data of Donaldson and Snedeker [26].

Figure 8 shows that in the free jet, away from the ground, $\sqrt{\overline{u^2}} \approx 4\sqrt{\overline{v^2}}$; and as the jet deflects near the ground $\sqrt{\overline{u^2}} \approx 2\sqrt{\overline{v^2}}$. The Reynolds-shear-stress profiles (Fig. 9) near the ground show substantial regions of zero shear stress ($Z/D = 0.2$ and below) despite the large velocity-gradients. The reasons for such a behavior are two fold: (1) the velocity gradients might cancel each other due to different signs, (2) it is also possible that the turbulence exhibits a more delayed response to the ground plate than the mean velocities, as also evidenced by the profiles of $\sqrt{\overline{u^2}}$ and $\sqrt{\overline{v^2}}$ near the ground. Limited measurements on the second jet show similar initial profiles and jet development.

Fountain Upwash Flow: Figure 10 illustrates the variations of the mean velocities (U and V) across the fountain. The distributions of V reveal that at $Z/D = 0.75$, the variation in V across the fountain is comparable to that at stations farther downstream and the magnitude of V is relatively close to zero at those stations. It should be observed that even at $Z/D = 0.2$, 0.3, and 0.5, the shape of the U profile resembles that typical of other downstream stations. The fountain formation region within which both U and V are significant (Fig. 1) extends $1.5D$ laterally and $0.5D$ vertically. In this region, $Z/D < 0.5$, $\sqrt{\overline{u^2}}$ was observed to be relatively small because of the stagnating flow. The symmetry in the data distributions about the centerline of the fountain is an indication of the quality of the present measurements; the symmetry of fountain data was a major problem in the majority of the earlier investigations.

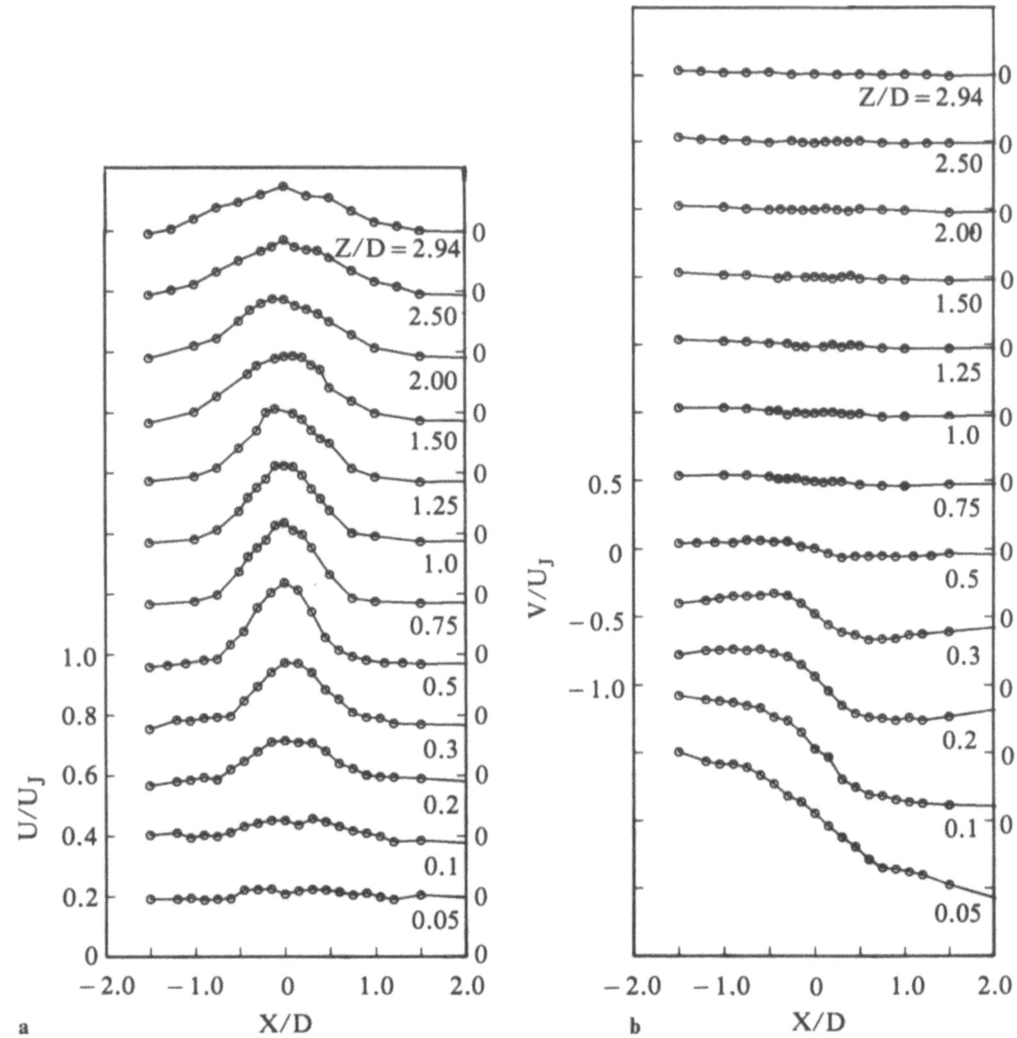

Fig. 10. Variation of the mean velocities (U, V) across the fountain for configuration (1); $S/D = 9.0$, $H/D = 3.0$

Similarity of the Fountain Measurements: Observation of the mean velocity (Fig. 10) and turbulence profiles in the fountain suggested self-similarity. The fountain profiles were normalized with U_{max} and $X_{1/2}$ where U_{max} is the local maximum velocity and $X_{1/2}$, the half-width, is the distance from the center of the fountain to the point where the velocity is $U_{max}/2$. Values of U_{max} and $X_{1/2}$ were obtained by curve-fitting the fountain mean velocity (U) profile with a least square curve of the form

$$U = A + B \exp[-(X - X_0)^2/2\,\delta^2], \tag{1}$$

as suggested by Gilbert [4]. This curve fit gives the symmetry coordinate X_0, the maximum velocity ($A + B$), and the parameter δ. The relation between the fountain half-width ($X_{1/2}$)

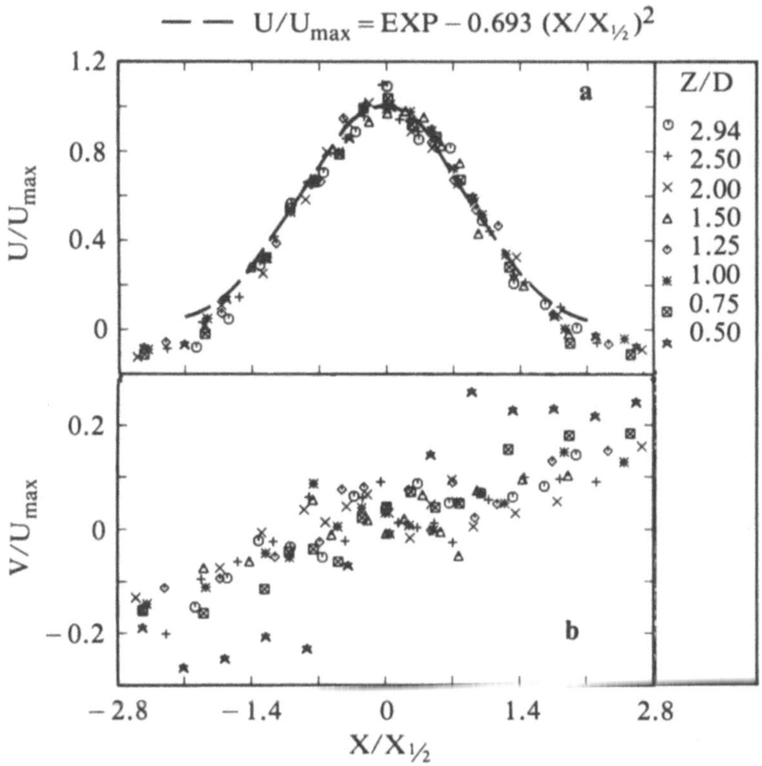

Fig. 11 a and b. Similar profiles of mean velocities (U, V) across the fountain for configuration (1); $S/D = 9.0$, $H/D = 3.0$

and the parameter δ is given in (2)

$$X_{1/2} = \left[2\ \mathrm{LOG}\left(\frac{2B}{B-A}\right)\right]^{1/2}\delta, \qquad \approx 1.177\,\delta \quad (\text{for } B \gg A); \tag{2}$$

also

$$U/U_{max} \approx \exp - 0.693\,(X/X_{1/2})^2 \quad \text{for } B \gg A. \tag{3}$$

The value of the X_0 for $Z/D \geqq 0.5$ was found to be very small establishing once again the inherent symmetry in the present data.

Figure 11 a shows the streamwise fountain velocity (U) profiles shifted to their symmetry point and normalized with respect to U_{max} and $X_{1/2}$, obtained by curve fitting. A striking similarity can be observed in the velocity profiles. Similar observations were made for a two-dimensional fountain in [4]. The profiles below $Z/D = 0.5$ were excluded because they fall definitely into the fountain-formation region, although the profiles at $Z/D = 0.3$ and 0.2 do not differ much from the similarity form shown in Fig. 11 a. The station $Z/D = 0.5$ falls as a borderline case. The profiles reach a similar form within a short distance ($Z/D = 0.5$) above the fountain-formation region, possibly because of greatly enhanced mixing at the base of the fountain.

Figure 11 b shows the variation of the cross-stream velocity (V) across the fountain in non-dimensional coordinates. The relatively large cross-stream velocities on the outer edges

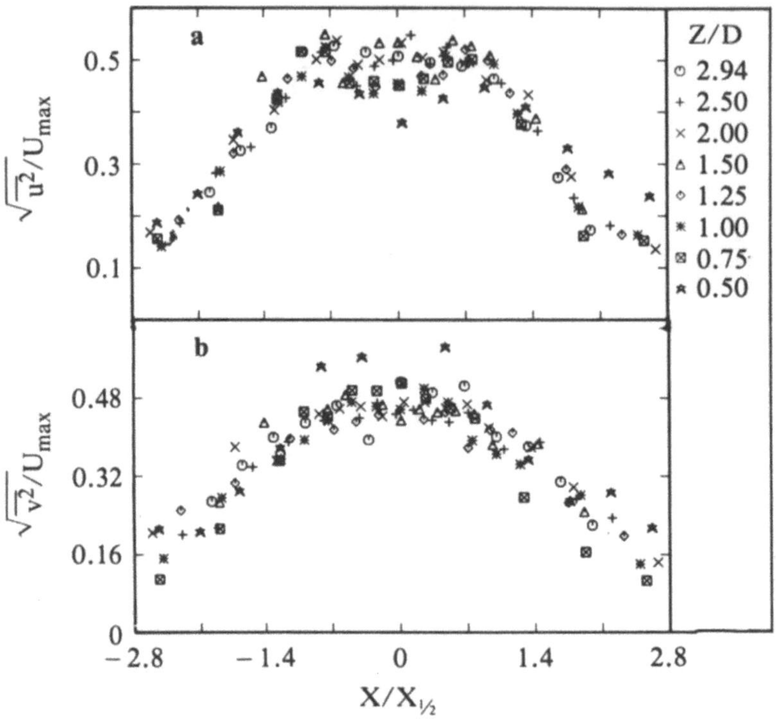

Fig. 12 a and b. Similar profiles of turbulence intensities in the fountain for configuration (1); $S/D = 9.0$, $H/D = 3.0$

of the fountain reflect entrainment. The entrainment velocities show the expected change in their direction from one side of the fountain to the other, and their maxima are of the order of 0.2 U_{max} at a given station. The cross-stream velocities within the fountain were observed to be very small and the expected change in the direction V within the fountain with zero crossing at the center of the fountain, was observed only at few stations because of the difficulty in resolving such small velocities. The values of V at $Z/D = 0.5$ are slightly higher compared to the other downstream stations because of this station's proximity to the fountain formation region.

Figure 12 shows the turbulence intensities across the fountain non-dimensionalized with the similarity variables U_{max} and $X_{1/2}$ obtained by curve fitting. Once again, similarity can be observed in the turbulence intensity profiles at various downstream stations. The profiles of $\sqrt{\overline{u^2}}$ and $\sqrt{\overline{v^2}}$ (Figs. 12a and b) show that they are generally of the same magnitude and shape. The maximum turbulence intensities are around 0.5, based on the local maximum velocity (U_{max}), suggesting the possibility of instantaneous flow reversal; therefore, the LDV is a more suitable instrument than the hot-film anemometer for measuring the highly turbulent fountain flows. The turbulence intensities are about twice of those reported in [4] for a planar fountain.

Figure 13 shows the distributions of Reynolds shear stress $-\overline{uv}$ and the correlation function $-\overline{uv}/\sqrt{\overline{u^2}}\sqrt{\overline{v^2}}$ non-dimensionalized with the similarity variables U_{max} and $X_{1/2}$. The similarity revealed by the Reynolds shear stress ($-\overline{uv}$) data (Fig. 13 a) is indicative of the accuracy and precision of the experiment because shear stress data are particularly sensitive to the measurement technique. The maximum shear stress occurs at $X/X_{1/2} \approx 0.7$ and amounts to $|\overline{uv}/U^2_{max}|_{max} \approx 0.08$, which is three times higher than the value reported in [4] for the planar fountain. Figure 13b shows the similarity exhibited by the correlation

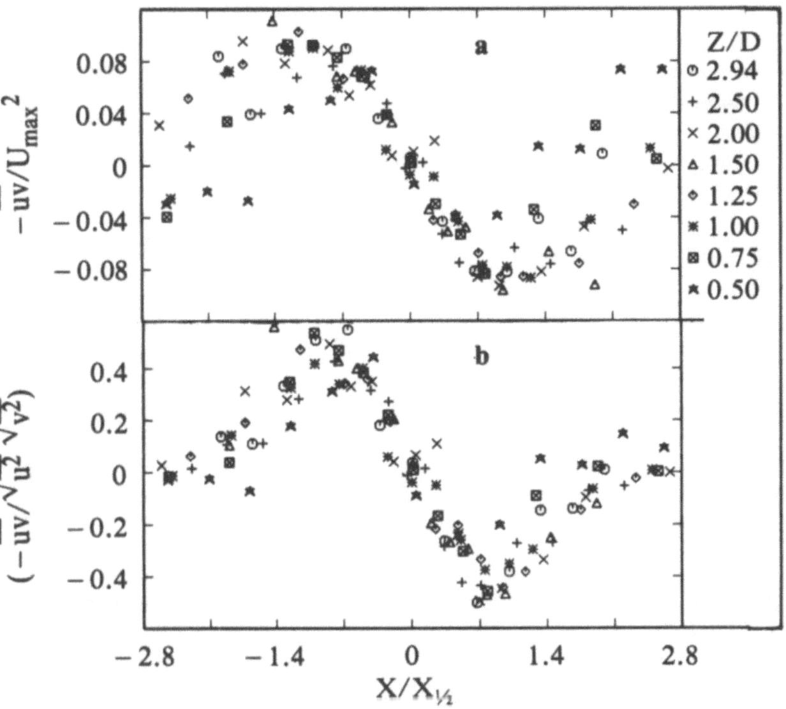

Fig. 13a and b. Similar distributions of (**a**) Reynolds shear stress $(-\overline{uv}/U_{max}^2)$ and (**b**) correlation function $(-\overline{uv}/\sqrt{\overline{u^2}}\sqrt{\overline{v^2}}$ across the fountain for configuration (1); $S/D = 9.0$, $H/D = 3.0$

function. The absolute maximum value of the correlation function is observed to be approximately 0.4. The variation of the Reynolds shear stress and the correlation function between the maxima is linear; they pass through zero at the centerline of the fountain ($X = 0$). The turbulence profiles at $Z/D = 0.5$ did not follow the similarity trends closely because of proximity to the fountain formation region. In general, the turbulence profiles in the fountain were observed to take a longer distance to attain their self-similar shape than the corresponding streamwise mean-velocity (U) profiles.

Configuration (2); $S/D = 14.0$, $H/D = 3.0$

In this configuration (2) (Fig. 6a), the wall jet is allowed to develop over a longer distance, resulting in different wall jet conditions at the fountain end compared to that in configuration (1). The jet-impingement region is essentially same as in configuration (1); hence, no data were taken in that region. Mean velocity and turbulence profiles in the fountain upwash were taken at ten stations ($Z/D = 0.1$ to 3.5) between the jet exit and the ground.

Figure 14 shows selected mean-velocity data for the fountain near the ground. The distributions of the cross-stream velocity (V) in the fountain reveal that the variation V at $Z/D = 1.0$ is comparable to that at further downstream stations and the magnitude of V is relatively close to zero at those stations. The fountain formation region extends $2D$ laterally and $0.75D$ vertically, which is higher than for configuration (1) since the wall jet in configuration (2) is thicker.

Figure 15a shows the streamwise mean velocity (U) profiles in the fountain normalized with U_{max} and $X_{1/2}$. These profiles were obtained by the curve fitting mentioned earlier for configuration (1) using Eq. (1). The data at various stations above the fountain formation region collapse on one another, indicating the similarity behavior once again. The non-

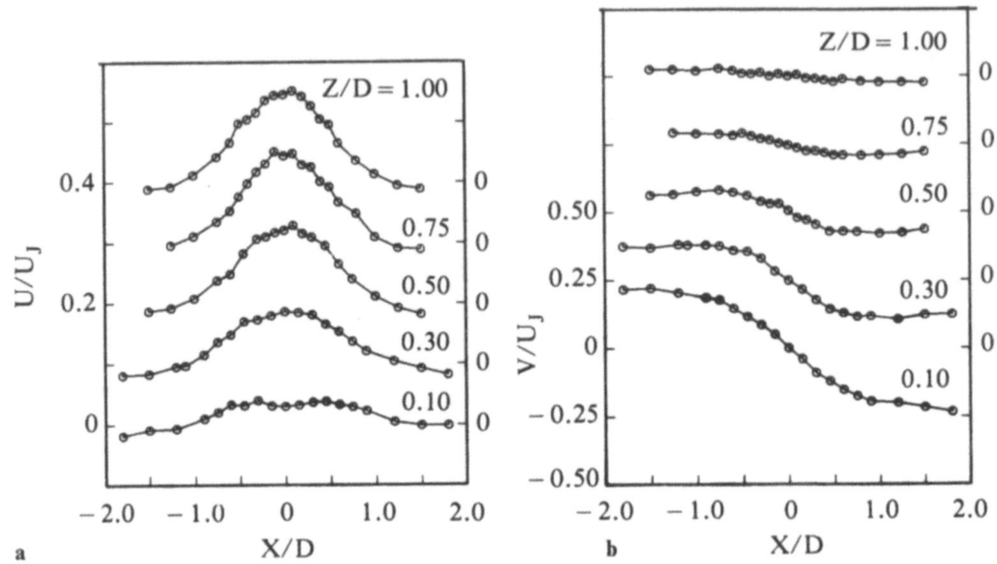

Fig. 14. Variation of the mean velocities (U, V) across the fountain in the near ground region for configuration (2); $S/D = 14.0$, $H/D = 3.0$

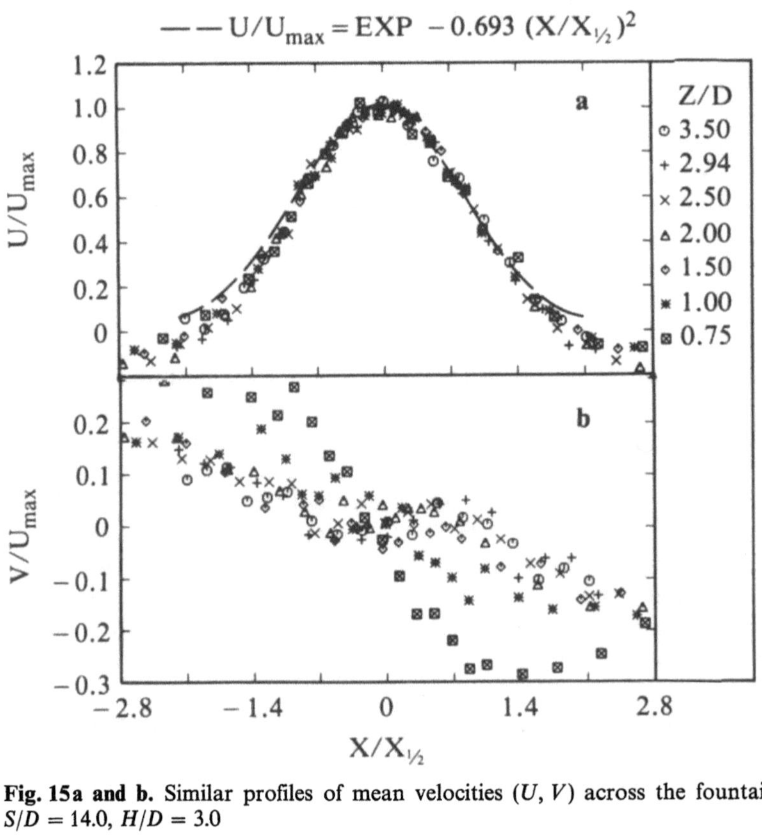

Fig. 15a and b. Similar profiles of mean velocities (U, V) across the fountain for configuration (2); $S/D = 14.0$, $H/D = 3.0$

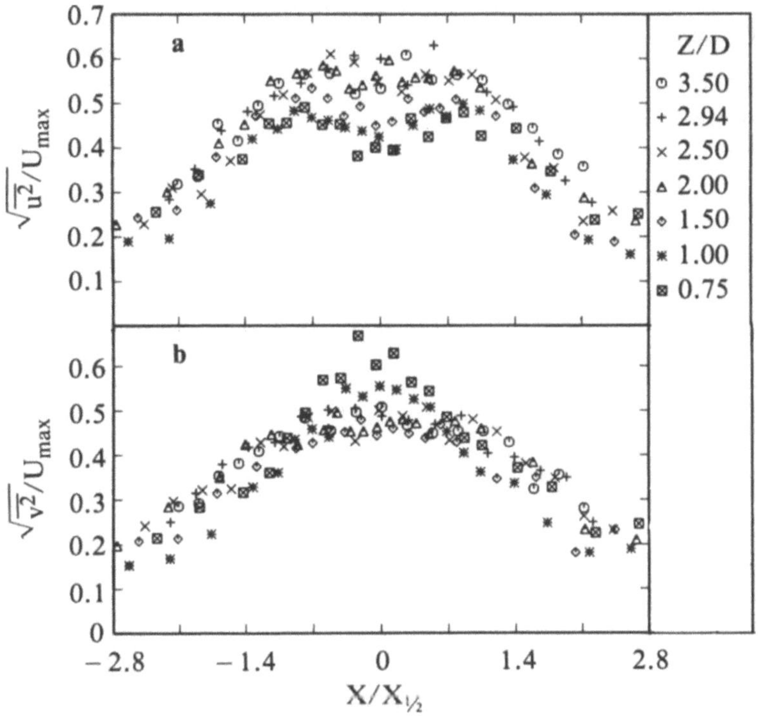

Fig. 16 a and b. Similar profiles of turbulence intensities in the fountain for configuration (2); $S/D = 14.0$, $H/D = 3.0$

dimensional mean-velocity (U) distribution matches that of configuration (1) exactly. It should be noted that the U profiles at $Z/D = 0.3$ and 0.5 follow the similarity distribution in Fig. 15a even though they belong to the fountain formation region. The cross-stream velocity (V) distributions shown in Fig. 15b are similar to those for configuration (1) with maximum entrainment velocities (0.2 U_{max}) of the same magnitude.

Figure 16 shows the turbulence intensities in non-dimensional variables. Figure 17 shows the corresponding variation of Reynolds shear stress and correlation function. The shapes of the non-dimensional turbulence distributions and the respective maximum values are typical of those for configuration (1). The turbulence profiles take a longer distance to reach the similarity form compared to the mean velocity (U) profiles, as evidenced by the turbulence profiles at $Z/D = 0.75$ and 1.0.

Configuration (3); $S/D = 9.0$, $H/D = 5.5$

The impinging free-jet in this configuration (3) (Fig. 6b) is allowed to take a longer distance to develop, resulting in different mean-velocity and turbulent conditions at the fountain end of the wall jet compared to that in configurations (1) and (2). Mean-velocity and turbulence data were taken at seven stations in the jet impingement flow and at twelve stations in the fountain upwash flow ranging from $Z/D = 0.1$ to 4.5.

Figure 18 shows selected data on the jet impingement flow near the ground. The jet impingement region extends 3.0D laterally and 0.5D vertically being larger compared to configuration (1). The influence of the ground, however, extends up to $Z/D = 0.75$ as evidenced by the cross-stream velocity profiles. The dip in U profiles at the center near the ground as observed in configuration (1), cannot be clearly seen here; it seems to be typical only of the interaction of potential core with the ground. Most of the remaining features of the jet impingement region are similar to those observed for configuration (1).

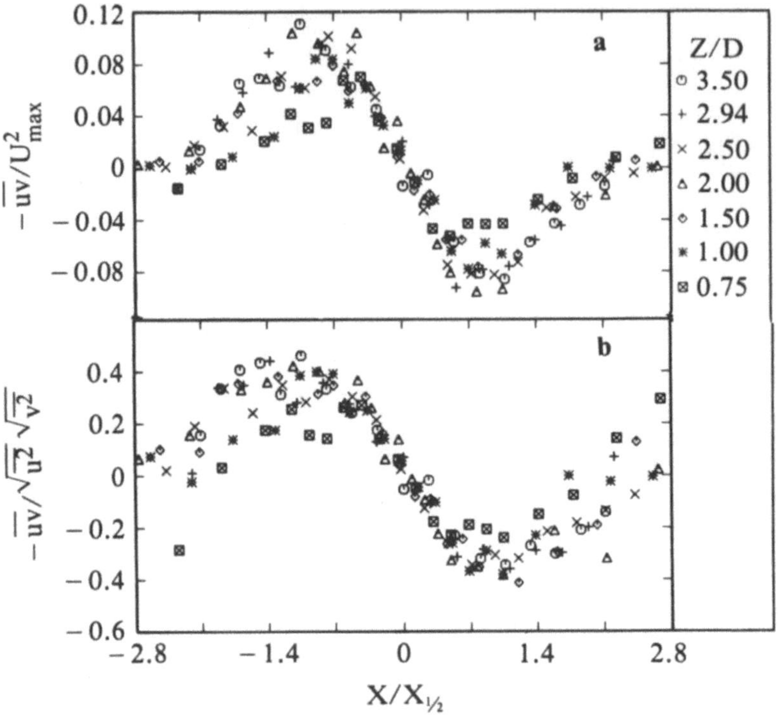

Fig. 17 a and b. Similar distributions of **(a)** Reynolds shear stress ($-\overline{uv}/U_{max}^2$) and **(b)** correlation function ($-\overline{uv}/\sqrt{\overline{u^2}}\sqrt{\overline{v^2}}$ across the fountain for configuration (2); $S/D = 14.0$, $H/D = 3.0$

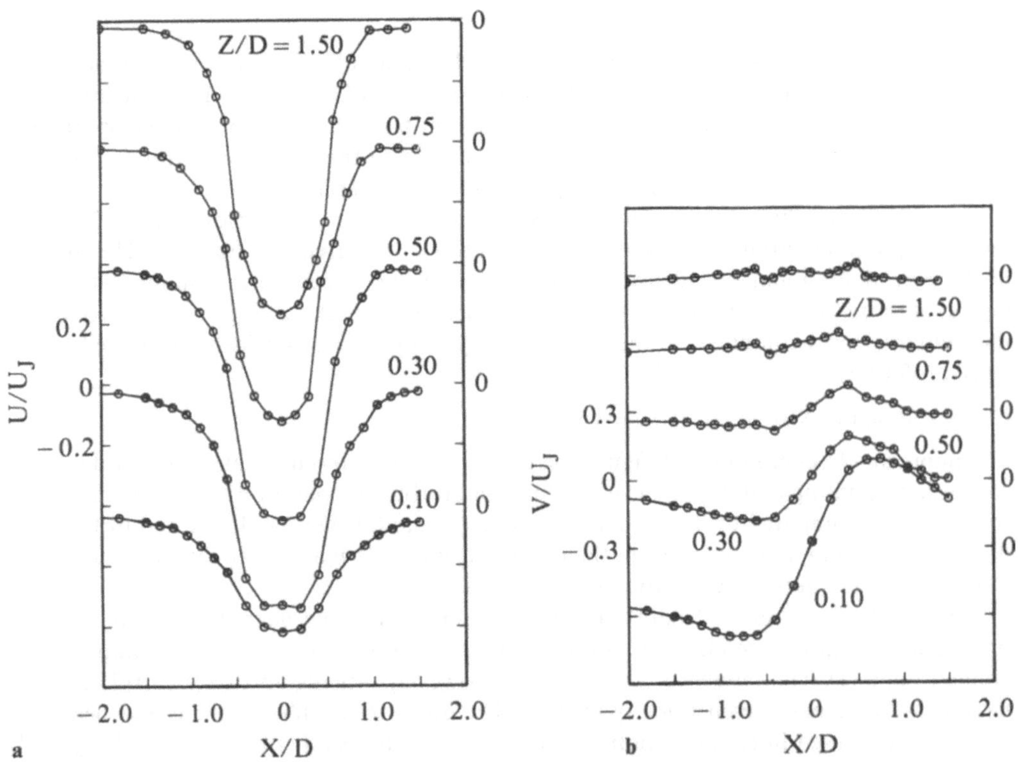

Fig. 18. Variation of the mean velocities (U, V) across the impinging jet in the near ground region for configuration (3); $S/D = 9.0$, $H/D = 5.5$

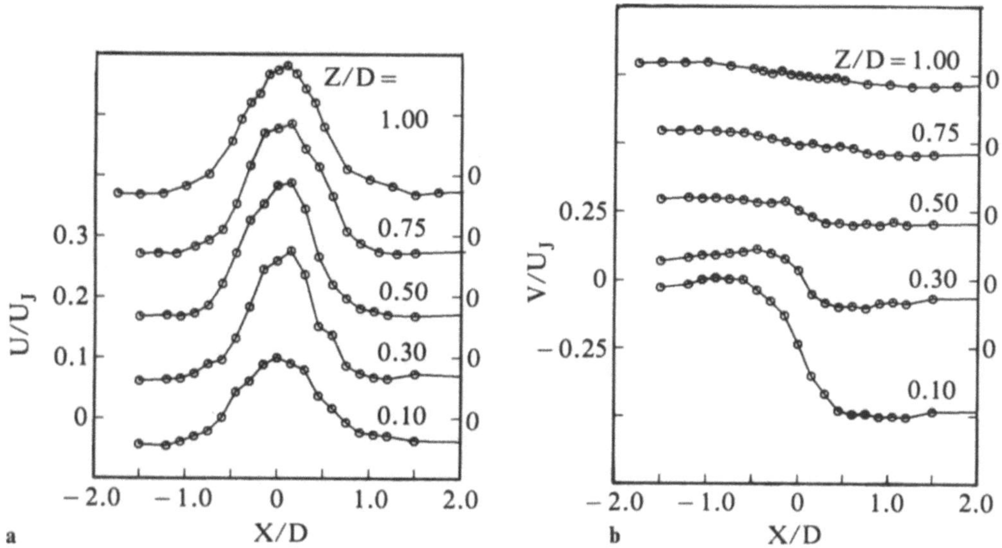

Fig. 19. Variation of the mean velocities (U, V) across the fountain in the near ground region for configuration (3); $S/D = 9.0$, $H/D = 5.5$

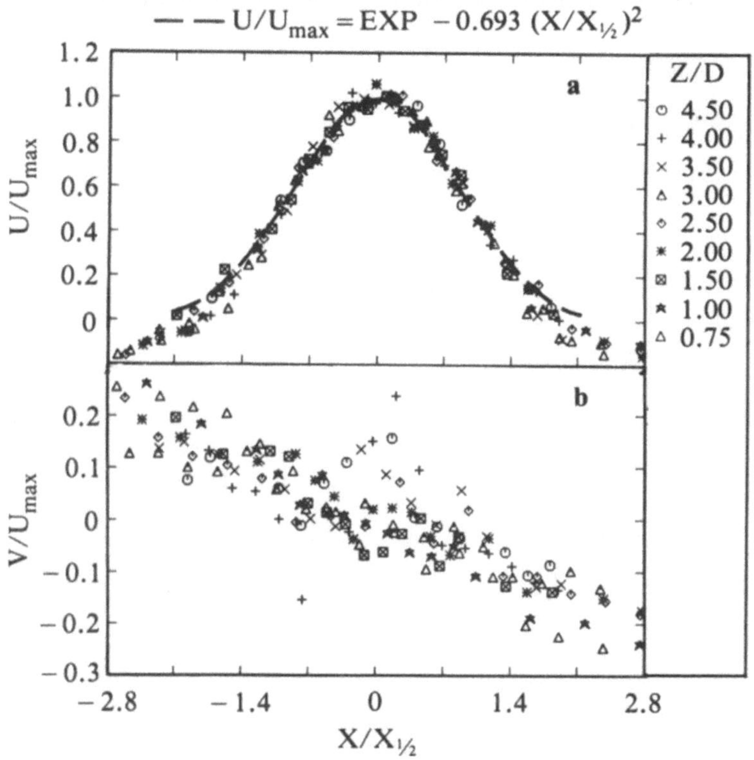

Fig. 20 a and b. Similar profiles of mean velocities (U, V) across the fountain for configuration (3); $S/D = 9.0$, $H/D = 5.5$

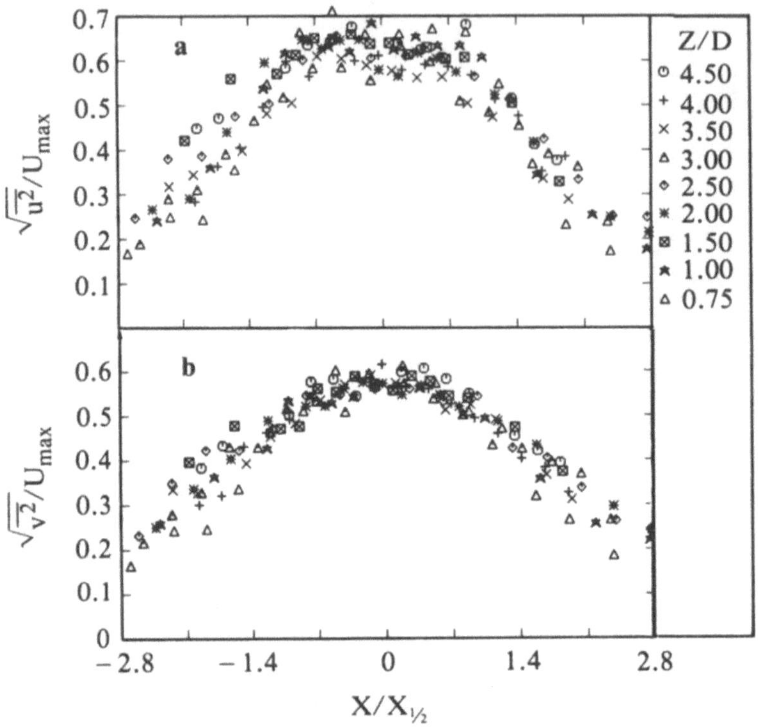

Fig. 21 a and b. Similar profiles of turbulence intensities in the fountain for configuration (3); $S/D = 9.0$, $H/D = 5.5$

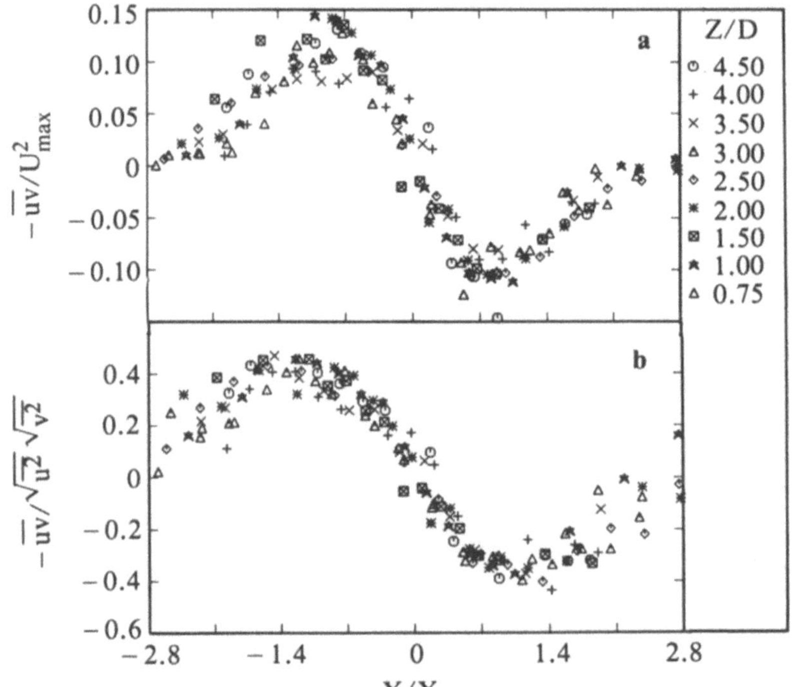

Fig. 22 a and b. Similar distributions of **(a)** Reynolds shear stress ($-\overline{uv}/U^2_{max}$) and **(b)** correlation function ($-\overline{uv}/\sqrt{\overline{u^2}}\sqrt{\overline{v^2}}$ across the fountain for configuration (3); $S/D = 9.0$, $H/D = 5.5$

Figure 19 shows selected mean-velocity profiles in the fountain flow near the ground. Here the fountain formation region extends $0.75D$ vertically and $2D$ laterally, which is the same for configuration (2) but larger than for configuration (1), indicating a thicker approaching wall jet at the fountain end.

Figure 20a shows the streamwise mean-velocity (U) profiles in the fountain non-dimensionalized with U_{max} and $X_{1/2}$ obtained by curve fitting, similar to that for configuration (1) using Eq. (1). Similarity is once again exhibited by these profiles above the fountain formation region, and the non-dimensional profile agrees exactly with those of the earlier configurations. As before, even some of the profiles in the fountain formation region follow the similarity distribution shown in Fig. 20a. The cross-stream mean-velocity distributions shown in Fig. 20b resemble those of earlier configurations. Figure 21 shows the variation of non-dimensionalized turbulence intensities and Fig. 22 shows the corresponding Reynolds shear stresses and correlation functions across the fountain and above the fountain formation region. All of the turbulence quantities exhibit the similarity behavior observed for the earlier configurations; however, the maximum turbulence intensities are relatively higher; $(\sqrt{\overline{u^2}}/U_{max})_{max} \approx (\sqrt{\overline{v^2}}/U_{max})_{max} \approx 0.6$. The maximum Reynolds shear stress is also higher; $|\overline{uv}/U^2_{max}|_{max} \approx 0.10$. The maximum $|\overline{uv}/\sqrt{\overline{u^2}}\sqrt{\overline{v^2}}|$ values still remain the same as those for other configurations. The turbulence in this case does not show a delayed response compared to streamwise mean velocity in attaining self similarity, in contrast to the observations made for the other configurations.

Growth and Decay Characteristics of the Fountain

The growth of the fountain with height above the ground is shown in Fig. 23 indicated by the increase in $X_{1/2}$ with Z for the three different configurations. $X_{1/2}$ is obtained by using (1) to curve fit the streamwise mean velocity (U) profiles. The fountain grows linearly with height above the ground for all three configurations. The growth rates of the fountain can be observed to be equal, $(dX_{1/2}/dZ) = 0.16$ for the different configurations. This spreading rate is higher than that (≈ 0.11) for a radial jet, [27, 28] but lower than that observed in [4]

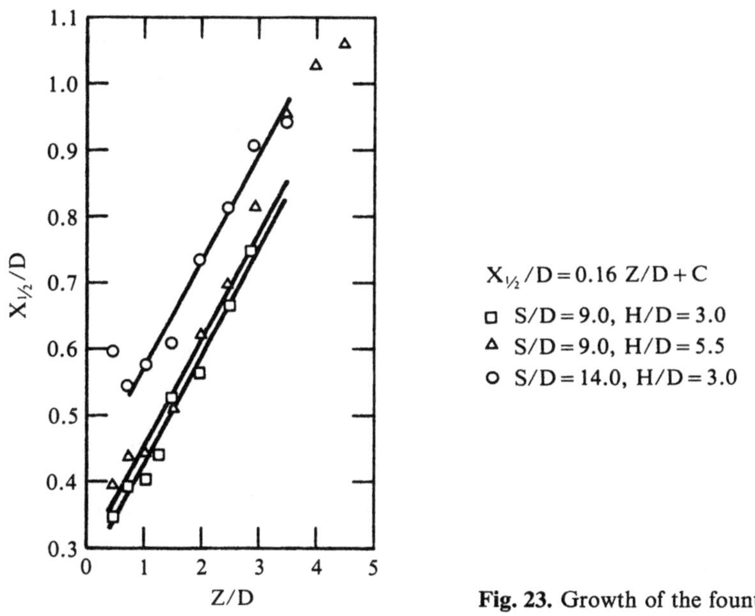

Fig. 23. Growth of the fountain half width $X_{1/2}$

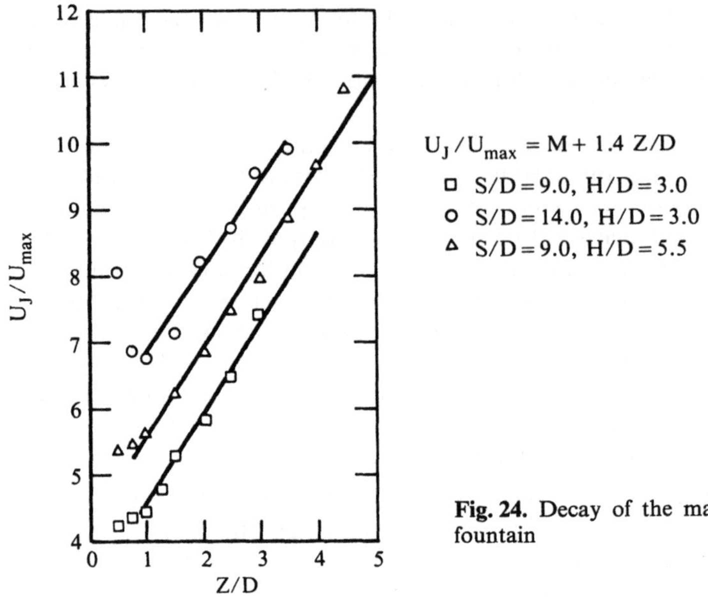

Fig. 24. Decay of the maximum velocity (U_{max}) in the fountain

for a planar fountain. The deviation from the linear distribution of the fountain growth at $Z/D = 3.5$ and beyond for configuration (3) can be partly attributed to the influence of the jet as the fountain and the jet are much closer (Fig. 6) at these heights.

The decay of the maximum velocity (U_{max}) in the fountain is shown in Fig. 24 for the three different configurations. U_{max} is obtained by using (1) to curve fit the streamwise mean velocity (U) profiles. The mean velocity decay can be seen to be inversely proportional to the height above the ground for all the configurations. The rate of decay $d(U_J/U_{max})/d(Z/D)$ can be observed to be equal (1.4) for all three configurations. These velocity decay characteristics are similar to those observed for a radial jet [27] where conservation of momentum dictates velocity decay to be inversely proportional to the downstream distance. Thus, there are indications of similarity between the fountain flow and radial jet. However, additional measurements away from the symmetry plane are needed to confirm this behavior. The recovery of the jet momentum was highest, for configuration (1), and lowest for configuration (2). The lowest momentum recovery for configuration (2), may be attributed to a larger amount of frictional losses on the ground.

Conclusions

1. Mean velocity and turbulence measurements are reported on three axisymmetric twin-jet impingement configurations with fountain formation, obtained by using the laser Doppler velocimeter. Emphasis is placed on the fountain upwash region, for which detailed measurements were taken. Data are also reported on the impinging jet. Measured turbulence quantities include turbulence intensities, Reynolds shear stresses, and correlation functions.

2. The distributions of mean velocity and turbulence quantities across the fountain exhibit remarkable self similarity for all configurations when non-dimensionalized with proper parameters except in the fountain formation region.

3. The fountain was observed to grow linearly; the growth rates for all the three configurations were identical.

4. The decay of the streamwise maximum velocity in the fountain was found to be inversely proportional to the height above the ground; the decay rates for all three configurations were identical.

5. The fountain flow was found to be highly turbulent (50–60% turbulence intensity), requiring use of the LDV measurement technique.

6. There are indications of similarity between the fountain flow and radial jet. However, additional measurements on the flow away from the symmetry plane connecting the nozzle centerlines should be made with a three-component LDV to confirm this behavior.

7. The turbulence and the near-ground mean-velocity measurements in the three-dimensional fountain are believed to be the first data reported on such flows.

Acknowledgement. This work was done under the McDonnell Douglas Independent Research and Development program. The author sincerely acknowledges the support provided by J. C. Kroutil in the design of the experiment and R. E. Price during data acquisition.

References

1. Kotansky, D. R., Glaze, L. W.: The Effects of Ground Wall-Jet Characteristics on Fountain Upwash Flow Formation and Development. Report ONR-CR212-216-1F (15 June 1980)
2. Jenkins, R. C., Hill, W. G., Jr.: Investigation of VTOL Upwash Flows Formed by Two Impinging Jets, Grumman Research Department Report RE-548 (November 1977)
3. Kind, R. J., Suthanthiran, K.: The interaction of two opposing plane turbulent wall jets. AIAA Paper 72-211, AIAA 10th Aerospace Sciences Meeting, San Diego, CA (Jan. 1972)
4. Gilbert, B. L.: An Investigation of Turbulence Mechanisms in V/STOL Upwash Flow Fields, Grumman Aerospace Report Re-688 (1984)
5. Foley, W. H., Finley, D. B.: Fountain jet turbulence. AIAA Paper No. 81-1293, AIAA Fluid and Plasma Dynamics Conf., Palo Alto, CA (June 1981)
6. Siclari, M. J., Hill, W. G, Jr., Jenkins, R. C., Migdal, D.: VTOL in-ground effect flows for closely spaced jets. AIAA Paper No. 80-1880 (Aug. 1980)
7. Migdal, D., Hill, W. G., Jr., Jenkins, R. C., Siclari, M. J.: VTOL In-Ground Effect Flows for Closely Spaced Jets. NASA CR-152321 (Dec. 1979)
8. Hall, G. R., Rogers, K. H.: Recirculation Effects Produced by a Pair of Heated Jets Impinging on a Ground Plane. NASA CR 1307 (May 1978)
9. Karemaa, A., Smith, C. W., Weber, H. A., Garner, J. E.: The Aerodynamic and Thermodynamic Characteristics of Fountains and Some Far-Field Temperature Distributions. Report ONR-CR 212-237-1F (July 1978)
10. Lummus, J. R.: The Criticality of Engine Exhaust Simulations on V/STOL Model – Measured Ground Effects. Report ONR-CR212-255-1F (Aug. 1979)
11. Siclari, M. J., Hill, W. G., Jr., Jenkins, R. C.: Investigation of stagnation line and upwash formation. AIAA Paper No. 77-615, AIAA/NASA, Ames V/STOL Conference (June 1977)
12. Durst, F., Melling, A., Whitelaw, J. H.: *Principles and Practice of Laser-Doppler Anemometry* (Academic, New York 1976)
13. Durrani, T. S., Greated, C. A.: *Laser Systems in Flow Measurement* (Plenum, New York 1977)
14. Saripalli, K. R., Kroutil, J. C.: A novel experimental facility for conducting jet-impingement studies related to VTOL aircraft. AIAA Paper 85-0052 (1985)
15. Saripalli, K. R.: Visualization of multi-jet impingement flow. AIAA Paper 81-1364 (July 1981)
16. Saripalli, K. R.: "Visualization Studies of Jet Impingement Flows at McDonnell Douglas Research Laboratories," Proceedings of the Third Intl. Symp. on Flow Visualization, Ann Arbor (Sept. 1983)
17. Saripalli, K. R.: Visualization of multi-jet impingement flow. AIAA J. **21**, 483 (1983)
18. McLaughlin, D. K., Tiederman, W. G.: Biasing corrections for individual realization laser anemometer measurements in turbulent flows. Phys. Fluids **16** (12), 2082–2088 (Dec. 1973)
19. Durão, D. F. G., Whitelaw, J. H.: Relationship between velocity and signal quality in a laser-Doppler anemometer, J. Phys. E: Sci. Instrum. **12**, 47–50 (1979)
20. Adams, E. W., Johnston, J. P., Eaton, J. K.: Experiments on the Structure of Turbulent Reattaching Flow. Report MD-43; Dept. of Mech. Eng. Stanford University (May 1984)

21. Adams, E. W., Eaton, J. K.: "An LDA Study of the Backward-Facing Step Flow, Including the Effects of Velocity Bias," Intl. Symp. on Laser Anemometry, ASME Winter Annual Meeting (Nov. 1985)
22. Johnson, D. A., Modarress, D., Owen, F. K.: An experimental verification of laser-velocimeter sampling bias and its correction. J. Fluids Eng. **106**, 5–12 (1984)
23. Nakayama, A.: Measurements of separating boundary layer and wake of an airfoil using laser Doppler velocimetry. AIAA Paper 85-0181 (Jan. 1985)
24. Panel Report on Velocity Bias, Intl. Symp. on Laser Anemometry, ASME Winter Annual Meeting (Nov. 1985)
25. Saripalli, K. R.: Laser Doppler velocimeter measurements in a 3-D impinging twin-jet fountain flow. AIAA Paper 85-4036 (Oct. 1985)
26. Donaldson, C., Snedeker, R. S.: A study of free jet impingement. Part I. Mean properties of free and impinging jets. J. Fluid Mech. **45**, 281 (1971)
27. Rajaratnam, N.: *Turbulent Jets,* Developments in Water Science, Vol. 5 (Elsevier, Amsterdam 1976)
28. Witze, P. O., Dwyer, H. A.: "Impinging Axisymmetric Turbulent Flows: The Wall Jet, The Radial Jet and Opposing Free Jets," First International Symposium on Turbulent Shear Flows, April 1977, University Park, PA

Part III

Wall Flows

Introduction

P. Bradshaw

Department of Aeronautics, Imperial College, London, England

The questions usually asked about turbulent boundary layers fall into three main groups: "what happens in the inner layer?", "what happens in the outer layer?", and "what is the response to perturbations?". Between them, the papers in this section address most of these questions.

One of the cornerstones of fluids dynamics is the "law of the wall", which states that the turbulent flow near a solid surface depends only on the distance from the solid surface, the surface shear stress (whose value is of course determined by outer flow conditions), the fluid properties, and the surface roughness if any. A corollary is the universality of von Karman's constant, $\varkappa$: another corollary is that the spectra of the shear-stress-producing part of the turbulent motion should also scale on y and on the shear stress or friction velocity. Any calculation method that is compatible with law-of-the-wall scaling – and virtually all of them are – ensures that something between half and two-thirds of the rise in velocity from zero at the solid surface to the free-stream value at the edge of the boundary layer will be accounted for, so that the dependence of predictions on the outer-layer model is comparatively small. This is the reason why quite poor outer-layer models give good results in simple, unperturbed, attached boundary layers at high Reynolds numbers. Near separation, or in strongly-perturbed flows, or on surfaces whose roughness length scale is not very small compared to the boundary layer thickness, the law of the wall becomes suspect. Unfortunately, the subgrid-scale models opf turbulence used in large-eddy simulations are tested most severely in the inner layer, where an increasing fraction of the shear stress is actually carried by the subgrid-scale motion. Full turbulence simulations, which resolve the whole spectrum of the turbulence, can in principle be used to derive and test subgrid-scale models for large eddy simulations, but it is currently not clear how small the grid size in full turbulence simulation must be in order to resolve all the turbulent eddies in the viscous sublayer. On all counts, therefore, measurements of inner-layer structure are still needed and the present papers provide welcome input.

The paper by Acharya and Escudier discusses wall roughness produced by an expanded-metal grid. The mesh size (major axis 6 mm) is not small compared to the boundary layer thickness, 44 mm, and is in fact of the order of the expected thickness of the inner layer. The equivalent sand roughness, 4.4 mm, is just one-tenth of the boundary layer thickness. It is therefore perhaps not entirely surprising that the von Karman constant should be slightly different from the conventional value of 0.41. Indeed, it seems possible that this kind of diamond-shaped cavity roughness may exhibit the "*d*-type" behaviour of Perry et al. (1969) so that its effects depend on the overall width of the flow and not merely on inner-layer conditions. The authors point out a rather large discrepancy between their carefully-executed measurements of surface shear stress and the value deduced from using Hama's (1954) velocity-defect profile. Escudier and the present writer later compared the Hama profile with the well-accepted Coles law of the wake, and found that the Hama profile implies a "wake parameter" P of only 0.47, while the value of the wake parameter P in Fig. 3a is

Turbulent Shear Flows 5

about 0.65. Coles' preferred value is 0.55, although the results of Erm et al. asymptote to at least 0.6 at high Reynolds number and several other experiments suggest that 0.55 is an underestimate. Using $P = 0.6$ and $\kappa = 0.41$, the value of c_f inferred from the profile in Fig. 3a is only about 6% higher than Acharya and Escudier's floating-element measurements. Therefore the paper implies that Coles' defect law, with $P = 0.6$ say, should be applicable with reasonable accuracy to rough walls as well as smooth, even for roughness whose effective height is as large as a tenth of the boundary layer thickness. Evidently Hama's profile is somewhat out of line with more modern data correlations. Acharya and Escudier also join with Perry (paper presented at the Fourth Symposium) and others in pointing out the unreliability of cross-hot-wire probes with the conventional wire sweep angle of only 45°, especially in flows where the turbulence intensity is as large as 25% of the mean velocity [compare Figs. 3a and 4a at $y/\delta = 0.1$].

Erm, Smits and Joubert consider a perturbation of the outer layer, namely that caused by low bulk Reynolds number. (It is generally agreed that the law of the wall is unaffected by bulk Reynolds number unless the Reynolds number is so low that streamwise deceleration of the flow near the wall leads to large stress gradients in the inner layer.) The variation of wake parameter with Reynolds number obtained by Erm et al. agrees very well with that of Coles, based on a survey of all data available in 1962. Their spectra show that inner-flow scaling, on the friction velocity and the distance from the surface, is not adequate, at least for the u-component fluctuations. At low wave number, the spectra collapse better on outer-flow variables, implying a large contribution from the "footprints" of outer layer eddies on the flow near the wall. At high wave number, viscosity is significant and only Kolmogorov scaling can collapse the spectra. However, with charity, it can be deduced that the shear-stress-producing motion does scale on inner-flow variables, as it must do if von Karman's constant is really to be constant.

Durst, Jovanovic and Kanevce present some very interesting data fits for the moments of the probability distribution of the u-component velocity fluctuation. The data fits are somewhat more sophisticated than the usual Gram-Charlier weighting of a Gaussian distribution. Naturally, the authors are not able to provide a simple explanation of the close relation between probability-distribution moments of different order, but empirical results like that in their Fig. 7, showing the relation between flatness factor and skewness, should be very useful. The authors went to considerable trouble to avoid the effects of heat conduction to the wall and other wall-proximity effects on the hot-wire signals (see Bhatia et al., 1982) but it is not immediately clear that their wire lengths are sufficiently small to avoid resolution difficulties in the viscous sublayer. Their measurements very close to the wall (Fig. 3) imply a limiting rms u-component fluctuation of 0.23 of the mean velocity, while Table 1 of the present Introduction shows values as large as 0.4: evidently more discussion is needed.

The limiting value of $\sqrt{\overline{u'^2}}/U$ at a solid surface is the same as the ratio of rms surface shear stress (strictly, the x-wise component only) to the mean surface shear stress: most of the values in Table 1 (based on the MSc project report of Cheung, Imperial College, 1985) are nominally surface shear stress measurements. Simple inner layer scaling implies that this quantity is an absolute constant, and pressure fluctuations imposed by the outer layer should not cause great changes from one case to another. The range of values in Table 1 is enormous, and the lower values immediately suggest poor spatial resolution or poor frequency response. Of course, it is not certain that the largest value is also the best, but (1) the spanwise length of a hot wire or wall stress probe must be less about 25 wall units if it is not to attenuate the streaky structure in the sublayer, while a 1 mm long hot wire in a boundary layer exceeds this limit at a *free stream* speed of only about 10 m/s. (2) it is well known that hot films on a solid surface have a high-frequency response significantly lower than the

Table 1. Intensity of wall shear stress fluctuations obtained with different methods (W. C. Cheung, MSc project report, Imperial College, 1985)

Authors	Method/flow	$\sqrt{\overline{\tau_w'^2}}/\tau_w$
Mitchell and Hanratty (1966)	electrochemical, aqueous soln., pipe flow	0.32
Sandborn (1979)	hot-wire, air zero p.g. t.b.l	0.3
Sirkar and Hanratty (1970)	electrochemical, aqueous soln. channel flow	0.29
Blinco and Simons (1974)	hot-film, water, zero p.g. t.b.l., (flush mounted)	0.244 to 0.394
Eckelmann (1974)	hot-film, oil, channel flow (flush mounted)	0.24
Sreenivasan and Antonia (1977)	hot-film, air, channel flow (flush mounted)	0.25
Madavan et al. (1985)	hot-film, water, zero p.g. t.b.l., (flush mounted)	0.2
Morrison (1983)	hot-wire, air, channel flow (using the gauge of Ajagu et al.)	0.2
Castro, Bradbury and Dianat	pulsed-wire, air, zero, p.g. t.b.l.	0.37
Durst et al.	hot wire air, p.g. t.b.l.	0.23
Spalart and Leonard	full turbulence simulation – t.b.l. in range of p.g.	0.36–0.42

mean calibration implies, because of heat transfer to the substrate: for a detailed review see Mathews and Poll (1985). Figure 3 of Durst et al. gives one of the lowest values of $\sqrt{\overline{u'^2}}/U$ but agrees qualitatively with some simulation results in implying that the maximum value of $\sqrt{\overline{u'^2}}/U$ occurs away from the wall. Clearly, a sensor mounted at, say, $y^+ = 3$ would indicate a *higher* value than a true surface measurement. However, in Spalart and Leonard's simulation the nearest point to the wall was typically at $y^+ = 0.08$, so their very high "wall" values of rms u'/u cannot be blamed on the non-linearity shown in Fig. 3 of Durst et al. (Their z-wise mesh spacing was 6.5 wall units, implying better spanwise resolution than most of the hot-wire measurements.)

The papers by Kim and by Spalart and Leonard mark the coming-of-age of computer simulations of turbulence. Both papers present the results of full turbulence simulations, without the use of any subgrid scale model: Kim's paper produces a quite spectacular demonstration of the evolution of a horseshoe vortex structure, which could in principle have been obtained by flow visualization experiments but which can be achieved much more straightforwardly on a computer. Clearly, the rather pitifully limited statistics obtainable with hot wire probes could also have been generated by the computer simulation. It must be admitted that Kim's calculations *started* with an ensemble-averaged vortical structure, obtained from the author's previous simulation work by an eduction process specially designed to reveal horseshoe vortices; and, as has been well pointed out by Sato (1984), selective quotation is at least as pernicious in turbulence studies as in literary criticism! However, the work of Moin and Kim (1985) and collaborators leaves no doubt about the prevalence of horseshoe vortices in all kinds of turbulent shear flow. Kim's paper also demonstrates, rather decoratively, the usefulness, and indeed indispensability, of computer graphics in presenting three-dimensional results on two-dimensional paper.

The paper by Spalart and Leonard, like that of Erm et al., considers low-Reynolds-number turbulent boundary layers: however, it could be argued that the latter authors did it on purpose whereas the former couldn't help it – that is, full turbulence simulations are at present confined to rather low bulk Reynolds numbers. With this limitation, the extension of turbulence simulations to a real spatially-growing flow, without making impossible demands on computing facilities, is a considerable achievement. The process by which the instantaneous velocity field at the downstream end of the computational domain is reduced in length scale, rescaled in amplitude, and fed back in at the upstream end is not uncontroversial but seems unlikely to be critical. A possible exception is the separating boundary layer, "Stratford's flow", case 9, where the two-dimensional structure is rather like that obtained by Metcalfe et al. (this volume) in a simulation of a temporally-growing mixing layer with deliberate two-dimensional forcing. However Spalart and Leonard obtained this flow by starting with a fully three-dimensional field at $t = 0$ and gradually increasing the pressure gradient until the surface shear stress fell to zero: that is, there was no forcing as such, only the above-mentioned feedback. As the authors say, experimental verification of the two-dimensional structure is necessary.

If full turbulence simulation on the best present day computers can only just manage to deal with the very restricted class of low-Reynolds-number equilibrium flows, it is clear that large-eddy simulation, which uses a coarser grid and represents the sub-grid-scale turbulence by an empirical model, offers the only hope of computer simulation of more practical flows, and it is to be hoped that the difficulties of sub-grid-scale modelling will not encourage people to abandon large eddy simulation for the aesthetically more pleasing full turbulence simulations.

An example of a flow that is currently inaccessible to time-dependent computer simulation is the boundary layer with an imbedded streamwise vortex, investigated experimentally by Westphal, Eaton and Pauley and calculated by Liandrat, Aupoix and Cousteix. However, even experiments on these flows would be almost unthinkable without the help of modern digital data processing and computerized data logging. Westphal et al., whose measurements unfortunately do not include the $\overline{vw}$ shear stress appearing in the longitudinal vorticity equation, show that the tendency of a vortex near a solid surface to flatten out is more noticeable in an adverse pressure gradient, where $\partial v/\partial y$ is positive, contrary to what might be expected. It is fairly clear from the studies of Liandrat et al. that even the best current Reynolds-stress transport models are not entirely satisfactory, even for the simplest possible case of vortices in constant-pressure boundary layers. This reviewer, who formulated one of the earliest shear-stress transport models about 20 years ago, is not altogether sure that Reynolds-stress models will reach engineering standards of accuracy and reliability in flows like these before they are superseded by the next generation of large eddy simulation methods.

Finally, Castro et al. discuss the response of the pulsed-wire anemometer as used for skin-friction measurement. This technique should be useful for measurements of skin-friction fluctuations as well as mean values, providing that the spanwise length of the pickup wires is less than about 30 wall units and providing that the effects of wall proximity that plague hot wires can be discounted. In view of the great current interest in unsteady and/or separated flows, and in view of the difficulties of making laser doppler velocimeter measurements close to a solid surface, it seems likely that this extension of the pulsed-wire technique will be much used in the next few years.

As a group, these papers show that even experiments on simple flows still have much to contribute to our knowledge of turbulence, although eddy simulations are starting to be competitive if not cost-effective. For the more complex flows, the experimenter's livelihood seems even more secure, even if the main application of experimental results, the development of Reynolds-average turbulence models for engineering calculations, is progressing at a disappointingly slow rate.

References

Ajagu, C. O., Libby, P. A., LaRue, J. C. (1982): Modified gauge for time-resolved skin-friction measurements. Rev. Sci. Instrum. **53**, 1920

Bhatia, J. C., Durst, F., Jovanovic, J. (1982): Correction of hot wire anemometer measurements near walls. J. Fluid Mech. **122**, 411
Blinco, P. H., Simons, D. B. (1974): Characteristics of turbulent boundary shear stress. J. Eng. Mech. Div., Proc. ASCE **100**, EM2, 203
Eckelmann, H. (1974): The structure of the viscous sublayer and the adjacent wall region in a turbulent channel flow. J. Fluid Mech. **65**, 439
Hama, F. R. (1954): Boundary layer characteristics for smooth and rough surfaces. Trans. Soc. Naval Arch Marine Engin. **62**, 333
Madavan, N. K., Deutsch, S., Merkle, C. L. (1985): Measurements of local skin friction in a microbubble-modified turbulent boundary layer. J. Fluid Mech. **156**, 237
Mathews, J., Poll, D. I. A. (1985): The theory and application of heated films for the measurement of skin friction. Cranfield Inst. of Tech. CoA Report 8515
Mitchell, J. E., Hanratty, T. J. (1966): A study of turbulence at a wall using an electrochemical wall shear stress meter. J. Fluid Mech. **26**, 195
Moin, P., Kim. J. (1985): The structure of the vorticity field in turbulent channel flow. Part 1. Analysis of instantaneous fields and statistical correlations. J. Fluid Mech. **155**, 441
Morrison, J. F. (1983): Unpublished work at Imperial College
Perry, A. E., Schofield, W. H., Joubert, P. N. (1969): Rough wall turbulent boundary layers. J. Fluid Mech. **37**, 383
Sandborn, V. A. (1979): "Surface Shear Stress Fluctuations in Turbulent Boundary Layers." Second Symposium on Turbulent Shear Flows (Imperial College, London) p. 361
Sato, H. (1984): Cognition and description of patterns in turbulent flows. Proc. 2nd Asian Congr. Fluid Mechanics (VNU Science Press, Utrecht)
Sirkar, K. K., Hanratty, T. J. (1970): Relation of turbulent mass transfer to a wall at high Schmidt numbers to the velocity field. J. Fluid Mech. **44**, 589
Sreenivasan, K. R., Antonia, R. A. (1977): Properties of wall shear stress fluctuations in a turbulent duct flow. J. Appl. Mech. **44**, 389

Turbulent Flow Over Mesh Roughness

M. Acharya and M. P. Escudier

Brown Boveri Research Center, 5405 Baden, Switzerland

Abstract

The surface shear stress, mean velocity, turbulence intensities and Reynolds shear stress have been measured in the boundary layer developing over a mesh type rough surface. The results confirm the conclusions of Perry et al. (1983) than an improved measurement of the Reynolds shear stress in the near-wall region is obtained by increasing the included angle of the X-wire probe from 90° to 120°. Direct measurement of the wall shear stress using a floating element also confirm the conclusion of Perry et al. that the value estimated from the Hama defect law is about 20% too high. The roughness function $\Delta u/u_\tau$ determined from mean velocity profiles indicates an equivalent sand grain roughness for the mesh surface of about ten times the centre line average roughness length.

Nomenclature

A, B	Constants in hot-wire calibration equation	$\overline{u'v'}$	Turbulent shear stress
B'	Constant in hot-wire calibration equation, $B' = B\cos^n\phi$	v'	Transverse turbulent fluctuation
e	Fluctuating voltage	y	Transverse coordinate, distance from wall
E	Instantaneous bridge voltage	y_m	Distance from measurement origin
k	Roughness length scale	ε	Roughness offset
n	Exponent in hot-wire calibration equation	δ	Boundary layer thickness
R	Pipe radius	κ	von Kármán constant
R_a	Centerline average roughness height	ϕ	Hot-wire yaw angle
S_{ij}	$i, j = 1, 2$; sensitivity coefficients	ν	Fluid kinematic viscosity
u	Velocity	ϱ	Fluid density
u_τ	Shear velocity, $\sqrt{\tau_s/\varrho}$	τ_s	Wall shear stress
Δu	Velocity profile shift	$\overline{(\ )}$	Mean quantity
u'	Streamwise turbulent fluctuation	$(\ )_{\text{eff}}$	Effective value
		$\langle\ \rangle$	Root-mean-square value
		$(\ )_1$	Relating to wire 1
		$(\ )_2$	Relating to wire 2

Introduction

The accurate determination of the wall shear stress in rough-wall turbulent boundary layer flows has proved to be a major problem for experimentalists. As discussed by Acharya and Escudier (1983), techniques such as the Clauser plot for smooth-wall boundary layers based upon similarity concepts cannot be extended with any reasonable degree of accuracy to rough-wall situations. Heated film gauges, Preston tubes and other devices which depend upon the existence of a similarity region close to the surface are also inapplicable, and it is well known that values of the shear stress obtained using the momentum integral equation are subject to large uncertainties. Potentially the most reliable technique appears to be the direct measurement of the shear force exerted on a small segment of the surface using, for example, a floating-element device. Although such an approach is not without its own

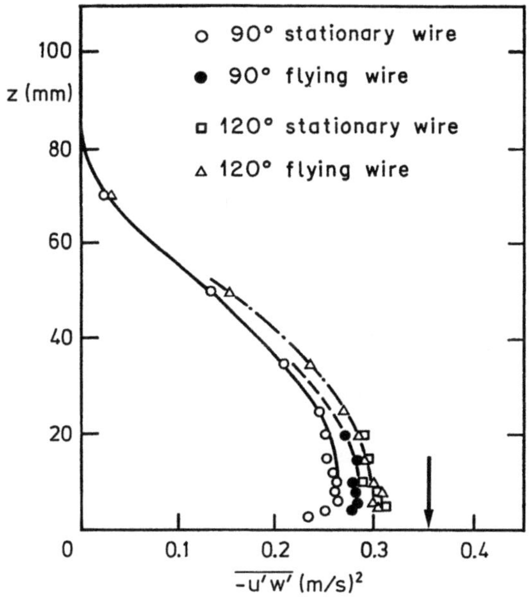

Fig. 1. Reynolds shear stress profiles (from Perry et al). Arrow indicates Hama defect law prediction

inherent difficulties, its successful use in a variety of both smooth- and rough-wall boundary layers has been demonstrated by the present authors (1983).

Perry et al. (1983) have suggested that the wall shear stress in a rough-wall boundary layer can be obtained by extrapolation of accurate X-wire measurements of the Reynolds shear stress in the near-wall region, provided that the geometry of the X-wire probe fulfils certain requirements. They carried out measurements of the Reynolds shear stress in the boundary layer developing over a mesh-type rough surface with X-wires of 90° and 120° included angles, both in the stationary and "flying" modes of operation. The results were compared with a wall shear stress value determined by fitting the mean-velocity data to the velocity-defect law that is generally presumed to be valid for both smooth and rough surfaces. Figure 1, taken from the paper of Perry et al., compares the different results obtained. It is concluded that the 120° probe yields better results than the 90° probe and that an additional improvement is obtained by "flying" the hot wires. The authors argue that close to a rough surface, the instantaneous velocity vector describes a significantly larger wedge angle than is the case for a smooth-wall boundary layer, and that errors in the measured Reynolds shear stress arise when the magnitude of this wedge angle approaches or exceeds the included angle of the X-wire probe. As a consequence, increasing this included angle results in an improved measurement. Further they show that large lateral velocity fluctuations normal to the plane of the X-wire probe introduce additional errors in the measurement and this effect is reduced by "flying" the wires through the boundary layer. It is however seen from Fig. 1 that the values of the wall shear stress obtained by extrapolating the X-wire data are 15–25% lower than the value obtained from a fit to the velocity-defect law. Perry et al. cast doubt on the validity of this law for rough-surface boundary layers.

From the above, it is apparent that an independent measurement of the wall shear stress would be of value in making a more definitive evaluation of the various alternatives considered. The present paper describes such measurements for a boundary layer developing over a mesh surface similar to that used by Perry et al. The experiments were motivated by the availability of a boundary-layer wind tunnel and a floating-element device for the direct measurement of the wall shear stress, both of which could be adapted to duplicate the measurements of Perry et al. A fully-developed smooth-wall pipe flow facility was also available for control measurements with the X-wires in a well defined flow using the same instrumentation, data acquisition and calibration procedures as for the boundary-layer experiments.

Experiment

Rough Surface Geometry

An expanded mesh with the geometry shown in Fig. 2 was used to build up the rough surface on the test plates and the floating element. The basic dimensions of the mesh (6.0 × 3.2 mm) may be compared with those of Perry et al. (29 mm × 10.5 mm). The use of a smaller mesh was dictated by the need to obtain a sufficiently large number of roughness elements on the floating element, a disc 20 mm in diameter. The test surface was 3 m in length and 0.8 m wide.

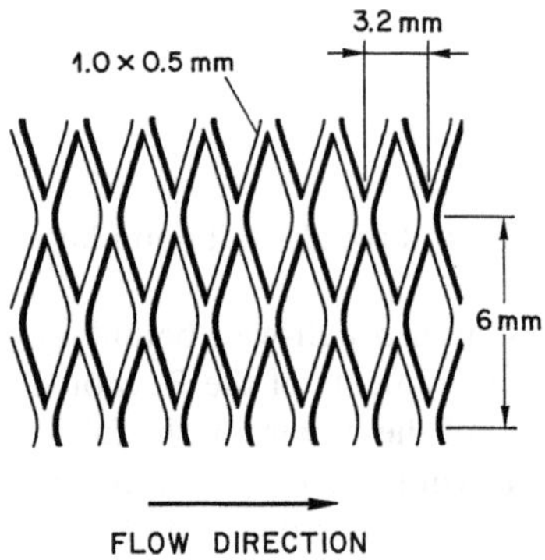

Fig. 2. Mesh geometry

Flow Facilities

The experiment was carried out in a conventional Bradshaw-type blower tunnel. The mesh surface formed the upper wall of the working section. Access to the boundary layer was through fitted plugs which were interchangeable with instrumentation plugs carrying the floating element and probe traverse. The tunnel was adjusted for zero pressure gradient in the test section and for most of the measurements reported here the free stream velocity and turbulence level were 20 m/s and 0.2 % respectively. Measurements were made at a location 1.2 m from the start of the mesh where the boundary-layer thickness was 44 mm and the momentum thickness Reynolds number 7750. The floating-element instrument is a nulling device which permits measurement of the average shear stress acting over a segment of the surface 20 mm in diameter. For zero pressure gradient, the accuracy of the instrument is about 5 %. The instrument has been described in more detail by Acharya et al. (1984). Profile measurements were made in the boundary layer over the mesh surface with 90° and 120° X-wire probes and the wall shear stress obtained from floating element measurements. Additional measurements of the mean velocity using a single hot wire (DISA probe 44 P15) were made at streamwise locations of 1.06 and 1.2 m for a range of free-stream velocities.

A 100 mm diameter 10 m long pipe-flow facility was used for the control measurements. The flow is fully developed in both the mean and turbulence properties over the last 4 m of this pipe. The same X-wires as for the boundary-layer measurements were used to obtain profile data at a location 8.4 m from the inlet for a centerline velocity of 17.6 m/s. The wall shear stress values obtained from pressure drop measurements and the Clauser plot technique agreed to within 2%.

Reynolds Shear Stress Measurements

The X-wire probes used for the Reynolds shear stress measurements were a standard DISA 55P61 probe with 5 μm platinum-plated tungsten sensors and an included angle of 90°, and a probe specially manufactured by DISA, identical to the 55P61 but with an included angle of 120°. The test surface could be approached to within 0.7 mm (measured from the center of the X configuration) by mounting the probes in a holder with an 8° inclination to the mean flow direction. The probes were operated in the constant temperature mode using DISA 55M10 units. The instantaneous bridge voltages were processed directly during calibration and measurements.

The X-wire probes were calibrated *in situ* against a pitot probe, over a range of about ten wind-tunnel speed settings, using a static calibration procedure. The relationship

$$\bar{E}^2 = A + B'\bar{u}^n, \tag{1}$$

where

$$B' = B \cos^n \phi$$

was assumed to be valid for each of the sensors and a least squares regression procedure was used to determine the appropriate calibration constants for each wire.

The fluctuating X-wire signals were related to the corresponding velocity fluctuations [see Perry (1982)] by

$$e_1' = S_{11}u' + S_{12}v',$$

$$e_2' = S_{21}u' - S_{22}v',$$

where the sensitivities are given by

$$S_{11} = \frac{n_1 B_1'}{2\bar{E}_1 \bar{u}^{1-n_1}}, \qquad S_{12} = S_{11} \tan\phi_{1\,\mathrm{eff}},$$

$$S_{21} = \frac{n_2 B_2'}{2\bar{E}_2 \bar{u}^{1-n_2}}, \qquad S_{22} = S_{21} \tan\phi_{2\,\mathrm{eff}}.$$

The effective angle ϕ_{eff} for each wire was determined using a technique suggested by Bradshaw (1971): the probe was yawed successively through a number of angles $-10° < \Delta\phi < 10°$ about the nominal position and the mean voltage was recorded. It can be shown using (1) that

$$\left(\frac{\bar{E}^2_{\phi+\Delta\phi} - A}{\bar{E}^2_{\phi} - A}\right)^{1/n} - \cos\Delta\phi = -\tan\phi_{\mathrm{eff}} \sin\Delta\phi.$$

A plot of the left hand side against $\sin\Delta\phi$ was used to obtain the slope $\tan\phi_{\mathrm{eff}}$.

The Reynolds normal and shear stresses were obtained as

$$\overline{u'^2} = \overline{(S_{22} e'_1 - S_{12} e'_2)^2}/D,$$

$$\overline{v'^2} = \overline{(S_{21} e'_1 - S_{11} e'_2)^2}/D,$$

$$\overline{u'v'} = (S_{21} S_{22} \overline{e'^2_1} + [S_{12} S_{21} - S_{11} S_{22}]\, \overline{e'_1 e'_2} - S_{11} S_{12}\, \overline{e'^2_2})/D,$$

where

$$D = S_{11} S_{22} - S_{12} S_{21}.$$

Typically 5000 data samples were taken to form the averages at each measurement point.

Discussion of Results

Mean velocity $\bar{u}$ and streamwise turbulence fluctuation $\langle u' \rangle$ profiles measured with the two X-wire probes are shown in Figs. 3 and 4 respectively for both the mesh boundary layer and the pipe flow. These measurements show little sensitivity to the probe geometry (i.e. yaw angle). The mean velocity profiles have been normalized to wall coordinates $\bar{u}/u_\tau$ and $y u_\tau/v$ with $u_\tau \equiv \sqrt{\tau_s/\varrho}$. The profiles show the usual semi-logarithmic variation near the surface, with a roughness function $\Delta u/u_\tau = 11.0$ for the mesh surface for this particular profile. The disappearance of the sharp peak in the $\langle u' \rangle$ profile close to the mesh surface is attributed to an extensive region of intense turbulent mixing which is commonly observed for rough surfaces. The transverse turbulence fluctuations, $\langle v' \rangle$, shown plotted in Figs. 5a and 5b are strongly influenced by the angle of yaw, a characteristic also observed by Sandborn (1968). The Reynolds shear stress profiles for the two flows are shown in Figs. 6a and 6b. Also shown in Fig. 6b is the expected variation of the Reynolds shear stress which, in fully developed pipe flow, is given by

$$\frac{\overline{u'v'}}{u_\tau^2} = 1 - \frac{y}{R} - \frac{v}{u_\tau^2} \frac{\partial \bar{u}}{\partial y}.$$

The rough-surface measurements from the 120° probe are systematically higher than those obtained with the 90° probe, as was observed by Perry et al. and much closer to the expected values in the near-wall region of the pipe than the 90° probe data. However, the differences in the results of the two probes are not as large as those found by Perry et al. (Fig. 1) and the wall shear stress values obtained by extrapolation of the X-wire data are within a few percent of that measured using the floating element. The skin friction coefficient determined by fitting the Hama defect law to mean velocity profiles measured using single hot wires is 20% higher than the floating element value. This discrepancy is comparable to that obtained by Perry et al.

The mean velocity profile in a rough-wall boundary layer expressed in wall coordinates $(\bar{u}/u_\tau, y u_\tau/v)$ is

$$\frac{\bar{u}}{u_\tau} = \frac{1}{\kappa} \ln \left[\frac{(y_m + \varepsilon)\, u_\tau}{v} \right] + A - \frac{\Delta u}{u_\tau}, \tag{2}$$

where y_m is the distance measured from a chosen origin and ε the incremental distance to the aerodynamic origin. The vertical shift $\Delta u/u_\tau$, generally termed the roughness function,

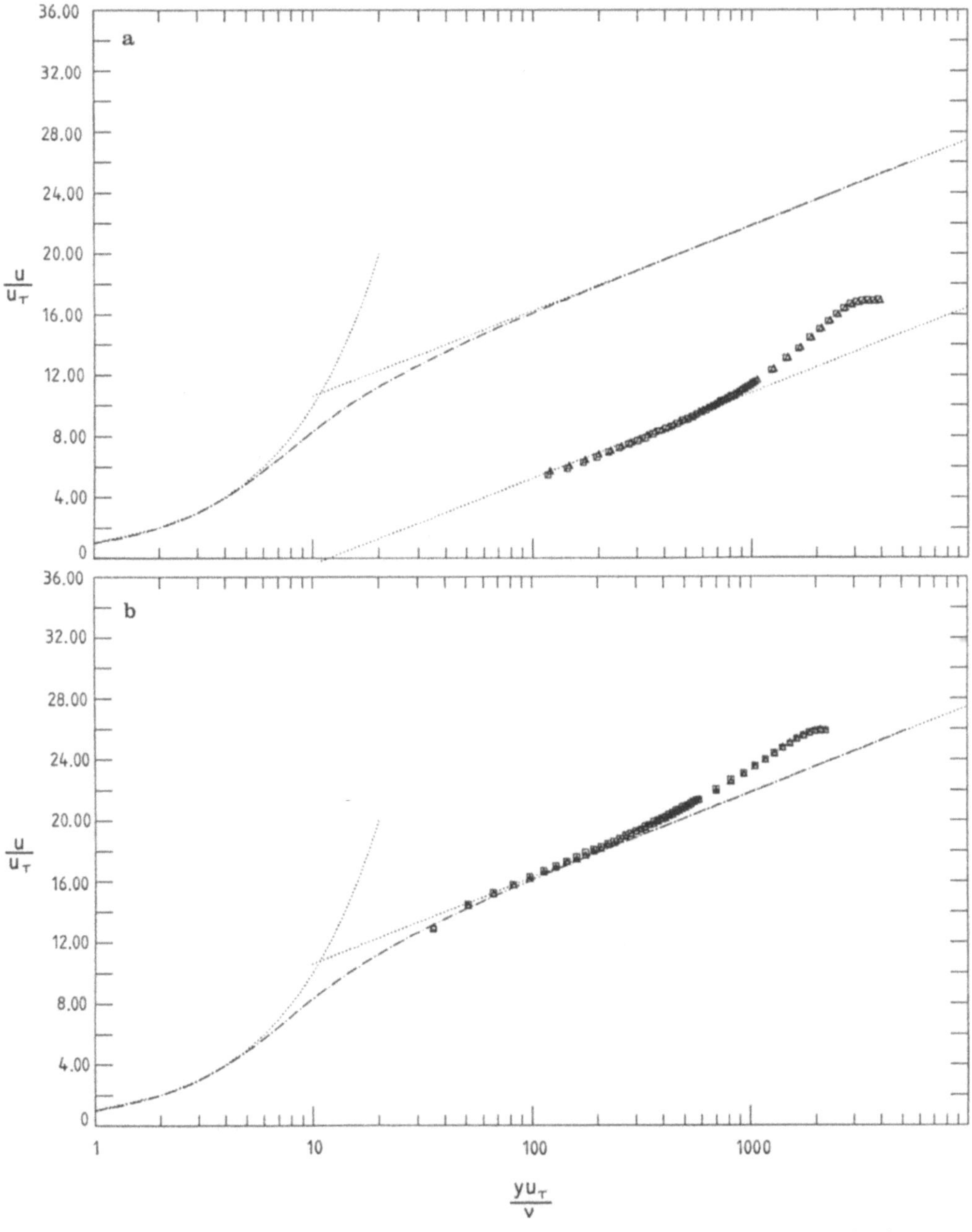

Fig. 3a. Mean-velocity profile in mesh-surface boundary layer: △ 90° probe; □ 120° probe; **b** Mean-velocity profile in pipe flow: △ 90° probe; □ 120° probe

is dependent upon a sublayer length scale u_τ/ν and the roughness geometry. The mean velocity $\bar{u}$ was measured using a single hot wire and the corresponding wall shear stresses determined using the floating element. The roughness function and offset ε were then determined from a fit of (2) to the experimental data.

Figure 7 shows the variation of the roughness function with the roughness Reynolds number $u_\tau R_a/\nu$ where R_a, the centerline average roughness height, is computed to be

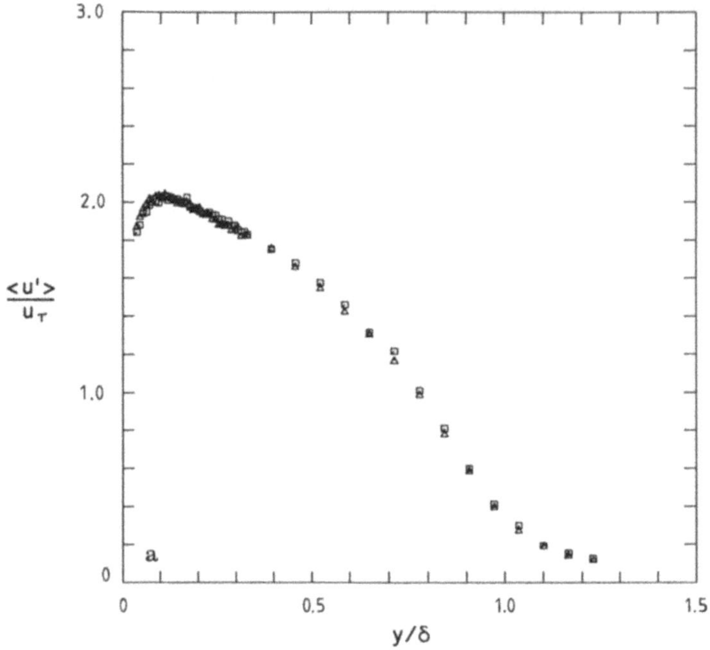

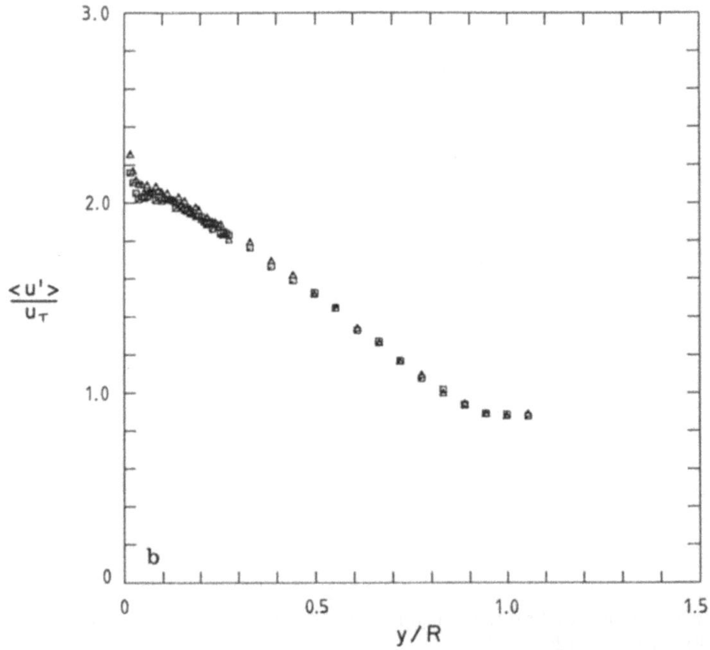

Fig. 4a. Streamwise turbulent fluctuation profiles in mesh-surface boundary layer: △ 90° probe; □ 120° probe; **b** Streamwise turbulent fluctuation profiles in pipe flow: △ 90° probe; □ 120° probe

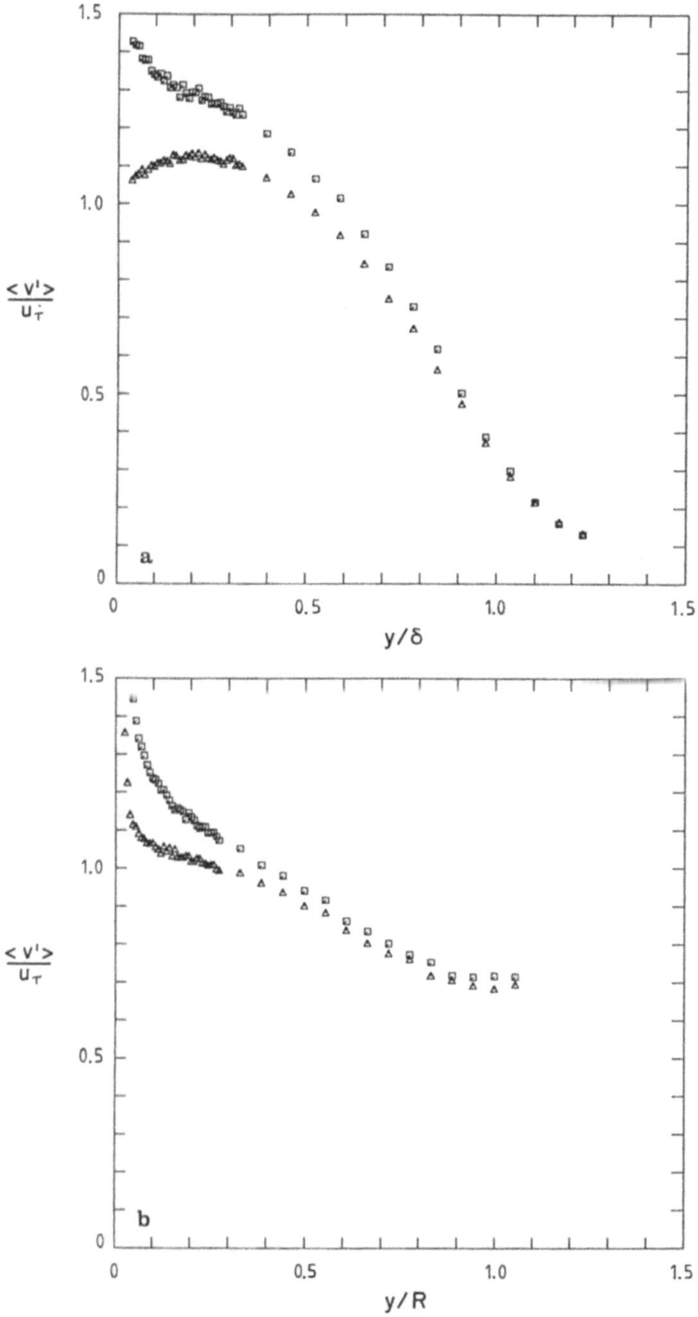

Fig. 5a. Transverse turbulent fluctuation profiles in mesh-surface boundary layer: △ 90° probe; □ 120° probe; **b** Transverse turbulent fluctuation profiles in pipe flow: △ 90° probe; □ 120° probe

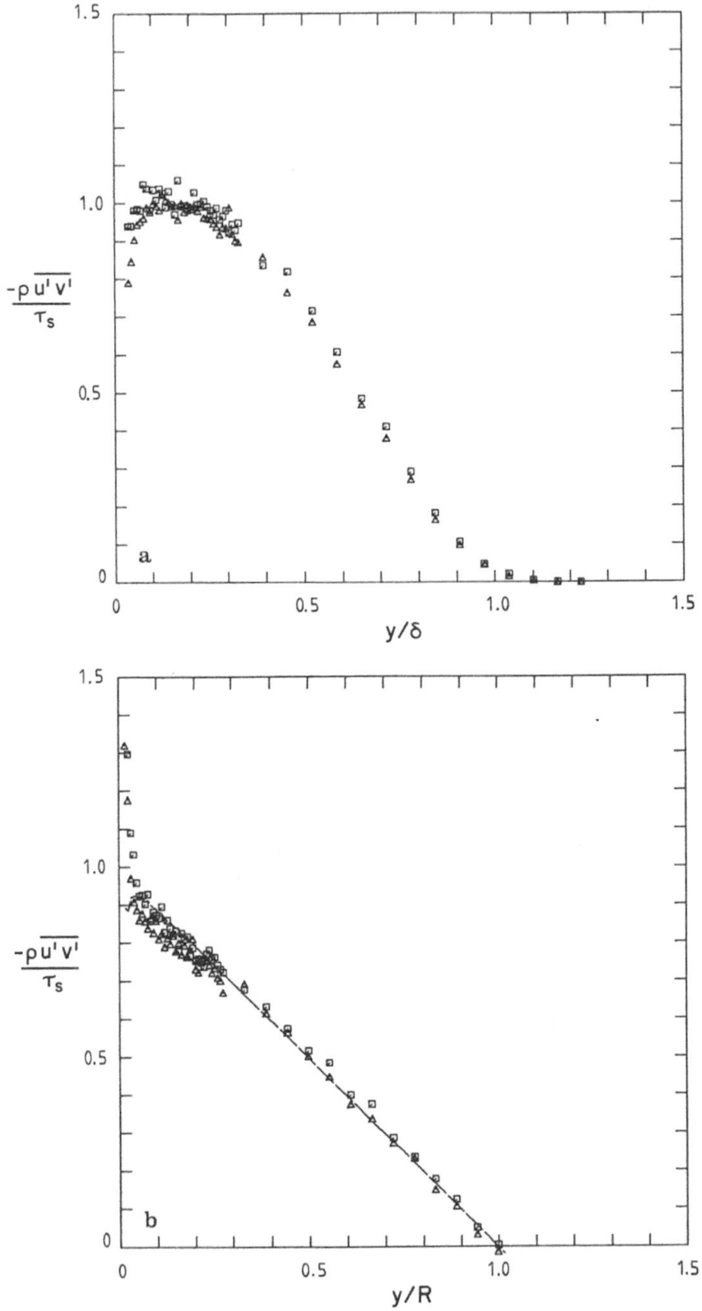

Fig. 6a. Reynolds shear stress profiles in mesh surface boundary layer: △ 90° probe; □ 120° probe; **b** Reynolds shear stress profiles in pipe flow: △ 90° probe; □ 120° probe

0.44 mm. Also shown are the results for sand-grain roughness with k, the sand-grain size, as the length scale. It is seen that a roughness length scale of 10 R_a brings the mesh data into line with the sand grain data and so 10 R_a may be regarded as the equivalent sand grain roughness for this surface. Also worth remarking is the fact that the mesh data exhibit a slope different from $1/\kappa$, which is the generally accepted value for data in the fully rough regime.

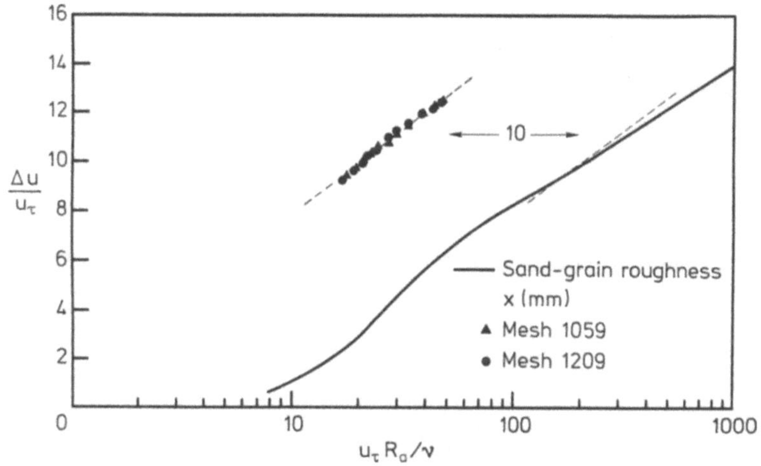

Fig. 7. Variation of Roughness function

Conclusions

Measurements have been made of the surface shear stress, mean velocity, turbulence intensity and Reynolds shear stress in a turbulent boundary layer developing over a mesh type rough surface. The surface shear stress distributions determined with a floating-element instrument were found to agree with the extrapolated Reynolds shear stress value, but to be 20% lower than the value estimated from the mean velocity profile using the Hama defect law. These results confirm the conclusions of Perry et al., and also their suspicion that the smooth-wall defect law may not be applicable to rough-wall boundary layers. Measurements to examine the roughness function indicate that the equivalent sand grain roughness scale for the mesh surface is about an order of magnitude larger than the centerline average roughness height.

References

Acharya, M., Escudier, M. P. (1983): "Measurements of the Wall Shear Stress in Boundary-Layer Flows", *Turbulent Shear Flows 4,* ed. by L. J. S. Bradbury et al. (Springer, Berlin, Heidelberg) pp. 277–286

Acharya, M., Bornstein, J., Escudier, M. P., Vokurka, V. (1984): Development of a floating element for the measurement of surface shear stress. AIAA J. **23**, 410–415

Bradshaw, P. (1971): *An Introduction to Turbulence and Its Measurement* (Pergamon, New York)

Perry, A. E. (1982): *Hot wire Anemometry* (Clarendon, Oxford)

Perry, A. E., Lim, K. L., Henbest, S. M., Chong, M. S. (1983): Presented at the Fourth International Symposium on Turbulent Shear Flows, Karlsruhe, FRG

Sandborn, V. A. (1968): "Hot Wire Anemometer Measurements in Large-Scale Boundary Layers, Advances in Hot-Wire Anemometry," *Proceedings of the International Symposium on Hot Wire Anemometry,* (University of Maryland)

Low Reynolds Number Turbulent Boundary Layers on a Smooth Flat Surface in a Zero Pressure Gradient

L. P. Erm *, A. J. Smits ** and P. N. Joubert *

* Department of Mechanical Engineering, University of Melbourne, Parkville, Victoria, 3052, Australia

** Mechanical and Aerospace Engineering Department, Princeton University, Princeton, New Jersey, 08544 U.S.A.

Abstract

An experimental investigation of low Reynolds number turbulent boundary layers ($R_\theta < 5000$) on a smooth flat surface in a nominally zero pressure gradient is described. Stimulator pins of three different heights were used to trip the boundary layer. Mean flow velocity profile measurements were analysed using the method of Coles [2]. This method involved fitting the data to a log-law line having constants $\kappa = 0.41$ and $C = 5.0$ and inferring values of Π and u_τ. Acceptable low Reynolds number turbulent boundary layers were shown to exist. Broadband turbulence and spectral measurements for $R_\theta = 5010$, the highest Reynolds number obtainable in these experiments, were compared, where possible, with other data and the agreement was good. Both mean flow and turbulence results were shown to be dependent on Reynolds number and apparently dependent on the size of the tripping device.

Nomenclature

C	Log-law constant
C'_f	Local skin friction coefficient
k_1	Streamwise wave number, 1/m
R_θ	Reynolds number based on momentum thickness
u	Longitudinal component of velocity, m/s
u'	Longitudinal fluctuating component of velocity, m/s
u_τ	Friction velocity, m/s
v'	Normal fluctuating component of velocity, m/s
w	Coles "Law of the wake"
x	Distance from tripping device in longitudinal direction, m
y	Distance from wall, m
δ	Boundary layer thickness where velocity has 99.5% of its freestream value, m
δ_H	Hama boundary layer thickness, m
$\delta_H u_\tau/\nu$	Karman number
ε	Turbulent energy dissipation, m^2/s^3
$\eta = (\nu^3/\varepsilon)^{1/4}$	Kolmogoroff length scale, m
θ	Boundary layer momentum thickness, m
θ_0	Momentum thickness for the most upstream profile, m
κ	Log-law constant
ν	Kinematic viscosity, m^2/s
Π	Profile parameter
$y = (\nu\varepsilon)^{1/4}$	Kolmogoroff velocity scale, m/s
$\Phi_{11}(k_1 y)$	Energy density per unit $k_1 y$
$\Phi_{11}(k_1 \delta)$	Energy density per unit $k_1 \delta$
$\Phi_{11}(k_1 \eta)$	Energy density per unit $k_1 \eta$

Introduction

A turbulent boundary layer is considered to be a low Reynolds number flow when the Reynolds number based on momentum thickness, R_θ, is less than 5000. Low Reynolds number turbulent boundary layers play an important part in many fluid flow problems and yet relatively little is known about them. One of the reasons for this is the difficulties involved in obtaining reliable experimental data. Low Reynolds number flows imply proximity to transition which means that the layers are very thin and consequently it is difficult to obtain accurate measurements. If freestream velocities are reduced so as to obtain thicker boundary

Turbulent Shear Flows 5

layers, then experimental difficulties are still encountered as the pressure differences to be measured are very small.

In 1956 Coles [1] carried out an extensive survey of mean velocity profile measurements at large Reynolds numbers, both with and without pressure gradients, and proposed that the velocity profile outside the viscous sublayer could be accurately described by the expression

$$\frac{u}{u_\tau} = \frac{1}{\kappa} \ln \frac{y u_\tau}{\nu} + C + \frac{\Pi}{\kappa} w\left(\frac{y}{\delta}\right),$$

where $\kappa = 0.40$ and $C = 5.1$. Π is a profile parameter and for flows at constant pressure it has a value of approximately 0.55. The function $w(y/\delta)$ is termed the "law of the wake". In 1962 Coles [2] extended this initial work by analysing virtually all of the published data on low Reynolds number turbulent boundary layers on smooth flat surfaces in nominally zero pressure gradients. He identified a normal state for the turbulent boundary layer at constant pressure and expressed this normal state in terms of a variation of the strength of the wake component, $\Delta u/u_\tau$, with R_θ. Figure 1 shows this relationship plotted in the form Π versus R_θ, where $\Delta u/u_\tau = 2\Pi/\kappa$. Coles [2] showed that the above expression is still valid provided Π varies with R_θ in this manner. In this later work Coles [2] changed the values of the log-law constants to $\kappa = 0.41$ and $C = 5.0$, and both were assumed to be independent of Reynolds number. The asymptotic value of $\Delta u/u_\tau$ given by Coles [2] in his 1962 paper is approximately 2.7. In 1968 Coles [3] used a slightly different analytic technique when reanalysing data and in this case he found that the asymptotic value of $\Delta u/u_\tau$ was approximately 3.0.

Since Coles undertook these extensive surveys, other workers have studied low Reynolds number flows and have arrived at various conclusions. Simpson [4] concluded that κ and C were not constant and for values of $R_\theta < 6000$ they varied with R_θ. He suggested that changes in the profile shapes at low Reynolds numbers, which Coles correlated with Π, could equally well be explained in terms of changes in the log-law constants. Huffman and Bradshaw [5] studied duct flow data and concluded that κ is probably constant but C is Reynolds number dependent. However, they indicate that even the variation of C is likely to be small in boundary layers unless the influence of the outer layer is extremely large.

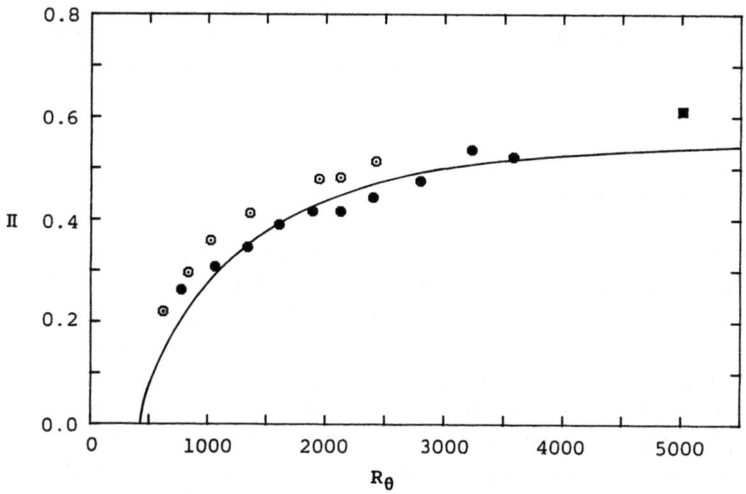

Fig. 1. Variation of profile parameter with Reynolds number
o 8 m/s, • 14 m/s, ▪ 20 m/s, —— Coles [2]

Granville [6] analysed previously published data and concluded that there is a lower limit to the value of u/u_τ corresponding to the disappearance of the log-law region. Support for the conclusion that κ and C are constant is given by Purtell et al. [7], but these authors found that the variation of Π with R_θ is different from that given by Coles [2]. The authors also presented broadband turbulence intensity measurements for zero pressure gradient flow. The results of Murlis et al. [8] indicated that observed changes in mean velocity profiles at low Reynolds numbers and at constant pressure arise because of a reduction in the wake component and not through a failure of the inner-layer logarithmic law. These authors also presented a large number of broadband turbulence measurements. A further study was undertaken by Smits et al. [9], who reported the results of an extensive experimental investigation into the mean flow properties of turbulent boundary layers with $R_\theta < 3000$ for both zero and favourable pressure gradients. They found a logarithmic region occurred in boundary layers with very low R_θ (354 for zero pressure gradient flow and 261 in the stronger of the two favourable pressure gradients tested) and they indicated that it seemed the log law is established almost immediately after transition.

The general consensus of most workers after Coles is that Coles' [2] criterion for a normal state for a turbulent boundary layer at constant pressure, as depicted in Fig. 1, is basically correct. Thus in any fundamental study of flow behaviour at low R_θ, it is essential that Coles' criterion be satisfied before proceeding. As well as this, it is essential that the flow satisfy the momentum integral equation balance. Only if these conditions are satisfied can any subsequent measurements, such as turbulence measurements or measurements in a pressure gradient, have any credibility attached to them.

For flow over models, the low Reynolds number region may be a significant part of the length of a particular body. However, even if this region is relatively small, it is still important since it is this region which sets up the initial conditions for subsequent layer development. As shown by Joubert et al. [10], prediction methods tend to break down, never to recover, at the nose of a body of revolution, where the flow has a low R_θ, a favourable pressure gradient, streamline divergence and lateral curvature.

From the above it is clear that despite some excellent contributions, there is still a scarcity of reliable experimental data for low Reynolds number flows. Virtually all the measurements have been taken in zero pressure gradient layers and the measurements comprise mainly mean flow characteristics such as velocity profiles taken with Pitot tubes or skin friction coefficients measured with Preston tubes. Very few broadband turbulence characteristics, especially Reynolds stresses, have been given and details of spectral behaviour are even more scarce. One of the aims of the current project was to help overcome this shortage of experimental data.

Experimental Apparatus and Techniques

The wind-tunnel used was an open-return suction type of conventional design having a working section 2.5 m long and inlet dimensions 309 mm × 613 mm. The smooth flat surface upon which measurements were made formed one of the walls of the working section. The opposite wall was adjustable so that different pressure gradients could be obtained. Transition was accomplished by means of round stimulator pins placed 8 cm downstream of the contraction outlet. The first and last measuring stations were 0.26 m and 1.94 m respectively, downstream from these pins. Measurements were taken at three different reference unit Reynolds numbers, corresponding to approximate reference velocities of 8, 14 and 20 m/s, and in each case a nominal zero pressure gradient existed. In order to obtain acceptable tripping of the boundary layers at these three velocities, it was necessary to use pins of three different heights. However, all pins had a fixed diameter of 3.0 mm and the spacing between pin centres was fixed at 9.0 mm. The pin heights corresponding to the above three velocities were 2.0, 0.8 and 0.6 mm respectively. Each set of pins was mounted on a metal insert that could be bolted flush into the smooth wall thereby enabling the different sets of pins to be changed quickly. The use of different tripping devices enabled any possible effects of tripping device on both the mean flow and turbulence characteristics of low Reynolds number flows to be investigated.

The mean flow results consist of velocity profiles taken using a round Pitot probe and skin friction coefficients measured with Preston tubes. All pressures were measured using an electronic manometer in conjunction with an electronic integrator.

Broadband turbulence measurements were taken using both single and crossed hot-wire probes which had 5 μm diameter etched Wollaston wires. The etched lengths were typically 1.0 mm, giving a length to diameter ratio of 200. The hote wire probes were calibrated using dynamic calibration techniques as developed by Perry and Morrison [11] and Perry [12]. Turbulence signals were fed into a digital computer via an analogue-to-digital converter and then processed.

The spectral measurements were all taken using a single uncalibrated hot wire of similar construction to that used for the broadband turbulence measurements. Once again the signal processing involved a digital computer and in this case use was made of a Fast Fourier Transform algorithm.

Analysis of Experimental Results

At this stage the experimental programme is still in progress and consequently all experimental data are not yet available.

Mean Flow Measurements

The mean flow velocity profile data were analysed in the same way as that used by Coles [2]. The log-law constants were assumed to have values of $\kappa = 0.41$ and $C = 5.0$ and values of Π and u_τ were inferred by fitting the data to the log-law line having these constants. For purposes of fitting the data, probe displacement corrections were removed.

It can be seen from Fig. 1 that for freestream velocities of 8 and 14 m/s the variation of Π with R_θ was substantially the same as that given by Coles [2]. A spot check at $x = 1.94$ m, the most downstream station, for 20 m/s also indicated good agreement. As well as the expected decrease in Π with decrease in Reynolds number, the size of the tripping device used also appeared to influence the results. A check on the balance of the momentum integral equation was also performed at 8 and 14 m/s by evaluating $(\theta - \theta_0)/\theta_0$ and $(1/\theta_0) \int (C'_f/2)\, dx$ for the developing flow and comparing the two terms at all stations. The cumulative imbalance at the last measuring station was less than 3% for both velocities although the imbalance went up to as high as about 14% for 14 m/s at earlier stations.

When the inferred values of C'_f associated with each profile were plotted against corresponding values of R_θ, they were found to follow Coles' [2] C'_f versus R_θ relationship reasonably closely.

An extensive region of log-law behaviour was found to exist even for $R_\theta = 617$, the lowest Reynolds number at which a velocity profile was taken. Figure 2 shows this profile. The agreement of the data with the log-law line is good. This also applied to other profiles. Thus the data of these experiments tend to confirm the values of the log-law constants to be $\kappa = 0.41$ and $C = 5.0$, as suggested by Coles [2].

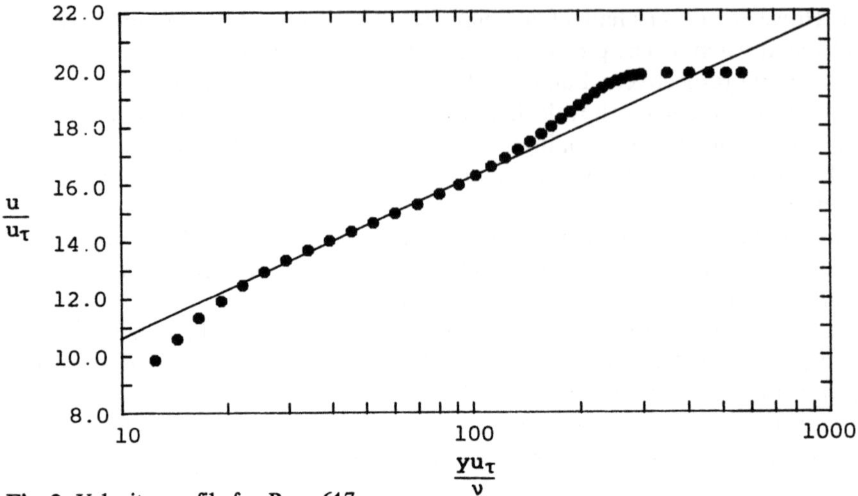

Fig. 2. Velocity profile for $R_\theta = 617$

From the above it is apparent that the flows have acceptable low Reynolds number behaviour as far as mean flow characteristics are concerned.

Broadband Turbulence Measurements

Since no criteria similar to the above ones have been produced for turbulence properties, checks were made where possible with established data at the highest R_θ which could be obtained in these experiments, viz. $R_\theta = 5010$.

The u' turbulence intensities taken with a single hot wire are shown plotted with inner flow scaling in Fig. 3, where the results for $R_\theta = 5010$ are compared with those of Purtell et al. [7] for a similar Reynolds number ($R_\theta = 5100$). As can be seen there is very close agreement.

Figure 4 shows the same data replotted with outer flow scaling on the abscissa. The data show a better collapse when plotted in this manner and a careful examination indicates that not only is there a Reynolds number dependence, but there also appears to be a dependence on tripping device as well. This apparent dependence on tripping device is not evident from Fig. 3. The results of Purtell et al. [7] are shown again and the rescaled results of Klebanoff [13] (R_θ approx. 8000) are also given for comparison. The results presented in Figs. 3 and 4 for different Reynolds numbers show remarkably similar trends to those of Purtell et al. [7].

The v' turbulence intensities are shown plotted in Fig. 5 using similar scaling to Fig. 4. Purtell et al. [7] did not give any results of this type but the rescaled results of Klebanoff [13] are again given for comparison. Once again there is a Reynolds number dependence and an apparent dependence on tripping device.

The kinematic Reynolds shear stress profiles are shown in Fig. 6 where they are normalised with u_τ^2. As for the previous data, there is a Reynolds number dependence and also an apparent dependence on tripping device. Superimposed on this plot are predicted distributions by J. E. Green, as given by Murlis [14], for $R_\theta = 500$ and 10,000 (approx. values). The data follow the predictions reasonably well. The rescaled data of Klebanoff [13] are also shown.

The decrease in normalised Reynolds shear stress with decreasing values of R_θ below $R_\theta = 5000$ has also been observed at the University of Melbourne by Watmuff et al. [15] in a different wind tunnel.

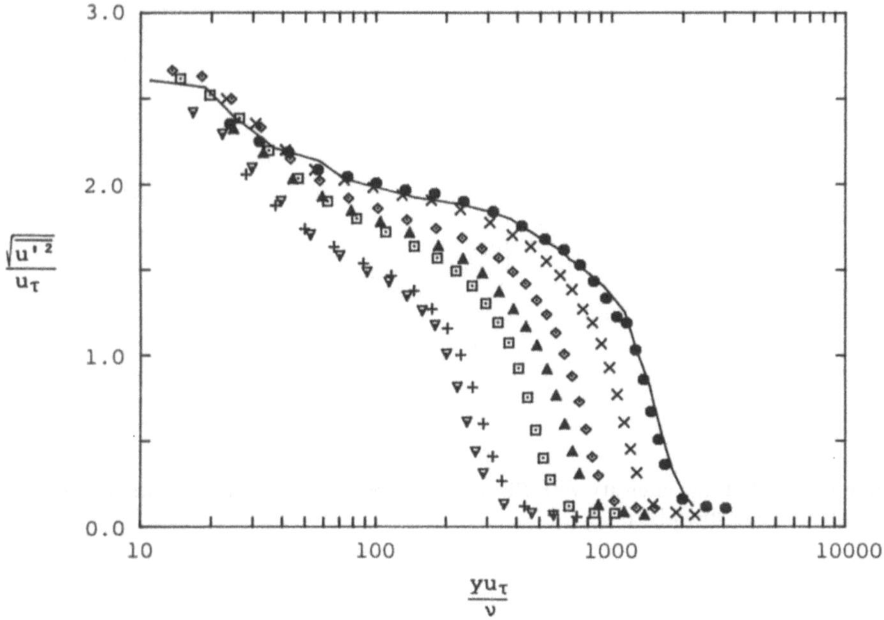

Fig. 3. Variation of u' turbulence intensity with Reynolds number – wall variables. Key to symbols showing values of R_θ and corresponding reference velocities (in curved brackets)
▽ 617 (8), + 772 (14), □ 1359 (8), ▲ 1890 (14), ◇ 2425 (8), × 3575 (14), ● 5010 (20), — Purtell et al [7]

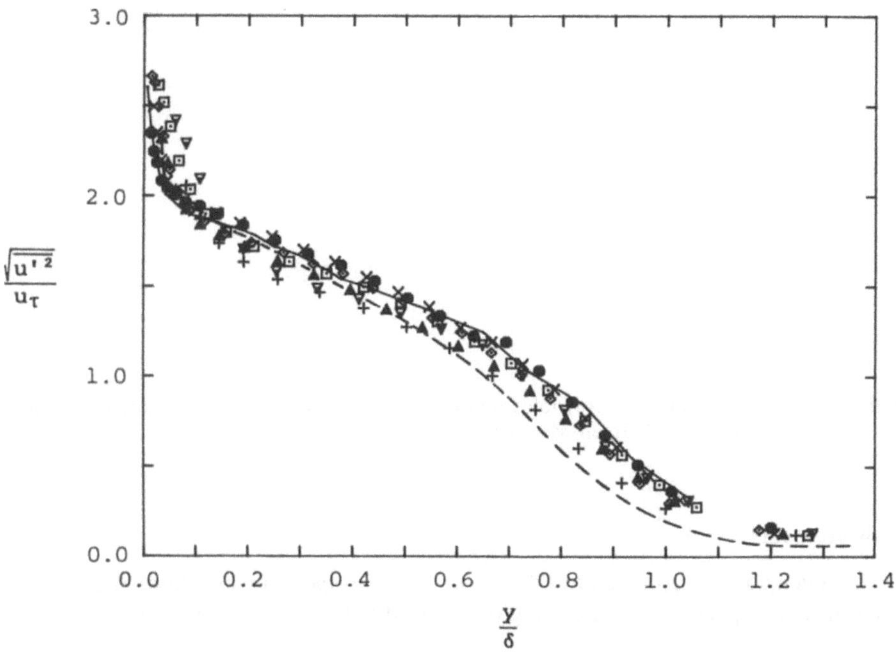

Fig. 4. Variation of u' turbulence intensity with Reynolds number – outer flow variables. For key to symbols see Fig. 3. ––– Klebanoff [13]

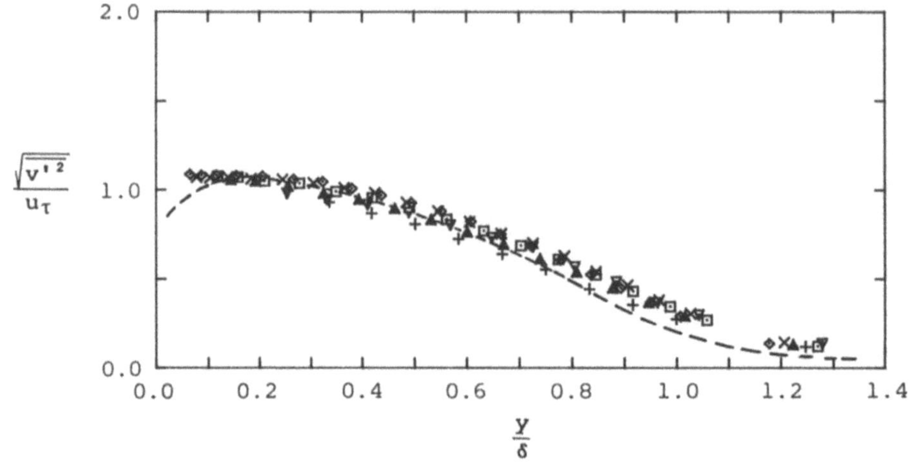

Fig. 5. Variation of v' turbulence intensity with Reynolds number – outer flow variables. For key to symbols see Fig. 3. --- Klebanoff [13]

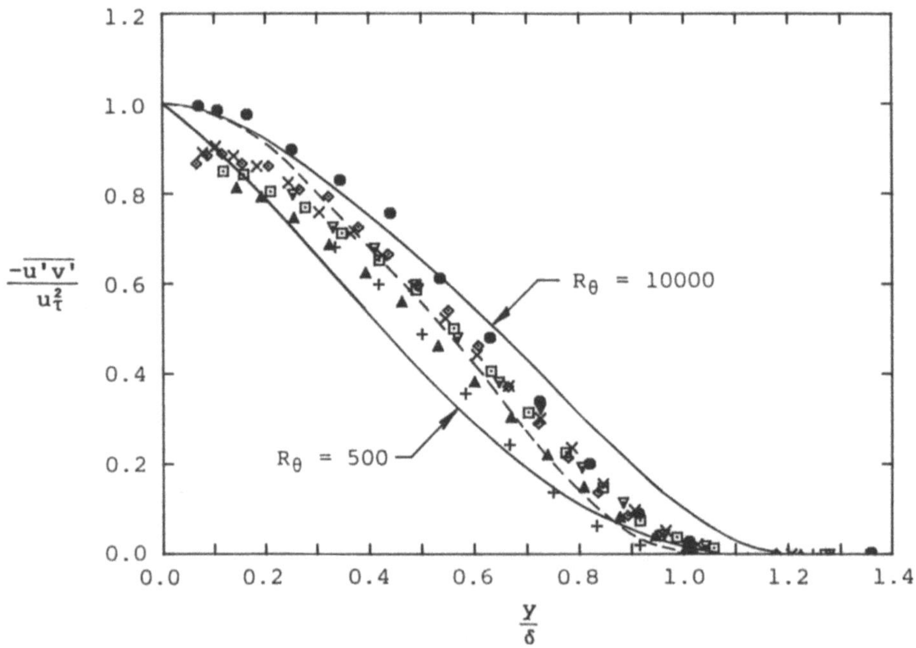

Fig. 6. Variation of Reynolds stress with Reynolds number – outer flow variables. For key to symbols see Fig. 3. --- Klebanoff [13], —— Green/Murlis [14]

Spectral Measurements

For each of the seven conditions listed in Fig. 3, a family of u' spectra was taken. Each family comprised seven to ten spectra corresponding to different values of y throughout the boundary layer.

Perry et al. [16], using dimensional analysis, have proposed spectral behaviour in the turbulent wall region which they defined as $yu_\tau/\nu \gg 1$, $y/\delta_H \ll 1$. These authors also plotted

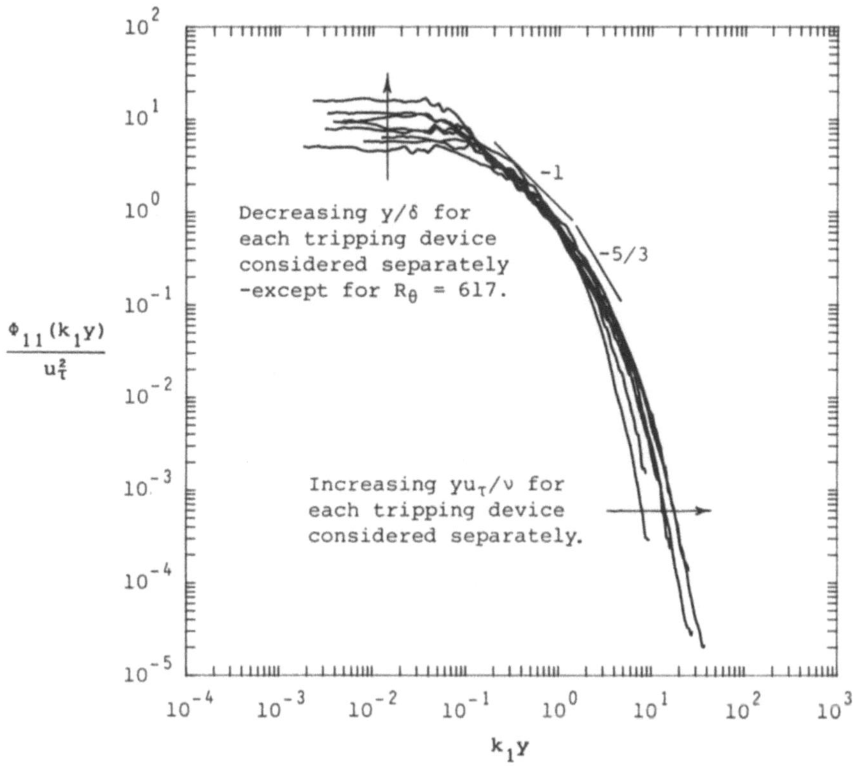

Fig. 7. u'-spectra, turbulent wall region, inner flow scaling

smooth wall, zero pressure gradient experimental spectral data using three different types of scaling, viz. "inner flow", "outer flow" and Kolmogoroff scaling, and their predictions were given encouraging support. All their plotted spectra satisfied the conditions $yu_\tau/v > 100$ and $y/\delta_H < 0.15$ and the lowest Reynolds number of their spectra was $R_\theta = 2791$ (private communication).

Since the present work is specifically concerned with low Reynolds number effects, it was decided to plot wall-region spectra using the scaling of Perry et al. [16], to observe the deviations, if any, from their model. Although the spectra for $R_\theta = 5010$ showed good agreement with the model, some problems were encountered. An analysis of the spectra indicated that for families having $R_\theta = 2425$, 1359, 772 and 617, the above turbulent wall region just did not exist (δ was used instead of δ_H). Notwithstanding this, the spectrum most closely corresponding to this region was selected from each of these families and plotted with others that satisfied the criterion as shown in Figs. 7–9. The figures correspond to "inner flow", "outer flow" and Kolmogoroff scaling respectively.

A close analysis of Figs. 7 and 8 indicated that the peeling off at the low and high wavenumber ends of the spectra followed the trends of the data of Perry et al. [16], provided that spectra corresponding to the different tripping devices were not always compared with each other and also provided that the spectrum corresponding to $R_\theta = 617$, the lowest Reynolds number, was disregarded in most cases. In other words, the size of the tripping device appeared to affect the results, as was the case for the broadband turbulence measurements, and also the spectrum at the very lowest Reynolds number did not fit the model. In Fig. 9 the spectra peel off in order of increasing Karman number and increasing Reynolds number at the low wavenumber end, as for the data of Perry et al. [16].

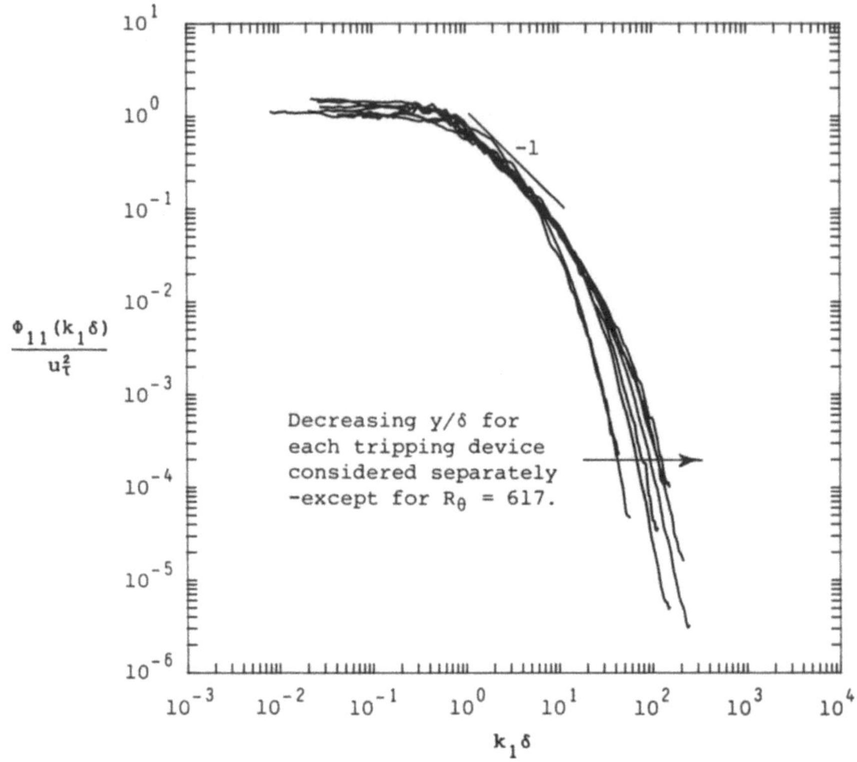

Fig. 8. u'-spectra, turbulent wall region, outer flow scaling

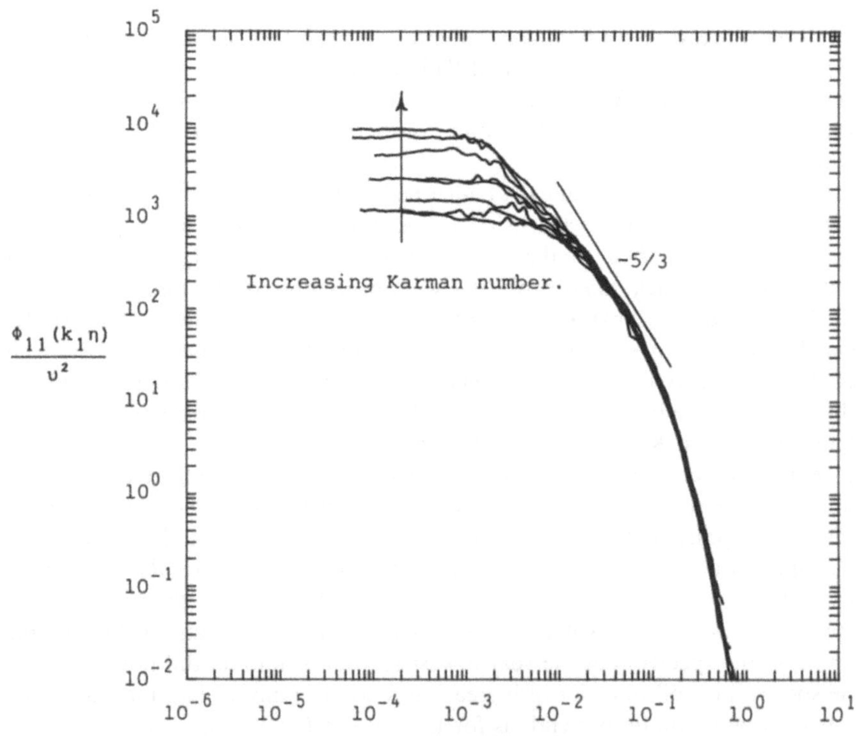

Fig. 9. u'-spectra, turbulent wall region, Kolmogoroff scaling

Considering the regions of collapse for the inner and outer flow scaling, Figs. 7 and 8 clearly show an inverse power law region and Fig. 7 seems to have a short inertial subrange region. These regions of collapse are all similar to the data of Perry et al. [16]. For the Kolmogoroff scaling, Fig. 9 indicates that the data do not quite match the line of slope $-5/3$ associated with the inertial subrange, the deviation most likely being due to low Reynolds number effects. All spectra show good collapse at the high wavenumber end as expected with this type of scaling.

Conclusions

When the mean flow velocity profile data were reduced using Coles' [2] method, the inferred values of Π and u_τ resulted in a similar Π versus R_θ relationship to that of Coles [2] and also an acceptable balance of the momentum integral equation. Extensive regions of log-law behaviour were found to exist even down to $R_\theta = 617$, the lowest Reynolds number at which a velocity profile was taken. The size of the tripping device appeared to affect the Π versus R_θ relationship.

When a u' turbulence intensity profile for $R_\theta = 5010$, the highest Reynolds number obtainable in these experiments, was compared with the results of Purtell et al. [7] for $R_\theta = 5100$, there was very close agreement. A Reynolds shear stress profile for $R_\theta = 5010$ was compared with the results of Klebanoff [13] for $R_\theta = 8000$ (approx.) and the agreement was reasonable. The turbulence intensity and Reynolds shear stress results were dependent on R_θ for R_θ less than 5010 and they also appeared to be dependent on tripping device as well.

Spectral measurements were compared with the model of Perry et al. [16], which was shown to be valid within the range $yu_\tau/\nu > 100$ and $y/\delta_H < 0.15$. Low Reynolds number spectra, including some from families in which this range did not exist, were shown to fit the model when "inner flow" and "outer flow" scaling was used. The order in which these spectra peeled off at the low and high wavenumber ends followed the model provided that spectra corresponding to the different tripping devices were not always compared with each other. The regions of collapse of all spectra for these two types of scaling were as predicted by the model. When Kolmogoroff scaling was used, all spectra peeled off at the low wavenumber end in order of increasing Karman number and increasing Reynolds number, irrespective of the tripping device. The spectra collapsed at the high wavenumber end but deviated slightly from the line of slope $-5/3$, possibly due to low Reynolds number effects.

References

1. Coles, D. E. (1956): The law of the wake in the turbulent boundary layer. J. Fluid Mech. **1,** 191–226
2. Coles, D. E. (1962): The turbulent boundary layer in a compressible fluid. Rand Report R-403-PR
3. Coles, D. E. (1968): "The Young Person's Guide to the Data", in *Proceedings Computation of Turbulent Boundary Layers, AFOSR-IFP-Stanford Conference,* Vol. 2, ed. by Coles, Hirst
4. Simpson, R. L. (1970): Characteristics of turbulent boundary layers at low Reynolds numbers with and without transpiration. J. Fluid Mech. **42,** 769–802
5. Huffman, G. D., Bradshaw, P. (1972): A note on von Karman's constant in low Reynolds number turbulent flows. J. Fluid Mech. **53,** 45–60
6. Granville, P. S. (1977): Drag and turbulent boundary layer of flat plates at low Reynolds numbers. J. Ship Res. **21,** 30–39
7. Purtell, L. P., Klebanoff, P. S., Buckley, F. T. (1981): Turbulent boundary layer at low Reynolds number. Phys. Fluids **24,** 802–811
8. Murlis, J., Tsai, H. M., Bradshaw, P. (1982): The structure of turbulent boundary layers at low Reynolds numbers. J. Fluid Mech. **122,** 13–56

9. Smits, A. J., Matheson, N., Joubert, P. N. (1983): Low Reynolds number turbulent boundary layers in zero and favourable pressure gradients. J. Ship Res. **127**, 147–157
10. Joubert, P. N., Sinclair, T. J., Hoffman, P. H. (1978): A further study of bodies of revolution. J. Ship Res. **22**, 54–63
11. Perry, A. E., Morrison, G. L. (1971): Static and dynamic calibrations of constant-temperature hot-wire systems. J. Fluid Mech. **47**, 765–777
12. Perry, A. E. (1982): *Hot-Wire Anemometry* (Oxford University Press, New York)
13. Klebanoff, P. S. (1955): Characteristics of turbulence in a boundary layer with zero pressure gradient. NACA Report 1247
14. Murlis, J. (1975): "The Structure of a Turbulent Boundary Layer at Low Reynolds Number", Ph.D Thesis, Imperial College, London
15. Watmuff, J. H., Witt, H., Joubert, P. N. (1985): Developing turbulent boundary layers with system rotation. J. Fluid Mech. **157**, 405–448
16. Perry, A. E., Lim, K. L., Henbest, S. M. (1985): "A Spectral Analysis of Smooth Flat-Plate Boundary Layers", Fifth Symposium on Turbulent Shear Flows, Cornell University, Ithaca, New York, USA

Probability Density Distribution in Turbulent Wall Boundary-Layer Flows

F. Durst *, J. Jovanovic ** and Lj. Kanevce **

* Lehrstuhl für Strömungsmechanik, Universität Erlangen-Nürnberg, Egerlandstr. 13, 8520 Erlangen, FRG

** Boris Kidric Institute of Nuclear Sciences, P.O. Box 522, 11001 Beograd, Yugoslavia

Abstract

The present paper provides a summary of a literature survey on statistical properties of wall boundary-layer flows. Existing knowledge on the subject and the authors' own investigations are described that were initially carried out to yield mean and fluctuating flow properties in turbulent wall boundary layers at high Reynolds numbers. From these investigations, instantaneous velocity information resulted which was processed to yield local velocity information in the form of mean velocity and probability density distributions of the turbulent velocity fluctuations. It is shown that the latter obey a general analytical distribution law. Hence, it is possible to describe probability density distributions in turbulent boundary-layer flows in an analytical form. In the sublayer as well as in the wake region of the boundary layer, a simple form of the general distribution, known as the hyperbolic function, describes the probability density distributions of the velocity fluctuations very well. The latter curve contains four parameters whereas the general distribution requires seven parameters for describing the entire probability density distribution of turbulent velocity fluctuations. The distributions of these parameters across a wall boundary-layer flow are given in the paper.

Nomenclature

A	Additive constant of log-law
$\bar{f}_B$	Mean frequency of occurrence of coherent structure
F	Flatness factor
L_τ	Integral time scale
n, k	Coefficients for hot-wire power-law calibration curve
$p(\kappa)$	Probability density distribution
R_θ	Reynolds number based on momentum thickness
S	Skewness factor
SS	Super-skewness factor
SF	Super-flatness factor
$\bar{T}_B$	Mean time duration of coherent structure
Δt	Sampling interval
u	Instantanious streamwise velocity component
u^+	Non-dimensional velocity $u^+ = u/u_\tau$
u_τ	Friction velocity $u_\tau = \sqrt{\tau_w/\varrho}$
U	Average velocity
U_∞	Free stream velocity
$\Delta u/u_\tau$	Coles's wake parameter
y^+	Non-dimensional distance from the wall $y^+ = y u_\tau/\nu$

Definitions of Greek Symbols

γ	Intermittency factor or parameter of general distribution function
δ	Boundary layer thickness or parameter of general distribution function
ϑ	Momentum thickness of investigated boundary layer
κ	Von Karman's constant
ν	Kinematic viscosity
ϱ	Fluid density
τ_w	Wall shear stress

Introduction

Turbulent wall boundary layers have been extensively studied by many research workers in fluid mechanics and a vast amount of information has become available on mean and fluctuating flow properties as well as spectral information. Through these data, it is nowa-

Turbulent Shear Flows 5

days well understood that the far-reaching wall influence on the flow is due to the fact that the integral length scale of the turbulent motion is usually of the order of the boundary-layer thickness and, hence, information on the damping of turbulence by the presence of the wall and information on turbulence production in the near-wall region are transferred to the entire boundary layer. It is this property of wall turbulence, in conjunction with the assumption of a hypothetical homogeneity, a constant shear stress, and the steadiness of the flow structure in a plane at a constant distance from the wall, which leads to the similarity consideration proposed by Prandtl (1927), (1932) and v. Kármán (1930). This similarity hypothesis yields the generally applicable laws for turbulent wall boundary layers when normalized with near-wall variables formed by the wall shear stress, the density and the viscosity of the fluid. Four characteristic regions result which are usually referred to as the viscous sublayer, the buffer region, the log-law region and the wake region. This subdivision is based on numerous experimental results of mean velocity distributions in turbulent wall boundary-layer flows and their presentation as "normalized wall boundary layer velocity profiles".

The above-mentioned general law for the normalized mean velocity profiles of turbulent wall boundary layers has been extensively used in closure procedures (turbulence models) for high-Reynolds-number wall-flow predictions to compute the flow properties for the wall-nearest grid point of a finite difference mesh. In this way, predictions very close to the wall were avoided and computational times reduced to acceptable values, e.g. see Launder and Spalding (1972). It is well known, however, that this approach prevents predictions of wall flows in which strong flow decelerations, or accelerations, flow separations, or drag reduction due to polymer additives, etc. cause deviations from the above mentioned law of the wall. Extensions of turbulence models have therefore been proposed, see Jones and Launder (1972). They suggested taking the low Reynolds number terms into the formulation of the differential equations of the turbulence model and performing flow predictions all the way down to the wall. In order to verify their model and also other turbulence models that are nowadays used in turbulent flow predictions, information on the turbulent fluctuations, such as the distribution of kinetic energy, correlations between the velocity fluctuations, turbulent length scales, etc. are needed over the entire boundary layer. Information like this is available in many publications reporting experimental investigations of turbulent boundary-layer flows. To date most experimental studies have concentrated on providing mean velocity information, and second order moments of turbulent velocity fluctuations in wall boundary-layer flows. Moments up to the 4th order have been reported, e.g. see Zaric (1974), and some papers have provided distributions of moments of up to 18th order, see Anselmet et al. (1984).

In more recent years, turbulence model developments have also used partial differential equations to predict the probability density distribution of turbulent velocity fields and the corresponding probability density distributions of scalar concentrations, e.g. see Pope (1980). This has emphasized the need to obtain experimental information on complete probability density distributions of velocity fluctuations across the entire boundary layer. Such measurements have become available, e.g. see Zaric (1974), and these clearly show the complexity of the flow field of turbulent wall boundary layers. The distributions show strong variations in their shape when compared at for different points in the inner and outer parts of the boundary layer. This finding has led to the belief that it is not possible to provide quantitative information on the complete probability density distribution of turbulent velocity fluctuation in a boundary layer in any other form than a table of the amplitude of the distribution as a function of the instantaneous deviation of the local velocity field from the mean velocity value at the point of consideration. Contrary to this general belief, the present paper finds that the probability density distribution of turbulent velocity fluctuations in wall boundary-layer flows can be described analytically by a single function referred to as the "general distribution function of turbulence flow properties". The normal distribution, the log-normal distribution, the log-hyperbolic distribution, etc. are special cases of this general distribution function. It is to be shown here that probability density distributions of turbulent velocity fluctuations in the sublayer of a turbulent wall boundary layer and in its wake region are described by the hyperbolic form of the general distribution function. Only in the buffer region and in the logarithmic region of the wall boundary-layer flow, the general distribution function has to be used to describe probability density distributions of turbulent velocity fluctuations.

Section 2 of the present paper provides a brief summary of a literature survey carried out to assess the existing knowledge on probability density distributions of velocity fluctuations in turbulent wall boundary-layer flows. The authors' studies revealed that complete information on probability density distributions exists in wall boundary-layer flows only for low Reynolds numbers. The authors carried out investigations of their own for high Reynolds numbers using hot-wire anemometry to obtain local instantaneous velocity measurements. These investigations are described in Sect. 3 where detailed information is also provided on the experimental equipment employed and on its use to obtain mean-flow properties and probability density measurements in high Reynolds number wall boundary-layer flows. In Sect. 4, the basic relationship for the proposed general distribution for turbulent velocity fluctuations is presented and its application to various parts of the boundary layer is shown. Conclusions and final remarks are presented in Sect. 5.

Amplitude Statistical Properties of Wall Boundary-Layer Flows: A Brief Literature Survey

In the past fifteen to twenty years, extensive studies of turbulent wall boundary-layer flows have become available providing evidence that turbulence in the near-wall region, and throughout the entire boundary layer possesses properties that deviate remarkably from those computed from hypothetical "Gaussian flow fields". Measurements of complete probability density distributions of turbulent velocity fluctuations, e.g. see Zaric (1974), and/or their integral properties, i.e. their higher order moments, clearly indicate strong deviations of the flow properties from those predicted from the normal distribution. One finds, for example, in the viscous sublayer and the wake region of the boundary layer strongly skewed probability density distributions also showing high values of flatness factor, e.g. see Johansson and Alfredson (1982), Andreopoulos et al. (1984). These deviations from Gaussian distribution profiles are usually attributed to the pronounced intermittent nature of the flow in these regions. In contrast, the regions of the boundary layer where turbulence production dominates, i.e. in the buffer and the logarithmic region of the flow, high rms-values of the velocity fluctuations are found and smaller values of the third and fourth moment of the velocity fluctuations. However, even for the buffer and the logarithmic region of a boundary-layer flow, the local velocity probability density distributions are far form being Gaussian. This has been reported in numerous publications for the longitudinal velocity fluctuations, e.g. see Zaric (1974), Eckelmann (1974) and Gupta (1971), but the finding can more or less be assumed to be also valid for the cross-flow velocity components perpendicular and parallel to the wall. This was shown in a recent study of Andreopoulos et al. (1984) who used a hot-wire wake sensor to measure local probability density distributions of the velocity fluctuations perpendicular to the wall. These distributions also revealed non-Gaussian shapes.

Non-Gaussian probability density distributions of other fluctuating flow properties such as the cross products $\overline{u_i u_j}$, the veloctiy derivative $(\partial u'/\partial t)$, the vorticity ω' etc. have also been found, e.g. see Zaric (1974), Gupta (1971) and Willmarth (1977). These quantities are assumed to reflect that strongly skewed probability density distributions are always found in flow regions where the flow exhibits strongly intermittent behaviour. In wall boundary-layer flows, it is found, however, that the time derivatives of velocity fluctuations and Reynolds shear stress also possess skewed distributions across the entire turbulent region of the flow. In general one finds more pronounced non-Gaussian probability density distributions for products of turbulent velocity fluctuations and for time derivatives of turbulence properties than for the turbulent fluctuations themselves.

It is often argued that the non-Gaussian characteristics of wall boundary-layer flows are the key to a better physical understanding of the self-maintenance of wall turbulence yielding a detailed understanding of turbulence production, convection, redistribution and dissipation, e.g. see Blackwelder (1978). Although not primarily developed for the task, many researchers in fluid mechanics have employed methods of structural flow analysis to study the believed cause of the non-Gaussian flow behaviour, the so-called coherent flow structures. The employed techniques for coherent structure detection postulate the existence of quasi-ordered patterns in the flow and employ detection schemes that utilize features of non-Gaussian probability density distribution as detection criteria of the structured and non-structured parts of the flow. This yields the strong coupling between the higher-order moments of velocity probability density distributions and the structural patterns of the most energetic fluctuations reported in recent studies by Johansson and Alfredson (1982) and Andreopoulos et al. (1984). The evolution of the reported flow patterns across the wall region from highly skewed profiles in the viscous sublayer to the nearly symmetrical shapes in the buffer and logarithmic region can be directly related to the measured values of the skewness factor. Strong connections like this have also been reported by Chen (1974). The results reported by Andreopoulos et al. (1984) reveal that the skewness factor can be related to the particular structure of coherent flow patterns whereas variations of the flatness factor across the boundary layer are inherently caused by variations of their frequency and their local time of duration. Due to the physics of the flow and the employed detection schemes for coherent structures, the flatness factor should be inversely related to the percentage of time occupied by the higher energy fluctuations of the local velocity field. This can be expressed by a so-called intermittency factor which expresses the product of frequency of occurrence of flow structures and their time of duration. Throughout the available data obtained by conditional sampling in boundary-layer flows, especially those of Wallace et al. (1977), the strong interrelation of the intermittency factor with the properties in the following expression:

$$\gamma = \bar{f}_B \bar{T}_B \sim \frac{1}{F}$$

can be directly related to the high values of the flatness factor in the viscous sublayer.

Considering the above-mentioned complexity of the structure of turbulent wall boundary-layer flows and the reflection of this complexity in the non-Gaussian probability density distribution of turbulent velocity fluctuations has led to the general belief that there is simply no proper way to describe the resultant distributions by a simple formula or even by a single analytical expression. The present paper shows that this beliefe is incorrect; a general distribution function is given that permits probability density distributions of turbulent velocity fluctuations to be described. This function is referred to as the *general distribution function of turbulent velocity fluctuations*. It has been previously used by Barndorff-Nielsen (1979) and coworkers for describing results of statistical processes. It is a function which embraces the normal, the log-normal, the hyperbolic, etc. distributions that have been extensively used in different fields of physics and engineering science to describe parameters of statistical processes. The authors show that this function also describes the amplitude statistics of the velocity field of turbulent wall boundary layers. In some regions a reduced form of the distribution, the so-called hyperbolic distribution, can be employed to adequately describe probability density distributions of turbulent velocity fluctuations.

Experimental Investigations

Wind Tunnel Measuring Section and Measuring Procedure

The present investigations were performed in the closed circuit wind tunnel of the Institut für Hydromechanik of the University of Karlsruhe, details of which are provided by Ermshaus and Naudascher (1977). It consists of a well-designed settling chamber with wire meshes followed by a contraction with an area ratio of 10:1 entering into an 8 m long working section. The cross-sectional shape of the wind tunnel is octagonal with an internal diameter of 1.5 m. The entire roof of the working section is adjustable to permit flow investigations at zero longitudinal pressure gradient. This experimental condition can be achieved without influencing the low turbulence intensity in the free-stream which was approximately 0.06% at 40 m/sec.

The investigated boundary layer developed along a flat plate made of formica which was installed horizontally inside the tunnel section 30 cm from the tunnel floor. The total length of the plate was 6 m and it was supported by strong vertical bars mounted over the entire plate length. This kind of support avoids plate deflections due to static pressure differences. The plate material was chosen to avoid the usually necessary hot-wire corrections close to the wall. This kind of wall material was suggested in a theoretical and experimental study performed by Bhatia et al. (1982) which explains the actual mechanism of the increased heat transfer from hot-wires close to walls. The study showed that heat losses due to wall proximity are negligible if highly heat-insulating wall materials are employed.

The traversing mechanism used in the present study for the hot-wire probes was installed inside the wind tunnel and was driven by an automatic, externally located position control. The actual position driving unit consisted of a stepping motor which permitted precise relative positioning of the hot wire in the three directions of a cartesian coordinate system with a step size of 25 μm in each direction. For measurements close to the wall, the relative motion of the probe was also checked independently by means of a precision dial gauge of 1 μm positioning accuracy. The absolute distance of the wire form the wall surface was obtained through an additional position calibration performed in still air outside the wind tunnel. This calibration was performed on the top of a wall consisting of the same material as the flat plate inside the wind tunnel. It was found that the hot-wire anemometer output voltage was a function of wire distance from the wall as shown in Fig. 1. The actual wire position in this distance calibration was obtained with a precision microscope with a traverse for the scaled objective lens system having a traverse accuracy of 2 μm. The measurements inside the tunnel were started without any flow and the starting wire position was obtained using the position calibration information of Fig. 1.

Special care was taken to position the prongs of the probe parallel to the wall in order to approach the surface as closely as possible and to minimize flow disturbances by the wire prongs and probe support. At higher flow velocities, the hot wire probes were inclined with respect to the wall surface by up to about 5° in order to permit measurements inside the

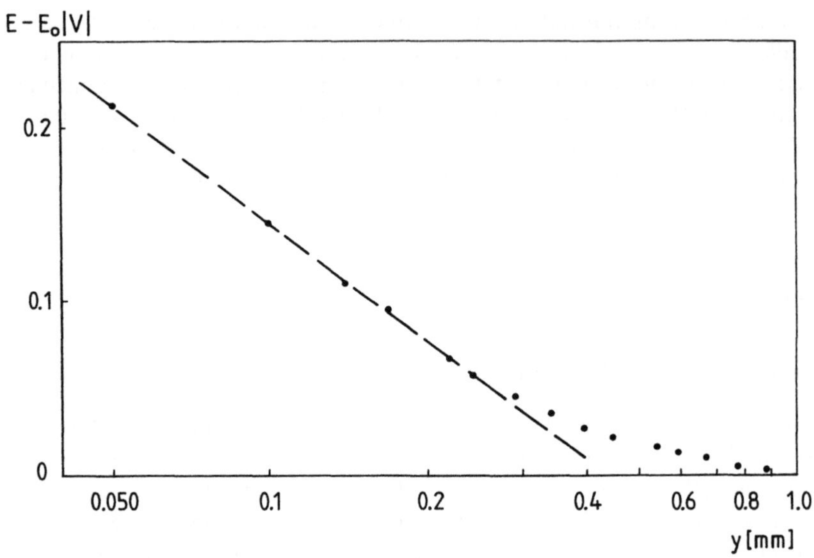

Fig. 1. Calibration of a hot-wire probe for determination of distance from the wall

viscous sublayer of the flow; this region becomes thinner as the Reynolds number of the flow increases.

Hot-Wire Equipment, Calibration and Measurement

Streamwise velocity measurements were performed with standard hot-wire anemometer equipment consisting of DISA P05 gold-plated boundary-layer probe operated with a DISA 55M01 constant temperature anemometer. The overheat ratio for the hot wire was 1.8 and the wire was calibrated for this fixed overheat in the range 0.3 m/s–40 m/s using the modified DISA 55D90 calibration equipment described by Brunn and Tropea (1979) together with a Baratron 3 mmHg pressure transducer.

Since the calibration data presented in Fig. 2 showed the well-known pronounced non-linearity, especially at low velocities, a non-standard method of data evaluation was used in conjunction with the calibration data shown in Fig. 2. Calibration data from Fig. 2 were curve-fitted for parts of the curve by the law proposed by Davies and Brunn (1969),

$$E^2 - E_0^2 = k u^n,$$

where the parameters k and n are velocity dependent, and E_0 corresponds to the anemometer output voltage for zero flow velocity. In order to evaluate the appropriate values of parameters k and n the calibration line from Fig. 2 was suitably subdivided, following the procedure described by Durst et al. (1971) to yield less than 1% deviations when computed by the above defined calibration law. Calibration constants obtained in this way are listed in Table 1. In the range of low velocities they differ considerably from those values defined by King's (1914) calibration law or by the law of Collis and Williams (1959).

After calibration, the hot wire was employed for velocity measurements as described in Sect. 3.1. All hot-wire data, including the calibration data, were recorded on-line by a data aquisition system based on a Hewlett-Packard minicomputer (Model 5451C) facilitated with

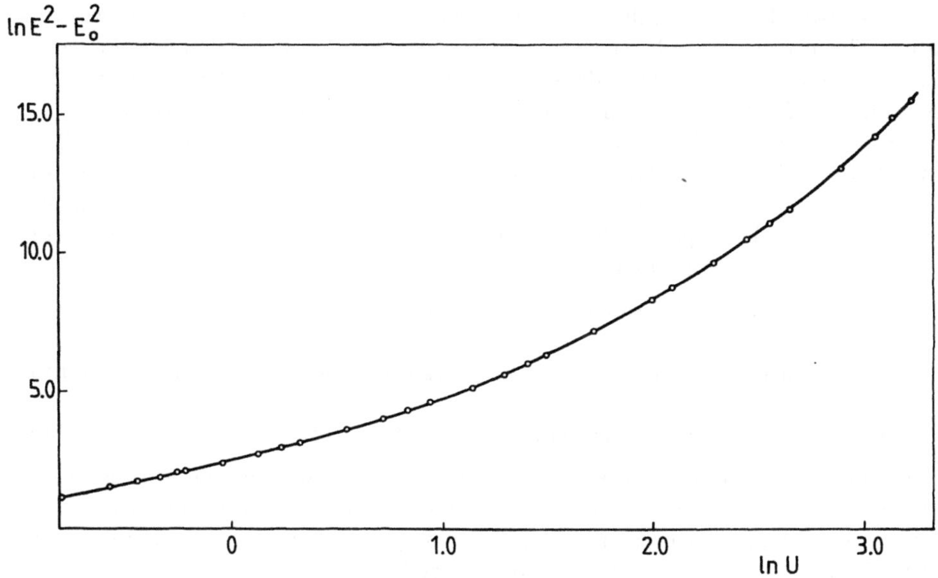

Fig. 2. Calibration data for DISA P05 gold-plated boundary layer probe

Table 1. Calibration constants k and n for DISA P05 gold-platted boundary layer probe

Segment i	Velocity range m/s u_{min}	u_{max}	k_i	n_i
1	0.0	0.6	2.920	1.161
2	0.6	0.8	2.533	0.849
3	0.8	1.4	2.467	0.731
4	1.4	2.5	2.555	0.625
5	2.5	4.4	2.667	0.579
6	4.4	7.3	2.795	0.547
7	7.3	14.0	2.985	0.514
8	14.0	40.0	3.145	0.494

a fast-Fourier analyser. The output of the hot-wire anemometers was first low-pass filtered at 10 kHz or 20 kHz and then digitized and recorded on magnetic tapes. For the actual measurement, signals between 20 sec. and 40 sec. total length were recorded at each measuring point with the recording length being dependent on the flow velocity and the measuring location from the wall; the longer samples in time were taken for the lower velocities and the closer wall positions. The sampling rate for the data was chosen to be between 5 to 20 kHz also depending on the Reynolds number of the flow. These sampling rates ensured the high frequency components of the velocity fluctuations were retained in the signal records. The recorded signals were subsequently analysed employing special software programs written for the UNIVAC 1108 digital computer of the University of Karlsruhe. On this computer the stored signals were converted into velocities and re-stored on tapes for further processing.

Mean Flow Properties

The present measurements have been carried out in the range of Reynolds numbers $3600 \lesssim Re_\theta \lesssim 15\,000$, where Re_θ is formed with the boundary-layer momentum thickness, the velocity far away from the wall and the kinematic viscosity of the gas at the working temperature. The lowest Reynolds number value was chosen in order to compare the present results with boundary-layer velocity data available in the literature. Most available data where provided for Reynolds numbers around this value or even for lower values.

The integral properties of the investigated boundary layers were measured with a Pitot tube connected to a differential Baratron 3 mm Hg pressure transducer. By numerical integration of the measured mean velocity profiles, these properties were found to have the well known values for fully developed turbulent boundary layers. For all investigated flows, skin friction coefficients where computed on the basis of three methods using:

- Mean velocity measurements inside the viscous sublayer obtained by hot-wire anemometry.
- Mean velocity measurements in the logarithmic region of the mean velocity distribution (Clauser-plot).
- Preston tube measurements together with Patel's (1965) calibration curve for such data.

For the higher Reynolds number boundary layers, the latter two methods were employed only because of the extremely small thickness of the viscous sublayer for the higher velocities. The resultant shear stress measurements were checked against von Karman's (1930) integral relationship for zero pressure gradient wall boundary-layer flows

$$\frac{\tau_w}{\varrho U_\infty^2} = \frac{d\theta}{dx},$$

i.e. the measured shear stresses were compared with the measured slopes of momentum thickness increase in the flow direction. The overall agreement among the results of the various skin friction measuring methods was very good yielding an accuracy of around 3–4% of the shear stress employed in this paper for normalizing the velocity information.

Mean-velocity measurements are presented in Fig. 3 for four of the investigated Reynolds numbers. The data are normalized with wall variables yielding the well-known general

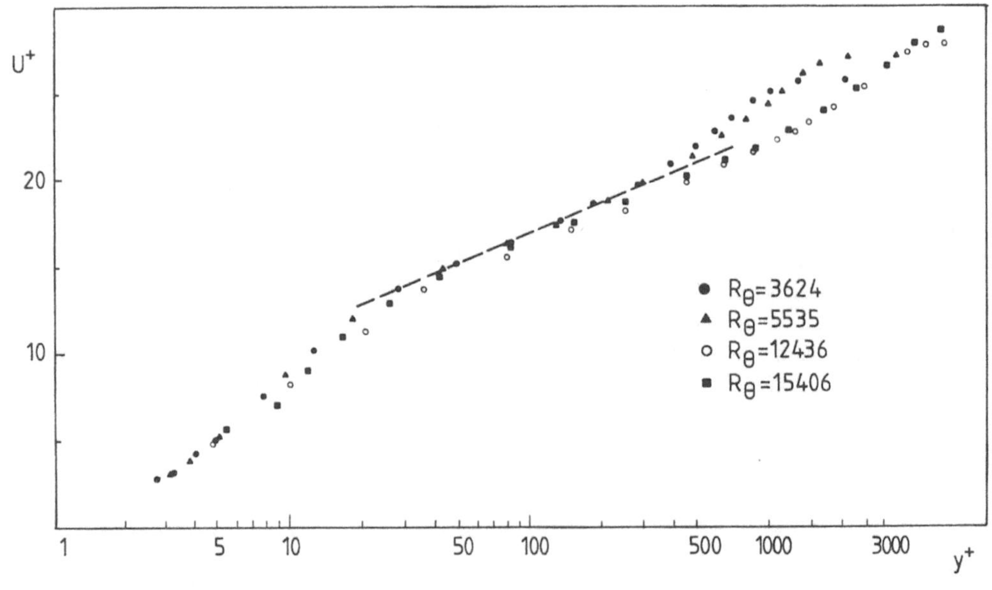

Fig. 3. Mean velocity profiles of investigated turbulent boundary layers

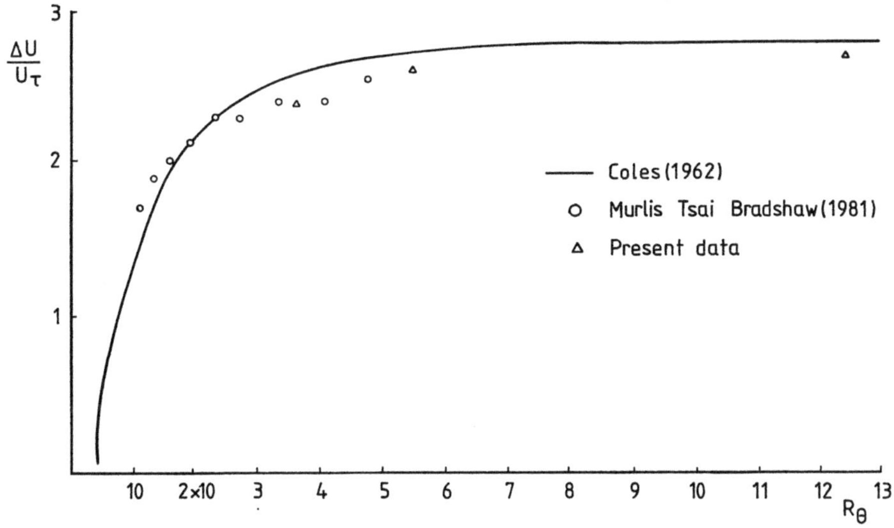

Fig. 4. Excess velocity of wake flow of the investigated boundary layers

distribution of mean velocity across the entire boundary layer. The data show a small Reynolds number dependence which was also noticed in the existing literature, see for example Comte-Bellot (1965) and Clark (1968). In the logarithmic region of the velocity distribution, the present data closely follow a log-law relationship

$$U^+ = \frac{1}{\kappa} \ln y^+ + A, \tag{1}$$

where κ is the von Karman's constant and this was found to have the value of $\kappa = 0.4$ for the data provided here. The additive constant A of the log-law turned out to depend on the Reynolds number and in the range of the present measurements was found to vary between 4.5 and 5.0.

The good agreement of the mean-flow properties of the boundary layers studied in this work is also demonstrated in Fig. 4 where the wake parameter $\Delta U/u_\tau$ as defined by Coles (1956) is plotted as a function of Reynolds number. The authors' data compare well with the data of Murlis et al. (1982) although there are differences between the additive constants A of Murlis et al. (1982) and the present results. This is thought to be due to the differences in the Reynolds numbers of the flows investigated.

The data presented in Fig. 3 readily suggest that the authors were able to eliminate wall influences on their hot-wire measurements by choosing the appropriate wall material, i.e. material with highly heat insulating properties, see Bhatia et al. (1982). This fact is also reflected by the near-wall measurements of rms-values of the longitudinal turbulent velocity fluctuations presented in Fig. 5. This figure shows the present data together with data of Eckelmann (1974) who carried out his measurements in an oil channel flow at an equivalent momentum thickness Reynolds number of $\mathrm{Re}_\theta \approx 300$. Very good agreement of the results is obtained down to $y^+ \approx 1.5$ and this indicates that in the region very close to the wall the measured fluctuations were also unaffected by the wall. The aggreement between the low-Reynolds-number values of Eckelmann (1974) and the high-Reynolds number values of this study also indicates that there is no Reynolds number dependence on the rms-values of the longitudinal velocity fluctuations in this region, if normalized with wall variables.

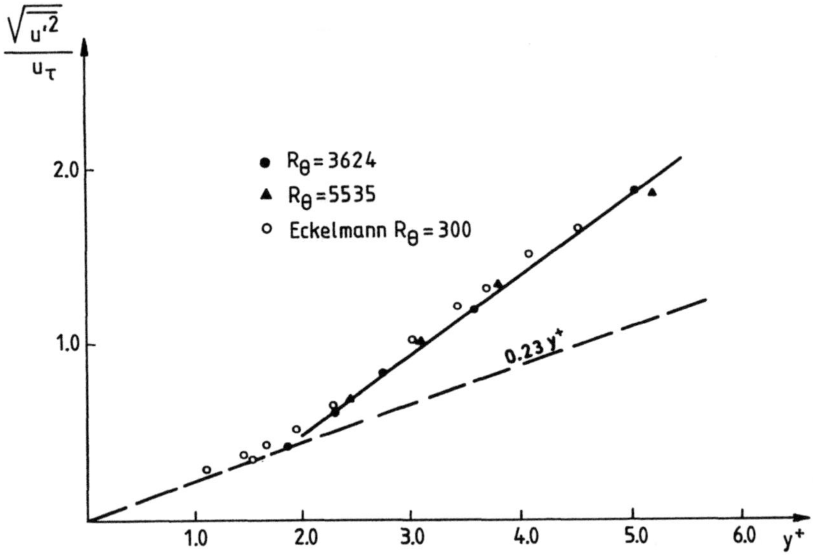

Fig. 5. Turbulence intensity distributions in viscous sublayer

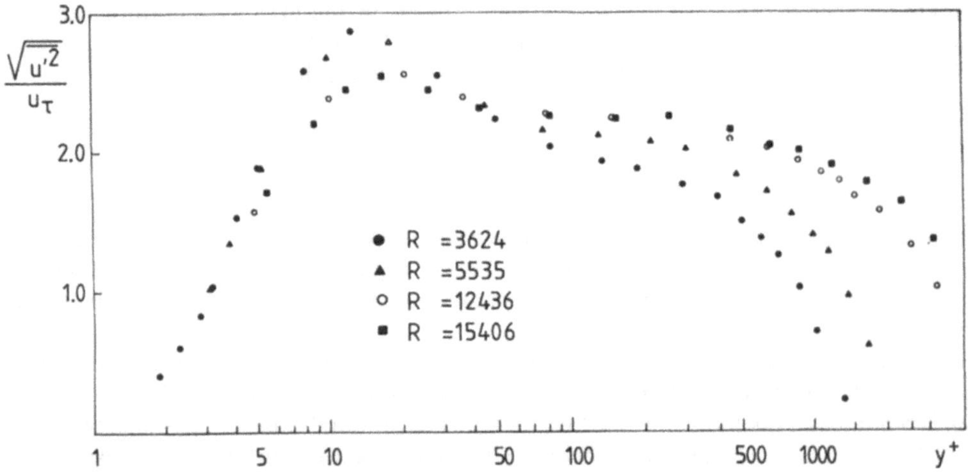

Fig. 6. Turbulent intensity measurements of investigated boundary layer

To complete the information on the rms-values of the longitudinal turbulent velocity fluctuations, Fig. 6 provides information of these values across the entire boundary layer. This figure reflects that there is a Reynolds number dependence on these data, at least in the buffer region. This may be caused by insufficient spatial resolution of the employed measuring sensors, as stressed by Blackwelder and Haritonidis (1983) and also by Johansson and Alfredson (1982).

Probability Density Measurements

From the authors' experimental investigations, continuous time records of the instantaneous, local, longitudinal velocity component, resulted in a time length for each Reynolds number corresponding to some thousands of boundary layer integral time scales δ/U_∞. For four Reynolds numbers, measurements were obtained for up to fifty points across the boundary layer. Using these time records, probability density distributions were computed by counting the relative number of occurrences (N_i/N) of the magnitude of the longitudinal velocity component Δu_i, where N_i was the number of occurrences in the velocity interval:

$$\Delta u_i = u_{i+1/2} - u_{i-1/2}. \tag{2}$$

To obtain the probability density distribution of the longitudinal turbulent velocity fluctuations, the following computations were performed:

$$\begin{aligned} \Delta u &= (u_{max} - u_{min})/NR \\ u_1 &= u_{min} + \Delta u/2 \\ u_i &= u_{i-1} + \Delta u \\ p(u_i) &= \frac{N_i}{N\,\Delta u} \\ i &= 1, 2, \ldots, NR, \end{aligned} \tag{3}$$

were u_{min} and u_{max} are the minimum and maximum values of the velocity in the recorded signals and NR is the total number of samples employed to compute the local probability density distribution. In the present analysis the number of subdivisions NR was chosen to be 25 in order to minimize the measuring uncertainties. Such uncertainties are inherently present and their rms-value can be computed according to Lumley (1970):

$$\overline{\varepsilon_i^2} = \frac{2}{N_i} \cdot \frac{L_\tau}{\Delta t}. \tag{4}$$

This formula shows that $\overline{\varepsilon_i^2}$ strongly depends on the ratio of the local integral time scale of turbulence and the sampling interval Δt, chosen for sampling the hot-wire data. $\overline{\varepsilon_i^2}$ also is inversely proportional to the number of samples that are available per velocity interval Δu_i. Hence, in terms of time and velocity intervals "highly resolved" probability density records show a large scatter of the measured distribution points relative to a mean distribution curve. The scatter also increases in the tail ends of the distribution, where N_i inherently decreases. This is shown in Figs. 7a and 7b for two particular examples of measurements, near the wall and at the outer edge of the boundary layer. An uncertainty analysis was performed and the results for $\overline{\varepsilon_i^2}$ are presented in Fig. 7. Integral time scales L_τ were calculated on the basis of autocorrelation measurements of Van Thinh (1969) and Zaric (1974), e.g. $y^+ = 3.179$, or from the global flow properties such as the boundary-layer thickness δ and the free-stream velocity U_∞, e.g. at $y/\delta = 0.845$. The lower parts of Figs. 7a and 7b are particularly suited to show the increase in the relative uncertainty of the calculated probability density values as one approaches the tail ends of the distribution.

The entire probability density distribution of instantaneous, local, longitudinal velocity fluctuations is shown in Fig. 8 in a form where the actual values are normalized by the local rms-value of the velocity fluctuation. Very near the wall and in the outer wake region of the boundary-layer flow, the probability density distributions show the well-known characteristic skewnesses and flatnesses corresponding to distribution widths exceeding, in some cases, ten rms-values. In these inner and outer regions of the boundary layer, the distributions are

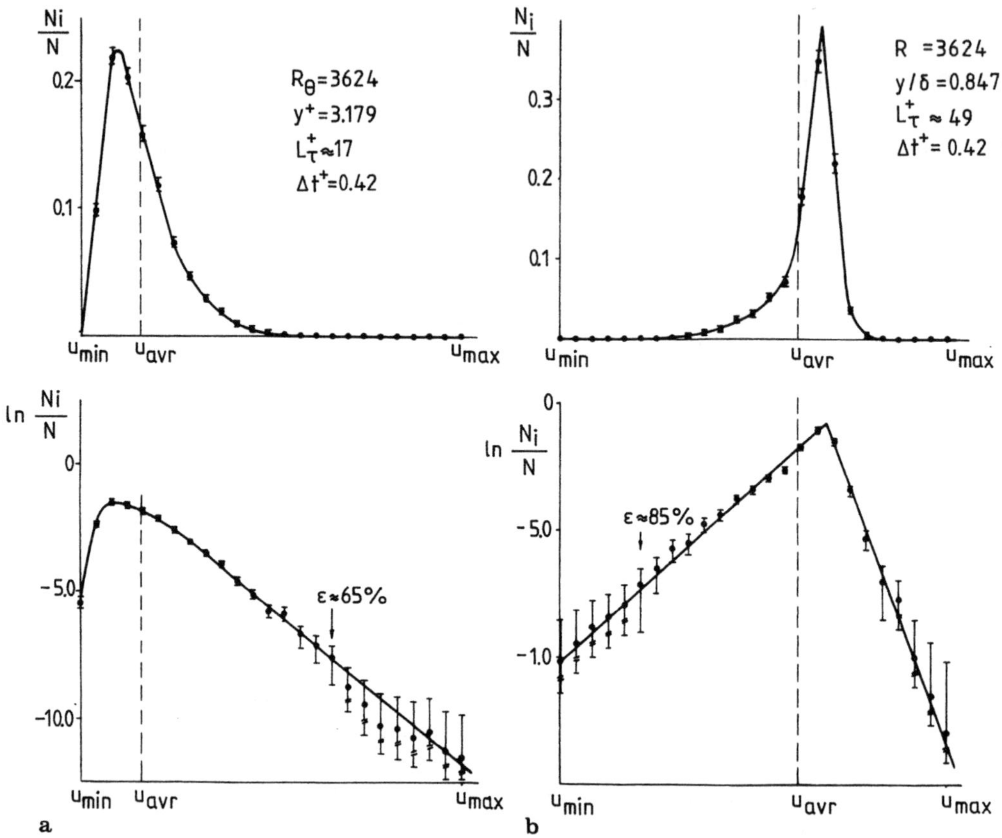

Fig. 7a. Error analysis of probability density measurements; **b** Error analysis of probability density measurements

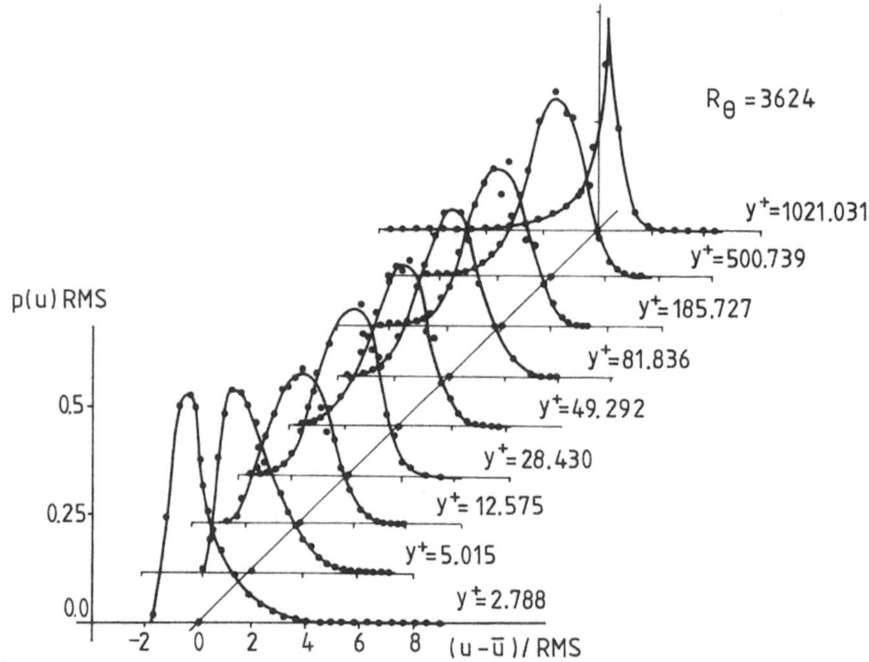

Fig. 8. Probability density distributions of streamwise velocity fluctuations

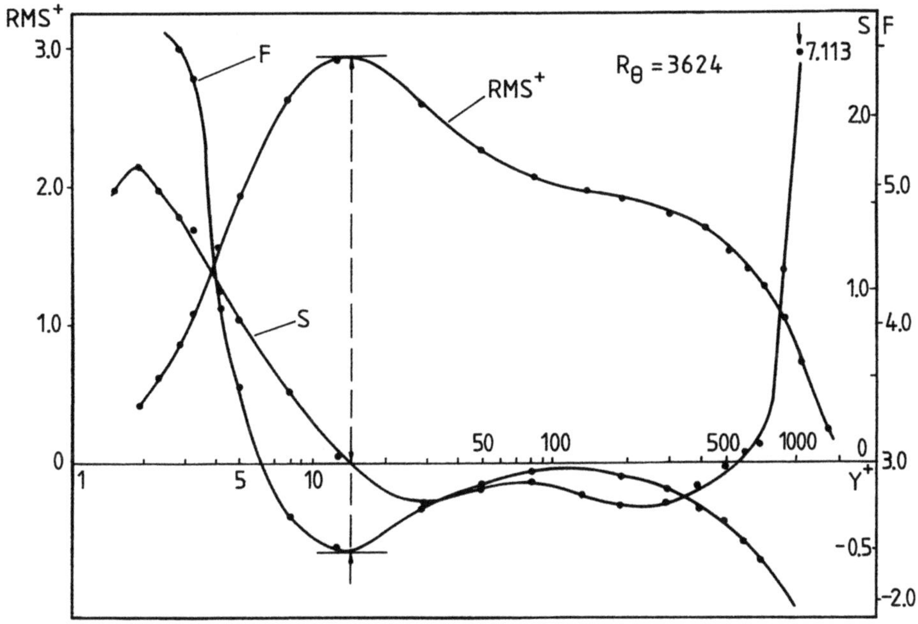

Fig. 9. Second, third and fourth moments of velocity probability density distributions

also characterized by narrow center regions corresponding to low flatness values of the distributions. These characteristic properties are summarized in Fig. 9 which shows the distributions of the second, third and fourth moments of the probability density distributions of Fig. 8.

For further analysis, the probability density distribution of Fig. 8 was replotted in Fig. 10 using semi-logarithmic coordinates. In this way characteristic features of the distributions become apparent, e.g. the fact that the tail ends of the distributions approach some asymptotic linear functions. Expecially in the buffer and the logarithmic region of the flow, the probability density distributions of Fig. 10 are self similar and are described by hyperbolic functions with characteristic tail shapes suggesting the following general behaviour to be valid for all distributions:

$$p(u_i) \to C \exp\{-a(u_i - b)^n\} \quad \text{for } u_i \to \pm\infty,$$

where n is found to lie in the range

$$1 \leqq n \leqq 3.$$

Continuing the studies of properties of measured probability density distributions, the authors have investigated relationships between moments of the distributions and have found that well defined relationships exist between the third and fourth moments, see Fig. 11, and the fifth and sixth moments, see Fig. 12. These figures also contain the appropriate moments of other authors and all data suggest that for subsequent order moments, there exists a remarkably ordered and regular dependence on each other. The resultant distributions also show an originally unexpected symmetry of the functional dependence of the even-number moments on the odd-number moments and this readily suggests that there might be analytical relationships available that describe the local probability density distri-

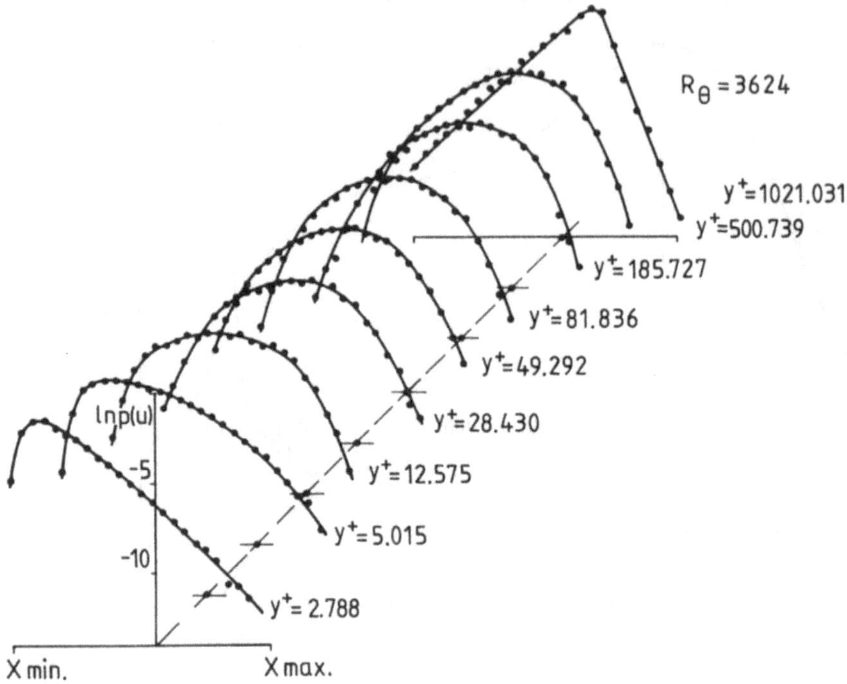

Fig. 10. Semi-logarithmic pot of probability density distributions across boundary-layer

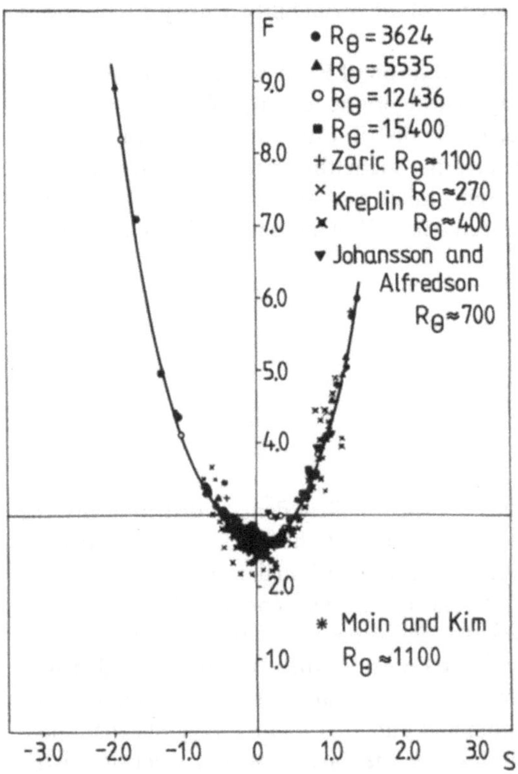

Fig. 11. Fourth moment of a probability density distribution and its dependence on third moment

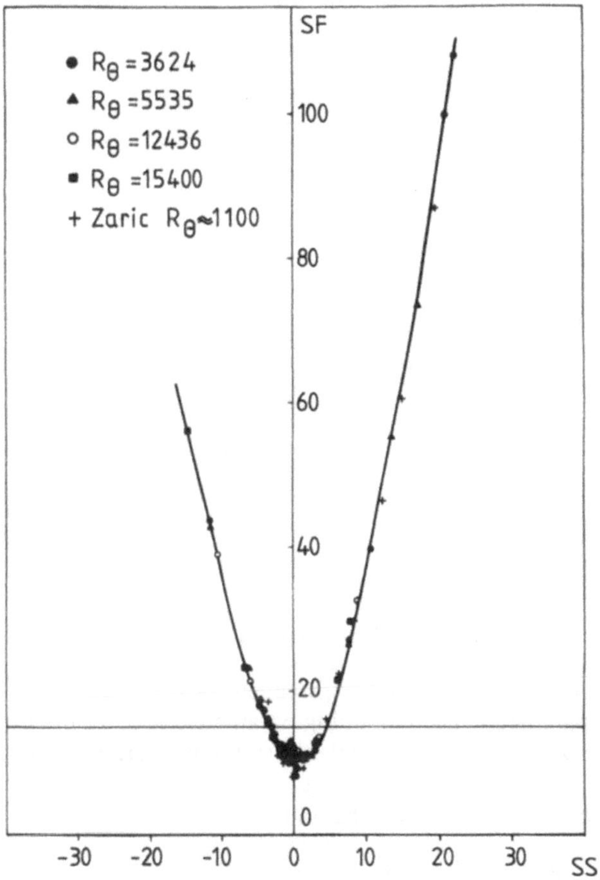

Fig. 12. Sixth moment of probability density distributions and its dependence on fifth moment

butions of turbulent velocity fluctuations. This has previously been postulated by Barndorff-Nielsen (1979) and co-workers who suggested that the probability density distribution of statistical processes should obey the general distribution function given below:

$$p(x) = \exp\{-\tfrac{1}{2}[\phi\{[\delta^2 + (x-\mu)]^{1/2} - (x-\mu)\}^{n_1} + \psi\{[\delta^2 + (x-\mu)]^{1/2} + (x-\mu)\}^{n_2}] + \gamma\}. \tag{5}$$

This is a function containing seven parameters that can be freely chosen to match experimental probability density distributions.

Analytical Description of Probability Density Distribution

Adaption of Distribution Function

Following the work by Barndorff-Nielsen (1979) and co-workers, the authors attempted to describe analytically their measured probability density functions of turbulent velocity fluctuations by the general distribution function given below:

$$p(x, \phi, \psi, \delta, \mu, n_1, n_2) = a \exp\{-\tfrac{1}{2}\phi[\sqrt{\delta^2 + (x-\mu)^2} - (x-\mu)]^{n_1} - \tfrac{1}{2}\psi[\sqrt{\delta^2 + (x-\mu)^2} + (x-\mu)]^{n_2}\} \quad \text{with } \gamma = \ln a, \tag{6}$$

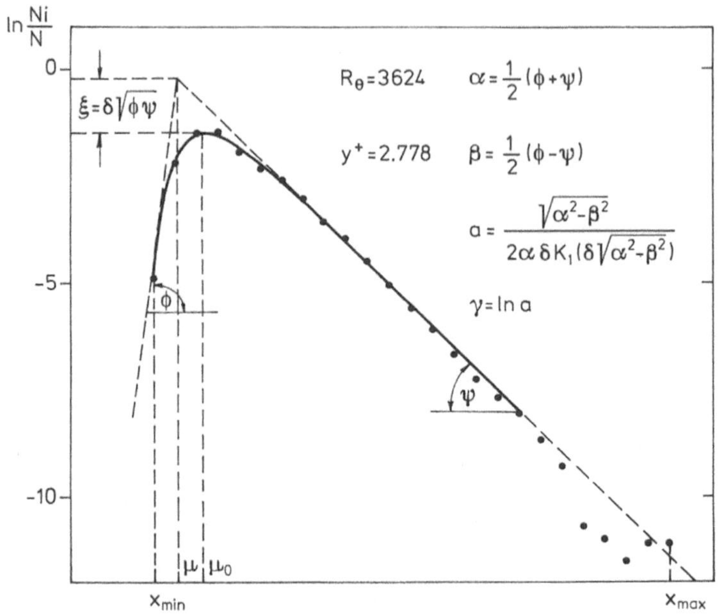

Fig. 13. Parameters of general distribution function and their physical meaning

where a $(\phi, \psi, n_1, n_2, \delta)$ is a normalizing constant, μ is the so-called location parameter ($\mu \in (-\infty, +\infty)$), and $\delta > 0$ the scale parameter. The remaining parameters ϕ, ψ, n_1 and n_2 are parameters with values larger than zero and all being related to the inclination of the asymptotes of the general distribution function sketched in Fig. 13.

The above function shows the following asymptotic behaviour:

$$\begin{aligned} p(x, \phi, \psi, \delta, \mu, n_1, n_2) &\to \exp\{-2^{n_1-1}\phi\,|x-\mu|^{n_1}\} \quad \text{as} \quad x \to -\infty, \\ p(x, \phi, \psi, \delta, \mu, n_1, n_2) &\to \exp\{-2^{n_1-1}\phi\,|x-\mu|^{n_2}\} \quad \text{as} \quad x \to +\infty. \end{aligned} \tag{7}$$

It is readily seen from (7) that symmetric distributions result if the following relationships hold:

$$\phi = \psi \quad \text{and} \quad n_1 = n_2.$$

It can also be deduced from (6) that the normal distribution with mean value μ and variance τ^2 results if the above parameters take on the following values:

$$\phi = \psi = (4\tau^2)^{-1}, \quad n_1 = n_2 = 2, \quad \delta = 0.$$

The hyperbolic distribution

$$p(x, \alpha, \beta, \delta, \mu) = a\exp[-\alpha\sqrt{\delta^2 + (x-\mu)^2} + \beta(x-\mu)] \tag{8}$$

is a special case of the above general distribution and results from the following set of parameters:

$$\alpha = \tfrac{1}{2}(\phi + \psi), \quad \beta = \tfrac{1}{2}(\phi - \psi), \quad n_1 = n_2 = 1.$$

Other known distributions of statistics can be derived if the appropriate parameter combinations are chosen, see Barndorff-Nielsen (1979).

To fit the general distribution function (6) to the authors' experimental probability density distributions, a computer code developed by Bhatia and Durst (1984) was employed with slight modifications to improve its computational performance. Through this code, the parameters of the general distribution function are found employing either a least-square approach or maximum-likelihood principles. For the least-square approach, the method of finding the set of parameters is similar to that developed by Jensen (1983) who introduced a weighting procedure for the experimental points to take into account that the data in the tails of the distribution have a smaller relative reliability than the data in the center of the distribution. Taking this into account involves minimizing the following function:

$$S = \sum_{i=1}^{k} \frac{N_i/N}{1 - N_i/N} \left\{ Y_i - \left[-\frac{1}{2} \phi \{\sqrt{\delta^2 + (x-\mu)^2} - (x-\mu)\}^{n_1} \right.\right.$$
$$\left.\left. - \tfrac{1}{2} \psi \{\sqrt{\delta^2 + (x-\mu)^2} + (x-\mu)\}^{n_2} + \gamma \right] \right\}^2 \tag{9}$$

where $(N_i/N)/(1 - N_i/N)$ is the weighting factor introduced by Jensen (1983) and $Y_i = \ln f_i$ the experimental values of the probability density distribution. The condition to minimize S reduces to a system of seven equations with γ being treated as a "free parameter" to simplify computations:

$$\sum_{i=1}^{k} \frac{N_i/N}{1 - N_i/N} [Y_i - F_i] \frac{\partial F_i}{\partial X} = 0, \qquad X \in \{\phi, \psi, \delta, \mu, n_1, n_2, \gamma\} \tag{10}$$
$$F = -\tfrac{1}{2} \phi [\sqrt{\delta^2 + (x-\mu)^2} - (x-\mu)]^{n_1} - \tfrac{1}{2} \psi [\sqrt{\delta^2 + (x-\mu)^2} + (x-\mu)]^{n_2} + \gamma.$$

To solve the above set of seven equations, the authors employed a Newton-Raphson iterative method. The solution was started with an unknown set of parameters X and successive increments ΔX were computed solving sets of linear equations of the following type:

$$\sum_{i=1}^{k} W_i \left[Y_i \frac{\partial^2 F_i}{\partial X \partial \phi} - \frac{\partial F_i}{\partial \phi} \frac{\partial F_i}{\partial X} - F_i \frac{\partial^2 F_i}{\partial X \partial \phi} \right] \Delta\phi$$
$$+ \sum_{i=1}^{k} W_i \left[Y_i \frac{\partial^2 F_i}{\partial X \partial \psi} - \frac{\partial F_i}{\partial \psi} \frac{\partial F_i}{\partial X} - F_i \frac{\partial^2 F_i}{\partial X \partial \psi} \right] \Delta\psi$$
$$+ \sum_{i=1}^{k} W_i \left[Y_i \frac{\partial^2 F_i}{\partial X \partial \delta} - \frac{\partial F_i}{\partial \delta} \frac{\partial F_i}{\partial X} - F_i \frac{\partial^2 F_i}{\partial X \partial \delta} \right] \Delta\delta$$
$$+ \sum_{i=1}^{k} W_i \left[Y_i \frac{\partial^2 F_i}{\partial X \partial \mu} - \frac{\partial F_i}{\partial \mu} \frac{\partial F_i}{\partial X} - F_i \frac{\partial^2 F_i}{\partial X \partial \mu} \right] \Delta\mu \tag{11}$$
$$+ \sum_{i=1}^{k} W_i \left[Y_i \frac{\partial^2 F_i}{\partial X \partial n_1} - \frac{\partial F_i}{\partial n_1} \frac{\partial F_i}{\partial X} - F_i \frac{\partial^2 F_i}{\partial X \partial n_1} \right] \Delta n_1$$
$$+ \sum_{i=1}^{k} W_i \left[Y_i \frac{\partial^2 F_i}{\partial X \partial n_2} - \frac{\partial F_i}{\partial n_2} \frac{\partial F_i}{\partial X} - F_i \frac{\partial^2 F_i}{\partial X \partial n_2} \right] \Delta n_2$$
$$+ \sum_{i=1}^{k} W_i \left[Y_i \frac{\partial^2 F_i}{\partial X \partial \gamma} - \frac{\partial F_i}{\partial \gamma} \frac{\partial F_i}{\partial X} - F_i \frac{\partial^2 F_i}{\partial X \partial \gamma} \right] \Delta\gamma = -\sum W_i [Y_i - F_i] \frac{\partial F_i}{\partial X}$$

$$W_i = \frac{N_i/N}{1 - N_i/N}, \qquad X \in \{\phi, \psi, \delta, \mu, n_1, n_2, \gamma\}.$$

In order to improve convergence for experimental probability density distributions for which only rough initial estimates were available, a convergence factor Ω was introduced as proposed by Bhatia and Durst (1984).

$$X^i = X^{i-1} + \Delta X\,\Omega, \qquad \Delta X \in \{\Delta\phi, \Delta\psi, \Delta\delta, \Delta\mu, \Delta n_1, \Delta n_2, \Delta\gamma\}. \tag{12}$$

At the start of computations, small values of Ω were chosen. As the computations progressed, Ω was increased up to $\Omega = 1$ and maintained at this value until convergence for S being a minimum was achieved. Thereafter, the parameter γ was reevaluated in order to satisfy the normalization condition:

$$\int_{-\infty}^{+\infty} f(x)\,dx = 1.$$

With the new value of γ, the other parameters were recomputed to yield the final values of all parameters. Only small modifications of all parameters were usually obtained during this repetition of the parameter computation.

Parameter Distribution Across Boundary Layer

It has been common practice to describe the turbulent flow field of wall boundary layers by the normalized mean velocity distribution $U^+ = f(y^+)$ and by the corresponding distribution of the rms-value of the local turbulent velocity fluctuations. To characterize the deviations of the probability density distribution of local, turbulent velocity fluctuations from the corresponding normal distribution ($\sigma^2 = \overline{u'^2}$), distributions of skewness and flatness factors are given in a number of papers, e.g. see Zaric (1974). Hence, distributions of about 4 parameters are commonly used to describe turbulent wall boundary flows. But the description is incomplete because no information is provided on higher moments of the probability density distributions of the velocity fluctuations.

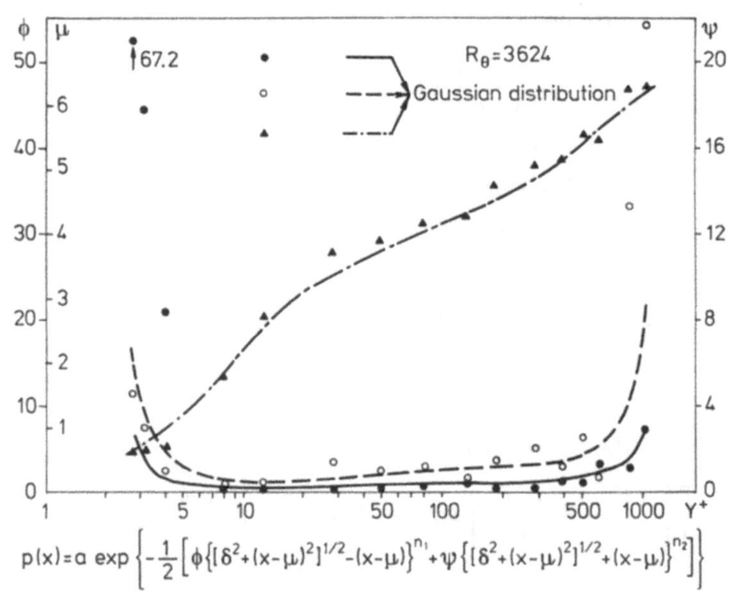

Fig. 14. Distribution of ϕ, ψ and μ across a turbulent wall boundary-layer

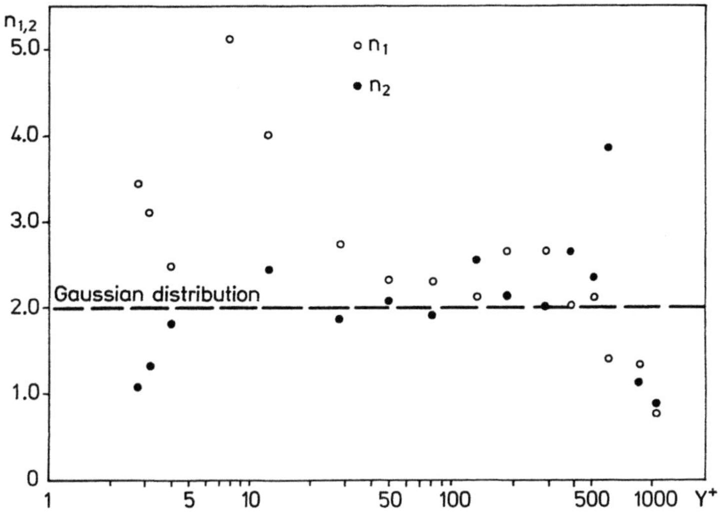

Fig. 15. Distribution of n_1 and n_2 across the turbulent boundary-layer

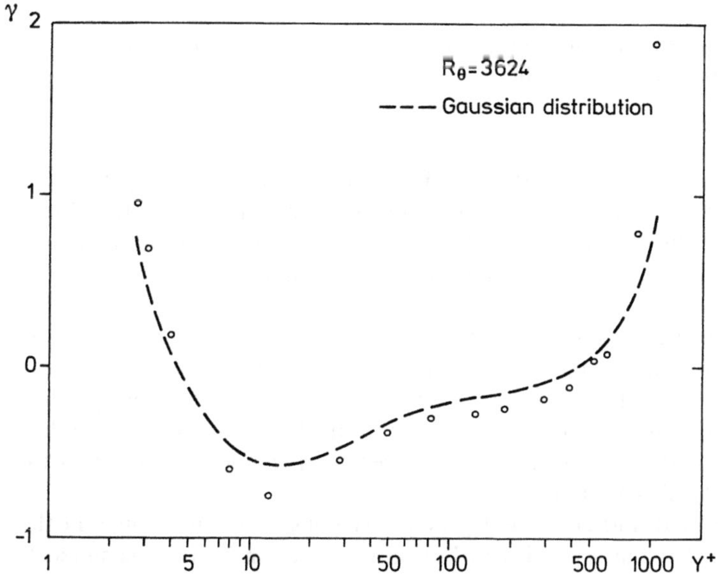

Fig. 16. Distribution of γ across a wall boundary-layer

Using the general distribution function given in Sect. 3.1, an alternative description of the velocity field of wall boundary-layer flows results in terms of the parameters of the general distribution function. Employing the computational procedure outlined in Sect. 3.1, the parameters were computed matching the general distribution function to the experimental data. These computations resulted in $\delta \approx 0$ over the entire boundary layer. The parameters ϕ, ψ, μ, n_1, n_2 and γ were computed to have non-zero values and their distributions are given in Figs. 14 to 16. The parameters are plotted as a function of the normalized wall distance y^+. Large variations occur for all the parameters where ϕ and ψ strongly increase close to

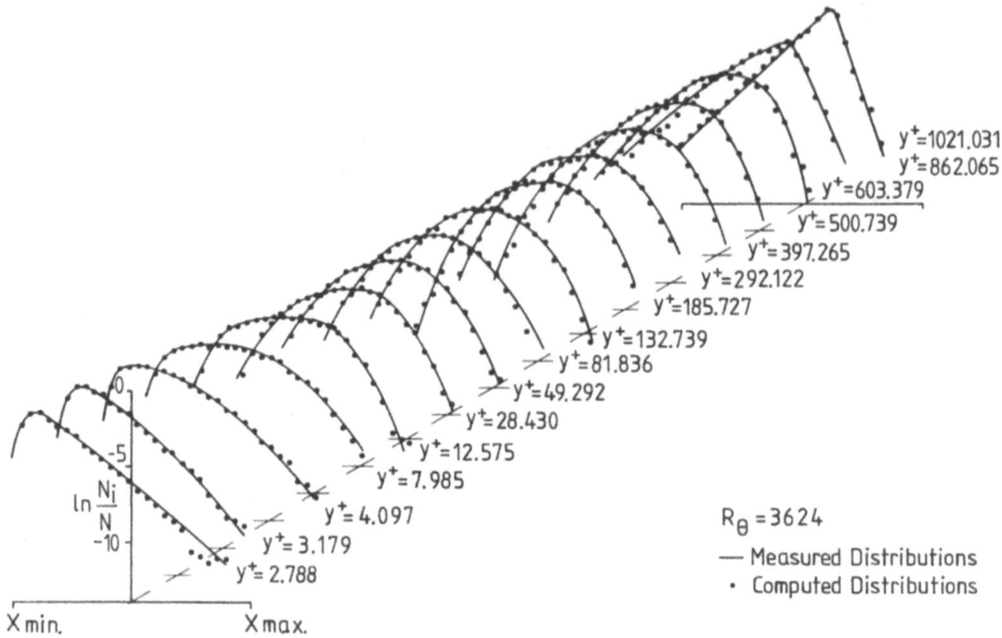

Fig. 17. Comparison of measured and computed probability density distribution. Computations employed the proposed general distribution function

the wall but show nearly constant values in the logarithmic region of the velocity profile of the boundary layer. The μ-distribution resembles features of the normalized mean velocity profile of wall boundary-layer flows. It is not equal to the mean velocity distribution because μ is the location parameter of the general distribution function and not the mean velocity itself.

In the present treatment of the data, the distributions of the parameters n_1 and n_2 did not come out to be very smooth. The reasons for this are due to the strong skewnesses of some of the probability density distributions the nature of the velocity fluctuations yielding only one or two points on the steep slopes of the distributions. Re-evaluations of the data are necessary with respect to these parameters using finer Δu_i increments in parts of the measured probability density distributions.

Figure 17 presents measured and computed probability density distributions and this figure clearly demonstrates that the general distribution proposed in Sect. 3.1 is suited to describe the probability density distributions of longitudinal velocity fluctuations in wall boundary-layer flows. Using the parameters in Figs. 14 to 16 allows higher order moments of the probability density distribution to be computed at each point of the flow. Hence, the parameters given in Figs. 14 to 16 represent a more complete description of the flow field than other plots of parameters that are presently available.

Conclusions and Suggestions for Further Research

Based on experimental investigations of wall boundary-layer flows and their evaluation to obtain higher order moments of local probability density distributions of turbulent velocity fluctuations, an unexpected regularity of distribution of the following quantities were ob-

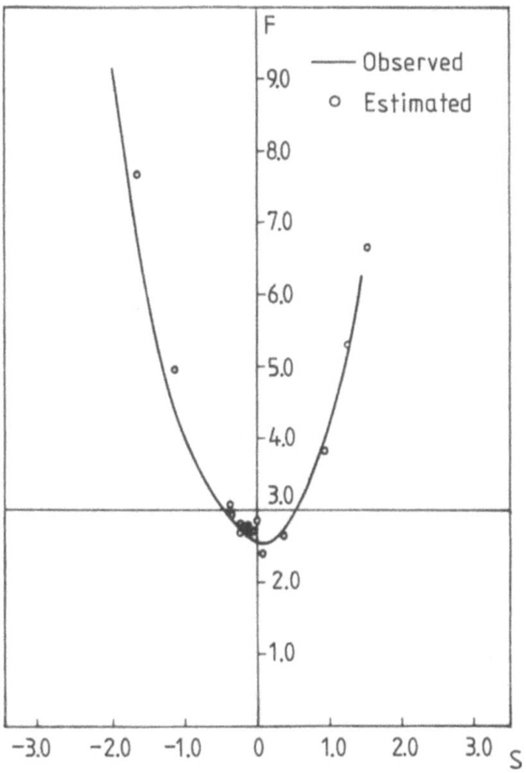

Fig. 18. Flatness factor as a function of skewness factor; prediction and measurements

served ($n > 2$):

$$F_{2n} = \frac{\overline{u'^{2n}}}{(\overline{u'^2})^n} \quad \text{versus} \quad S_{2n-1} = \frac{\overline{u'^{(2n-1)}}}{(\overline{u'^2})^{(n-1/2)}}.$$

The authors' findings compare well with corresponding data of other investigations, e.g. see Zaric (1974), Kreplin (1979), Johannsson and Alfredson (1982), Moin and Kim (1982). A general distribution function for probability density fuctions of turbulent velocity fluctuations is suggested in the paper and the parameters are computed from the experimental data presented. Using these parameters, the third to sixth moments of probability density distribution were re-computed at the locations for which measurements were available. Figures 18 and 19 show the re-computed values in comparison to the mean line through the measurements. The findings suggest that the proposed general distribution function is capable of satisfactorily describing the experimentally found dependence of the even numbered moments on the presequent odd numbered moments. This dependence seems to be Reynolds number independent, at least in the range 1 : 20 covered by the data in the literature and the data provided here.

There have been other attempts to treat correlations in turbulent flows, e.g. the treatment by Frenkiel and Klebanoff (1967 a) (1967 b), (1973), or by Van Atta and Chen (1968) and Van Atta and Yeh (1970) in terms of truncated Gram-Charlier distributions. These distributions describe probability density distributions as infinite series expansions yielding terms of higher order correlations and Hermite polynomials with the first term of the expansion being the Gaussian probability density distribution. Truncating such a series expansion at its fourth order term allows the fifth order moment (super skewness) to be related to the third

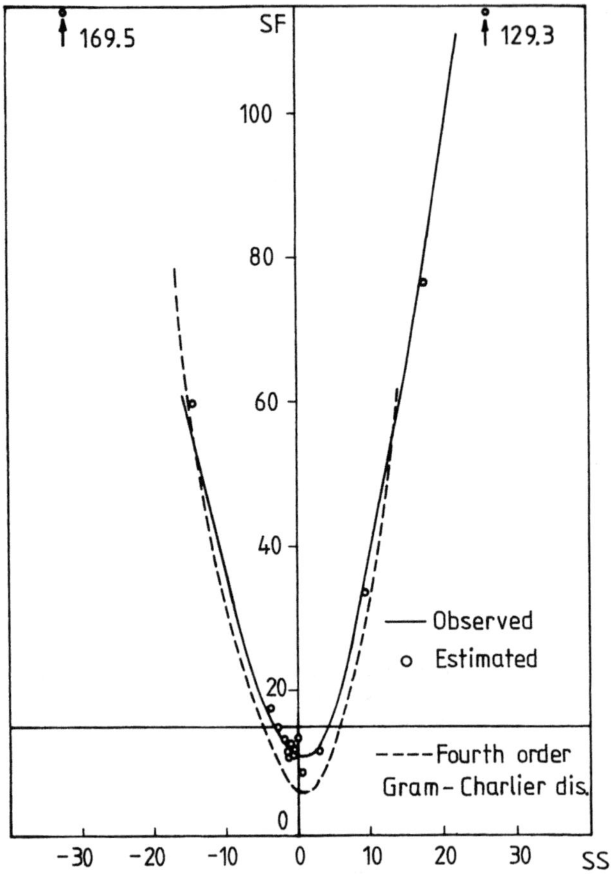

Fig. 19. Super flatness factor as a function of super skewness; predictions and measurements

order moment

$$SS = 10\,S$$

and the sixth order (super flatness) moment to the fourth order moment

$$SF = 15\,F - 30.$$

Using these relationships and the experimental results of Fig. 18 allowed the broken line in Fig. 19 to be computed. Figure 19 readily suggests that the general distribution function suggested in this paper presents better agreement with experimental findings than the Gram-Charlier series expansion. In addition, the general distribution function uses a finite set of parameters to describe probability density distributions of turbulent velocity fluctuations and is therefore easier to apply than the infinite number of terms of the Gram-Charlier series expansion.

Further studies in this field should take into account the fact that the general distribution function, used in the present study, belongs to a family of distribution functions that can be thought of as being made up of "mixtures" of simpler distributions. Of particular interest are those that consist of a mixture of normal distributions and may be defined as

$$H(x;\ldots) = N(\xi, \sigma^2)\,\widehat{\sigma^2}\,M(\sigma^2, \ldots),$$

where $N(\xi, \sigma^2)$ represents the normal distribution with mean value ξ and σ^2 as standard deviation. The function M is the so-called mixing function and $\widehat{\sigma^2}$ expresses the fact that the mixing is carried out with respect to σ^2. Applying this approach to a mixture of normal distributions obeying the following linear relationship between the mean value ξ and the variance σ^2

$$\xi = \mu + \sigma^2$$

together with the inverse Gaussian mixing distribution for σ^2,

$$M = \frac{1}{2} \frac{(\alpha^2 - \beta^2)^{1/2}}{\delta k_1 [\delta(\alpha^2 - \beta^2)^{1/2}]} \exp\left\{-\frac{1}{2}\left[\frac{\delta^2}{\sigma^2} + (\alpha^2 - \beta^2)\,\sigma^2\right]\right\},$$

results in the hyperbolic distribution. This distribution can be shown to satisfy all the requirements of the central limit theorem of statistics. It is therefore an "asymptotic distribution" of statistical systems. Hence, if the turbulent flow field at a point in space is made up of a sum of contributions of flow fields that are described by normal distributions, the local turbulent flow field might deviate from the normal distribution and still satisfy all the requirements of the central limit theorem.

Acknowledgements: The authors would like to acknowledge the support of this work by the Deutsche Forschungsgemeinschaft and by the Bundesministerium für Forschung und Technologie. The second author recieved also support from the Alexander-von-Humboldt-Foundation to participate in this work. In addition, the authors would like to thank Prof. R. S. L. Lee of the State University of New York at Stony Brook for his reading of the manuscript.

Miss A. Messner, Miss M. Laternik and Mr. Kaschak from the LSTM-Erlangen contributed to the completion of the final manuscript.

References

Andreopoulos, J., Durst, F., Jovanovic, J., Zaric, Z. (1984): Influence of Reynolds number on characteristics of turbulent wall boundary layers. Exp. Fluids **2**, 7–16

Anselmet, F., Gagne, Y., Hopfinger, J. E., Antonia, A. R. (1984): High-order velocity structure functions in turbulent shear flows. J. Fluid Mech. **140**, 63–89

Barndorff-Nielsen, O. (1979): Models for non-Gaussian variation with applications to turbulence. Proc. R. Soc. London **A 368**, 501–520

Bhatia, J. C., Durst, F., Jovanovic, J. (1982): Corrections of hot-wire anemometer measurements near walls. J. Fluid Mech. **122**, 411–431

Bhatia, J. C., Durst, F. (1984): On the estimation of parameters for two non-Gaussian distributions. Report LSTM 37/T/84, University of Erlangen-Nürnberg

Blackwelder, R. F., Haritanidis, H. J. (1983): The bursting frequency in turbulent boundary layers. J. Fluid Mech. **132**, 87–103

Blackwelder, R. F. (1978): "Pattern Recognition of Coherent Eddies," Proc. Dynamic Flow Conference – Dynamic Measurements in Unsteady Flows, Marseille-Baltimore

Blaesild, P. (1978): The shape of generalized inverse Gaussian and hyperbolic distributions. Research Report, No. 37, Dep. of Theoretical Statistics, University of Aarhus

Brunn, H. H., Tropea, C. D. (1979): The SFB 80 nozzle calibration facility. Report 80/M/14, University of Karlsruhe

Chen, C. H. F. (1974): "The Large Scale Motion in a Turbulent Boundary Layer: A Study Using Temperature Contamination;" Ph.D. Thesis, University of Southern California

Clark, J. A. (1968): A study of incompressible turbulent boundary layers in channel flow. Trans. ASME, Journal Basic Eng. D90, 455–468

Coles, D. (1956): The law of the wake in the turbulent boundary layer. J. Fluid Mech. **1**, 191–226

Collis, D. C., Williams, M. J. (1959): Two-dimensional convection from heated wires at low Reynolds numbers. J. Fluid Mech. **6**, 357

Comte-Bellot, G. (1965): Ecoulement turbulent entre deux parois paralleles, Pub. Sci. et Tech. No. 419, Paris
Davies, P. O. A. L., Brunn, H. (1969): The performance of yawed hot wire. ISVR Report No. 284
Durst, F., Melling, A., Whitelaw, J. H. (1971): The interpretation of hot-wire signals in low-turbulence flows. Report Department of Mechanical Engineering, Imperial College, London S.W. 7
Eckelmann, H. (1974): The structure of the viscous sublayer and the adjacent wall region in a turbulent channel flow. J. Fluid Mech. **65**, 439–459
Ermshaus, R., Naudascher, E. (1977): Der Niedergeschwindigkeitskanal des Instituts für Hydromechanik an der Universität Karlsruhe. Z. Weltraumforsch. **1**, 419–425
Frenkiel, F. N., Klebanoff, P. S. (1967): Higher-order correlations in a turbulent field. Phys. Fluids **10**, 507–520
Frenkiel, F. N., Klebanoff, P. S. (1967): Correlation measurements in a turbulent flow using high-speed computing methods. Phys. Fluids **10**, 1737–1747
Frenkiel, F. N., Klebanoff, P. S. (1973): Probability distributions in a turbulent boundary layer. Phys. Fluids **16**, 725–737
Gupta, A. K. (1971): "An Experimental Investigation of the Viscous Sublayer Region in a Turbulent Boundary Layer;" Ph.D. Thesis University of Southern California
Jensen, J. L. (1983): Maximum likelihood estimation of the hyperbolic parameters from grouped observations. Research Reports No. 91, Department of Theoretical Statistics, University of Aarhus
Johansson, V. A., Alfredson, P. H. (1982): On the structure of turbulent channel flow. J. Fluid Mech. **122**, 295–314
Jones, W. P., Launder, B. E. (1972): The prediction of laminarization with a two equation model of turbulence. Int. J. Heat Transfer **15**, 301
King, V. L. (1914): The convection of heat from small cylinders in a stream of fluid. Determination of the convection constants of small platinium wires with application to hot-wire anemometry. Philos. Trans. Roy. Soc. London **A 214**, 373–432
Kreplin, H., Eckelmann, H. (1979): Behavior of the three fluctuation velocity components in the wall region of a turbulent channel flow. Phys. Fluids **22**, 1233–1239
Launder, B. E., Spalding, D. B. (1972): Mathematical Methods of Turbulence (Academic, London, New York)
Lumley, J. (1970): Stochastic Tools in Turbulence (Academic, London, New York)
Murlis, J., Tsai, M. H., Bradshaw, P. (1982): The structure of turbulent boundary layers at low-Reynolds numbers. J. Fluid Mech. **122**, 13–56
Moin, P., Kim, J. (1982): Numerical investigation of turbulent channel flow. J. Fluid Mech. **118**, 341–377
Patel, V. C. (1965): Calibration of the Preston tube and limitations on its use in pressure gradients. J. Fluid Mech. **23**, 185–208
Prandtl, L. (1927): Über den Reibungswiderstand strömender Luft. Ergeb. AVA Göttingen, III. Lfg. 1–5
Prandtl, L. (1932): Zur turbulenten Strömung in Rohren längs Platten. Ergeb. AVA Göttingen IV. Lfg. 18–29
Pope, S. B. (1980): "Probability Distributions of Scalars in Turbulent Shear Flow." in Turbulent Shear Flows 2, ed. by L. J. S. Bradbury et al. (Springer, Berlin, Heidelberg) p. 7
Van Atta, C. W., Chen, W. Y. (1968): Correlation measurements in grid turbulence using digital harmonic analysis. J. Fluid Mech. **34**, 497–515
Van Atta, C. W., Yeh, T. H. (1970): Some measurements of multi-point time correlations in grid turbulence. J. Fluid Mech. **41**, 169–178
Van Thinh, J. (1969): "Auto-correlation Measurements of Turbulent Velocity Fluctuations in the Viscous Sublayer Using Real Time Digital Techniques," International Seminar Herceg Novi-Yugoslavia
Von Karman, T. (1930): Mechanische Ähnlichkeit und Turbulenz, Nachr. Gesellsch. Wissensch. Göttingen, Math.-Phys. Klasse 1930, 58–76
Wallace, J., Bordkey, S. R., Eckelmann, H. (1977): Pattern recognized structures in bounded shear flows. J. Fluid Mech. **83**, 673–693
Willmarth, W. W. (1977): Survey on new measurements of turbulence structure near the wall. Phys. Fluids Suppl. **20**, No. 10, 9–21
Zaric, Z. (1974): Etude statistique de la turbulence parietale. Report Institut Boris Kidric-Vinca

Evolution of a Vortical Structure Associated with the Bursting Event in a Channel Flow

John Kim

NASA Ames Research Center, Moffett Field, California 94035, USA

The temporal evolution of a horseshoe vortex structure in a channel is investigated by a numerical simulation. A spectral numerical method is employed to integrate the time-dependent, three-dimensional Navier-Stokes equations. The initial vortical structure is obtained by applying a conditional sampling technique to a data base generated from a direct simulation of a turbulent channel flow. The evolution of this vortical structure under the influence of the self-induced motion and the mean shear is presented. It is shown that the initial sheet-like vortical structure rolls up into a vortex tube as it is convected downstream. Turbulence characteristics associated with the vortex are investigated. Production of vorticity due to vortex stretching is high inside the vortex legs, although it is also substantial in the tip region and above the legs. High Reynolds shear stress is produced near the tip of the vortex.

Introduction

Since Theodorsen [1] proposed a horseshoe vortex as the primary structure responsible for the creation of turbulence, both experimentalists and theoreticians have made extensive investigations of this structure (see [2] for a detailed review). The flow-visualization work of Head and Bandyopadhyay [3], in particular, provided a strong evidence for the presence of a horseshoe vortex in turbulent boundary layers. Recently Acarlar and Smith [4] showed a striking similarity between a laminar flow in which synthetically generated horseshoe vortices were embedded and a turbulent flow during a bursting process, thus providing an association of the horseshoe vortex with the bursting process in wall-bounded turbulent shear flows. Moin and Kim [5] conducted an investigation of the vortex structure using a data base obtained from a large-eddy simulation (LES) of a turbulent channel flow [6]. Detailed statistical analyses of the computed three-dimensional vorticity field showed the presence of a horseshoe vortex structure in a turbulent channel flow. It was shown that the vortical structure is generated from deformation (or roll-up) of sheets of transverse vorticity.

To investigate the organized structures associated with the bursting process, Kim and Moin [7] applied various conditional sampling techniques to data bases obtained from numerical simulations (direct simulation as well as LES). It was found that there exist two distinct horseshoe-shaped vortical structures associated with the bursting process: one associated with the ejection event and the other with the sweep event. The ensemble-averaged vortical structures have generally the same characteristics as the instantaneous structures, but they are smeared out by the averaging process. It was shown that the vortical structures are associated with high Reynolds shear stress, indicating that these structures make a significant contribution to turbulent energy production.

In this paper, we present some preliminary results from the investigation of a time-evolution of a horseshoe vortex structure in a channel flow. The objective of this work is to

Turbulent Shear Flows 5

study dynamics of the vortex structure in order to to gain a better understanding of its role in turbulent transport processes. For this purpose one has to follow the development of such a structure dynamically to observe the processes by which it is created, evolved, and annihilated. Performing this experiment in a real turbulent flow situation – numerically as well as in a laboratory experiment – is a rather difficult task. Instead, as a first step, we set out to examine the evolution of an already existing vortical structure in a nonturbulent environment. This is done by using one of the ensemble-averaged velocity fields of Kim and Moin [7] as an initial field. This field contains an ensemble-averaged vortical structure that is associated with the bursting process. Starting from this initial velocity field, time-dependent, three-dimensional Navier-Stokes equations are integrated in time. The resulting temporal evolution of this vortical structure and the turbulence characteristics associated with it are presented in this paper.

Numerical Method and Initial Field

A computer code developed for a direct simulation of turbulent channel flows [7] was used for this study. Details of the numerical method will be reported elsewhere, but a brief description is given below. A spectral method – Fourier series in the streamwise and spanwise directions and Chebyshev polynomial expansion in the normal direction – was used for spatial derivatives. The time advancement was made by a semi-implicit method: the Crank-Nicolson scheme for viscous terms and the Adams-Bashforth method for the nonlinear terms. The computation was carried out with about 2×10^6 grid points ($128 \times 129 \times 128$, in x, y, z) for a Reynolds number of 3200, based on the centerline velocity U_c, and channel half-width δ, (180 based on the wall shear velocity, u_τ). The grid spacings in the steamwise and spanwise directions were $\Delta x^+ \approx 18$, and $\Delta z^+ \approx 6$ in wall units. Nonuniform meshes were used in the normal direction; the first mesh point away from the wall was at $y^+ \approx 0.05$, and the maximum spacing (at the centerline of the channel) was about 4.4 in wall units. The streamwise and spanwise extents of the computational box were about 2300 and 750 in wall units, respectively ($12.8\,\delta$ and $4.2\,\delta$). No subgrid-scale model was used in the computation, since the grid resolution was sufficiently fine to resolve all the essential turbulent scales. The same calculation was carried out with about 4×10^6 grid points ($192 \times 129 \times 160$, in x, y, z) for a short period for a comparison, and essentially no difference was found between the two results.

The data base generated from the direct simulation of Kim and Moin [7] was used to construct an initial field. The quadrant analysis of Lu and Willmarth [8] was applied to the data base, and the initial field was obtained by ensemble-averaging over the events in the second quadrant ($u' < 0$, $v' > 0$) with the Reynolds shear stress higher than 10 times the mean value at $y^+ = 100$ (see [7] for detail). The ensemble-averaged velocity field contains an organized vortical structure as a perturbation to the mean velocity field. The initial velocity field in the channel was set such that the lower half of the channel contains the ensemble-averaged field and the upper half of the channel contains the time-averaged velocity field.

Results

One way of identifying three-dimensional vortical structures is by tracing vortex lines in three-dimensional space, $\boldsymbol{x}$. The lines are defined by

$$\frac{d\boldsymbol{x}}{ds} = \frac{\boldsymbol{\omega}}{|\boldsymbol{\omega}|}, \tag{1}$$

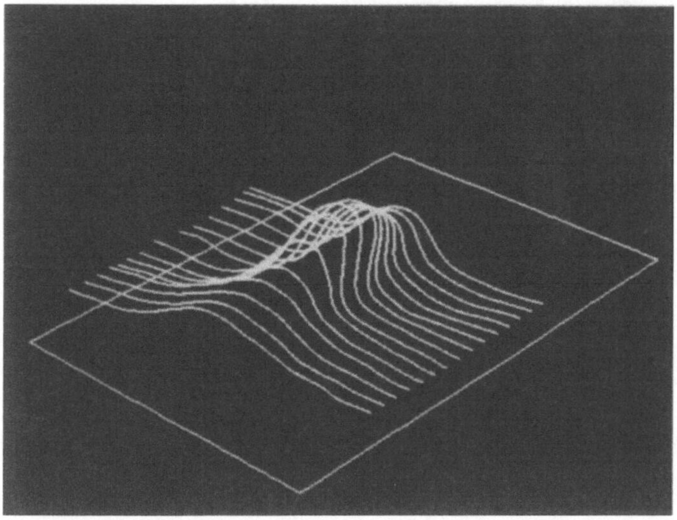

Fig. 1. Vortex lines displaying the initial vortical structure. This is a perspective view of a three-dimensional field. The mean flow direction is from lower left to the upper right, and the box shown in the figure represents a portion of the lower wall of the channel. The streamwise and spanwise extents of the wall shown are $3.2\,\delta$ and $2.1\,\delta$, respectively, or 575 and 375 in wall units

where s is the distance along the vortex line, and $\boldsymbol{\omega}$ is the vorticity vector. In Fig. 1, vortex lines displaying the initial vortical structure are shown. The box shown in the figure represents a small portion of the lower wall of the channel; its dimensions are given in the figure caption. Note that a *time-averaged* flow field contains only the spanwise vorticity component, and the resulting vortex lines would display a planar vortex sheet parallel to a (x, z)-plane. Any departure from the planar vortex sheet indicates a perturbation to the mean. The presence of a horseshoe-shaped vortical structure is evident in Fig. 1.

A disadvantage in identifying a vortical structure with vortex lines as above is that the vortex lines alone do not provide information about the strength of vorticity except when those vortex lines form a vortex filament, in which case one can infer the strength of vorticity from the diameter of the filament. A different method of illustrating the three-dimensional vortical structure is used in Fig. 2. The same vorticity field is visualized by a three-dimensional contour surface. Here, the shaded surface represents a surface of constant enstrophy [one half of the squared vorticity, i.e., $\frac{1}{2}(\omega_x^2 + \omega_y^2 + \omega_z^2)$]. Several contour levels can be drawn to display different shapes associated with each level (see Fig. 5 b, for example). In Fig. 2, however, only one contour surface that demonstrates an identifiable structure is shown for clarity. Higher levels of enstrophy exist either inside the horseshoe-shaped structure or underneath the flat surface. The high enstrophy underneath the flat surface is due mainly to the spanwise vorticity that forms the mean shear, whereas the high enstrophy inside the vortical structure is primarily a result of the perturbation vorticity. Thus the structure displayed in Fig. 2 represents a region associated with high vorticity relative to its surroundings. Visualized in this way, the presence of a horseshoe-shaped vortical structure is also apparent.

The initial vortical structure evolves under the influence of the self-induced motion and the mean shear as it is convected downstream. The effect of the induced motion is to lift the vortex away from the wall, whereas the effect of the shear is to stretch the vortex and to rotate it back toward the wall. The self-induced motion is high at the tip because of the high

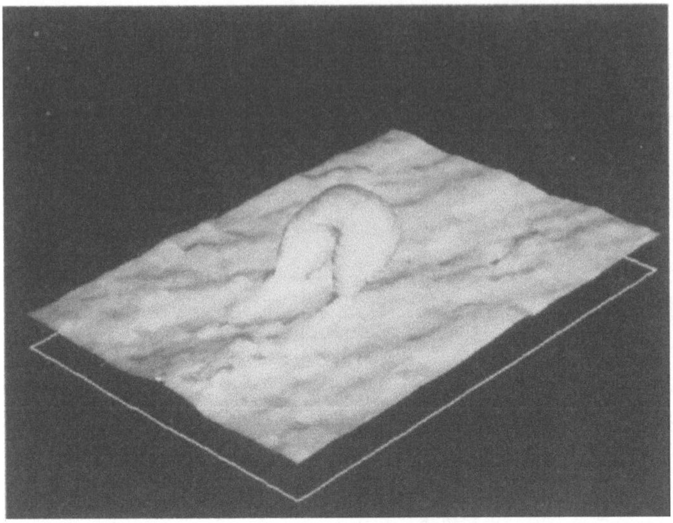

Fig. 2. Perspective view of a surface of constant enstrophy of the initial field, $t^* = 0$. See the caption for Fig. 1 for flow direction and other details

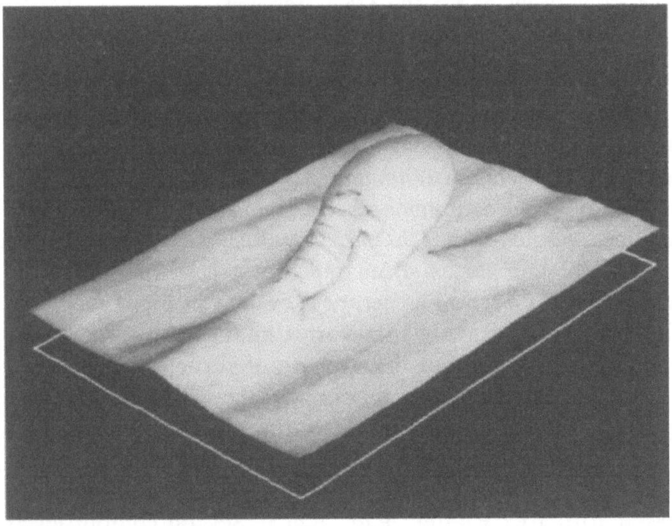

Fig. 3. Perspective view of a surface of constant enstrophy at $t^* = 4.3$. See the caption for Fig. 1 for flow direction and other details

curvature of the vortex lines; the stretching and rotating actions are high near the wall. Figures 3 through 6 show the time-evolution of the vortical structure. They are taken at $t^* = 4.3$, 8.6, 17.2, and 28.8, where t^* is a dimensionless time nondimensionalized by the channel half-width δ, and by the centerline velocity U_c (t can also be nondimensionalized by the wall shear velocity u_τ, from $U_c/u_\tau \approx 18$). Note that Figs. 3 through 6 are presented such that the tip of the vortex is located approximately in the center of each figure (each figure being a close-up of a small portion of the total computational box). The streamwise locations of the tip in each figure corresponds to $x \approx 3.4\,\delta$, $7.4\,\delta$, $15.2\,\delta$, and $27.0\,\delta$, respectively, where $x = 0$ corresponds to the initial location of the tip. In Fig. 5 b, which corresponds to the same

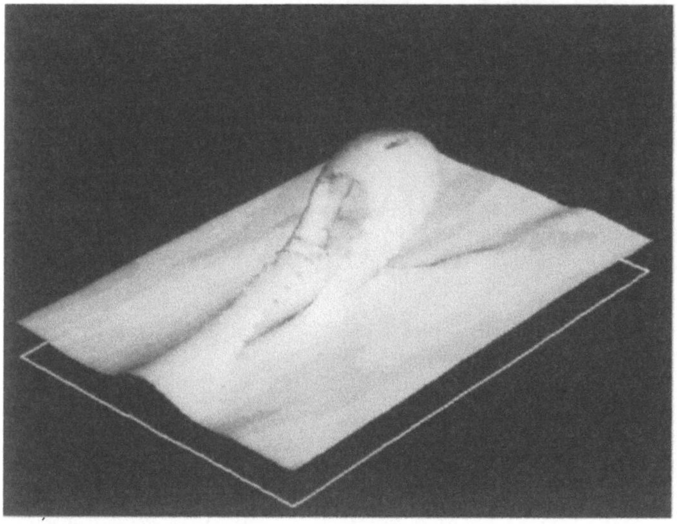

Fig. 4. Perspective view of a surface of constant enstrophy at $t^* = 8.6$. See the caption for Fig. 1 for flow direction and other details

time as Fig. 5a, three contour levels with different colors are shown to illustrate the variation of the enstrophy inside the contour surface shown in Fig. 5a. Here, the red lines represent the highest contour level and the blue lines represent the lowest that are at the same level as the surface shown in Fig. 5a. The yellow lines are the average of the levels represented by the red and blue lines.

An approximate convection velocity of the tip can be calculated from these tip locations. For example, the convection velocity from $t^* = 0$ to $t^* = 4.3$ (Fig. 2 to Fig. 3) is about 0.8 U_c, which is approximately the local mean velocity at the y-location of the tip. As the vortex moves downstream, the tip moves toward the centerline of the channel owing to the self-induced motion, and the convection velocity increases as the local mean velocity increases; for example, it is about 0.95 U_c from $t^* = 8.6$ to $t^* = 28.8$ (Fig. 4 through Fig. 6).

Once the tip of the vortex approaches the centerline (Figs. 5 and 6), the effect of the shear diminishes, but that of the self-induced motion continues. This results in a higher angle between the vortex and the wall; it also makes the tip of the vortex progressively Ω-shaped (see also [5] and [9]). It will eventually pinch off to form a vortex ring unless viscous diffusion takes over before such an event can occur [10]. It is also interesting to note, for the channel flow considered here, that the effect of the mean shear will be a compression rather than a stretching once the tip progresses beyond the centerline. It appears that this effect is somewhat reflected in the shape of the contour surface in Fig. 6; the shaded surface shows that the round, cigar-like structure in the leg region becomes a flattened surface (the cross section in a (x, y)-plane is an ellipse) at the tip region.

Figure 7 shows vortex lines near the vortical structure at $t^* = 28.8$. By comparing the vortex lines in Fig. 7 with those in Fig. 1, one can see the effect of the vortex stretching caused by the mean shear. The initial vortex structure visualized by the vortex lines is recognized as layers of bulged-out vortex sheets because of the smearing effect of the averaging process. As the structure evolves in time, this sheet-like structure rolls up into a vortex tube, as indicated in Fig. 7. Examination of the strength of the total vorticity indicates that the region where the vortex lines coalsece is associated with high vorticity, consistent with the kinematics of a vortex tube. Figure 7 also shows that several vortex lines have become Ω-shaped because of the self-induced motion.

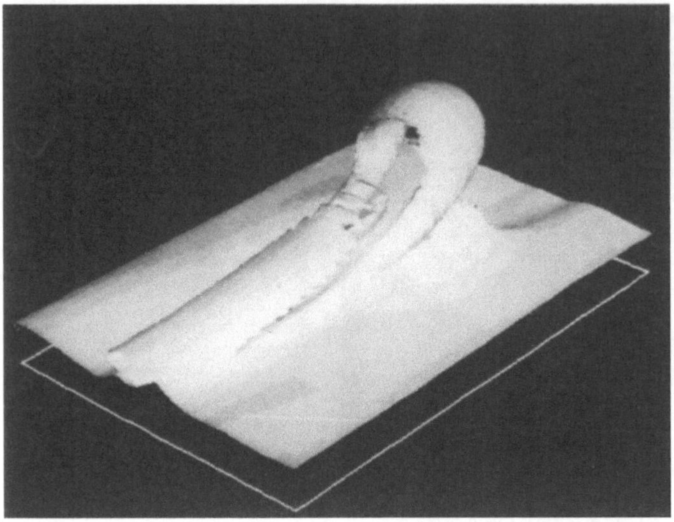

Fig. 5 a. Perspective view of a surface of constant enstrophy at $t^* = 17.2$. The tip of the vortex is located near the centerline of the channel. See the caption for Fig. 1 for flow direction and other details

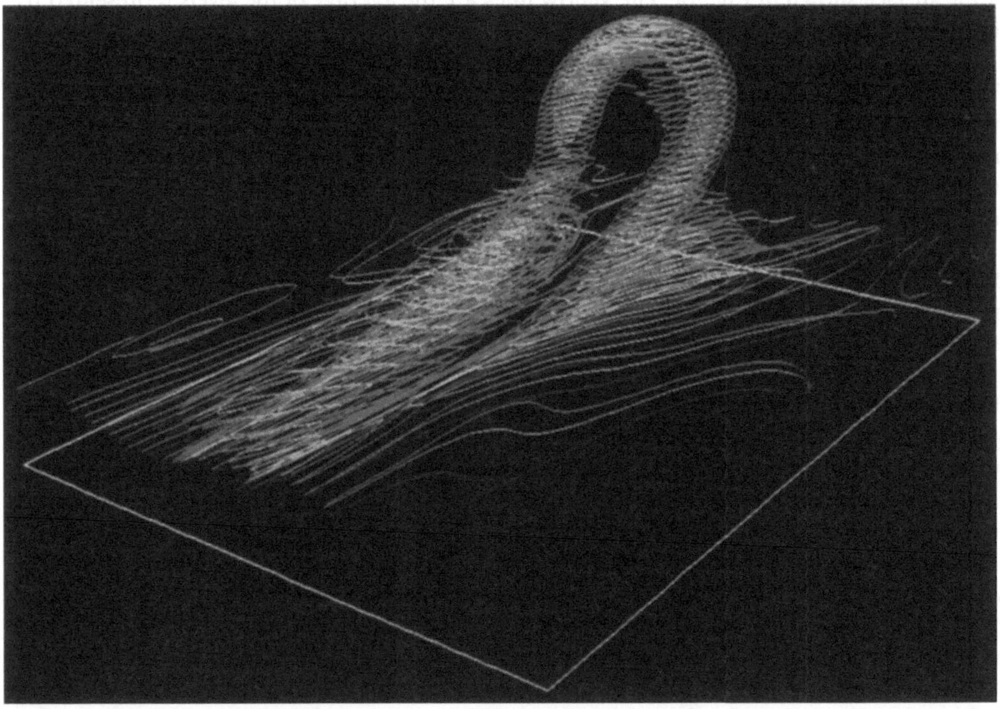

Fig. 5 b. Perspective view of contour lines of constant enstrophy at $t^* = 17.2$. The highest and lowest levels are represented by the red and blue lines, respectively. The intermediate level is represented by the yellow lines. The tip of the vortex is located near the centerline of the channel. See the caption for Fig. 1 for flow direction and other details

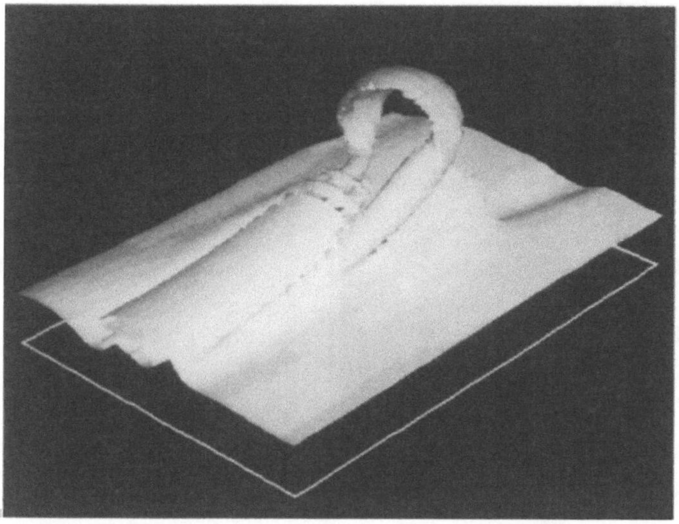

Fig. 6. Perspective view of a surface of constant enstrophy at $t^* = 28.8$. The tip of the vortex is located above the centerline of the channel. See the caption for Fig. 1 for flow direction and other details

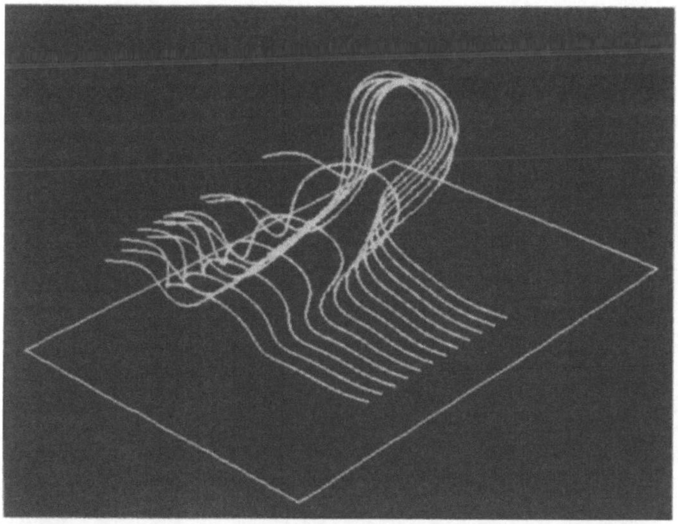

Fig. 7. Perspective view of Vortex lines displaying the vortical structure at $t^* = 28.8$. See the caption for Fig. 1 for flow direction and other details

The production of vorticity by vortex stretching – $\omega_i \omega_j s_{ij}$, where s_{ij} is the strain rate, $\frac{1}{2}(u_{i,j} + u_{j,i})$, and the summation is implied – is the dominant term in the equation for the vorticity budget. In Fig. 8, contour surfaces of $\omega_i \omega_j s_{ij}$ at $t^* = 17.2$ (all the figures shown henceforth correspond to $t^* = 17.2$, unless stated otherwise) are shown to illustrate the regions associated with an intense vortex stretching. In Fig. 8 a, one level of contour is shown for clarity; in Fig. 8 b, three contour levels of the vortex stretching term are displayed to illustrate the variation inside the surface shown in Fig. 8 a. Some of the surfaces shown in the figures are not associated with horseshoe vortex; instead, they represent regions near the wall associated with the high vorticity production. As anticipated, the most intense stretching

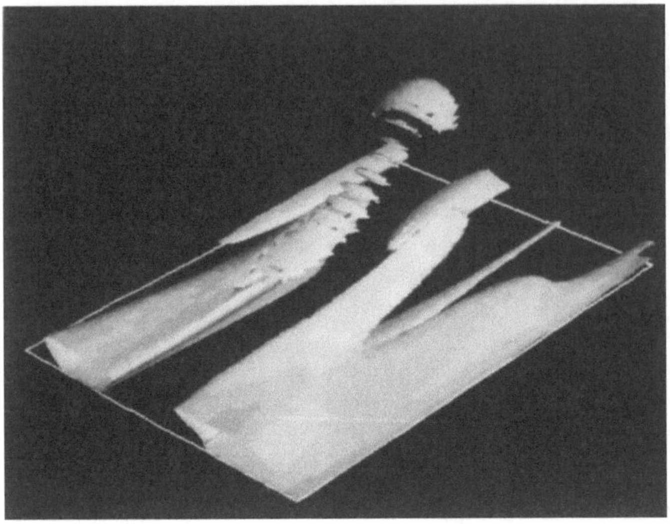

Fig. 8a. Perspective view of surfaces of constant vorticity production owing to vortex stretching, $\omega_i \omega_j s_{ij}$, at $t^* = 17.2$. See the caption for Fig. 1 for flow direction and other details

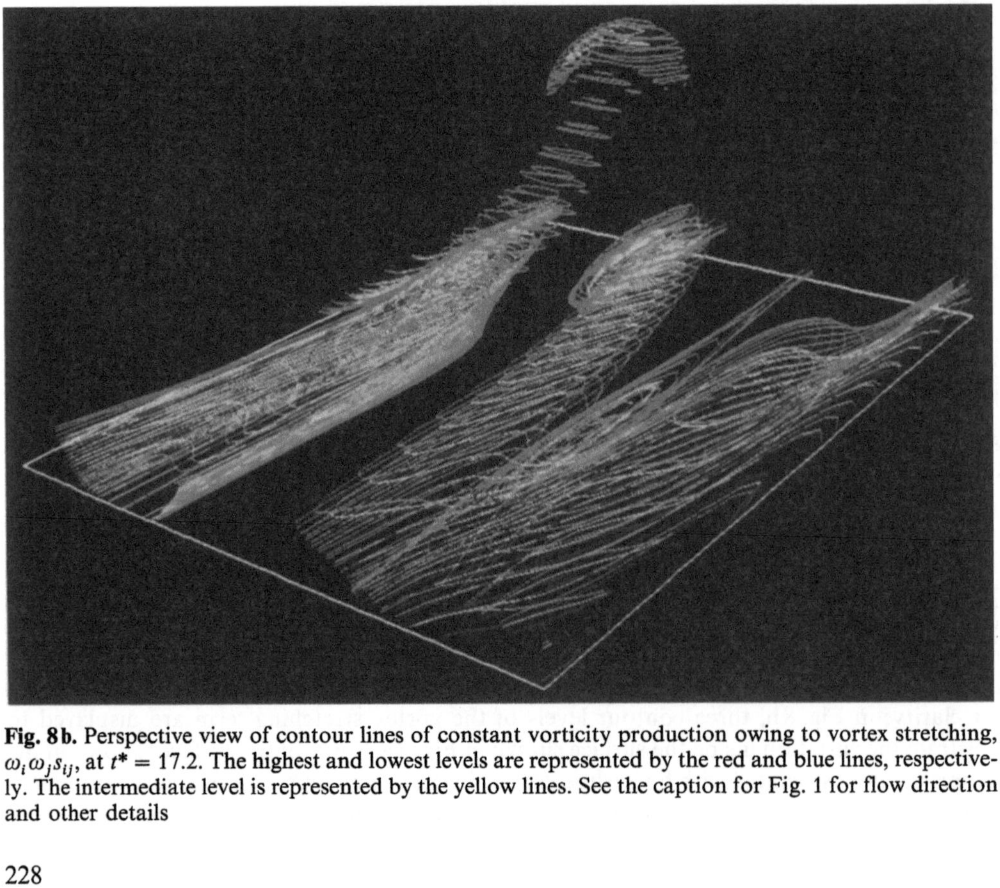

Fig. 8b. Perspective view of contour lines of constant vorticity production owing to vortex stretching, $\omega_i \omega_j s_{ij}$, at $t^* = 17.2$. The highest and lowest levels are represented by the red and blue lines, respectively. The intermediate level is represented by the yellow lines. See the caption for Fig. 1 for flow direction and other details

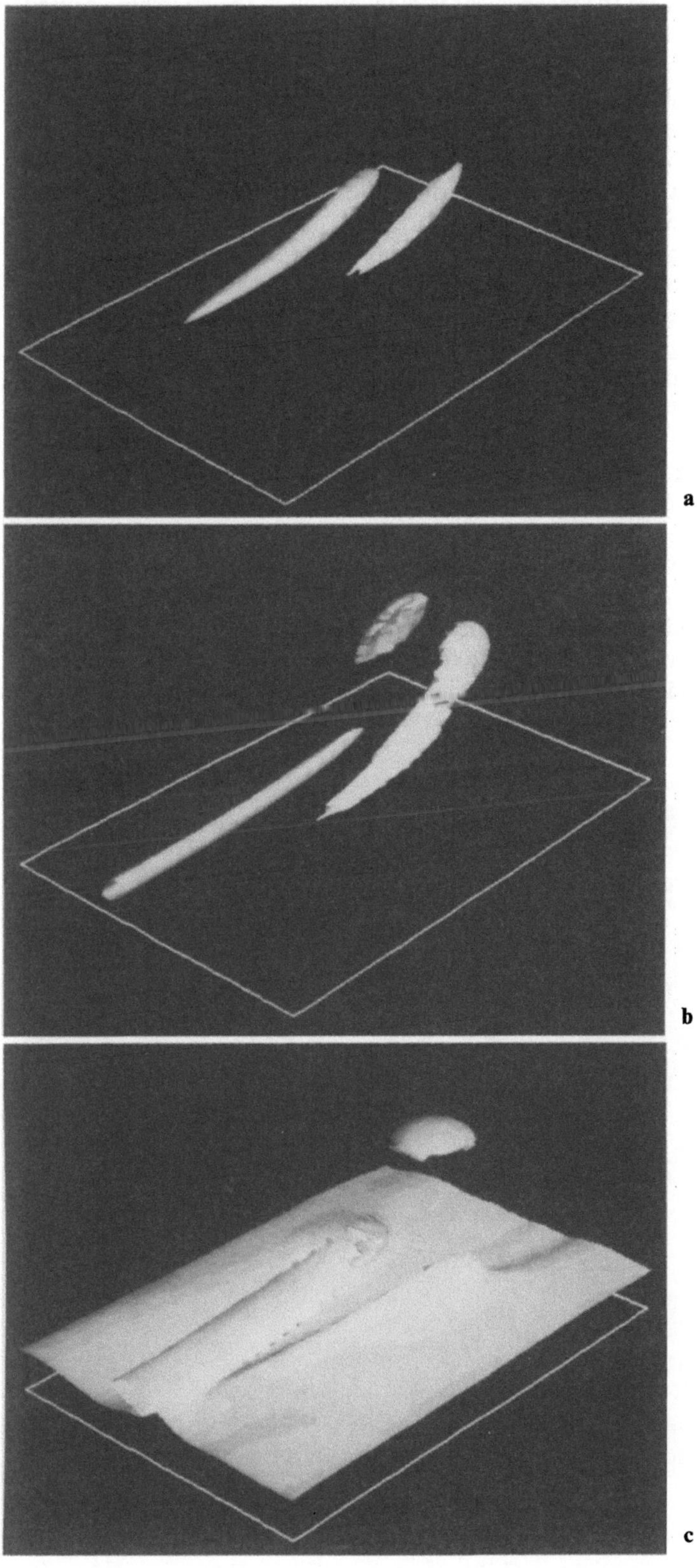

Fig. 9a–c. Perspective view of surfaces of constant vorticity at $t^* = 17.2$: **a** streamwise, ω_x; **b** normal to the wall, ω_y; **c** spanwise, ω_z. See the caption for Fig. 1 for flow direction and other details

takes place in the vortex legs, where the mean shear is high, although the stretching caused by the self-induced motion in the tip region and the region just above the legs is also substantial.

Turbulence structures visualized by other turbulence quantities such as streamwise, normal to the wall, and spanwise vorticity; Reynolds shear stress; and pressure are shown in Figs. 9 through 11. The contour surface of the streamwise vorticity (Fig. 9a) displays the two legs of the horseshoe vortex, and the contour surface of the spanwise vorticity (Fig. 9c) emphasizes the tip of the vortex. Note that some portion of the contour surface of the spanwise vorticity shown in Fig. 9c is not a part of the horseshoe vortex structure; it is a secondary structure generated by the horseshoe vortex. The self-induced motion owing to the two legs of the vortex causes the vortex lines above the legs to stretch, thus, increasing the spanwise vorticity in that region. This secondary structure is shown in Fig. 9c as a region of high spanwise vorticity; it is also shown in Figs. 3 through 6 as a region of high enstrophy just above the vortex legs.

Figure 9 illustrates that one cannot obtain a complete picture of a three-dimensional vortical structure by examining one particular component of vorticity. Nishioka et al. [11], for example, suggested the presence of a horseshoe vortex structure at the five-spike stage of transition to turbulence, by examining contours of spanwise vorticity component. This, however, may lead to an erroneous conclusion with regard to the vortex legs since they primarily contain streamwise and normal vorticity. To illustrate this point, contour lines of the spanwise vorticity in the mid-plane of the vortical structure are shown in Fig. 10. In this plane, contour lines of the spanwise vorticity look like a horseshoe vortex, but they represent the secondary structure created by the horseshoe vortex instead of the horseshoe vortex itself.

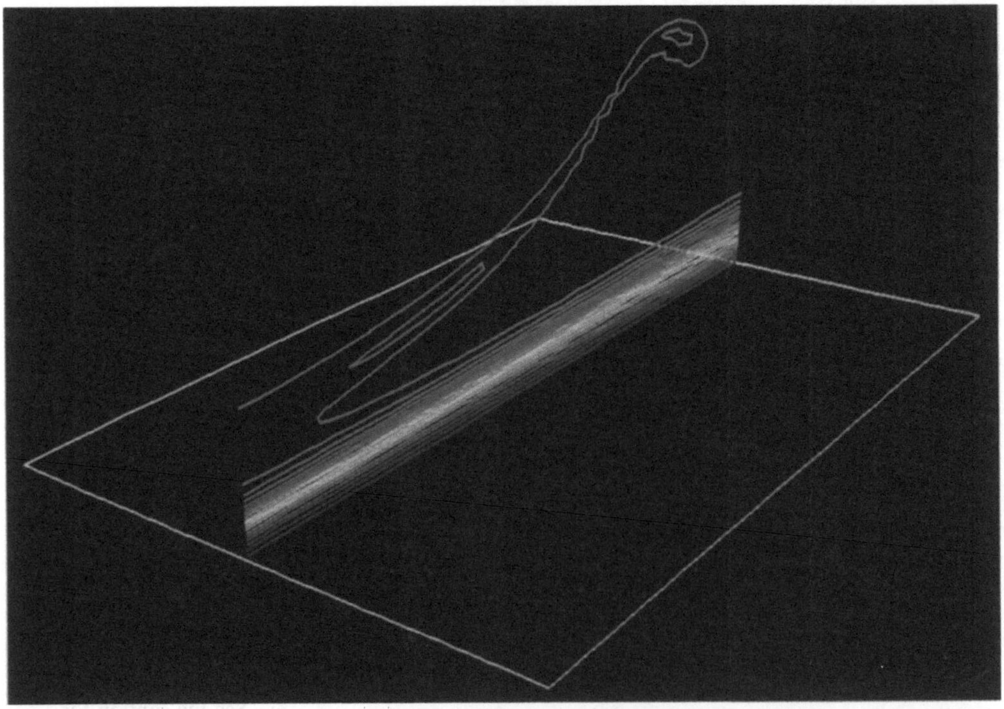

Fig. 10. Perspective view of contour lines of constant spanwise vorticity in the midplane of Fig. 9c. Red, yellow and blue represent contour levels that are progessively decreasing from high to low levels

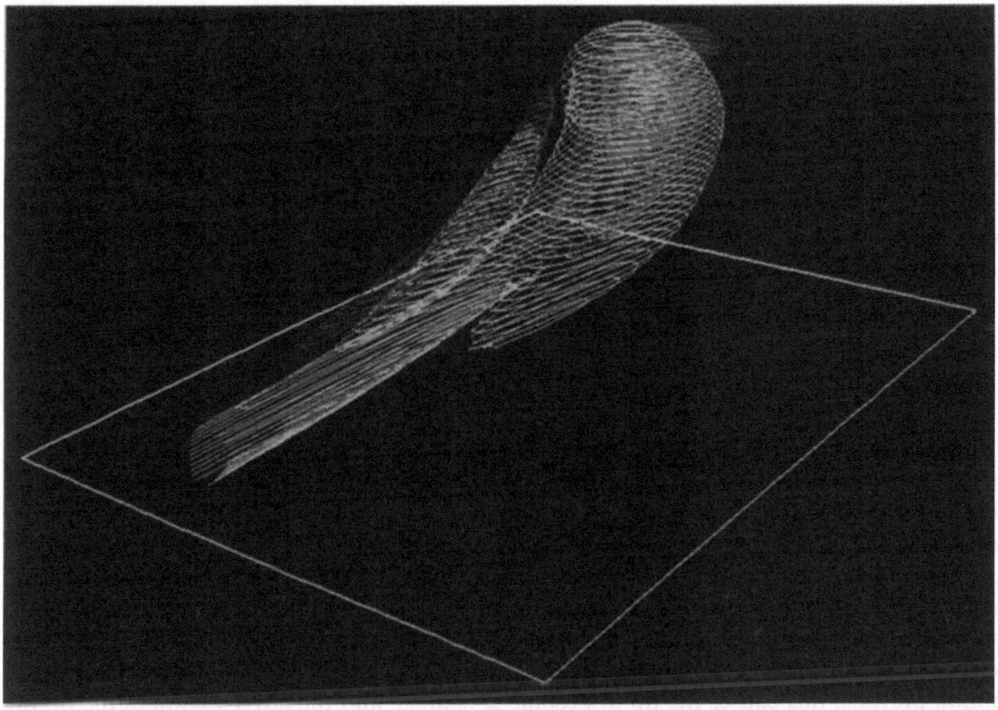

Fig. 11. Perspective view of contour lines of constant Reynolds shear stress at $t^* = 17.2$. White, blue, green and yellow and red represent a progression of contour levels from high to low, and the red represents a negative contour level of $-\overline{u'v'}$. See the caption for Fig. 1 for flow direction and other details

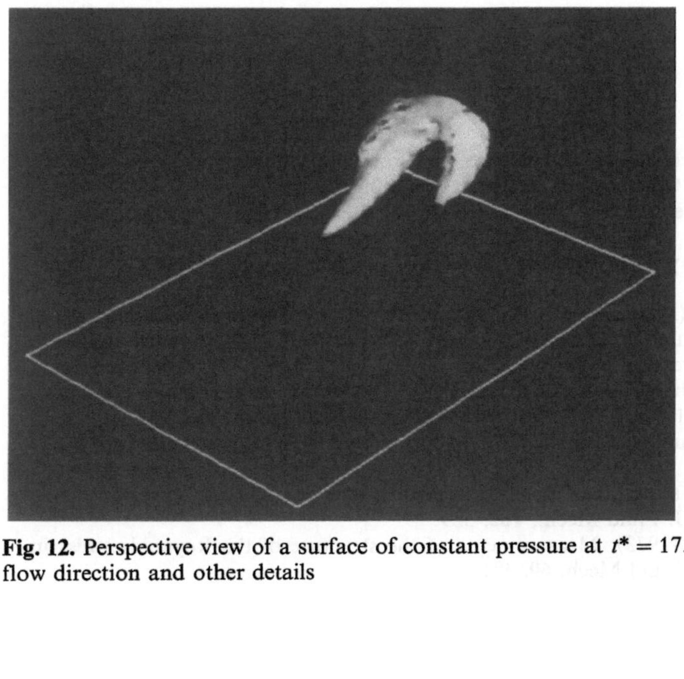

Fig. 12. Perspective view of a surface of constant pressure at $t^* = 17.2$. See the caption for Fig. 1 for flow direction and other details

The contours of Reynolds shear stress (Fig. 11) show that the most intense Reynolds shear stress is produced approximately in the middle of the tip, and a slightly lower level of shear stress is found near both legs. It is interesting to note that there exist regions of positive Reynolds stress (i.e., negative production) near the tip and above the left leg. The contour surfaces of pressure – the contour level shown in Fig. 12 represents a low-pressure region – indicate that pressure is low inside the tip of the vortex. A slightly higher level (but still lower than the surroundings) exists along the leg regions, indicating that the pressure is lower inside the entire horseshoe vortex consistent with the pressure distribution of an isolated vortex.

Summary

The temporal evolution of a horseshoe vortex structure in a channel was investigated by integrating the time-dependent, three-dimensional Navier-Stokes equations. The initial vortical structure was constructed by applying a conditional sampling technique to the data base generated from a direct simulation of a turbulent channel flow. This sheet-like vortical structure rolls up into vortex filaments as it is convected downstream at the local mean velocity. The self-induced motion of the vortex causes the vortex to move away from the wall, whereas the presence of shear near the wall causes the legs to stretch. Once the tip of the vortex approaches the centerline, the effect of shear diminishes, and only the self-induced motion prevails; this causes the tip to become Ω-shaped, and the cross section of the tip of the vortex becomes an ellipse. It was shown that the production of vorticity is high inside the legs of the horseshoe vortex; the stretching caused by the self-induced motion in the tip and above the legs is also substantial. The stretching due to the induced motion above the legs is found to create a secondary structure above the legs. High Reynolds shear stress is produced near the tip of the vortex because of the self-induced motion.

Acknowledgement. The author is grateful to Drs. A. Leonard and P. Moin for reviewing the draft of this manuscript, and to Drs. P. G. Buning and P. R. Spalart for the three-dimensional contour programs used in the paper.

References

1. Theodorsen, T. (1952): "Mechanism of Turbulence", in Proceedings of the 2nd Midwestern Conference on Fluid Mechanics, Ohio State University, Columbus, Ohio
2. Wallace, J. M. (1982): On the structure of bounded turbulent shear flow: a personal view. Dev. Theor. Appl. Mech. **11**, 509
3. Head, M. R., Bandyopadhyay, P. (1981): New aspects of turbulent boundary layer structure. J. Fluid Mech. **107**, 297
4. Acarlar, M. S., Smith, C. R. (1984): An experimental study of hairpin-type vortices as a potential flow structure of turbulent boundary layers. Report FM-5, Department of Mechanical Engineering and Mechanics, Lehigh University, Bethlehem, PA
5. Moin, P., Kim, J. (1985): The structure of the vorticity field in turbulent channel flow. Part 1. Analysis of instantaneous fields and statistical correlations. J. Fluid Mech. **155**, 441
6. Moin, P., Kim, J. (1982): Numerical investigation of turbulent channel flow. J. Fluid Mech. **118**, 341
7. Kim, J., Moin, P. (1986): The structure of the vorticity field in turbulent channel flow. Part 2. Study of ensemble-averaged fields. J. Fluid Mech., **162**, 339
8. Lu, S. S., Willmarth, W. W. (1973): Measurements of the structure of the Reynolds stress in a turbulent boundary layer. J. Fluid Mech. **60**, 481

9. Hama, F. R. (1962): Progressive deformation of a curved vortex filament by its own induction. Phys. Fluids **5**, 1156
10. Moin, P., Leonard, A., Kim, J. (1986): Evolution of a curved vortex filament into a vortex ring, Phys. Fluids **29**, 955
11. Nishioka, M., Asai, M., Iida, S. (1981): "Wall Phenomena in the Final Stage of Transition to Turbulence", in Transition and Turbulence, ed. by R. E. Meyer (Academic, New York)

Direct Numerical Simulation of Equilibrium Turbulent Boundary Layers

Philippe R. Spalart and Anthony Leonard*

NASA Ames Research Center, USA

Abstract

This paper describes the simulation of turbulent boundary layers by direct numerical solution of the three-dimensional time-dependent Navier-Stokes equations, using a spectral method. The flow is incompressible, and $Re_{\delta^*} = 1000$ for most cases. The equations are written in the similarity coordinate system and normalized by the local friction velocity. Periodic streamwise and spanwise boundary conditions are then imposed. A family of nine "equilibrium" boundary layers, from the strongly accelerated "sink" flow to Stratford's separating flow, is treated. Good general agreement with experiments is observed. The effects of the pressure gradient and of the Reynolds number are discussed.

Nomenclature

a	$(X - X_1)\, U_{\infty X}/U_\infty$ Bradshaw's pressure gradient parameter
A	Envelope function, see (6)
C	Intercept, see (12a)
D	$u_\tau U_{\infty X}/(U_\infty u_{\tau X})$ Coles' pressure gradient parameter
G	$(H-1)\, U_\infty/(H u_\tau)$ Clauser's shape factor
H	δ^*/θ shape factor
Re_{δ^*}	$\delta^* U_\infty/\nu$ Reynolds number, based on δ^*
S	$\partial y/\partial x\|_\eta$ slope of the similarity lines
t	Time
u, v, w	Velocity components
U, V	Mean velocity components
u', v', w'	Fluctuating velocity components
u_p, v_p, w_p	Periodic normalized fluctuating components, see (6)
U_∞	Free-stream velocity
u_τ	Friction velocity
u^+	u/u_τ wall units
x, y, z	Streamwise, vertical, spanwise cartesian coordinates
X	Distance along the boundary layer
X_1	Virtual origin in Bradshaw's equilibrium boundary layer
y^+	$y u_\tau/\nu$ wall units
α	$d\tau/dy$ stress derivative at the wall in case 9
β	$-U_\infty U_{\infty X}\delta^*/u_\tau^2$ Clauser's pressure gradient parameter
δ	Characteristic thickness, see (1b)
δ^*	Displacement thickness
ΔU^+	Wake component
η	Vertical coordinate in similarity system
κ	Karman constant, see (12)
Λ_x, Λ_z	Periods in x and z directions
ν	Kinematic viscosity
ω	Vorticity
Ω	Over-relaxation parameter
θ	Momentum thickness
τ	Shear stress
$\langle . \rangle$	Average taken over x, z and t
ϕ_t, ϕ_x, etc.	Derivatives of ϕ with respect to t, x, etc.
∇^2	$\partial_{xx} + \partial_{yy} + \partial_{zz}$ Laplace's operator

Introduction

Successful numerical simulations of several turbulent flows have been performed in the past few years: homogeneous turbulence with and without mean strain, shear or rotation, plane channel and curved channel flows. These simulations, either "direct" or "large-eddy",

* Present address: California Institute of Technology, USA

Turbulent Shear Flows 5

showed good general agreement with experiments and can now be relied upon to provide information experiments could not provide [1]. This paper describes the simulation of turbulent boundary layers. Unlike homogeneous or channel flows, boundary layers have an outer inviscid flow-turbulent interface. In addition, they exist with a wide range of streamwise pressure gradients, while the channel flow demands a weak favorable gradient.

A numerical method was designed to handle the semi-infinite domain accurately. It was described in [2] and was used in [3] to simulate transition induced by a vibrating ribbon in a Blasius boundary layer. The method has now been applied to the direct simulation of fully turbulent boundary layers, with minor changes. The new element in the present study is the use of self-similar coordinates to account for the growth of the boundary layer.

Unlike fully-developed channel flow, for example, boundary layers are not homogeneous in the streamwise direction so that periodic conditions are not applicable a priori. For the transition simulations [3] the mean flow was treated as parallel and the fluctuations as periodic. A forcing term was introduced to approximate the non-parallel effects, thus controlling the evolution of the mean flow. For turbulent cases a prescribed forcing term cannot be used, because the velocity profile is not known a priori as it was in the laminar case. Periodic conditions are still used, but in a non-cartesian "similarity" coordinate system. Furthermore the velocities are normalized by the friction velocity before periodicity is applied.

The main objectives of this work are to study the structure of wall-bounded turbulence, and to provide accurate and complete statistical data for the calibration of turbulence models. Equilibrium boundary layers were chosen as the first subject [4–7]. These boundary layers are characterized by having mean profiles, properly normalized, that are independent of X. Such boundary layers are rare in practice, but they have the advantage of being well defined and of providing good samples for averaging. In addition, the wall region of a boundary layer is often close to equilibrium even when the wake region is not. Thus, the results obtained near the wall will be applicable to more general cases.

Mathematical Formulation

Motivation

The important aspect of the formulation is the choice of boundary conditions. The y and z directions are easy to deal with. The y-domain will be $[0, \infty]$ with the no-slip condition, $u(x, 0, z) = 0$, and the freestream condition, $(u - U_\infty) \to 0$ exponentially as $y \to +\infty$. The boundary layers are assumed to be statistically homogeneous in the z direction. This motivates the use of periodic conditions, provided that the period is sufficiently large compared with the macro-scales of the flow.

The x direction poses a problem, because in general the flow is not homogeneous in that direction; the velocity and stress profiles slowly vary. Ideally one would simply simulate the spatially-developing boundary layer, and the growth would happen naturally. One would either simulate the full transition process or use turbulent inflow conditions, which would probably be rather crude. In either case one would need to simulate a very long stretch of the boundary layer (tens or hundreds of boundary-layer thicknesses) in order to eliminate the influence of the inflow conditions. This would be prohibitively expensive on current computers. In this work we focus on equilibrium boundary layers and take advantage of the fact that their growth is both slow and self-similar to reduce the domain of integration. The assumptions made are largely empirical and based on experimental evidence: we assume that the mean velocity and the Reynolds-stress tensor obey similarity laws [4, 5].

Similarity Assumptions

The streamwise evolution of the boundary layer is governed by three quantities: the edge velocity $U_\infty(X)$, the friction velocity $u_\tau(X)$, and the characteristic thickness $\delta(X)$. In order to emphasize the different length scales let us introduce a "slow" variable X and a "fast" variable x. Typically, the long scale is the distance from the leading edge while the short scale is the boundary-layer thickness. The quantities U_∞, u_τ, δ and all the statistics are slowly varying, functions of X.

The mean velocity profile U is assumed to obey the "law of the wall" in the wall region,

$$U(X, y) = u_\tau(X)\, f_1\left(\frac{y u_\tau(X)}{\nu}\right), \tag{1a}$$

and the "defect law" in the outer or wake region,

$$U(X, y) = U_\infty(X) - u_\tau(X)\, f_2\left(\frac{y}{\delta(X)}\right). \tag{1b}$$

The non-dimensional functions f_1 and f_2 are unknown.

The Reynolds stresses are assumed to scale with u_τ^2 in both regions:

$$\langle u_i' u_j' \rangle\,(X, y) = u_\tau^2(X)\; g_{ij}\left(\frac{y u_\tau(X)}{\nu}\right), \tag{2a}$$

$$\langle u_i' u_j' \rangle\,(X, y) = u_\tau^2(X)\; h_{ij}\left(\frac{y}{\delta(X)}\right), \tag{2b}$$

where the functions g_{ij} and h_{ij} are unknown. This property has not been verified, experimentally, as thoroughly as (1). Equations (2a) and (2b) can overlap only in a constant-stress region, a behavior which is not strongly supported by available experimental or numerical results. A more complex model may be needed in the future to make the assumptions fully consistent. However (2) is satisfactory as a first approximation.

Similarity Coordinates

The normal coordinate y is replaced by a coordinate η that is adapted to (1) and (2). In the wall region η is chosen to be a function of y^+, so that the η = constant lines coincide with the y^+ = constant lines. These are the "similarity lines". In the wake region η is a function of y/δ. The only important quantity is the slope S of the similarity lines with respect to the wall. S is defined as $\partial y/\partial x$ (η constant) and is equal to $-\, y u_{\tau X}/u_\tau$ and $y \delta_X/\delta$ in the wall and wake regions, respectively. Note that only the ratios, $(u_{\tau X}/u_\tau)$ and (δ_X/δ), enter the equations (thus δ is defined only up to a multiplicative constant). These quantities, and S, are both small and slowly-varying. Terms containing higher order combinations like S^2 or S_X will be neglected.

Let us focus on a station of the boundary layer, say at X_0. In order to make the formulas lighter η is normalized so that at X_0, η and y coincide. Let u, v and w designate the contravariant velocity components in the (x, η, z) system. The continuity condition is

$$u_x + v_\eta + w_z = -\, S_\eta u. \tag{3}$$

The non-zero right-hand side is due to the non-parallel, fan-like character of the coordinate lines. Similar terms appear in the momentum equation; for instance the u-equation is

$$u_t + u u_x + v u_\eta + w u_z = - p_z + S p_\eta \\ + \nu (u_{xx} + (1 + S^2) u_{\eta\eta} + u_{zz} - 2 S u_{x\eta} - S_X u_\eta + S S_\eta u_\eta). \quad (4)$$

Normalization of the Velocity

The normalization of the mean velocity is given by (1): the velocity U itself scales with u_τ near the wall, and the defect $(U_\infty - U)$ scales with u_τ away from the wall. This allows one to express the X-derivative U_X (η constant) solely in terms of the profile at X_0, U:

$$U_X = \frac{u_{\tau X}}{u_\tau} U, \quad (5a)$$

$$U_X = U_{\infty X} + \frac{u_{\tau X}}{u_\tau} (U - U_\infty), \quad (5b)$$

in the wall and wake regions respectively. Equations (1) and (5) are equivalent; (1) made the assumptions more clear, but (5) is a more usable form.

The treatment of the turbulent fluctuations u', v' and w' is more involved. According to (2), as one follows a similarity line (η = constant, z = constant) the fluctuating component has fast random-like variations, with a local rms that is proportional to $u_\tau(X)$. This suggests representing u', for instance, in the following manner:

$$u'(X, x, \eta, z) = A(X)\, u_p(x, \eta, z), \quad (6)$$

where $A(X)$ is an "envelope" function proportional to $u_\tau(X)$ and u_p is periodic in x.

Equation (6) expresses the key assumption that made possible meaningful simulations of turbulent boundary layers within a short domain in the streamwise direction. Compared with a simulation that would include a long stretch of the boundary layer and capture the growth, this procedure not only allows simulations within a much shorter domain, it also obviates the need for inflow/outflow conditions. Such conditions are crucial both from a physical and a numerical point of view.

Let us evaluate the validity of (6). Clearly, a product of that type can generate a turbulent-like signal, and the slow variation in amplitude of the fluctuation is taken into account. On the other hand, by making u_p periodic one is imposing uniform length and time scales from the inlet to the outlet of the domain. The streak spacing, Kolmogorov scale and burst frequency are examples of relevant scales. The assumption can be a good one only if these scales do not vary significantly over one period Λ_x. In the case of the boundary layer the length and time scales vary slowly, on the X scale, while Λ_x is a few boundary-layer thicknesses. Therefore assumption (6) is satisfactory. In contrast, this assumption would be inappropriate in a simulation of a spatially-developing mixing layer: in that flow the length and time macro-scales double for each pairing and a proper simulation must cover at least one or several pairings.

Equation (6) is an elementary form of the method of multiple scales. The function $A(X)$ can be normalized so that $A(X_0) = 1$, and A_X/A is equal to $u_{\tau X}/u_\tau$ since A is proportional to u_τ. Differentiating (6) in the x direction then yields

$$\frac{\partial u'}{\partial x} = u_{p_x} + \left(\frac{u_{\tau X}}{u_\tau}\right) u'. \quad (7)$$

The first term is the "fast" derivative associated with the fast, periodic variable x. The second term is the "slow" derivative associated with the slow variable X and is expressed solely in terms of values at X_0 as in (5). Again only the ratio $(u_{\tau X}/u_\tau)$ enters the equation. From here on u_x will designate the fast derivative and u_X the slow derivative. The total derivative is $u_x + u_X$. The mean components only have slow derivatives, like U_X.

With this new notation the continuity (3) is written

$$u_x + v_\eta + w_z = -u_X - S_\eta u. \tag{8}$$

Using the definition of S and u_X one obtains

$$V_\eta = 0, \tag{9a}$$

$$u'_x + v'_\eta + w'_z = 0 \tag{9b}$$

near the wall, as a result of a fortuitous cancellation. Away from the wall the equations are more complex:

$$U_{\infty X} + \frac{u_{\tau X}}{u_\tau}(U - U_\infty) + \left(\frac{\delta_X}{\delta}\right) U + V_\eta = 0, \tag{10a}$$

$$u'_x + v'_\eta + w'_z = -\left(\frac{u_{\tau X}}{u_\tau} + \frac{\delta_X}{\delta}\right) u'. \tag{10b}$$

With the existing numerical method, including the (small) right-hand-side of (10b) is not possible. This term had to be neglected. It is then consistent to neglect the corrections in the pressure term as well, so that it is properly coupled with the continuity equation. On the other hand, all the terms in (9) and (10a) are retained. The mean x-momentum equation becomes a one-dimensional integro-differential equation, similar to the Blasius equation. The integral terms are introduced into the equation by V, which is given by (9a) and (10a).

Numerous terms appear in the momentum equation when the slow derivatives are introduced, and only the leading terms are retained. Let ε represent first order small terms. u_τ is $O(\varepsilon)$. The boundary layer grows under the effect of u_τ^2, so S and the X-derivatives are $O(\varepsilon^2)$. The corrections in the viscous terms are neglected. The expansion of (4) and of the other components of the momentum equation up to order ε^3 is

$$\begin{aligned} u'_t + (U + u')\, u'_x + v'(U + u')_\eta + w' u'_z &= -p_x + \nu \nabla^2 (U + u') \\ &\quad - \left[U U_X - U_\infty U_{\infty X} + V U_\eta + U_X u' + V u'_\eta + U\left(\frac{u_{\tau X}}{u_\tau}\right) u' \right], \end{aligned} \tag{11a}$$

$$v'_t + (U + u')\, v'_x + v' v'_\eta + w' u'_z = -p_\eta + \nu \nabla^2 v' - \left[V_\eta v' + V v'_\eta + U\left(\frac{\delta_X}{\delta} + \frac{u_{\tau X}}{u_\tau}\right) v' \right], \tag{11b}$$

$$w'_t + (U + u')\, w'_x + v' w'_\eta + w' w'_z = -p_z + \nu \nabla^2 w' - \left[V w'_\eta + U\left(\frac{u_{\tau X}}{u_\tau}\right) w' \right]. \tag{11c}$$

The terms that are associated with the spatial growth, as opposed to the usual Navier-Stokes terms, are between square brackets. The most significant of these "growth terms" is the mean term, $(U U_X - U_\infty U_{\infty X} + V U_\eta)$, in the x-momentum equation (11a). This term supplies momentum to the boundary layer and allows it to maintain a statistically steady state, in spite of the momentum loss due to the shear stress at the wall. It includes the mean pressure gradient $-U_\infty U_{\infty X}$. All the other terms act on the fluctuations. Terms like $-Vu'_\eta$ and $-U(u_{\tau X}/u_\tau)\,u'$ represent transport, or "advection", of the turbulence by the mean flow. The terms $-U_X u'$ and $-V_\eta v'$ represent straining of the turbulence by the mean flow.

In practice one obtains the state of statistical equilibrium by a time-like relaxation process; a term ΩU_t is added to the left-hand-side of (11 a), where Ω is a constant. One can accelerate the relaxation without causing stability problems by setting Ω to a value smaller than 1. Values as low as 0.1 have been used. The slow response of the turbulence to the mean shear is the limiting factor. Once equilibrium has been obtained U_t vanishes and one recovers (11 a).

Behavior of U_∞, δ and u_τ

In order to apply the system of equations (5, 9–11) one finally needs three equations to prescribe $U_{\infty X}/U_\infty$, δ_X/δ and $u_{\tau X}/u_\tau$. One uses the current values of u_τ, κ and U to compute the values of $U_{\infty X}/U_\infty$, δ_X/δ and $u_{\tau X}/u_\tau$ for the next time step.

General Case

No fully satisfactory theory exists to predict the variations of U_∞, u_τ and δ along an equilibrium boundary layer. Dimensional arguments are not sufficient because there are two velocity scales, U_∞ and u_τ. There is only experimental evidence that with carefully tailored pressure distributions boundary layers can be generated that satisfy (1) closely. Clauser's empirical definition of an equilibrium boundary layer is that the non-dimensional parameter β is independent of X [4]. Bradshaw's definition is that a is independent of X [6]. In Coles' definition it is D that is independent of X [5]. The quantities β, a and D are all non-dimensional measures of the pressure gradient.

The present study does not allow any evaluation of these various definitions, because only one station of the boundary layer is simulated. Even when several stations (with different Reynolds numbers) are simulated separately, the differences between the definitions are likely to be fine and to be dominated by low-Reynolds number effects. Clauser's definition is used in this study mostly because the value of β was given for all the experiments that will be used for comparison. Prescribing the value of β gives the first equation.

A second equation is derived by matching (1 a) and (1 b), [5]. The functions f_1 and f_2 are unknown, but experimental and numerical evidence shows that there is a region of overlap between (1 a) and (1 b), and that the velocity profile is logarithmic:

$$f_1(y^+) = C + \frac{1}{\kappa}\log(y^+), \tag{12a}$$

$$f_2\left(\frac{y}{\delta}\right) = B - \frac{1}{\kappa}\log\left(\frac{y}{\delta}\right), \tag{12b}$$

for large enough y^+ and for small enough y/δ, respectively. Equations (1) and (12) imply

$$\frac{U_\infty}{u_\tau} = C - B + \frac{1}{\kappa}\log\left(\frac{\delta u_\tau}{\nu}\right). \tag{13}$$

The values of κ, C and B are assumed not to depend on X since f_1 and f_2 do not. Differentiating (13) yields a linear equation relating the unknowns:

$$\left(\frac{U_{\infty X}}{U_\infty}\right) = \left(1 + \frac{u_\tau}{U_\infty \kappa}\right)\left(\frac{u_{\tau X}}{u_\tau}\right) + \frac{u_\tau}{U_\infty \kappa}\left(\frac{\delta_X}{\delta}\right). \tag{14}$$

The third equation is provided by the conservation of momentum. The integration of (11 a) over the whole domain yields $\Omega(U_\infty \delta^*)_t = u_\tau^2 - (U_\infty^2 \theta)_X - U_\infty U_{\infty X} \delta^*$. Its right-hand-side involves various integrals of the velocity profile and is linear in $U_{\infty X}/U_\infty$, δ_X/δ and $u_{\tau X}/u_\tau$. Thus one can choose values for U_∞ and ν, and then directly control δ_t^* during the relaxation period until the desired value of $R_{\delta*}$ is obtained. This procedure roughly amounts to surveying a spatially-developing boundary layer to find the station at which $R_{\delta*}$ takes the desired value. After the relaxation period δ^* is constant and the usual momentum balance equation $u_\tau^2 - (U_\infty^2 \theta)_X - U_\infty U_{\infty X} \delta^* = 0$ is satisfied.

Special Cases

The first special case is that of the "sink" flow in which U_∞ is proportional to $1/(X_1 - X)$ [5, 6]. Dimensional analysis shows that $1/\delta$ and u_τ must be proportional to U_∞ as functions of X. Full similarity is ensured. One then needs only one equation, the momentum balance. Equation (14) is trivial, and β is not specified. The second special case is Stratford's separating boundary layer [8, 9]. The friction velocity u_τ is 0 so that (1 a), (12 a), etc., do not apply. Instead, $U_{\infty X}/U_\infty$ (the pressure gradient) is adjusted to obtain $u_\tau = 0$ and δ_X/δ is used to control δ^*. Following Coles [5], D is assumed to be equal to 1.

Numerical Method

The numerical method used is spectral in all directions [2]. Fourier series are used in the x and z directions. In the η direction the spectral basis functions are based on an exponential mapping and Jacobi polynomials [2]. Divergence-free vector functions are constructed in the manner of Leonard and Wray [10]. Leray's weak formulation of the incompressible Navier-Stokes equations is then used [16]. It eliminates the pressure term and reduces the memory requirements. The time-advance scheme is Crank-Nicolson for the Stokes terms and a low-storage third-order Runge-Kutta scheme for the convection and growth terms. Aliasing is eliminated by the "2/3" rule in all directions.

Table 1

Case	β	$\frac{\delta^* \cdot U_{\infty X}}{U_\infty}$	$\frac{\delta^* \cdot \delta_X}{\delta}$	$\frac{\delta^* \cdot u_{\tau X}}{u_\tau}$	$\frac{dp^+}{dx^+}$
1	−0.58	0.0015	−0.0015	0.0015	−0.012
2	−0.53	0.0012	−0.0009	0.0012	−0.010
3	−0.35	0.0008	−0.0007	0.0007	−0.007
4	0.00	0.0000	0.038	−0.0004	0.000
5	0.90	−0.0021	0.0120	−0.0032	0.019
6	1.80	−0.0061	0.0295	−0.0088	0.037
7	5.40	−0.0124	0.0509	−0.0172	0.150
8	8.00	−0.0149	0.0636	−0.0213	0.199
9	∞	−0.0264	0.1200	−0.0264	∞

Results

The Cases

Nine cases have been treated, with increasingly adverse pressure gradients. Case 1 is the sink flow. Cases 2 and 3 have favorable gradients and are to be compared with Herring and Norbury's experimental results [7]. Case 4 is the constant-pressure flow. Cases 5 to 8 have adverse gradients and correspond with Bradshaw's and Clauser's flows [4, 6]. Case 9 is Stratford's flow [8, 9]. Table 1 summarizes the characteristics of these flows.

The Reynolds number $R_{\delta*}$ is 1000. This the highest Reynolds number at which one can undertake a series of simulations, using a Cray X-MP computer. This Reynolds number is sufficient for well-developed turbulence to be generated (turbulence can be maintained down to $R_{\delta*} \approx 500$). The periods are $\Lambda_x = 50\,\delta^*$ and $\Lambda_z = 17\,\delta^*$, except for case 9, where $\Lambda_z = 67\,\delta^*$. There are $132 \times 30 \times 85$ spectral modes. Typically, the first collocation point is 0.08 wall units from the wall, and the spacing between points in the x and z directions is 13 and 6.5 wall units, respectively. Cases 2 to 8 were simulated using the same program and parameters except of course for the value of β, and for the "switching function" from (1 a) and (2 a) to (1 b) and (2 b). This switching is done, progressively, within the log region, and the results show that the log region moves to lower values of y^+ with adverse pressure gradient.

General Description

The aspect of the flows will first be described using case 4. Contours of the streamwise velocity fluctuation u', in a plane near $y^+ = 10$, are shown in Fig. 1. This figure reveals the well-known "streaks": long regions of alternating low and high velocity. This is in agreement with experiments, and with numerical simulations in channels. Figures 2 a and b are side- and endviews of the instantaneous vorticity-magnitude contours. They reveal fine structures and an irregular interface between the vortical fluid near the wall and the irrotational fluid away from it. The intermittent character of the turbulence in this region is reproduced. The skewness and flatness factors of u' are plotted in Fig. 3 and their rapid variation near the edge of the boundary layer suggests a region of intermittent turbulence, with isolated lumps of low speed fluid. This is in agreement with visualizations and with Klebanoff's flatness measurements [11].

Figures 4 a and b show the two-point correlation coefficients of the three velocity components in the x and z directions, at $y^+ = 10$. They indicate that the value chosen for Λ_z is sufficient, but that Λ_x is only marginally sufficient. The presence of streaks with a spacing of about 100 wall units is confirmed by the negative peak of the u'-correlation, around

u_τ	H	G	D	κ	C	ΔU^+	z_p^+	$u'^+_{\max}$
0.050	1.40	5.7 (6.0)	1.00	0.42	5.8	0.0	32.8	2.40
0.049	1.44	6.4 (5.5)	1.03	0.39	5.3	0.2	31.9	2.40
0.050	1.45	6.3 (6.0)	1.23	0.40	5.2	0.3	32.8	2.45
0.049	1.52	6.7 (7.0)	0.00	0.39	4.4	1.3	31.9	2.55
0.049	1.55	7.2 (9.0)	0.64	0.38	3.7	1.9	31.1	2.70
0.046	1.65	8.6 (10.0)	0.69	0.31	1.6	2.8	30.1	2.95
0.048	1.86	11.0 (14.0)	0.84	–	–	–	28.4	4.20
0.044	1.92	12.0 (19.0)	0.83	–	–	–	28.4	4.80
0.000	2.30	∞	1.00	–	–	–	–	∞

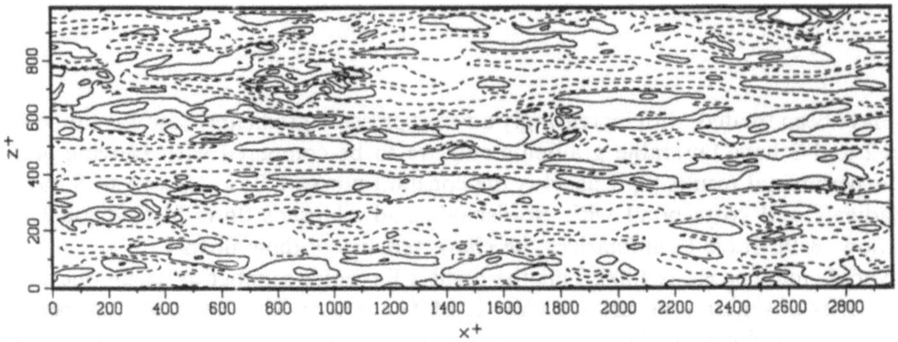

Fig. 1. u' contours. Case 4, $y^+ \approx 10$. —— high speed; --- low speed

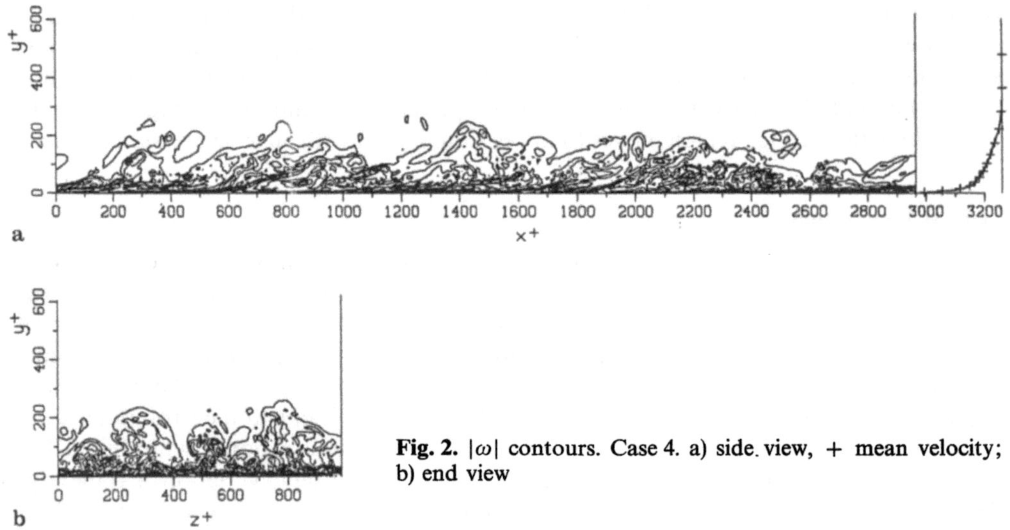

Fig. 2. $|\omega|$ contours. Case 4. a) side view, + mean velocity; b) end view

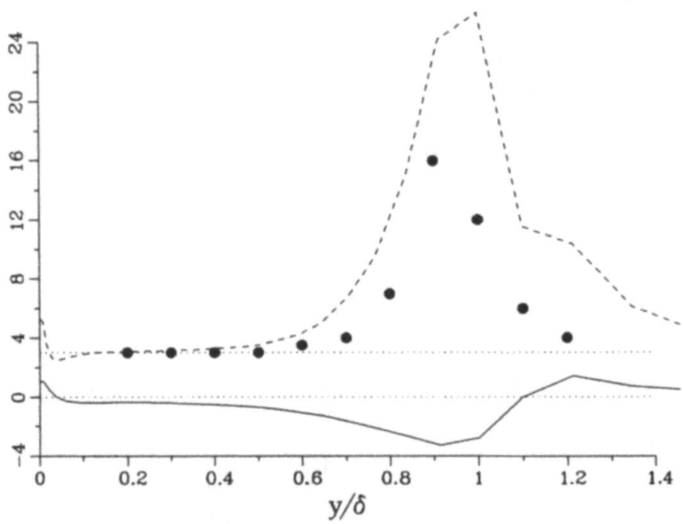

Fig. 3. Skewness and flatness factors of u'. Case 4. —— skewness; --- flatness; ● experimental flatness [11]

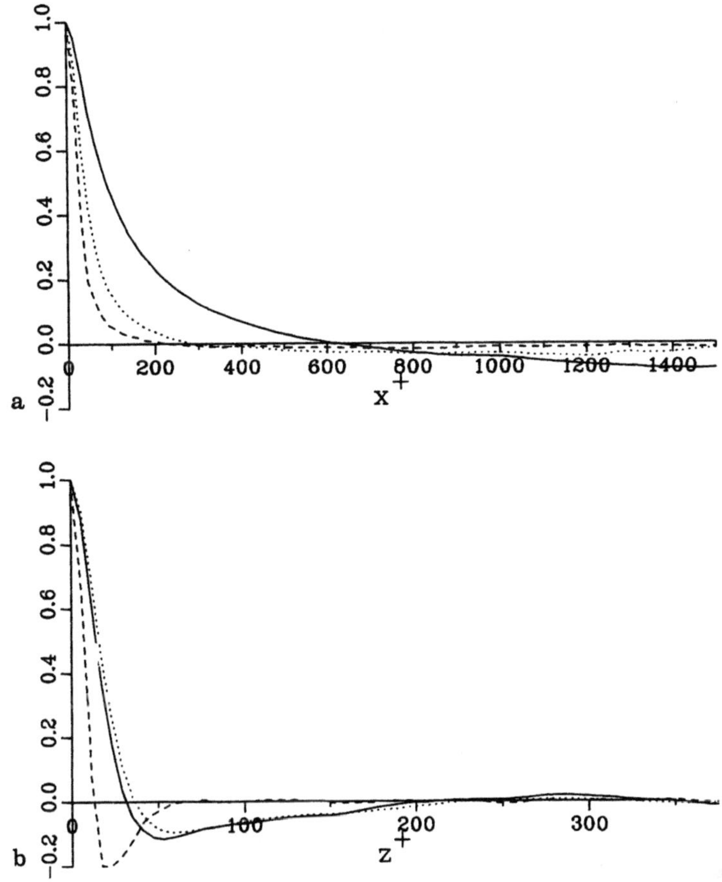

Fig. 4a and b. Two-point correlation coefficients. Case 4, $y^+ = 10$. —— u'; --- v'; ..... w'. **a** x direction; **b** z direction

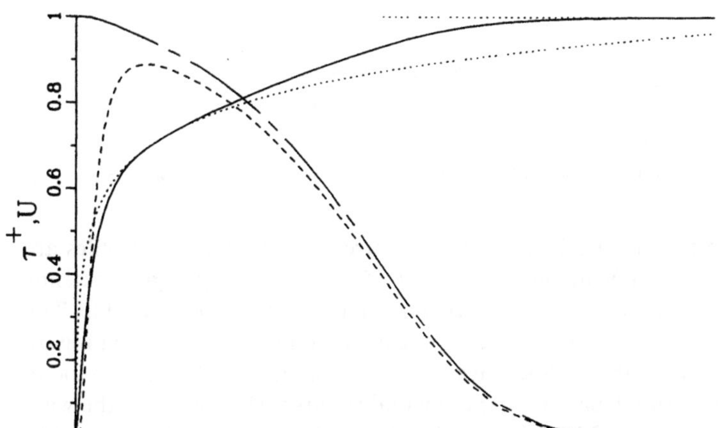

Fig. 5. Mean quantities. Case 4. —— velocity; --- Reynolds stress; -—- total stress; ..... log law; ..... U_∞

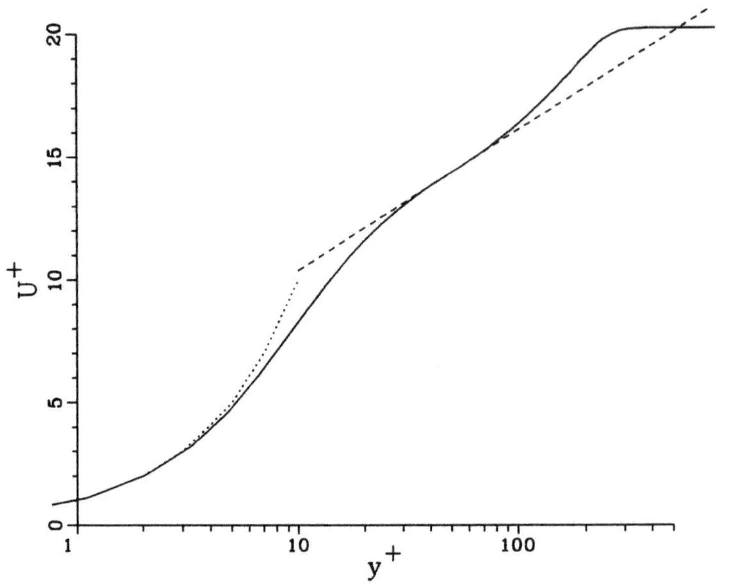

Fig. 6. Velocity profile. Case 4. —— velocity; --- log law; ..... $u^+ = y^+$

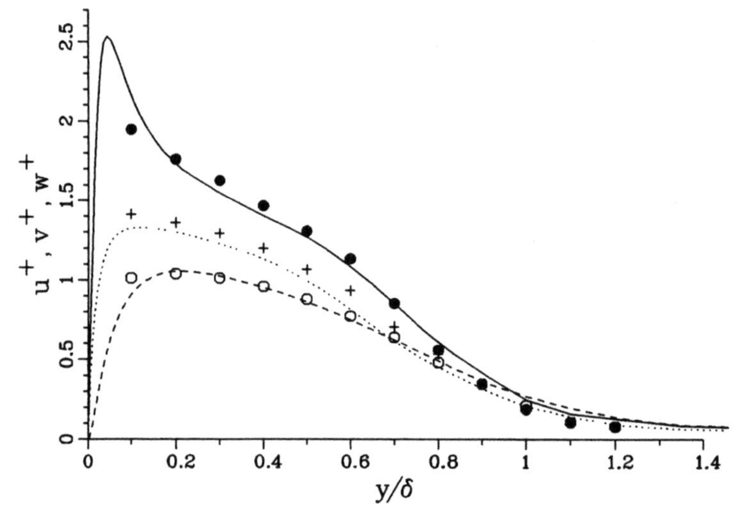

Fig. 7. rms profiles. Case 4. Present results: —— u'; --- v'; ..... w'. Experiments: [11]: ● u'; ○ v'; + w'

$z^+ = 50$ in Fig. 4b. In Fig. 5 the mean velocity, Reynolds stress and total stress profiles are plotted. The viscous stress is significant up to about $0.15\,\delta$, and the log region extends roughly from $0.1\,\delta$ to $0.2\,\delta$. Figure 6 is a semi-logarithmic plot of the velocity profile. The viscous sublayer, buffer layer, log layer and wake layer can all be identified. The log region is short, because the Reynolds number is low. Figure 7 shows the rms of the three velocity components. The agreement with Klebanoff's experimental results [11], away from the wall, is good. The disagreement at $y/\delta = 0.1$ is due to the Reynolds number difference [1]. The anisotropy and the presence of a "step" near $y = 0.7\,\delta$ are reproduced.

[1] Agreement with the lowest Reynolds number data of *Erm, Smits* and *Joubert* on pp. 191–192 of this Chapter, Figs. 4 and 5, is excellent Editor.

Pressure-gradient Dependence

Figures 8 and 9 are linear and semi-log plots of the mean velocity profiles in the various cases. Figure 8 reveals the expected change in shape of the profile and increase in both shape factors G and H under the effect of the pressure gradient. For zero and adverse gradient cases the experimental value of G (in parenthesis in the table) is consistently higher than the computed value, even though G was designed to be Reynolds-number independent [4]. The disagreement shows that the numerical results do not satisfy the high-Reynolds number defect law, and is at least partly explained by low-Reynolds number effects. These effects were shown by Coles [12] to be present up to $R_{\delta^*} \approx 10^4$, in case 4. A fully consistent study of low-Reynolds-number boundary layers will require a more general model than (1 b), possibly Coles' model with variable π [12].

Figure 9 shows that the function f_1 is significantly different for cases 1 to 8. As the pressure gradient gets more adverse the apparent value of κ decreases and C increases. These

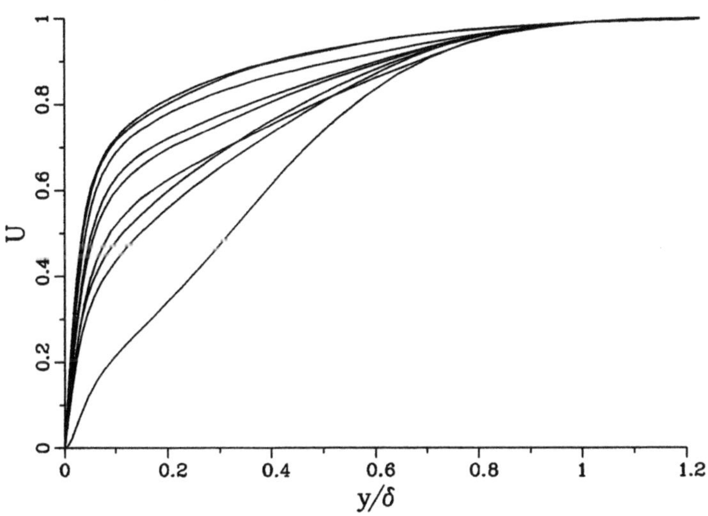

Fig. 8. Mean velocity profiles. Cases 1 (highest curve) to 9 (lowest curve). $U(\delta) = 0.99$

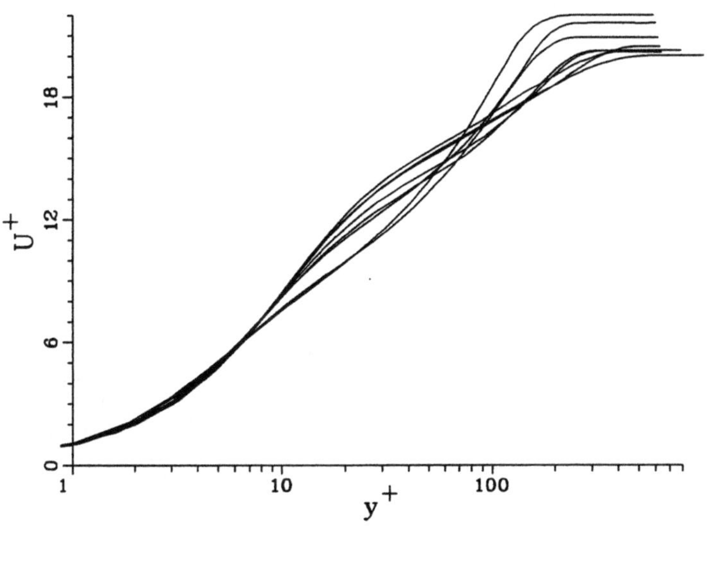

Fig. 9. Mean velocity profiles. Cases 1 to 8

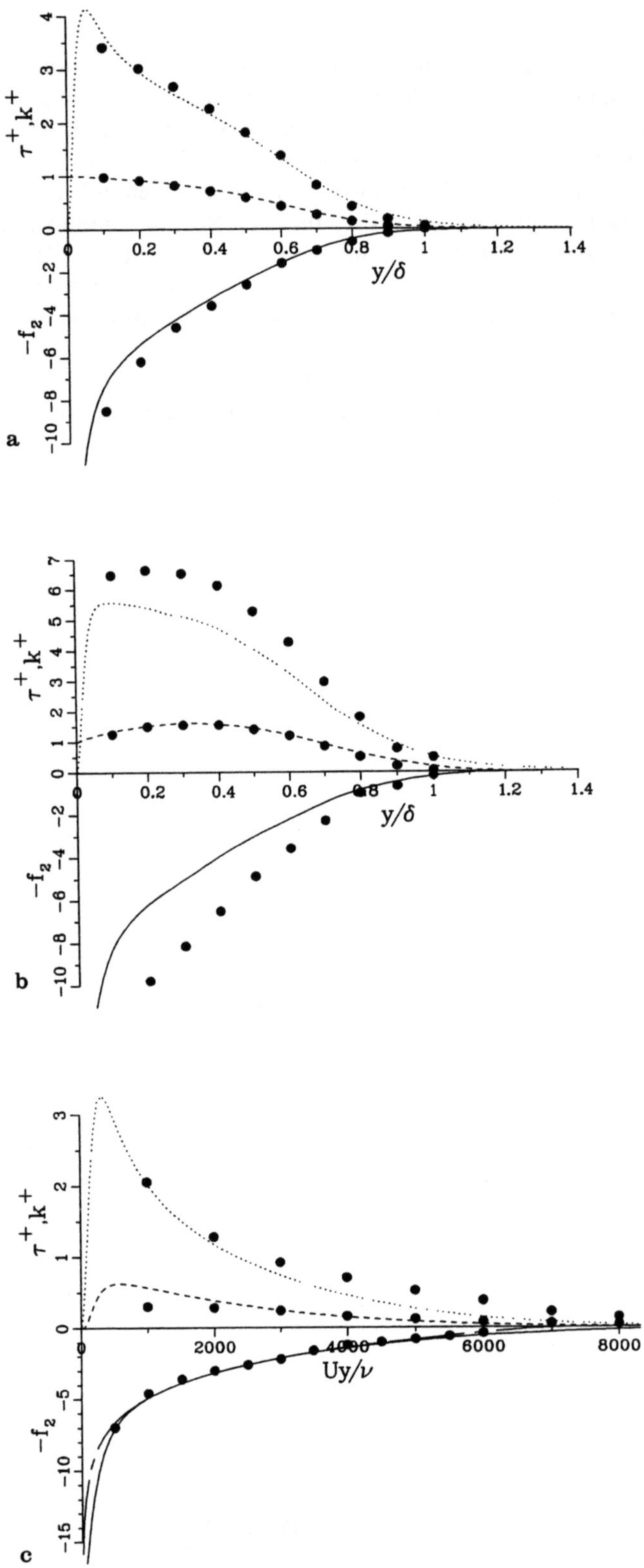

Fig. 10. Mean quantities. —— defect; --- total stress; ..... turbulent energy. • experiments [6], [11], [13]. a) case 4, $U(\delta) = 0.996$; b) case 5, $U(\delta) = 0.99$; c) case 1; —–— log law

values were computed by seeking the point of minimum slope in Fig. 9. This procedure has the merit of being systematic, but κ will be slightly over-estimated if any waviness is present on the curve. On the other hand the log regions are short, especially for cases 5 and 6, which might lower the apparent value of κ. The values of κ are not firmly established. In cases 7 and 8 the log region has disappeared altogether.

The position z_p^+ of the negative peak in the z-correlation of v', is rather well defined (see Fig. 4b). It was chosen to reflect possible changes in the flow structure near the wall. The table shows that z_p^+ is essentially insensitive to the pressure gradient. The peak value of the rms of $u^{+\prime}$, $u'^{+}_{\max}$, increases steadily from case 1 to 8.

Figures 10a and b compare the velocity defect, shear stress and turbulent kinetic energy in cases 4 and 5 with experimental results. In case 4 the stress and energy are in very good agreement with Klebanoff's high-Reynolds-number experiments, but the defect is not. In case 5 the agreement is good for the stress and acceptable for the energy, but poor for the defect. The pressure gradient is moderate: $\beta = 0.9$. One could improve the defect by comparing flows with the same G instead of β, but then the stresses would differ. The computed and experimental functions $G(\beta)$ do not agree for adverse-gradient cases. This may be due to the low-Reynolds-number effects being more intense with adverse pressure gradient. It may also be due to the breakdown of the assumptions that were made, especially the assumption of slow growth (notice the much larger values of $\delta^* \delta_X/\delta$). These assumptions are closely related to the boundary-layer assumptions.

Special Cases

The sink flow results compare well with the experimental results of Jones and Launder for $U_{\infty X}\nu/U_\infty^2 = 1.5 \times 10^{-6}$ [13], see Fig. 10c. The computed values of u_τ, H and Re_θ are 0.05, 1.42 and 690; the experimental values are about 0.05, 1.43 and 690. Good agreement is also observed for the other favorable-gradient cases. In the sink flow the velocity profile is logarithmic almost up to the edge of the boundary layer. Millikan's argument to support the existence of a log region is that the overlap layer must be logarithmic in order to "stretch" as δ^+ increases [17]. Here, δ^+ is independent of X. Another justification uses dimensional

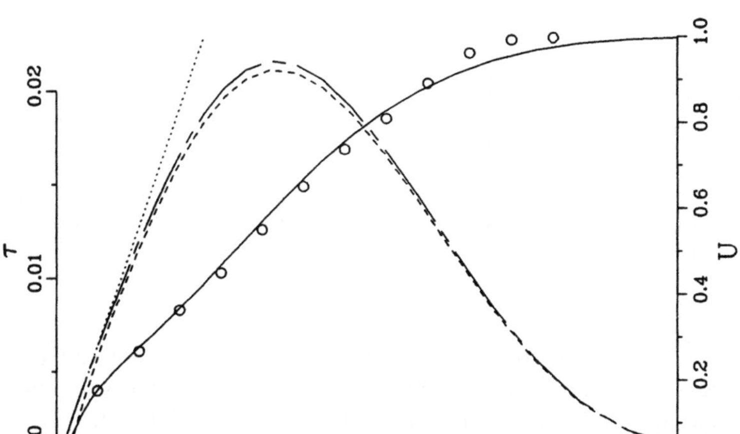

Fig. 11. Mean quantities. Case 9. $U(\delta) = 0.98$. —— velocity, --- Reynolds stress; –-–- total stress; ..... αy. o experiments [8]

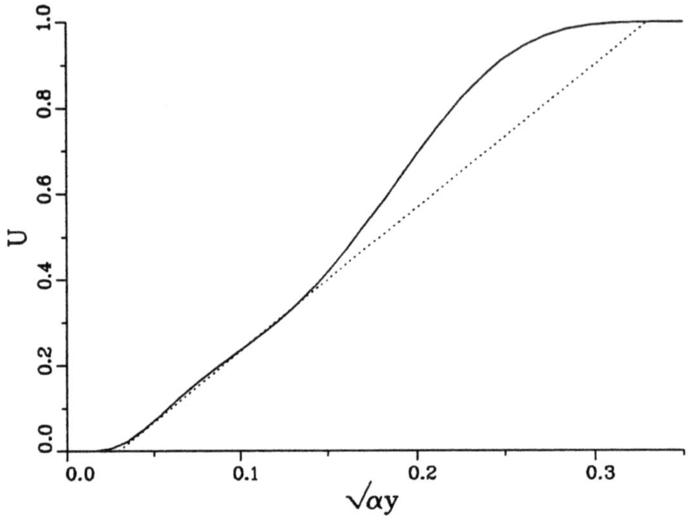

Fig. 12. Mean velocity. Case 9. ——velocity, ... $U = \sqrt{\alpha y}/0.6 - 0.1$

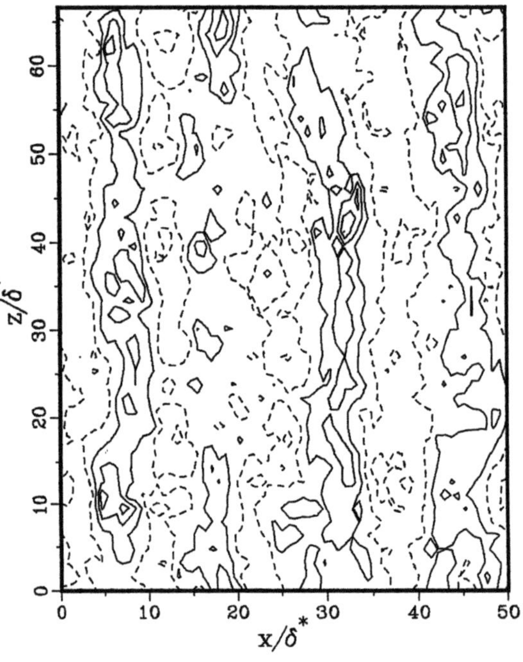

Fig. 13. u' contours. Case 9, $y/\delta \approx 1$. —— high speed; --- low speed

arguments and applies only to a constant-stress layer. Here, the stress has fallen to 20% of its wall value and the profile is still logarithmic. The same behavior is observed in a channel. Thus the log layer is a much more general feature than has been explained so far.

Stratford's flow is expected to be less dependent on Reynolds number than the others, because the viscous stress is small even near the wall, see Fig. 11. The computed and measured velocity profiles are in good agreement. The computed values $H = 2.3$ and $a = -0.22$ compare well with the experimental values. On the other hand if one assumes linear growth of the thickness with $(X - A_1)$ the computed value of $\delta/(X - X_1)$ is about 0.35

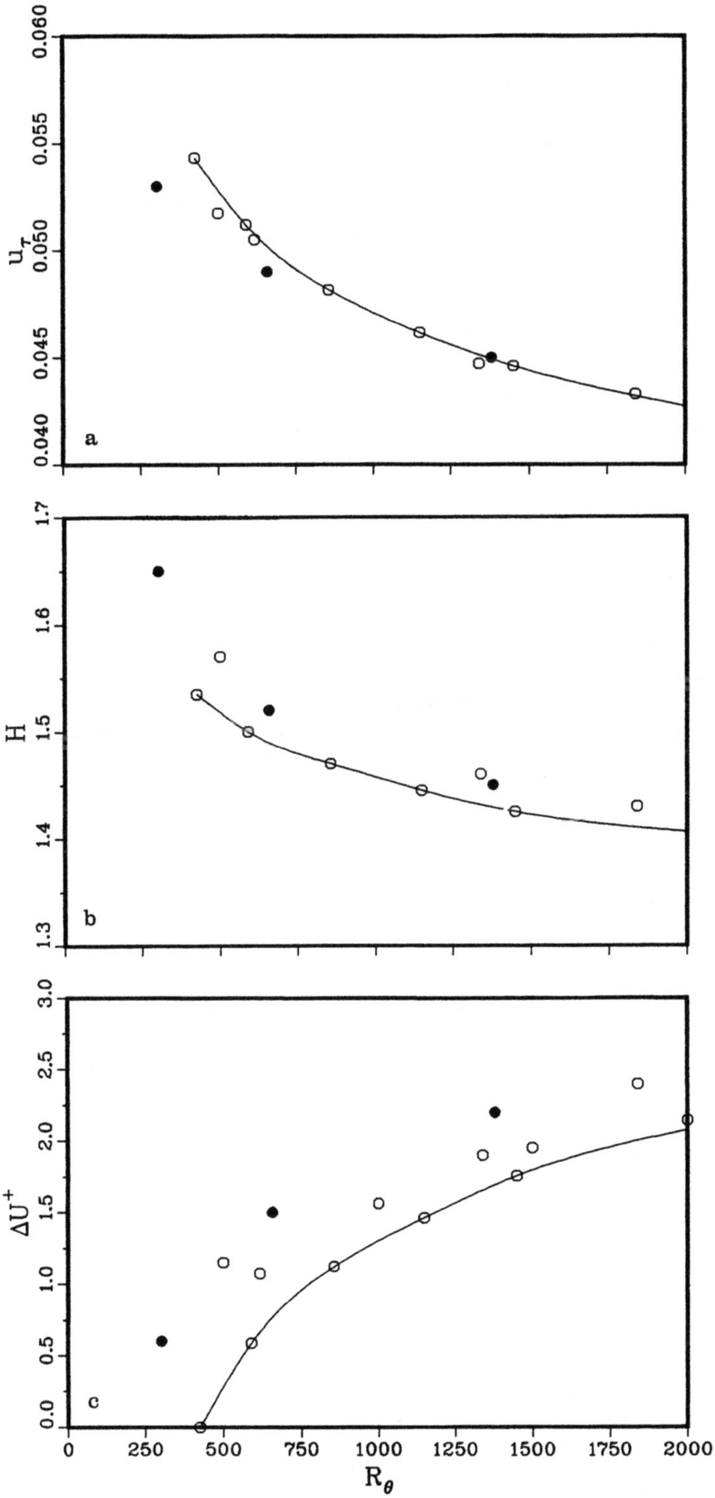

Fig. 14. Global quantities in zero-gradient boundary layer. • present results; ○ experimental results [14, 15]; Coles' compilation of experiments [12]. a) friction velocity; b) shape factor; c) wake component

as opposed to about 0.12 in the experiments: the growth angle is much too large [8, 9]. The velocity profile is plotted versus $\sqrt{\alpha y}$ in Fig. 12, revealing a "square root" layer for $\sqrt{\alpha y}$ between about 0.03 and 0.13. Such a behavior was predicted on dimensional grounds by Stratford [8] and Townsend [9], when the stress distribution is linear. The computed value of the constant K_0 (as defined by Stratford) is 0.6, in reasonable agreement with the experimental value of 0.5.

Velocity contours for Stratford's flow are plotted in Fig. 13. They are strongly correlated in the z direction, unlike in Fig. 1. The spanwise coherence remained strong even when Λ_z was quadrupled, from $17\delta^*$ to $67\delta^*$. The flow contains large spanwise two-dimensional vortices, spaced by $12.5\delta^*$, similar to the ones found in a free mixing layer. The similarity between the coherent structures in Stratford's flow and in the free mixing layer might not come as a surprise since this flow is a shear layer that is moving away from the wall. Its mean velocity profile has a long region with very little curvature, suggesting an inflectional instability. One could also expect this flow to thicken by vortex pairing, like the mixing layer. If this is true Stratford's flow is, strictly speaking, outside the range of validity of (6).

Reynolds-number Dependence

The boundary layer with zero pressure gradient (case 4) was simulated at three Reynolds numbers, $R_{\delta^*} = 500$, 1000 and 2000, in order to investigate the low-Reynolds number effects. Figure 14 compares the computed dependence of u_τ, H and ΔU^+ on R_θ with experimental results [12, 14, 15]. The strength of the wake component is defined by Coles [12] as ΔU^+, the maximum overshoot over the log law. Its deviation from the high-Reynolds number value of 3.2 is a measure of the severity of the low-Reynolds number effects. The numerical results fall within the experimental scatter and exhibit the proper dependence on the Reynolds number. The rather high computed values of ΔU^+ may be attributed to the fact that the free-stream turbulence level is effectively zero in the simulations. Note that ΔU^+ does not

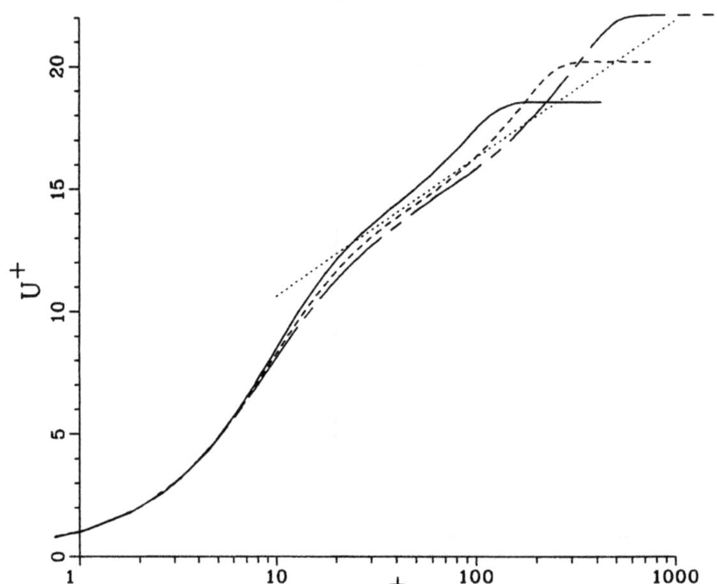

Fig. 15. Velocity profiles. —— $R_{\delta^*} = 500$, --- $R_{\delta^*} = 1000$, —-— $R_{\delta^*} = 2000$; ..... $U^+ = \log(y^+)/0.41 + 5$

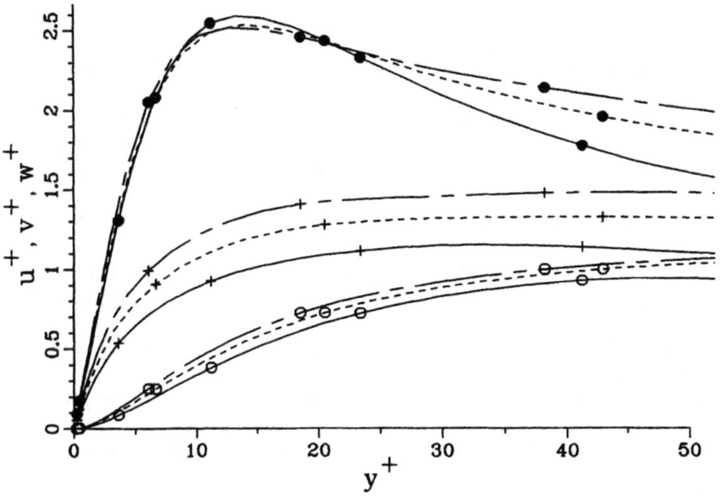

Fig. 16. rms profiles. —— $R_{\delta *} = 500$, --- $R_{\delta *} = 1000$, —-— $R_{\delta *} = 2000$. • u'; ○ v'; + w'

reach 0 even for R_θ as low as 300. This is in agreement with the results of Purtell et al. [14] and Erm et al. [15].

Figure 15 shows the velocity profiles; the log region increases in length as the Reynolds number increases. It also seems to move down, but it is too early to draw any conclusions[2] because the resolution of the high-Reynolds number case is marginal. Figure 16 shows the rms profiles near the wall. Even when expressed in wall units they exhibit a rather strong and complex dependence on the Reynolds number. This partial failure of the scaling laws may have been concealed by the fortuitous fact that the most visible and easily measured quantity, the peak value of the u' rms, is almost independent of R_θ. This dependence was also observed by Purtell et al. [14]; they remarked that the u' rms was independent of R_θ up to $y^+ \approx 15$. The variation of the w' rms is the most remarkable, in that it is appreciable even for very low values of y^+. Thus it cannot be attributed to the impingment of the wake region on the wall region. These results are inconsistent with (2) and seem to show that the present understanding of the turbulence behavior in the wall region is incomplete.

Conclusions

The results reported in this paper show that meaningful direct simulations of equilibrium turbulent boundary layers, at $Re_{\delta *} = 1000$, are possible. The Reynolds number effects are appreciable in the wake region and make comparisons with experiments difficult, especially with adverse pressure gradients. On the other hand large values of the pressure gradient, in wall variables, could be obtained. Two highlights of the results are the very long log layer in the sink boundary layer, which calls for a theoretical explanation, and the two-dimensional mixing-layer-like character of Stratford's flow, which ought to be verified experimentally. The behavior of the velocity and Reynolds-stress profiles indicates the need for improved scaling laws, in both the wall and wake regions.

Acknowledgments. The authors worked in close cooperation with Drs. J. Kim, P. Moin, R. Moser and A. Wray at NASA Ames Research Center.

[2] A shift as indicated would be consistent with the behaviour in channel flows at low Reynolds number but is a feature not found by Erm et al. Editor.

References

1. Rogallo, R. S., Moin, P. (1984): Numerical simulation of turbulent flows. Ann. Rev. Fluid Mech. **16**, 99–138
2. Spalart, P. R. (1984): A spectral method for external viscous flows. Contemp. Math. Vol. **28**, 315–335
3. Spalart, P. R. (1984): Numerical Simulation of Boundary-Layer Transition, in Proceedings of the 9th International Conference on Numerical Methods in Fluid Dynamics, Paris, Soubbaramayer and Boujot, Springer-Verlag, Heidelberg
4. Clauser, F. (1954): Turbulent boundary layers in adverse pressure gradients. J. Aeron. Sci. **21**, 91–108
5. Coles, D. (1957): Remarks on the equilibrium turbulent boundary layer. J. Aeron. Sci. **24**, 495–506
6. Bradshaw, P. (1967): The turbulence structure of equilibrium boundary layers. J. Fluid Mech. **29**, 625–645
7. Herring, H. J., Norbury, J. F. (1967): Some experiments on equilibrium boundary layers in favorable pressure gradients. J. Fluid Mech. **27**, 541–549
8. Stratford, B. S. (1959): An experimental flow with zero skin friction throughout its region of pressure rise. J. Fluid Mech. **5**, 17–35
9. Townsend, A. A. (1959): The development of turbulent boundary layers with negligible wall stress. J. Fluid Mech. **8**, 143–155
10. Leonard, A., Wray, A. (1982): A new numerical method for the simulation of three-dimensional flow in a pipe. NASA T.M. 84267
11. Klebanoff, P. S. (1954): Characteristics of turbulence in a boundary layer with zero pressure gradient. NACA T.N. 3178
12. Coles, D. (1962): A manual of experimental boundary-layer practice for low-speed flow. RAND Corp. Report. No. R-403-PR (Appendix A)
13. Jones, W. P., Launder, B. E. (1972): Some properties of sink-flow turbulent boundary layers. J. Fluid Mech. **56**, 337–351
14. Purtell, L. P., Klebanoff, P. S., Buckley, F. T. (1981): Turbulent boundary layer at low Reynolds number. *Phys. Fluids* **24(5)**, 802–811
15. Erm, L. P., Smits, A. J., Joubert, P. N. (1985): Low Reynolds Number Turbulent Boundary Layers on a Smooth Flat Plate in a Zero Pressure Gradient, this volume, pp. 186–196
16. Temam, R., Navier-Stokes Equations, Theory and Numerical Analysis. 2nd ed. North-Holland, Amsterdam, 1979
17. Millikan, C. B., A critical discussion of turbulent Flows in channels and circular tubes. Proc. 5th Int. Cong. For Applied Mechanics, 386–392, Wiley, 1939.

Calculation of Longitudinal Vortices Imbedded in a Turbulent Boundary Layer

J. Liandrat, B. Aupoix and J. Cousteix

Onera/Cert, Dept. of Aerothermodynamics, 2 avenue Edouard Belin, 31055 Toulouse Cedex, France

Abstract

Numerical simulations of the interaction between longitudinal vortices and a flat plate turbulent boundary layer have been performed.

Diffusive phenomena can be neglected in the longitudinal direction and the governing equations (Navier-Stokes and continuity) can be parabolized along this direction. A numerical forward marching procedure is then developed. Moreover, different levels of closure for the turbulent quantities have been compared. Simple models, based on the Boussinesq hypothesis provide good estimates of overall flow properties but more elaborate models using second-moment closures are required for detailed prediction. A new time-saving treatment of transport terms in the Reynolds stress governing equations is presented and provides satisfactory estimates of mean and turbulent quantities.

The different computations are checked by comparison with two available experiments.

Introduction

The present paper is devoted to numerical computations of mean and turbulent properties of decaying longitudinal vortices interacting with a flat plate turbulent boundary layer. Such interactions are encountered in many flows of practical engineering importance.

In many cases, vortices are generated by an otherwise fixed geometry and their interaction with a downstream boundary layer cannot be avoided. On an airplane, for instance, a canard-generated vortex can imbed itself in the downstream wing boundary layer and alter its aerodynamic properties. Another example is given by the horseshoe vortex wrapping around the junction of an axial flow turbomachine blade and a hub that impinges on the blade boundary layer. Other interactions may be intentionally generated, e.g. in the use of vortex generators to prevent or delay boundary layer separation on airfoils.

Measurements of mean and turbulent quantities have been performed by Bradshaw et al., (1982) in the case of a single vortex or two contrarotating vortices merging with an otherwise two-dimensional boundary layer in zero pressure gradient. In the case of a single vortex flow, the decay of the structures under the action of Reynolds stresses (stress-induced secondary flow) and main flow gradients (skew-induced secondary flow) is very slow; consequently, the vortex influences the flow for a very long distance downstream.

In the vortex-pair configuration, the interaction phenomena are much stronger due to the presence of a strong secondary flow between the two vortices and the decay of the structures may be more rapid.

Although in the case of a single vortex, the mean flow is only weakly three-dimensional, Bradshaw et al. demonstrate clearly, in the two above-mentioned experiments, a complex response of the turbulent quantities within the boundary layer.

Such flows, combining a very simple geometry with intricate turbulent and threedimensional phenomena, provide good test cases for numerical procedures and turbulence models.

Turbulent Shear Flows 5

Basic Hypotheses

The flow is assumed incompressible and statistically steady.

Streamwise vortices merging with a flat-plate turbulent boundary layer are characterized by much larger velocity gradients in both transverse (y and z) directions than in the streamwise (x) direction.

Asymptotic analysis of the governing equations under such hypotheses shows that diffusion is effective in the (y, z) crossflow plane but can be neglected in the x-direction. Moreover the pressure gradient term $\partial P/\partial x$ that initially ensures the elliptic nature of the Navier-Stokes equations is a prescribed quantity.

Thus, the governing equations are parabolized in the streamwise direction. However, they remain elliptic in the crossflow plane since a perturbation field $p(y, z)$ induced by the secondary flow v and w and ensuring continuity, is calculated in each x-cross section.

Mean Flow Equations

The equations governing the mean velocity (U, V, W) and the pressure perturbation term p, then read as follows:

Continuity equation: $\dfrac{\partial U}{\partial x} + \dfrac{\partial V}{\partial y} + \dfrac{\partial W}{\partial z} = 0.$

Momentum equations:

$$\frac{DU}{Dt} = -\frac{1}{\varrho}\frac{\partial}{\partial x}P(x, z) \qquad + \frac{\partial}{\partial y}\left(\nu\frac{\partial U}{\partial z} - \overline{u'v'}\right) \qquad + \frac{\partial}{\partial z}\left(\nu\frac{\partial U}{\partial z} - \overline{u'w'}\right),$$

$$\frac{DV}{Dt} = -\frac{1}{\varrho}\frac{\partial}{\partial y}(P(x, z) + p(y, z)), \qquad + \frac{\partial}{\partial y}\left(\nu\frac{\partial V}{\partial y} - \overline{v'^2}\right) \qquad + \frac{\partial}{\partial z}\left(\nu\frac{\partial V}{\partial z} - \overline{v'w'}\right),$$

$$\frac{DW}{Dt} = -\frac{1}{\varrho}\frac{\partial}{\partial z}(P(x, z) + p(y, z)) + \qquad + \frac{\partial}{\partial y}\left(\nu\frac{\partial W}{\partial y} - \overline{v'w'}\right) \qquad + \frac{\partial}{\partial z}\left(\nu\frac{\partial W}{\partial z} - \overline{w'^2}\right),$$

↑ y-diffusion terms aligned   ↑ z-diffusion terms aligned

where $D./Dt$ stands for the substantial derivative

$$U\frac{\partial .}{\partial x} + V\frac{\partial .}{\partial y} + W\frac{\partial .}{\partial z}.$$

Numerical Solution Procedure

The parabolic feature of the equations in the streamwise direction is amenable to a forward-marching solution procedure presented by Cousteix et al. (1985). Before discretization, the equations governing the flow are written in a quasi-conservative form. Then, the discretized approximations are obtained from a finite-volume technique on a staggered grid.

The computation advances step by step in the x direction.

As they are strongly coupled, the four governing equations are solved simultaneously at each x station. The x-advection terms are approximated using the known values of the upstream flow. For the sake of stability, an exponential scheme developed by Patankar (1980) is employed to estimate the y and z convection and diffusion terms.

The Reynolds stresses are expressed as

$$\overline{u_i' u_j'} = -2\nu_t S_{ij} + \overline{u_i' u_j'}^{\mathrm{Suppl}},$$

where $S_{ij} = \frac{1}{2}\left[\frac{\partial U_i}{\partial x_j} + \frac{\partial U_j}{\partial x_i}\right]$ and ν_t is the eddy viscosity.

$\overline{u_i' u_j'}^{\mathrm{Suppl}}$ is a source term but $-2\nu_t S_{ij}$ contributes to the implicit differencing terms that strengthen the stability of the scheme.

Finally, the discretized equations can be written at each x-station as a system of matrix equations relating at any grid point P the values of (U, V, W, p) to the values of the same variables at each neighbour N and S in the y direction, E and W in the z direction and U upstream.

$$A_N X_N + B_S X_S + C_E X_E + D_W X_W + E_P X_P = F_P,$$

where X is the (U, V, W, p) vector, A_N, B_S, C_E, D_W, E_P are 4×4 matrices and F_P is a four-coordinate vector standing for explicit source and upstream terms.

The solution of this matrix equation system is achieved by an iterative process. Each equation is written as

$$C_E X_E + D_W X_W + E_P X_P = F_P - A_N X_N^* - B_S X_S^*,$$

where X^* stands for the last iterative estimation of X. At each iteration, the system then reduces on every line to independent systems that can be solved using a block-tridiagonal matrix-inversion algorithm.

The number of iterations in each crossflow plane must be high enough to ensure the propagation of boundary conditions from the wall to the external flow. It is obviously connected with the number of grid points in the y direction. In addition, when transport equations for turbulent quantities are implemented, these extra equations are solved separately. Stability conditions require the use of a smaller step. Practically, for one step for the mean flow calculation, ten steps are necessary for the transport equations using the previously computed mean velocity gradients.

Finally, this numerical solution procedure allows large x integration steps (about one third of the average boundary layer thickness) and the limited number of iterations in each crossflow plane (about two times the number of grid points in the y direction) reduces the computation time to an acceptable level.

Turbulence Models

The Reynolds stresses $\overline{v'^2}$, $\overline{w'^2}$ and $\overline{v'w'}$ govern the evolution of the secondary motion and thus of the longitudinal vorticity. Special attention is therefore paid to the modelling of these stresses in the different turbulence models outlined below.

Boussinesq Viscosity Hypothesis

The simplest way to express the Reynolds stresses is provided by the Boussinesq hypothesis: $\overline{u_i' u_j'} - 2/3\,\delta_{ij} k = -2\nu_t S_{ij}$ where k is the turbulent kinetic energy.

Two models are tested to evaluate the eddy viscosity ν_t:

- The mixing length model, with the effective boundary layer thickness $\delta(z)$ at a given x being the value of y for which the local streamwise velocity equals 99% of the external streamwise velocity $U_{ext}(z)$. The mixing length L is obtained from $L = 0.085\,\delta \tanh(\kappa y/0.085\,\delta)$ and $\nu_t = F^2 L^2 (2 S_{ij} S_{ij})^{1/2}$ where F is the classical van Driest wall damping function and $\kappa = 0.40$. The turbulent kinetic energy, required to compute the normal stresses, is derived from the Prandtl-Kolmogorov formula:

$$\nu_t = \sqrt{2a_1 k}.L \quad \text{with } 2a_1 = 0.3.$$

- The $k-\varepsilon$ model based on the computation of the turbulent kinetic energy k and its dissipation rate ε adapting standard transport equations and the relation: $\nu_t = C_\mu (k^2/\varepsilon)$ with $C_\mu = (2a_1)^2$.

Reynolds Stress Transport Equations

Exact transport equations governing the Reynolds stresses are derived from the Navier-Stokes equations:

$$\begin{aligned}\text{Advection of } \overline{u_i' u_j'} &= \text{Production } (P_{ij}) - \text{dissipation } (\varepsilon_{ij}) \\ &\quad + \text{pressure strain } (\phi_{ij,1} + \phi_{ij,2} + \phi_{ijw}) + \text{diffusion } (T_{ij}).\end{aligned}$$

The production $P_{ij} \equiv -\overline{u_i' u_k'}\dfrac{\partial U_j}{\partial x_k} - \overline{u_j' u_k'}\dfrac{\partial U_i}{\partial x_k}$ is exact but the other terms need modelling. The dissipation is estimated with the local isotropy assumption as:

$$\varepsilon_{ij} = 2/3\,\delta_{ij}\varepsilon.$$

The non-linear part of the pressure-strain correlation $\phi_{ij,1}$ is modelled as proposed by Rotta:

$$\phi_{ij,1} = -C_1 \frac{\varepsilon}{k}\left(\overline{u_i' u_j'} - \frac{2}{3}\delta_{ij} k\right).$$

For the linear part of the pressure strain term, two models, proposed by Launder et al. (1975), are compared:

$$\phi_{ij,2} = \alpha'(P_{ij} - \tfrac{2}{3}\delta_{ij} P) + 2\beta' k S_{ij} + \gamma' D_{ij}$$

with

$$D_{ij} \equiv -\overline{u_i' u_k'}\frac{\partial U_k}{\partial x_j} - \overline{u_j' u_k'}\frac{\partial U_k}{\partial x_i}$$

and

$$\alpha' = -\frac{C_2 + 8}{11}, \quad \beta' = -\frac{30\,C_2 - 2}{55}, \quad \gamma' = -\frac{8\,C_2 - 2}{11}$$

with $C_2 = 0.4$; or, a simplified form with $\alpha' = -0.6$, $\beta' = \gamma' = 0$.

Moreover, wall reflexion terms are needed for near-wall regions. Gibson and Launder (1978) proposed the following expression of ϕ_{ijw} in the case of the above simplified model ($\alpha' = 0.6$, $\beta' = \gamma = 0$):

$$\phi_{ijw} = \left[C_1' \frac{\varepsilon}{k} \left(\overline{v'^2}\, \delta_{ij} - \frac{3}{2}\, \overline{u_i' v'}\, \delta_{j2} - \frac{3}{2}\, \overline{u_j' v'}\, \delta_{i2} \right) \right.$$
$$\left. + C_2' \left(\phi_{22,2} \delta_{ij} - \frac{3}{2}\, \phi_{2i,2} \delta_{j2} - \frac{3}{2}\, \phi_{2j,2} \delta_{i2} \right) \right] f$$

with $C_1' = 0.5$ and $C_2' = 0.3$.

A similar study has been done for the complete model with $C_2 = 0.4$ following the same basic ideas. That is, C_1' and C_2' are optimized to obtain the experimental ratios $\overline{u_i' u_j'}/k$ in the logarithmic region of a two-dimensional boundary layer

$$\left(\frac{\overline{u'^2}}{k} \approx 1.1;\ \frac{\overline{v'^2}}{k} \approx 0.25;\ \frac{\overline{w'^2}}{k} \approx 0.65 \text{ and } -\frac{\overline{u'v'}}{k} \approx 0.26 \right).$$

The values $C_1' = 0.35$ and $C_2' = 0.15$ have been selected.

The effects of the wall reflection terms are damped by the surface proximity function f as the influence of the wall decreases. The ratio l'/y, where l' is a characteristic length scale of turbulence, is used as the f parameter. Gibson et al. first proposed $f = l'/\kappa y$. This linear formulation provides $f = 1$ at the wall. Meanwhile, the classical boundary layer estimate $l' = 0.085\,\delta$ gives $f = 0.2$ at $y = \delta$. The quadratic expression $f = (l'/\kappa y)^2$ proposed by Naot and Rodi (1981) provides, with the same estimate for l', $f = 0.04$ at $y = \delta$ and seems more satisfactory. Finally, with $k^{3/2}/\varepsilon$ as the turbulence length scale, the surface proximity function reads:

$$f = \left[\frac{[2 a_1 k]^{3/2}}{\varepsilon \kappa y} \right]^2.$$

Treatment of the Diffusion Terms and Integration of the Reynolds Stress Transport Equations

Two expressions for the diffusion terms leading to different integration schemes of the Reynolds stress transport equations have been compared:

- The first estimate of the Reynolds stress diffusion term is given by the classical gradient closure:

$$T_{ij} = \frac{\partial}{\partial x_1} \left[(\nu + \nu_t) \frac{\partial \overline{u_i' u_j'}}{\partial x_1} \right].$$

Without any other approximations the Reynolds stress equations, can then be discretized and solved with the numerical procedure presented above for the mean flow equations. However, due to the complexity of a simultaneous solution of the seven equations for the six Reynolds stresses and the dissipation rate ε, the solution is iterated separately for each unknown.

- An alternative treatment of the Reynolds stress transport equations is possible by making an algebraic estimate of the diffusive transport terms T_{ij}.

After computation of k and ε with the above-mentioned procedure, T_{ij} is expressed as proposed by Launder (1982) as

$$T_{ij} = T_k[(1+\beta)\,\overline{u_i'u_j'}/k - \tfrac{2}{3}\,\delta_{ij}\beta],$$

where T_k is the diffusive transport for k.

Owing to this expression, the Reynolds stress transport equations are now hyperbolic and can be directly integrated along a streamline without iteration.

These equations then read

$$V_s \frac{\overline{du_i'u_j'}}{ds} = a\,\overline{u_i'u_j'} + C_{ij},$$

where s is the curvilinear coordinate, $V_s = \sqrt{U^2+V^2+W^2}$, $a = (T_k/k)\,(1+\beta) - C_1\,\varepsilon/k$ and

$$C_{ij} = P_{ij} - \varepsilon_{ij} + \tfrac{2}{3}\,C_1'\,\varepsilon\,\delta_{ij} + \phi_{ij,2} + \phi_{ijw}.$$

At each grid point P, if Δs_P stands for the streamline length from the upstream plane (u) to P, a direct integration provides

$$\overline{u_i'u_j'} = \left[-\frac{C_{ij}}{a} + \overline{u_i'u_j'}_u\right] \exp\left[\frac{a}{V_s}\,\Delta s_P\right] + \frac{C_{ij}}{a}.$$

Near-Wall Treatment

The (k, ε) and Reynolds stress transport equation models presented above are only valid for large turbulent Reynolds numbers and cannot be applied in the wall region.

In all the cases, the wall sublayer is computed with a mixing length model. Moreover, the turbulent kinetic energy k and its dissipation rate ε are expressed with a simplified version of the model developed by Hassid and Poreh (1975) as

$$\nu_t = \sqrt{2a_w \cdot k}\;\kappa y[1 - \exp(-0.0255\,y^+)],$$
$$\varepsilon = (2a_w k)^{3/2}\,[1 - \exp(-0.0255\,y^+)]/\kappa y + 2\nu k/y^2,$$

where $2a_w = 0.26$ and y^+ is the normal wall coordinate

$$y^+ = \frac{y\sqrt{\dfrac{\tau_w}{\varrho}}}{\nu}\;.$$

According to experimental measurements of near-wall turbulence, the normal stresses are expressed as $\overline{u'^2}/k = 1.1$, $\overline{v'^2}/k = 0.25$, $\overline{w'^2}/k = 0.65$ and the off-diagonal stresses are derived from a Boussinesq hypothesis. The matching of the wall-layer model with the high-Reynolds-number model occurs in the logarithmic region.

Comparison between Computation and Experiment

Two experiments from Bradshaw et al. have been simulated: a single vortex experiment and a double vortex one with flow, between the two vortices, upwards. In conformity with the experimental results presented by Bradshaw et al., the computations are presented on selected crossflow planes as maps of mean velocity and isocontours of turbulent quantities.

Influence of Initial Data

Special attention has been paid to initial mean velocity data as they strongly influence the downstream development of the computed flow.

As the crossflow circulation level evolves downstream very slowly in the case of a single vortex, its initial estimate must be given accurately. Moreover, initial values of secondary flow on the computation volume boundaries affect mass conservation and can produce lateral inflows or outflows that lead to an erroneous decay of the computed vortices. On the other hand, inadequate longitudinal mean velocity initial data affect the streamwise advection phenomena.

In fact, an initial secondary flow with a strong enough circulation in the crossflow plane and an initial distribution of longitudinal velocity reproducing exactly the main vortices are required. Moreover, the computation domain is extended to allow two-dimensional conditions on both sides and to have an undisturbed potential flow above.

Boussinesq Hypothesis (Figs. 1, 3–5)

The vortex/boundary-layer interaction drastically reorganizes the Reynolds stresses. Experiments bring into evidence large departures from the eddy viscosity hypothesis, so classical, simply-behaved, eddy viscosity models should fail.

However, in the case of the single vortex flow, even the mixing-length model provides satisfactory estimates of the main flow properties.

Mixing-length model

The general evolution of the single vortex flow is fairly well reproduced by this computation. The slow decay of the vortex strength under the action of Reynolds stresses and mean velocity gradients, the growth by diffusion of the vortex affected cross section and the spanwise trajectory of the vortex core due to the vicinity of the wall, are in agreement with experiment.

However, the computed iso-U lines are less distorted than the measured ones. An overestimate of the eddy viscosity due to the crude definition of the boundary layer thickness δ in the vortex interaction region can generate a too strong turbulent diffusion. The other probable reason is the rough estimate of the Reynolds stresses. Although the intensity contours of $\overline{u'^2}$, $\overline{v'^2}$ and $\overline{w'^2}$ look like the mean streamwise velocity contours, as in the experimental results, the velocity correlations are badly balanced. $\overline{v'w'}$ and $\overline{v'^2} - \overline{w'^2}$ which govern the secondary motion are underpredicted. $\overline{u'v'}$ is well estimated in the two-dimensional regions. However, $\overline{u'v'}$ and $\overline{u'w'}$ are miscalculated in the imbedded part of the vortex where the strong vortex/boundary-layer interaction leads to important perturbations of the shear stress field.

These defects are much more severe in the case of the vortex pair.

In the experiment, the two downward-moving sides of the vortices are very twisted. With the mixing length model, the use of the boundary-layer thickness δ induces an overestimate of the turbulent diffusion and provides a strong smoothing of the distortion. For the same reason, the iso-velocity contours of turbulent quantities profiles are badly reproduced.

$(k-\varepsilon)$ Transport Equation Model

For the single vortex flow, the results are very similar to those obtained with the mixing-length model. In the case of the vortex pair, however, (k, ε) transport equations provide noticeable improvements. The distortion of isolines is stronger and the estimate of turbulent quantities is improved in the region affected by vortex/boundary layer interactions. Nevertheless, the overall calculation of the flow is not entirely satisfactory as the estimate of $\overline{v'w'}$ and $\overline{u'w'}$ remain too weak.

The mixing-length formulation connecting the turbulence length scale to the mean flow properties using a boundary layer thickness $\delta(z)$ is not able, in the double vortex case, to generate the intricate behaviour of the turbulent quantities. The local estimate of this scale by means of k and ε, using two transport equations, refines the Reynolds stress computation. Nevertheless, the complete calculation must be improved by discarding the Boussinesq hypothesis.

Reynolds Stress Transport Equations

So far computations have been performed only for the single-vortex flow.

Classical Treatment of Reynolds Stress Transport Equations (Figs. 1–4)

The main characteristic feature of this computation is the appearance of two secondary vortices in the wall region of the down- and upward-moving sides of the main vortex.

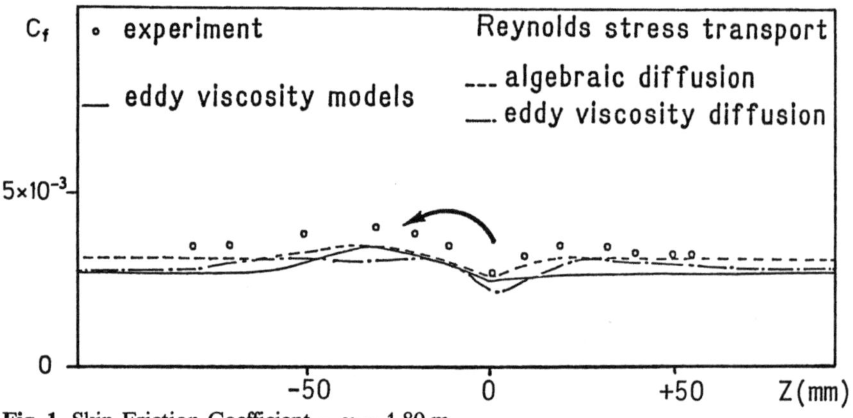

Fig. 1. Skin Friction Coefficient – $x = 1.80$ m

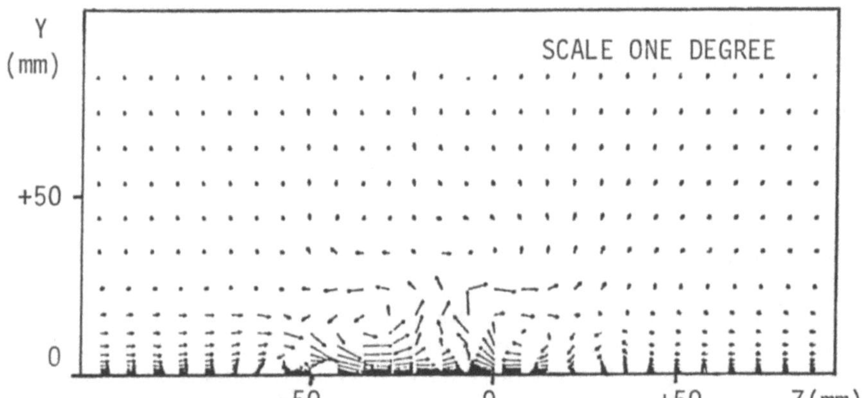

Fig. 2. Secondary Flow – $x = 0.60\ m$ – Reynolds Stress Transport-First Gradient Diffusion

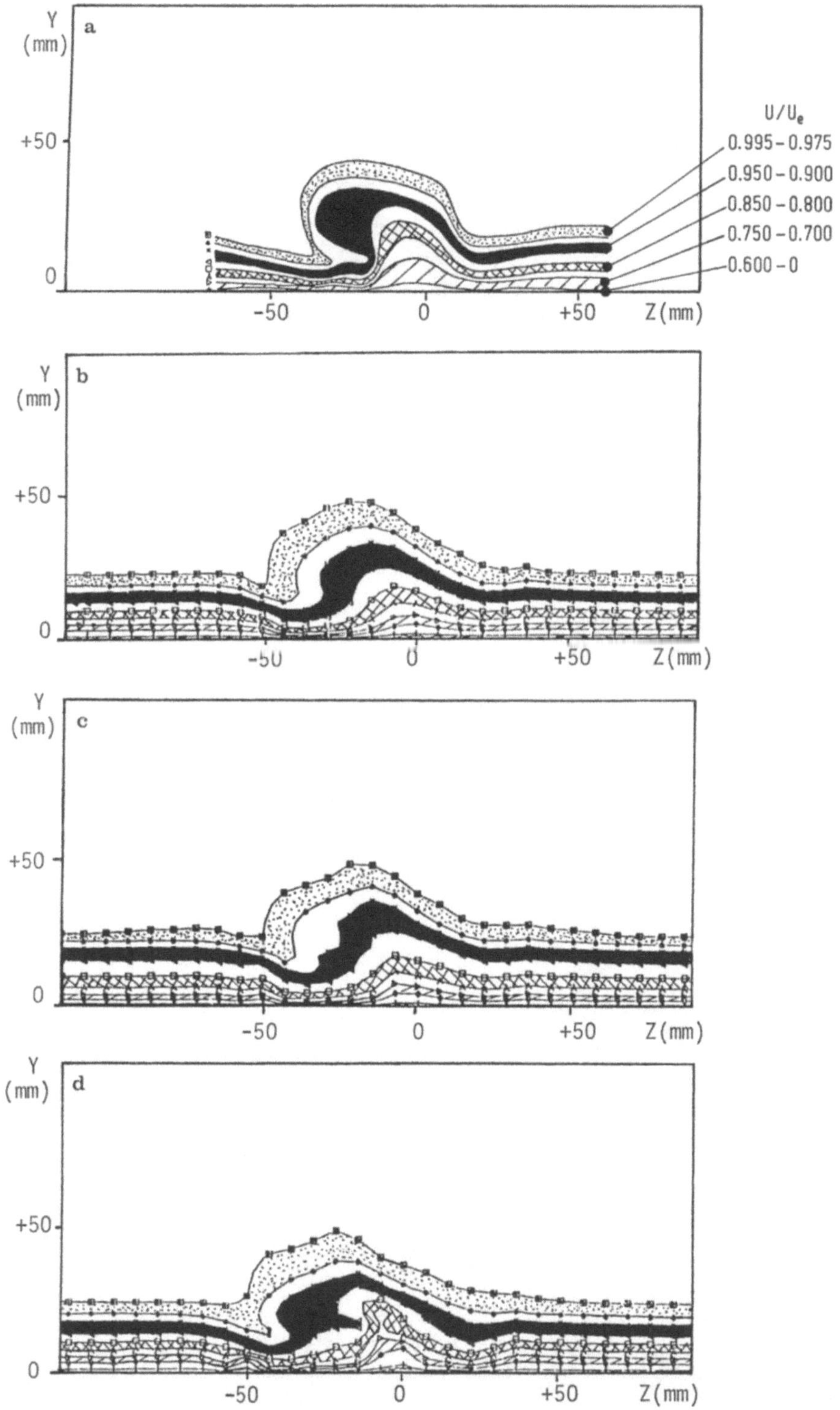

Fig. 3a. ISO-U Contours – $x = 0.60$ m – Experiment; **b** ISO-U Contours – $x = 0.60$ m – Boussinesq Hypothesis; **c** ISO-U Contours – $x = 0.60$ m – Reynolds Stress Transport – Algebraic Diffusion; **d** ISO-U Contours – $x = 0.60$ m – Reynolds Stress Transport – First Gradient Diffusion

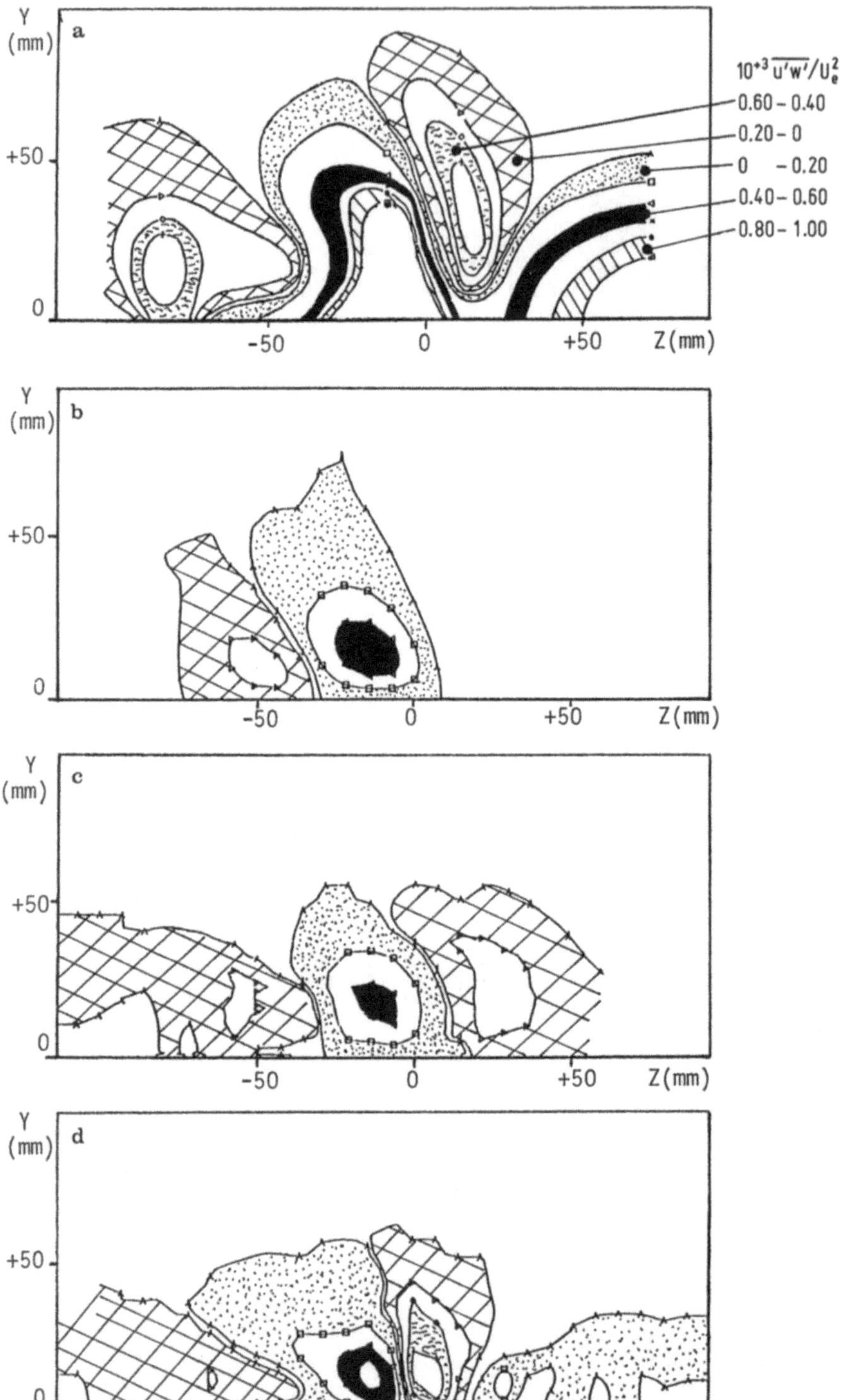

Fig. 4a. ISO-$\overline{u'w'}$ Contours – x = 0.60 m – Experiment; **b** ISO-$\overline{u'w'}$ Contours – x = 0.60 m – Boussinesq Hypothesis; **c** ISO-$\overline{u'w'}$ Contours – x = 0.60 m – Reynolds Stress Transport – Algebraic Diffusion; **d** ISO-$\overline{u'w'}$ Contours – x = 0.60 m – Reynolds Stress Transport – First Gradient Diffusion

These induced vortices are brought into evidence by the spanwise variation of the skin friction coefficient and by the secondary flow. Experimental estimates of skin friction bring into evidence an increase in wall shear stress near the upward-moving side of the vortex which can be linked to the presence of negative values of v associated with a secondary vortex. Nevertheless, according to Bradshaw et al., this secondary structure is not resolved by the hot-wire probes they used and a quantitative comparison between computational and experimental secondary flow estimates is not possible in that region. A direct study of the transport equation for the longitudinal vorticity indicates that secondary vortices are generated by the interaction of the main vortex with Reynolds stresses and mean velocity gradients in the near-wall region. Available experimental visualisations of vortex/boundary layer interactions (Mehta and Lim, 1984) also show the existence of these secondary structures.

The iso-U line distortion is improved near the downward-moving side of the vortex and the secondary flow is predicted better.

This is presumably due to the fact that Reynolds-stress transport equations can provide a larger difference $\overline{v'^2} - \overline{w'^2}$ than the Boussinesq hypothesis. However, the prediction of the Reynolds stresses is not yet completely satisfactory. Levels of $\overline{u'^2}$, $\overline{v'^2}$ and $\overline{w'^2}$ are well reproduced but iso-contours lines are less distorted than in the experiment. $\overline{u'w'}$ and $\overline{v'w'}$ are in better agreement with experiment, but still underpredicted.

Alternative Treatment of Reynolds Stress Transport Equations (Figs. 1, 3, 4)

The Reynolds stress transport equations are here solved along streamlines with the help of a simple algebraic model for the diffusion term. The influence of the governing parameter β is now discussed.

With $b_{ij} = \overline{u_i' u_j'}/k - \frac{2}{3}\delta_{ij}$, T_{ij} reads:

$$T_{ij} = T_k((1 + \beta)\, b_{ij} + \tfrac{2}{3}\delta_{ij})$$

and can be split onto two terms. $T_k(1 + \beta)\, b_{ij}$ links T_{ij} directly to the large-scale motion anisotropy while $2/3\, T_k \delta_{ij}$ relates the diffusion to isotropic small scales. As it is commonly assumed that diffusion is mainly connected with small scale motions, negative values of β that weaken the influence of the first term, sound more physical.

Computations performed on two-dimensional and three-dimensional cases with negative values of β showed that, for $-0.2 \leqq \beta \leqq 0$, realisability conditions on Reynolds stresses ($\overline{u_i' u_i'} \geqq 0$ and $|\overline{u_i' u_j'}| \leqq (\overline{u_i'^2}\, \overline{u_j'^2})^{1/2}$) cannot always be secured. In fact, the diffusion of the off-diagonal stresses $\overline{u'v'}$, $\overline{u'w'}$ and $\overline{v'w'}$ is unacceptably increased and can generate an unrealistic propagation of three-dimensional effects in initially two-dimensional regions. A parametric study on the case of vortex/boundary layer interaction provides an optimal value of -0.8, in agreement with Launder's proposal.

With this model, the computed secondary flow is less important than previously. Only one secondary vortex can form near the upward-moving side of the main vortex as can be seen from the spanwise distribution of the skin-friction coefficient. Moreover, the simplified expression for pressure strain term $\phi_{ij,2}$ has been tested. Naot and Rodi (1981) have pointed out that this expression was not suitable for flow with secondary motion mainly governed by the difference $\overline{v'^2} - \overline{w'^2}$. In our case, the use of this simplified model alters only slightly the estimation of $\overline{w'^2}$, $\overline{v'^2}$ and $\overline{u'^2}$. The main feature of this computation is to provide a weaker level for the off-diagonal stresses, $\overline{v'w'}$ and $\overline{u'w'}$ in disagreement with the experimental results. The prediction of $\overline{u'v'}$ is very similar to the previous one and the mean flow computation is not strongly affected.

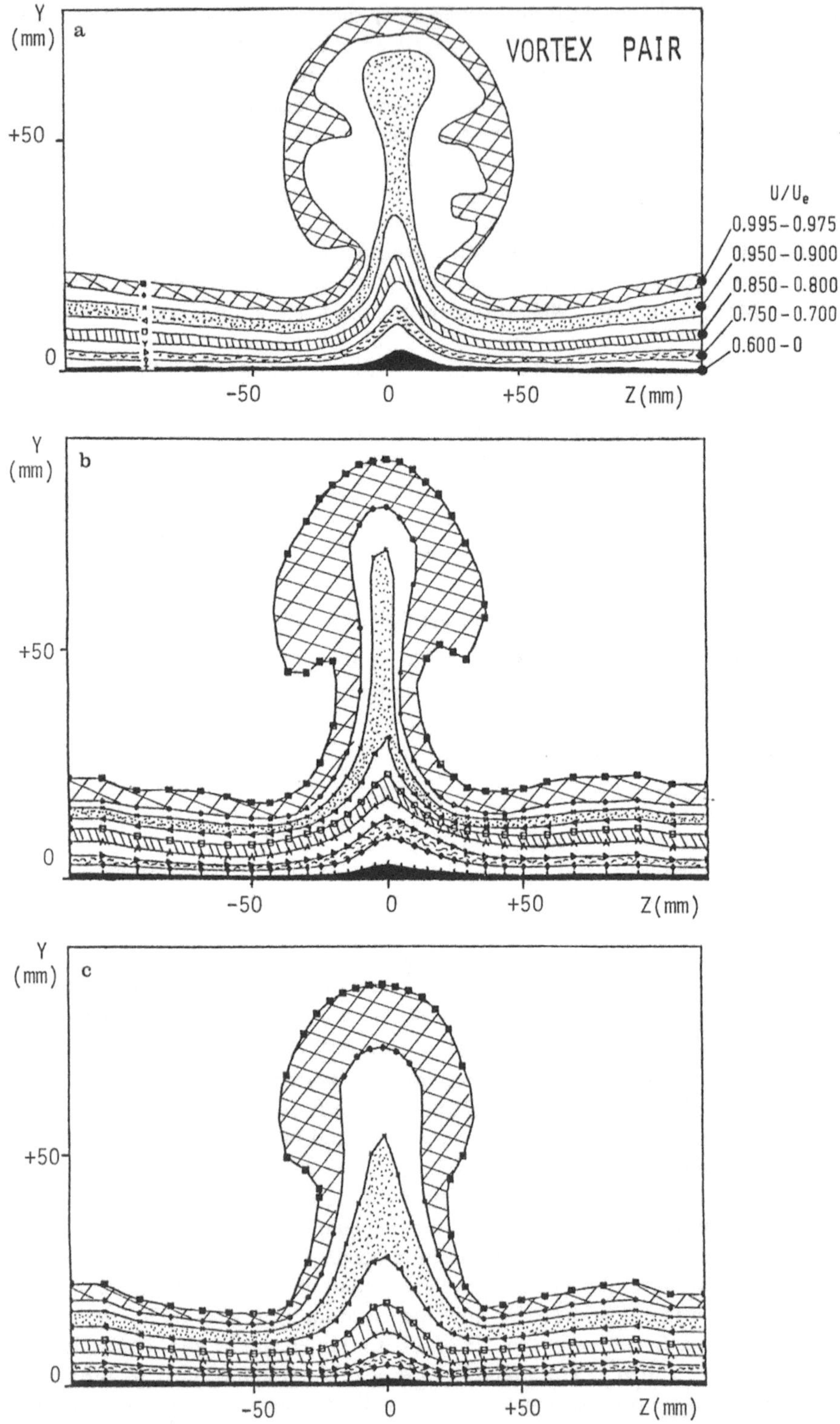

Fig. 5a. ISO-U Contours – $x = 0.45$ m – Experiment; **b** ISO-U Contours – $x = 0.45$ m – $(k - \varepsilon)$ Model; **c** ISO-U Contours – $x = 0.45$ m – Mixing Length Model

Conclusions

A forward marching numerical procedure has been developed to compute the interaction of streamwise vortices with flat plate turbulent boundary layers.

Simple models of turbulence, based on the Boussinesq hypothesis, provide satisfactory estimates of the main features of the flow in the case of a single vortex but partly fail in the more intricate situation of a vortex pair.

Reynolds stress transport equations are required to improve the predictions. As $\overline{v'^2} - \overline{w'^2}$ is better estimated, the evolution of the secondary motion and in turn the longitudinal vorticity is better computed. The main-flow transverse distortion is also better simulated. Moreover, estimates of $\overline{u'w'}$ and $\overline{v'w'}$ are improved but these stresses are still underpredicted.

The Reynolds stress transport equations can be integrated along streamlines with the help of a simple algebraic diffusion model. The value $\beta = -0.8$ for the diffusion governing coefficient has been selected. This method gives satisfactory results and a substantial reduction of computation time.

References

Cousteix, J., de Saint-Victor, X., Houdeville, R. (1985): "X-Marching Methods to Solve the Navier-Stokes Equations in Two- and Three-Dimensional Flows", *Third Symposium on Numerical and Physical Aspects of Aerodynamic Flows*, Long Beach

Bradshaw, P., Shabaka, I. M. M. A., Mehta, R. D. (1982): Turbulent vortex flows, Imperial College London, IC Aero TN 82–103

Gibson, M. M., Launder, B. E. (1978): Ground effects on pressure fluctuations in the atmospheric boundary layer. J. Fluid Mech. **86**, 491–511

Hassid, S., Poreh, M. (1975) A turbulent energy model for flows with drag reduction. Trans. ASME, vol 97, 234–241

Launder, B. E. (1982): A generalized algebraic stress transport hypothesis. AIAA J. **20**, 436–437

Launder, B. E., Reece, G. J., Rodi, W. (1975): Progress in the development of a Reynolds stress turbulent closure. J. Fluid Mech. **68**, 537–566

Mehta, R. D., Lim, T. T. (1984): Flow visualization study of a vortex/wing interaction. NASA Technical Memorandum 86656

Naot, D., Rodi, W. (1981): Numerical simulation of secondary current in open channel flow with an algebraic stress turbulence model. SFB, 80/T/187, Karlsruhe University

Patankar, S. V. (1980): Numerical Beat Transfer in Fluid Flow, Series in Computational Methods in Mechanics and Thermal Sciences (McGraw Hill, New York)

Rodi, W. (1972): The Prediction of Free Turbulent Boundary Layers by Use of a Two-Equation Model of Turbulence, Ph.D. Thesis, University of London

Shabaka, I. M. M. A., Mehta, R. D., Bradshaw, P. (1985) Longitudinal vortices imbedded in turbulent boundary layers. Part 1 – Single vortex. J. Fluid Mech. **155**, 37–57

Interaction Between a Vortex and a Turbulent Boundary Layer in a Streamwise Pressure Gradient

R. V. Westphal*, J. K. Eaton** and W. R. Pauley**

* NASA Ames Research Center, Moffett Field, California

** Stanford University, Stanford, California

Abstract

The effect of a moderate adverse pressure gradient on the interaction between a single streamwise vortex and a turbulent boundary layer is investigated experimentally. Quantitative characterization of vortex properties based on measurements of the mean cross-flow velocity components is attained. Growth of the vortex core is observed, followed by a flattening of the core shape which occurs when the core radius becomes comparable to the distance of the vortex center from the surface. The adverse pressure gradient causes an increase in the rate of core growth and, therefore, a stronger distortion of the core shape. Turbulence properties are even more strongly disturbed by an adverse pressure gradient than by constant pressure.

Nomenclature

C_f	Skin friction coefficient
C_P	Pressure coefficient; $C_P = [P(X) - P(X = 10\,\text{cm})]/[(1/2)\,\varrho U_0^2]$
P	Static pressure (measured with a wall static tap)
R_Y, R_Z	Vortex core radial dimensions in vertical and spanwise directions, respectively
R_Γ	Vortex circulation Reynolds number; $R_\Gamma = \Gamma/\nu$
u, v, w	Velocity components in X, Y, Z directions
U, V, W	Mean velocities; shorthand notation for $\bar{u}, \bar{v}, \bar{w}$
X, Y, Z	Right-hand Cartesian streamwise, vertical, and spanwise coordinate directions, respectively
Γ	Overall circulation
δ	Boundary-layer thickness
ν	Air kinematic viscosity
ϱ	Air density
Ω_x	Streamwise vorticity; $\Omega_x = \partial W/\partial Y - \partial V/\partial Z$
$-$	Time average
$'$	Turbulence component, e.g., $u = U + u'$

Subscripts

c	Refers to vortex center
e	Refers to local free-stream conditions
max	Maximum value for a particular crossflow plane
0	Reference value (measured in free stream at $X = 10$ cm)

Introduction

Streamwise vortices often interact with turbulent shear layers in practical engineering systems; an example is the tip vortex of one helicopter blade impinging on a trailing blade. An important application of an induced interaction is the use of vortex generators to prevent or delay stall (separation) on airfoils or in diffusers. In all such applications, a vortex or an array of vortices can interact with a turbulent boundary layer in the presence of a severe adverse pressure gradient. The combined effect of boundary-layer turbulence and adverse pressure gradient on a longitudinal vortex has not been investigated previously. The overall aim of the present research was to investigate the development of a fairly weak streamwise

Turbulent Shear Flows 5

vortex embedded within a turbulent boundary layer in the presence of a moderate pressure gradient.

There is considerable evidence that a strong adverse pressure gradient alone can have a catastrophic effect on a discrete vortex. Batchelor [1] considered changes in the structure of an axisymmetric, Rankine vortex caused by changes in far-field velocity. His analysis indicates that depending on the relative vortex strength, deceleration of the external stream beyond a certain critical value cannot occur without a total change in the structure of the vortex core. Vortices generated by a delta wing have been observed to undergo severe structural changes [2]; it has been postulated that this "vortex breakdown" may be due to the effect of a strong, adverse streamwise pressure gradient on the leading-edge vortex [3].

Shabaka et al. [4] and Mehta et al. [5] studied a single, weak vortex just above – and later embedded within – the boundary layer in a constant-pressure flow. Their results showed a strong distortion of turbulent stresses within the boundary layer – in spite of the fact that the mean flow was only weakly three-dimensional. These studies did not address the issue of the effect of the boundary layer on the vortex itself.

Several studies [3, 6, 7] have indicated that free vortices at constant pressure evolve very slowly with streamwise distance. For example, a very strong adverse pressure gradient was required to produce much effect on the free vortex studied by Leuchter and Solignac [3]. Because of the combined influence of boundary-layer turbulence and the impermeability constraint owing to the wall, it is not expected that vortex behavior near a wall would be similar to that observed for a free vortex. The additional influence of an adverse pressure gradient would cause rapid growth and would increase turbulent stresses within the boundary layer; therefore, it would also be expected to affect vortex development.

The current research objective was to characterize a vortex as it interacts with the turbulent boundary layer with and without an adverse pressure gradient. The present study was purposefully designed to yield parameters for vortex and boundary-layer properties similar to those used in [4] and [5], which are summarized in Table 1 and discussed further below.

Table 1. Parameters of present experiment compared with those of Refs. 4 and 5

Experiment	Inlet velocity U_0, m/sec	Test section cm	Trip
Present	27	20 × 80 × 300	Wire, 0.4 mm at $X = 90$ cm
Refs. 4, 5	30	13 × 76 × 290	Wire, 1.0 mm at $X = 0$

Experiment	Vortex generator	Vortex circulation Γ/U_0, cm	Vortex core height Y_c, cm	Data
Present	Half delta wing in test section at $X = 90$ cm	0.5	1.7 $X = 200$	Mean cross-flow Reynolds stresses Direct C_f
Refs. 4, 5	Half delta wing in contraction	0.2	2.5 $X = 194$	Reynolds stresses Triple products Preston tube C_f

Experimental Apparatus

The experiments were carried out in the Boundary Layer Wind Tunnel of the Aerodynamic Research Branch at NASA Ames Research Center (see Fig. 1). This facility has a test section with dimensions of 20 by 80 by 300 cm and inlet free-stream turbulence of less than 0.2%. Inlet mean-velocity uniformity at the test speed of 27 m/sec was within 1% across the center half-span of the tunnel. A round wire trip (0.4 mm diam) was positioned at $X = 20$ cm on the test surface. The facility has an adjustable wall opposite the flat test plate for pressure-gradient control; the wall was slotted for probe access.

A single vortex was generated with a delta wing (3.8 cm root chord length); the wing was mounted on the test surface at $X = 90$ cm and at an angle of attack of 12 deg. The generator height (1.8 cm) was chosen to position the vortex center near the undisturbed boundary-layer edge at station $X = 150$ cm. The vortex circulation obtained using this generator was about 0.13 when normalized on inlet free-stream velocity and generator root chord. The resulting vortex Reynolds number R_Γ was approximately 10^4.

Measurement of all three components of mean velocity and of five of the six independent components of the Reynolds stress tensor was achieved by using an automated, crossed hot-wire anemometer system [8]. The probe was rolled about its axis to obtain both $u - v$ and $u - w$ plane measurements. Additional mean velocity measurements were made using a four-hole pressure probe [9, 10] whose output was corrected for the effects of velocity gradient in both the Y- and Z-directions. Skin friction was measured and visualized beneath the vortex, using the oil-film interferometry technique [11, 12]. A special test plate with a partially reflective, metal-coated glass insert was used for the latter work. Silicone oil with a nominal viscosity of 50 centistokes (cS) was used.

Two test conditions were selected. For the first, denoted case A, the control wall was adjusted for constant pressure over the entire test section ($0 < X < 3$ m). For the second, case B, the control wall was adjusted to give a moderate adverse pressure gradient over the region $1 < X < 2$ m, followed by a region of "recovery" at constant pressure. Figure 2 shows the measured pressure distributions for each case. An inlet velocity of $U_0 = 27$ m/sec was used, yielding a boundary-layer thickness of about 1.3 cm at the vortex generator location ($X = 90$ cm). Note that the inlet velocity was used to normalize all quantities in the presentation of data below. The thicknesses of the boundary layers without the vortex for cases A and B are tabulated below.

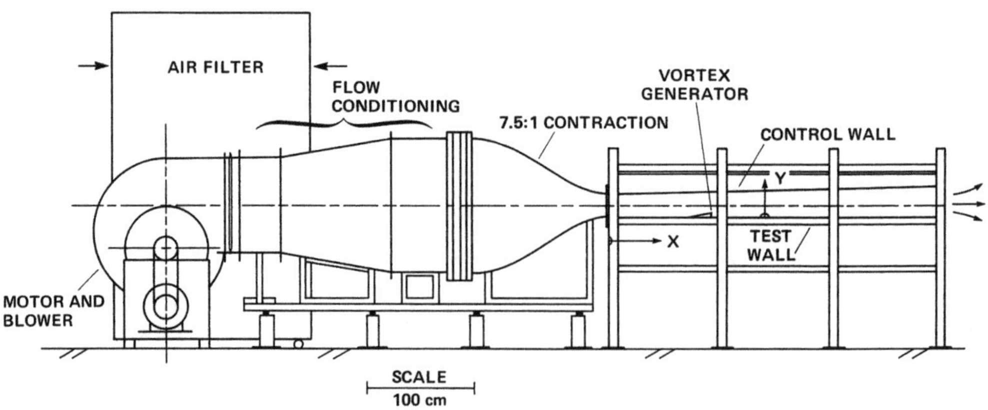

Fig. 1. Boundary Layer Wind Tunnel

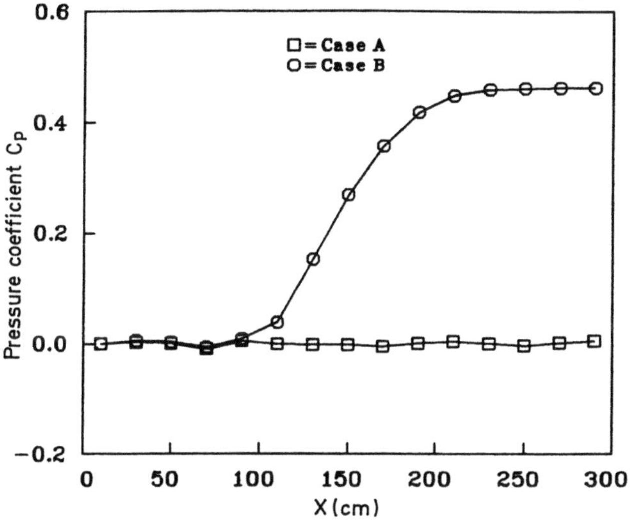

Fig. 2. Measured wall static pressure distributions: cases A and B

δ, cm		X, cm		
		100	300	300
Case A:	constant pressure	1.4	2.8	4.0
Case B:	adverse gradient	1.7	4.3	6.2

Results and Discussion

Vortex Scalar Descriptors

A key to the quantification of vortex evolution was to establish a model for vortex size and strength. Figure 3a shows mean cross-flow velocity vectors measured at $X = 199$ cm for zero pressure gradient. The importance of the proximity of the wall (image vortex) is apparent in the strong spanwise velocities measured beneath the vortex center. The tangential velocity component appears to decay approximately as the inverse distance from the vortex center outside the core. The cross-flow has the characteristics of a potential vortex with a rotational core with the image velocity field superimposed. Note that because of the importance of the image flow, the vortex center position and vortex strength cannot be easily defined directly from the cross-flow velocity vector plot.

Figure 3b shows the computed contours of streamwise vorticity Ω_x, using the data of Fig. 3a. The vorticity was computed by interpolating the measurements of V and W, using cubic splines, then evaluating the spline derivatives analytically. Since the image field is irrotational in the region of interest, the vorticity contours for a vortex near a wall would be expected to be the same as those for an isolated vortex. A real "core" must blend smoothly with the irrotational far-field, so the contours of constant vorticity would be expected to be approximately concentric, closed curves. The vorticity contours of Fig. 3b show precisely this behavior. The contours also point out the presence of a weak vortex of opposite sign adjacent to the main vortex.

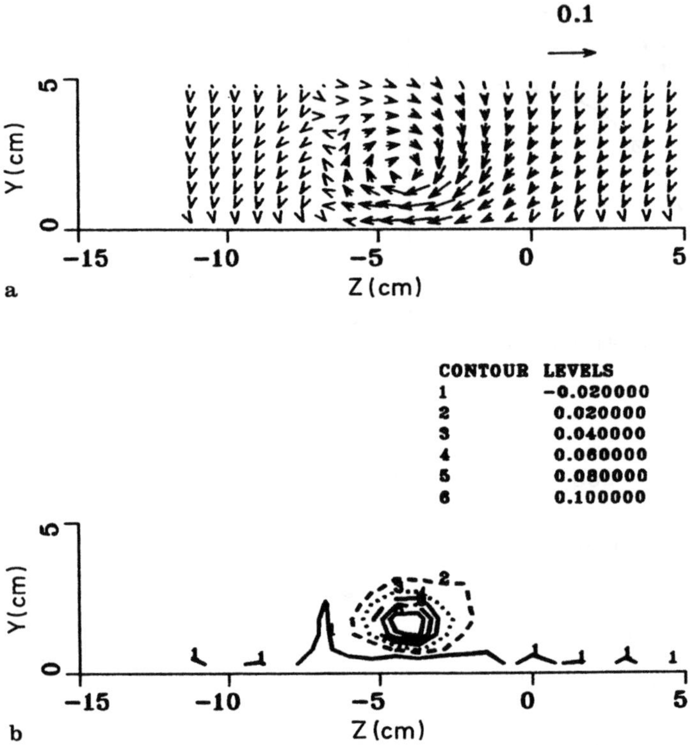

Fig. 3a and b. Mean cross-flow properties: case A, $X = 199$ cm. **a** Velocity vectors; **b** streamwise vorticity contours, Ω_x/U_0, cm^{-1}

The initial roundness of the vorticity contours suggests that the vortex does not display "meander" at this location, in spite of the suggestion [13] that meander is an inherent feature of flows with discrete vortices. Further evidence (from the skin-friction and turbulence measurements) that the vortex remains stationary is presented below.

A simple, objective characterization of the main vortex was suggested, based on the vorticity contours. Vortex position was easily located as the geometric center of the concentric vorticity contours. Vortex core strength was characterized by the peak value of vorticity attained in the core $\Omega_{x,\max}$. Overall circulation Γ was simply defined as the integral of the vorticity over the cross-flow plane; to provide some noise immunity in performing this integration, vorticity levels less than 10% of the peak were taken as zero vorticity. Because of the elliptical vorticity contours that were observed in some cases, vortex size had to be defined using two length scales. Those scales were chosen as the vertical and spanwise radial dimensions measured from the vortex center where the vorticity had decreased to 50% of the peak value of $\Omega_{x,\max}$. Other contours could just as well be used to define the vortex core length scale, but values less than about 30% of the peak are probably within the resolution in determining vorticity by the method used. Generally, the 20% contour radius was about 40% larger than the value from the 50% contour.

Streamwise Development of Vortex Properties

Mean velocity data were obtained at seven axial stations for each case. Overall circulation Γ/U_0, evaluated as described above, was observed to decrease steadily in both cases from

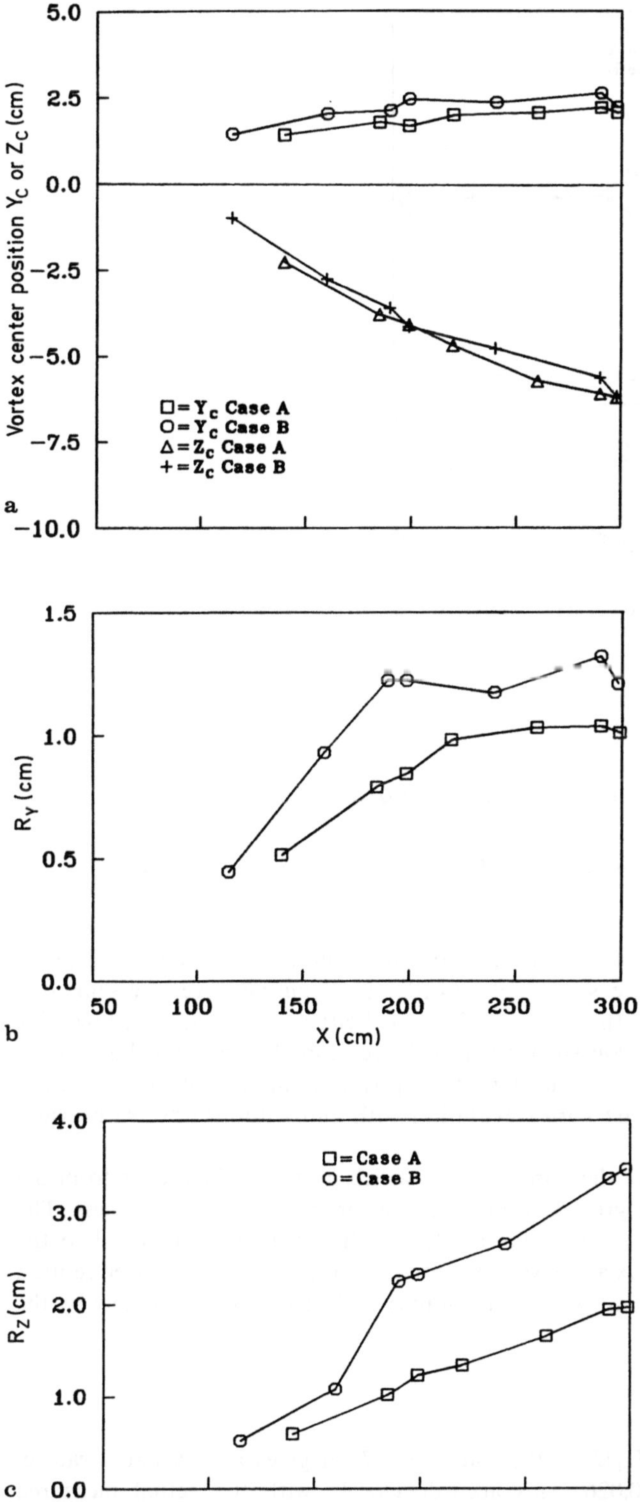

Fig. 4a–c. Streamwise evolution of vortex properties: cases A and B. **a** Vortex position Y_c and Z_c, cm; **b** vortex core vertical dimension R_Y, cm; **c** Vortex core spanwise dimension R_Z, cm

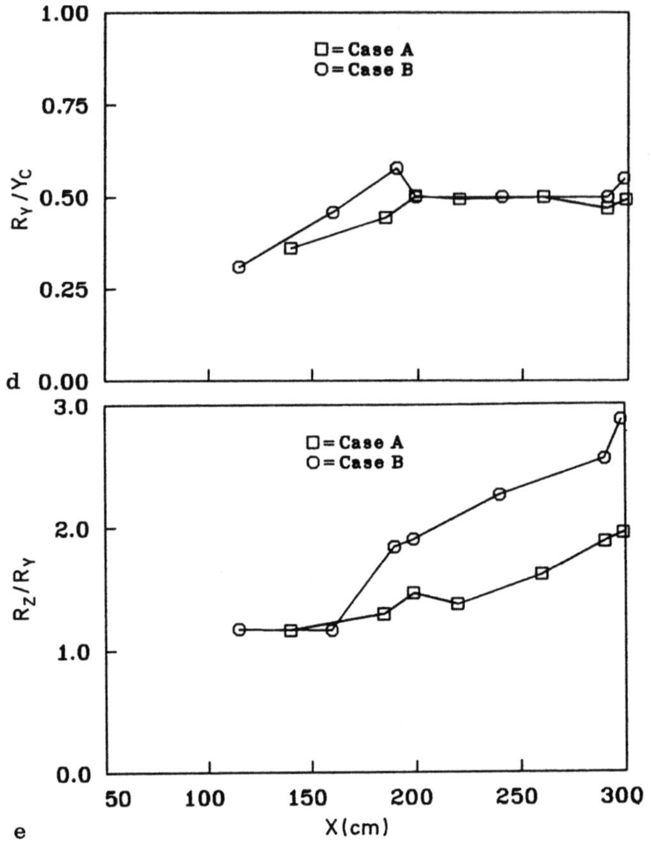

4d nondimensional vortex core height R_Y/Y_c; **4e** nondimensional core aspect ratio R_Z/R_Y

a value of about 0.6 cm immediately downstream of the generator to a value of 0.45 cm at $X = 298$ cm. Peak vorticity decreased about an order of magnitude over the same distance – from about 60 sec^{-1} to less than 6 sec^{-1}. Peak vorticity was substantially lower for case B (adverse pressure gradient), consistent with the larger core diameter for this case, as explained below. Other parameters characterizing the position and size of the vortex are shown in Figs. 4a–e, which contrast vortex evolution with and without adverse pressure gradient.

Figure 4a depicts the motion of the vortex center, which moves slightly away from the wall, following approximately the growth in boundary-layer displacement thickness. The observed spanwise motion of the core in the negative Z-direction was attributed to the velocity induced by the image vortex at the vortex center position. The total induced motion caused at the vortex center by its image, over a distance X, is given approximately by the following relation:

$$\Delta Z_c/\Delta X = \Gamma/(4\pi Y_c U_c).$$

Using values of $\Gamma/U_0 = 0.5$ cm, $U_c/U_0 = 0.9$, and $Y_c = 1.7$ cm gives an estimated rate of spanwise motion of $\Delta Z_c/\Delta X = 0.026$. This accords closely with the actual measured spanwise-motion value of 5 cm between $X = 115$ and $X = 298$ cm, which yields a value of 0.027.

The most important difference in the vortex properties between the two cases is demonstrated in Figs. 4b and c, which show how the vortex-core dimensions evolve. The vertical dimension of the core R_Y shows a decreasing rate of growth, while the core "width" R_Z displays an approximately constant growth rate.

It appears that the core grows symmetrically at first, then becomes increasingly elliptical. This observation suggests that the distance from the wall may limit core growth and be responsible for the onset of ellipticity. The core dimension R_Y divided by center height Y_c is plotted in Fig. 4d. It appears that once the value $R_Y/Y_c = 0.5$ is reached, the core cannot grow further in the vertical dimension. At this point, ellipticity of the core becomes more pronounced. Figure 4e shows the ratio of the major and minor axes of the ellipse, R_Z/R_Y, and indicates that after the core ratio R_Y/Y_c reaches 0.5, the ellipticity becomes more pronounced, finally reaching a value of nearly 3 for the case of adverse pressure gradient.

Figures 5a and b compare the actual vorticity contours for the two cases at $X = 290$ cm. The striking effect of the adverse pressure gradient in causing accelerated core growth and stronger ellipticity is brought out by comparing these contours.

It was initially thought that the ellipticity observed for the vortices after substantial downstream development might really be just a manifestation of meander or "snaking" of the vortex center – that is, a spanwise motion of the vortex center. However, by comparing the turbulence data upstream (where the core is circular) with the downstream data (with

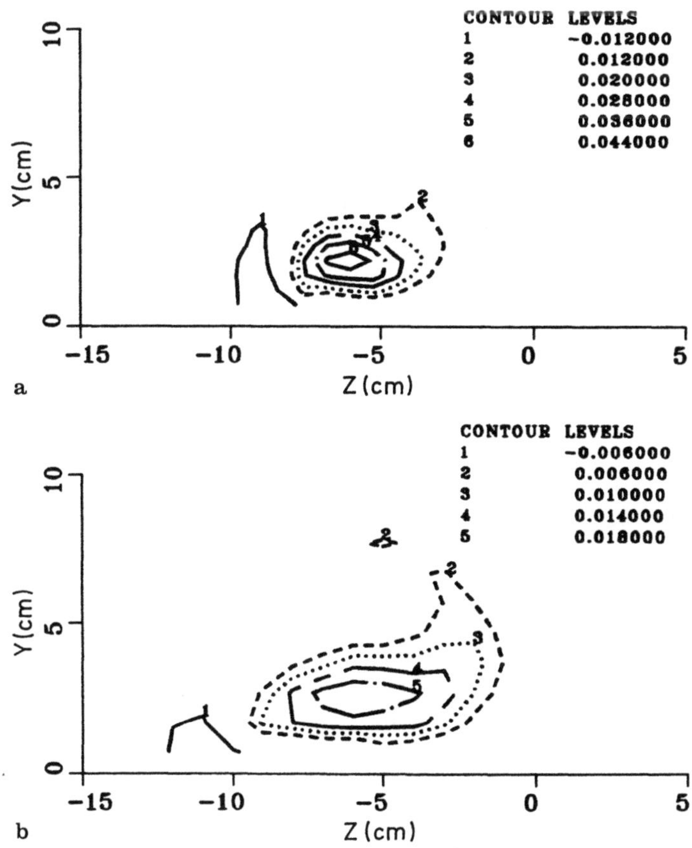

Fig. 5a and b. Streamwise mean vorticity contours Ω_x/U_0, cm^{-1}. **a** Constant pressure (case A); **b** adverse pressure gradient (case B)

elliptical mean-vorticity contours) we concluded that meander was not evident. This is discussed further below.

The increased core growth caused by the adverse pressure gradient is only partially accounted for by Batchelor's [1] simple model in which vortex core growth depends solely on changes in streamwise velocity. Using our values for vortex diameter and circulation yields a nondimensional vortex strength "ka" of less than 0.3, and application of (7.5.25) of [1] with a value of the velocity ratio of 0.73 yields a core growth of 18%, owing to free-stream velocity variation alone. This is in contrast to the experimental results, which indicated that the diameter of the vortex core affected by an adverse pressure gradient was some 50% greater by the end of the test section than that of the core in the constant-pressure case.

Mean Velocity and Skin Friction

The streamwise mean velocity contours are strongly affected by the presence of the vortex. Figures 6a (case A) and 6b (case B) show contours of velocity U which indicate how the vortex causes the boundary layer to be effectively thickened on the upwash side and thinned in the downwash region. The wake of the generator is quite small and is not visible in these plots, all of which are quite far (more than 10 times the root chord) downstream. It was also

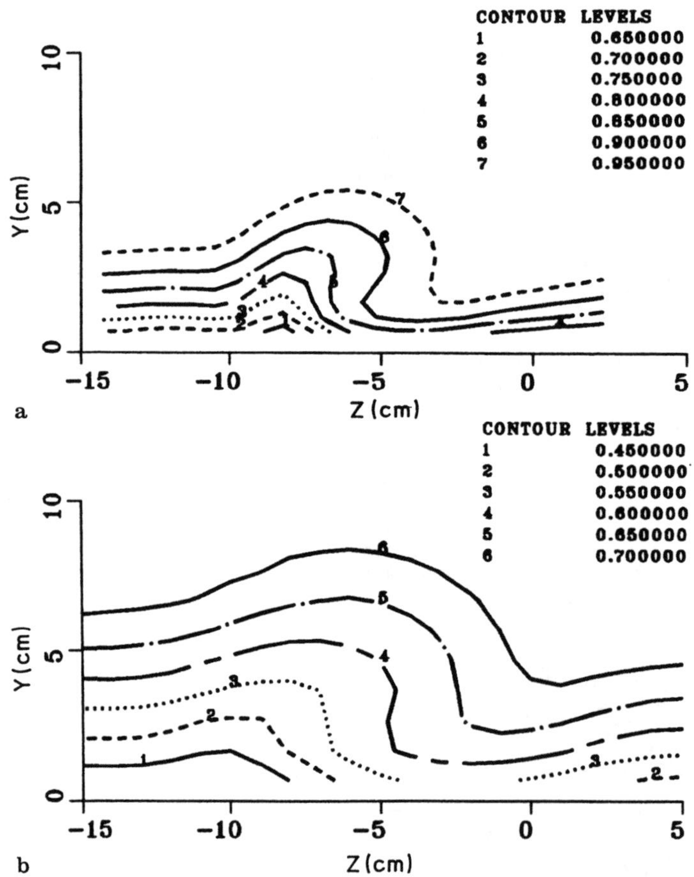

Fig. 6a and b. Streamwise mean velocity contours at $X = 290$ cm: U/U_0, dimensionless. **a** Constant pressure (case A); **b** adverse pressure gradient (case B). Note: $U_e/U_0 = 0.73$

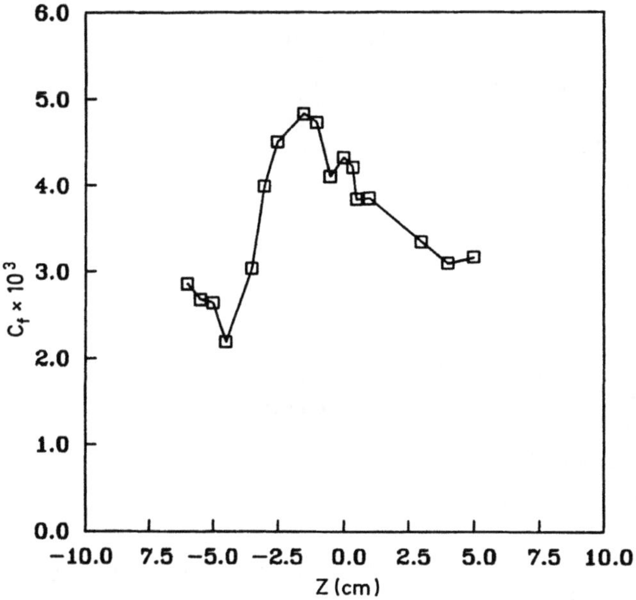

Fig. 7. Skin-friction coefficient C_f measured with the oil-film technique: case A at $X = 144.5$ cm. $C_f = 3.12 \times 10^{-3}$ in the absence of the vortex; $Z_c = 2.1$ cm

observed that the contour shapes at the various downstream stations display only a slight variation in spanwise nonuniformity of the mean-velocity field.

The skin-friction coefficient was measured beneath the vortex using the oil-film technique. These data are shown in Fig. 7 at one streamwise location for case A. As observed from Preston tube data [4, 5], there is a large variation of skin friction beneath the vortex, with peak values more than twice the minimum. The existence of a very sharp minima also suggests that the vortex does not meander at this station, in agreement with the inference drawn from the vorticity contours. The local validity of the law of the wall could not be assessed because of the coarse data mesh used for the anemometry measurements; this question is one subject of our current research.

Turbulence Reynolds Stresses

Strongly disturbed Reynolds-stress distributions were measured in the presence of the vortex, in general agreement with those shown in [4] and [5]. In Figs. 8a (case A) and 8b (case B) contours of $\overline{u'v'}/U_0^2$ are compared at $X = 199$. Note that a region of positive correlation is observed on the downwash side of the core for both cases, this region being much larger in extent for the adverse-pressure-gradient case.

Results for the spanwise component of the normal Reynolds stress $\overline{w'^2}$ are given in Fig. 9 at two streamwise locations for case A. These data further suggest that the vortex does not meander. For example, a side-to-side motion of the vortex should show up directly as increased levels of $\overline{w'^2}$; however, the values of this component were found to be similar to the upstream values. And production of $\overline{u'w'}$ would accrue from meander owing to the action of the mean strain dU/dZ with the term $\overline{w'^2}$; however, no significant differences were noted between the contours of $\overline{u'w'}$ when the contours of mean vorticity were circular and when they were more elliptical.

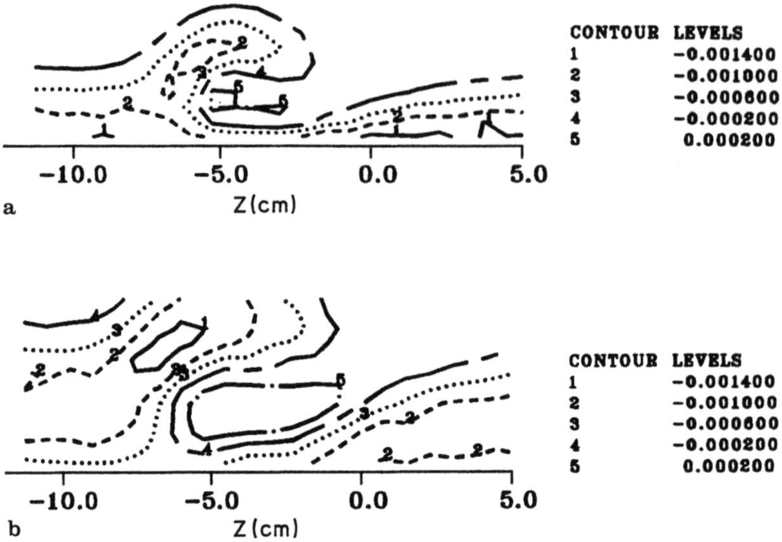

Fig. 8a and b. Turbulence shear stress contours at $X = 199$ cm: $\overline{u'v'}/U_0^2$, dimensionless. **a** Constant pressure (case A); **b** adverse pressure gradient (case B)

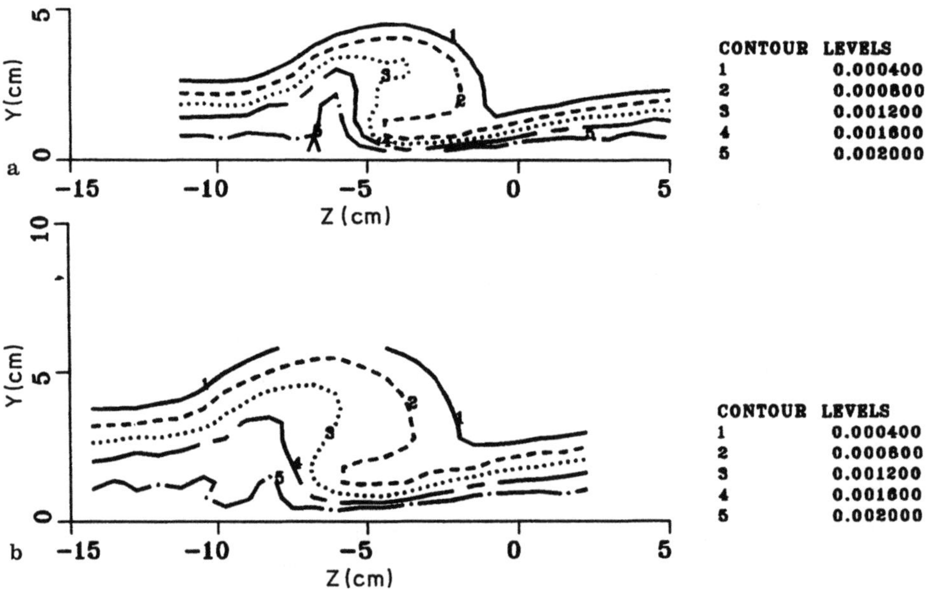

Fig. 9a and b. Turbulence normal stress contours, case A: $\overline{w'^2}/U_0^2$, dimensionless. **a** $X = 199$ cm; **b** $X = 298$ cm

Concluding Remarks

An interaction between a turbulent boundary layer and a relatively weak embedded streamwise vortex in the presence of an adverse pressure gradient has been investigated experimentally. Mean-velocity measurements have shown that the vortex generated by a delta wing mounted at angle of attack on the test plate may be characterized simply and objectively using the contours of streamwise vorticity computed from the mean cross-flow velocity distribution.

Substantial vortex core growth, relative to that expected for a similar free vortex [3, 6, 7], was shown for both the constant pressure and the moderate-adverse-gradient cases. Accelerated core growth was noted because of the adverse pressure gradient; the increase in core growth over that found for the constant-pressure case was not fully accounted for by Batchelor's [1] simple theory. When the diameter of the vortex core grew to a sufficient fraction of the height of the vortex center from the wall, the vorticity contours became increasingly elliptical in shape. The onset of ellipticity was accelerated in the case of adverse pressure gradient as a result of the higher core growth rate.

The vorticity contours were initially round, suggesting that the vortex does not meander. The hypothesis that the apparent ellipticity of the vortex was really a manifestation of meander was tested farther downstream by examining the turbulence measurements. There was not evidence of meander in these data.

Distortion of the Reynolds stresses was more pronounced in the case of the adverse pressure gradient relative to the flat-wall boundary-layer values. Modeling difficulties even greater than those in the constant-pressure case are anticipated because of the larger regions of strongly disturbed stress levels.

Acknowledgments. The contributions of Charles Hooper in writing some of the necessary data-acquisition and data reduction software are gratefully acknowledged. The authors would also like to thank Rabi Mehta for many useful discussions.

Funds for the support of this study were furnished by NASA Ames Research Center, Moffett Field, California, under Interchange No. NCA2-1R745-405.

References

1. Batchelor, G. K. (1967): *Introduction to Fluid Dynamics* (Cambridge University Press) 543–555
2. Leibovich, S. (1984): Vortex breakdown and stability: survey and extension. AIAA J. **22**, 1192–1206
3. Leuchter, O., Solignac, J. L. (1984): "Experimental Investigation of the Turbulent Structure of Vortex Wakes," in *Turbulent Shear Flows*, ed. by L. J. S. Bradbury et al. (Springer, Berlin, Heidelberg) 156
4. Shabaka, I. M. M. A., Mehta, R. D., Bradshaw, P.: (1985): Longitudinal vortices imbedded in turbulent boundary layers. Part I. Single vortex. J. Fluid Mech. **155**, 37
5. Mehta, R. D., Shabaka, I. M. M. A., Shibl, A., Bradshaw, P. (1983): Longitudinal vortices imbedded in turbulent boundary layers. AIAA Paper 83-0378, Reno, Nevada
6. Hoffman, E. R., Joubert, P. N. (1963): Turbulent line vortices. J. Fluid Mech. **16**, 395–411
7. Phillips, W. R. C., Graham, J. A. H. (1984): Reynolds-stress measurements in a turbulent trailing vortex. J. Fluid Mech. **147**, 353–371
8. Westphal, R. V., Mehta, R. D. (1984): Cross hot-wire anemometry data acquisition and reduction system. NASA TM-85871
9. Shepherd, I. C. (1981): A four-hole probe for fluid flow measurements in three dimensions. Transactions of the American Society of Mechanical Engineers, J. Fluids Eng. **103**, 590–594
10. Youssefmir, P. (1982): "Flow Studies of Full-Coverage Film Cooling on a Convexly Curved Surface;" Ph.D. Thesis, Department of Mechanical Engineering, Stanford University, Stanford, CA
11. Tanner, L. H. (1981): The application of fizeau interferometry of oil films to the study of surface flow phenomena. Opt. Lasers Eng. **2**, 105–118
12. Westphal, R. V., Bachalo, W. D., Houser, M. H. TM-882/6: Improved skin friction interferometer. NASA Technical Memorandum
13. Bushnell, D. M. (1984): Body-turbulence interaction. AIAA Paper 94-1527, Snowmass, CO

The Pulsed Wire Skin-Friction Measurement Technique

I. P. Castro, M. Dianat and L. J. S. Bradbury*

Department of Mechanical Engineering, University of Surrey, Guildford, UK
* Department of Mechanical Engineering, Plymouth Polytechnic, UK

Abstract

Studies of the convection and diffusion of a heat tracer close to a wall are described in the particular context of the use of a pulsed wire skin friction gauge. It is shown that the flight time of a heat tracer, measured as the time required for the front of the temperature pulse to reach some downstream station, is critically dependant on the form of the upstream velocity profile. The difference between flight times obtained in cases of different velocity profiles having the same surface velocity gradient is greater than that anticipated solely on the basis of the difference in convection velocity at the height of the heat source. Furthermore, this difference is a non-monotonic function of the Peclet number (based on shear velocity).

The implications of these results for the use of the pulsed wire wall probe for making skin friction measurements in highly turbulent flows is discussed.

Nomenclature

A, B, C	Constants in wall probe calibration function	T_s	Sensor wire time constant
C_P	Specific heat	x, y	Axial and vertical coordinates
l_s	Length of sensor wires	α	Parameter, s/h
H_c	Channel half-height	β	Diffusion parameter. $\kappa v/u_\tau^2 s^2$
h	Distance between probe wires and the surface	κ	Thermal diffusivity
U	Velocity	ϱ	Density
Pe	Peclet number	τ_w	Wall shear stress
s	Spacing between pulsed and sensor wires	θ	Temperature
		ω	Parameter, s/H_c

Introduction

The pulsed wire anemometer is now a well-tried and proven instrument for making velocity and turbulence measurements in highly turbulent flows. From its inception it was recognised that a similar technique might be useful for the measurement of surface shear stress in situations where other methods, like Preston tubes, Stanton fences, log-law fits etc., are practically useless. Early work by Ginder and Bradbury [1] proved partially successful. They used surface mounted thin film elements but encountered two major difficulties. Firstly, largely because of the unavoidable influence of the substrate, the signal/noise ratios were very much lower than for the standard pulsed wire velocity probe. Secondly, calibrations of the device in a laminar flow channel led to skin friction measurements in a turbulent boundary layer that were too high by as much as 20%. No firm conclusions regarding the reason for this behaviour were reached.

More recently, workers at Stanford University and the University of Surrey have taken up the original suggestion of Bradbury and used wires mounted just above the surface, rather

Turbulent Shear Flows 5

than films *on* the surface, in order to increase the signal/noise ratio. With the substantive improvements in the pulsed-wire signal-processing electronics that have been made over recent years, the signal/noise ratios obtainable from such a device are very much higher than they are for a surface-film gauge and this has led to a revival of the technique. Both the Stanford group (e.g. Eaton et al., [2], Westphal et al., [3]) and ourselves (Castro and Dianat, [4], Dianat and Castro [5]) have used a pulsed wire wall probe to make mean and fluctuating surface shear stress measurements in a variety of highly turbulent separated flows and the potential of such a technique seems considerable.

However, like Ginder [1, 6], both groups have found that laminar channel calibrations of the device lead to turbulent flow measurements which are significantly too high. One way to surmount this difficulty is to perform calibrations in a known turbulent flow, like a turbulent channel flow or a zero-pressure gradient boundary layer. There seems to be no guarantee that such an approach will always yield the correct results in a flow very different from that in which the probe is calibrated.

Leaving aside for the moment the whole question of the possible effects of spatial and time variations on the instrument's response when mounted in the linear sub-layer of a turbulent boundary layer, *some* difference in calibration is to be expected if a laminar channel flow is used and the probe is calibrated directly in terms of C_f. In the latter the velocity profile is parabolic whereas in a turbulent boundary layer it is linear, so for a given surface shear stress (proportional to du/dy at the wall) the convection velocity at the wire height will be different in the two cases. Ideally, the wires need to be as close to the wall as possible, both to minimise this difference and to reduce the possible effects of the turbulent motions in the inner region of the turbulent flows. Practically, 50 microns seems to be a reasonable height at which to mount the wires – this ensures that in a typical turbulent flow the wires are at a y^+ value (normalised by the usual wall units) of $0(1)$. Actual measurements of the heat tracer flight time show differences which are significantly higher that can be accounted for solely on the basis of the differences in convective velocity at the wire height. Westphal et al. [3] found that for their probe, in which the wires were rather further from the wall [at y^+ of $0(10)$], the difference in laminar and turbulent calibrations was generally less than about 5%, whereas we have found rather larger differences (see later). For surface films Ginder [6] found differences in excess of 20%. Clearly these various differences cannot be accounted for solely on the basis of the slightly different convective velocities at the wire height in the laminar and turbulent velocity profiles having the same surface shear, for then they would become smaller as the wire height decreased; the above trends have the opposite behaviour. Since diffusion effects are likely to be relatively significant (compared with the case of the ordinary pulsed-wire, see Bradbury and Castro, [7]) it is not at all obvious how the heat tracer actually arrives at the sensor wire. It is at least possible that the velocity curvature effects (see Westphal et al., [3]), in acting over a distance large compared with the wire height, could be magnified by the effects of lateral and longitudinal diffusion.

We have therefore studied the response of a pulsed wire wall probe in some detail, in order to identify the reasons for the different laminar and turbulent calibrations and hence, hopefully, to develop a rather more satisfactory calibration procedure.

Numerical Computations

We have developed a computational technique which essentially solves the two-dimensional heat convection/diffusion equation for the case of a line-source of heat near a boundary, with

$$\frac{\partial \theta}{\partial t} + U \frac{\partial \theta}{\partial x} = \kappa \left(\frac{\partial^2 \theta}{\partial x^2} + \frac{\partial^2 \theta}{\partial y^2} \right) + \frac{\mu}{\varrho c_p} \left(\frac{\partial U}{\partial y} \right)^2 . \tag{1}$$

As an initial step, the steady flow problem was considered. Ignoring the viscous heating term, which was subsequently shown to be always negligible, (1) becomes

$$U \frac{\partial \theta}{\partial x} = \frac{1}{\text{Pe}} \cdot \left(\frac{\partial^2 \theta}{\partial x^2} + \frac{\partial^2 \theta}{\partial y^2} \right), \tag{2}$$

where now all variables are normalised using U_h, the velocity at the height of the heat source and s, the spacing between the source and recepter positions; Pe is $U_h s/\kappa$. U depends on the upstream velocity profile: for a linear and a parabolic shear it is $(1 + \alpha y)$ and $(1 - \omega/2\alpha + (\alpha - \omega)\, y - \alpha\omega y^2/2)$, respectively, where $\alpha = s/h$ and $\omega = s/H_c$. h is the distance between the surface and the heat source (in practice, the wires) and H_c is the half-height of the channel providing the parabolic velocity profile (see Fig. 1). Note that $x = 0$, $y = 0$ is the location of the source (the pulsed wire).

Equation (2) was solved numerically using standard finite difference techniques. Grids of typically, 70 × 80 nodes were used and mesh expansion ratios were kept below 1.1 to minimise errors. This still allowed initial non-dimensional mesh sizes of around $10^{-4} - 10^{-3}$ near the source, with the mesh extending to $x = 1.5$ ($x = 1$ is the sensor wire position).

Assuming a non-conducting wall, the appropriate boundary conditions are straightforward, except for that defining the source temperature distribution. The simplest approach is to assume a Gaussian distribution of defined half-width, since that is the form of the analytic solution of (2) in the absence of longitudinal diffusion and in the case of a uniform velocity distribution. As might be expected, this did not lead to adequate solutions near the source. Replacing this distribution by the appropriate Bessel function solution of (2) with longitudinal diffusion *included* gave very satisfactory agreement with the analytic solution for a uniform velocity profile, throughout the region of interest, thus validating our finite-difference code.

Solutions were then obtained for cases with linear and parabolic velocity profiles, over a range of Peclet numbers. Figure 2 shows the ratio $(\theta_{\text{lin}}/\theta_{\text{par}})$ of the temperatures at the sensor wire location ($x = 1$, $y = 0$) obtained with these two velocity profiles, as a function of the 'diffusion' parameter, β, defined by $\beta = 1/(\alpha \cdot \text{Pe}) = \kappa\nu/u_\tau^2 s^2$. The results have been adjusted to ensure the same source heat flux in the two cases. Increasing Peclet number leads to an asymptotically constant ratio of the two temperatures, as expected, but as Pe falls,

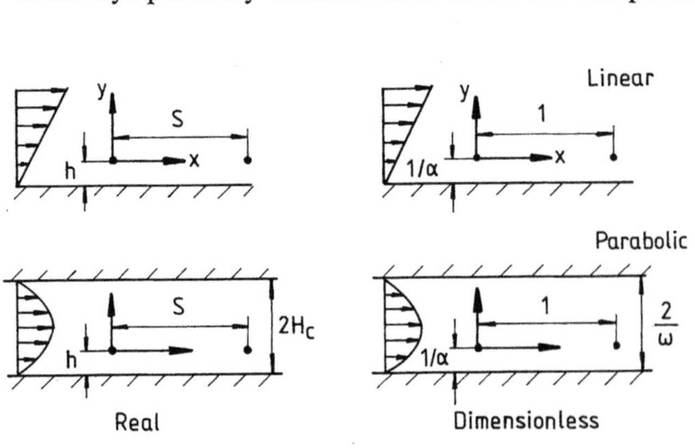

Fig. 1. Notation

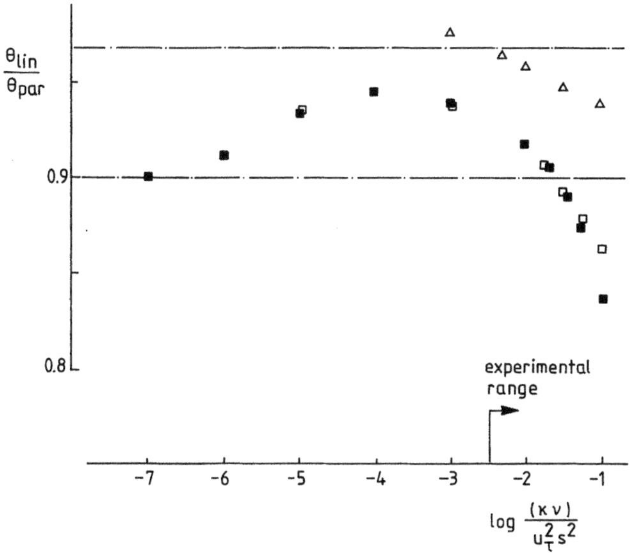

Fig. 2. Ratio of temperatures at $x = 1$, $y = 0$ for linear and parabolic velocity profiles with the same surface stress. ▵, $\alpha = 10$, $\omega = 0.63$; □, ■, $\alpha = 5$, $\omega = 1$; closed symbol, longitudinal diffusion neglected (results obtained from a 'marching' solution). ——, no diffusion results $(1 - \omega/2\alpha)$

diffusion effects become more significant. In particular, over at least part of the typical experimental Peclet number range, *longitudinal* diffusion is important. The results for the case in which the longitudinal diffusion term was ignored were obtained using a standard (parabolic) marching algorithm. In this case the grid was expanded laterally as the solution marched downstream. (The agreement at high Peclet number between the solutions with and without longitudinal diffusion was a further confirmation of the adequacy of the full elliptic code). $\theta_{\text{lin}}/\theta_{\text{par}}$ is precisely what would be anticipated from earlier experimental results; a higher value of θ would lead to earlier triggering of the sensor wire signal and hence a higher measured shear stress. Note particularly that this ratio is less than the 'convection only' limit, as experiments have suggested – calibrations in a channel lead to too high a shear stress measurement in the linear sub-layer of a turbulent boundary layer.

The full time-dependent problem was then solved by using a standard alternating direction implicit two-step method. Spatial boundary conditions were the same as for the steady case and at $t = 0$ an appropriate temperature profile was imposed at the source, followed by an exponential decay whose form depended on the wire Reynolds number. The approach was similar to that described by Bradbury and Castro [7].

Figure 3a shows typical computations of the variation with time of the temperature θ_f at the sensor location, along with the resulting variation of the sensor temperature itself. This is related to the flow temperature by

$$\theta_s = 1/T_s \exp(-t/T_s) \int_0^t \theta_f \exp(-t/T_s)\, dt,$$

where T_s is the sensor wire time constant. At higher values of Peclet number the sensor temperature is essentially the integral of the flow temperature, since then T_s is much higher than typical convection or diffusion times. At low Peclet numbers, the reverse is true so that θ_s essentially follows θ_f.

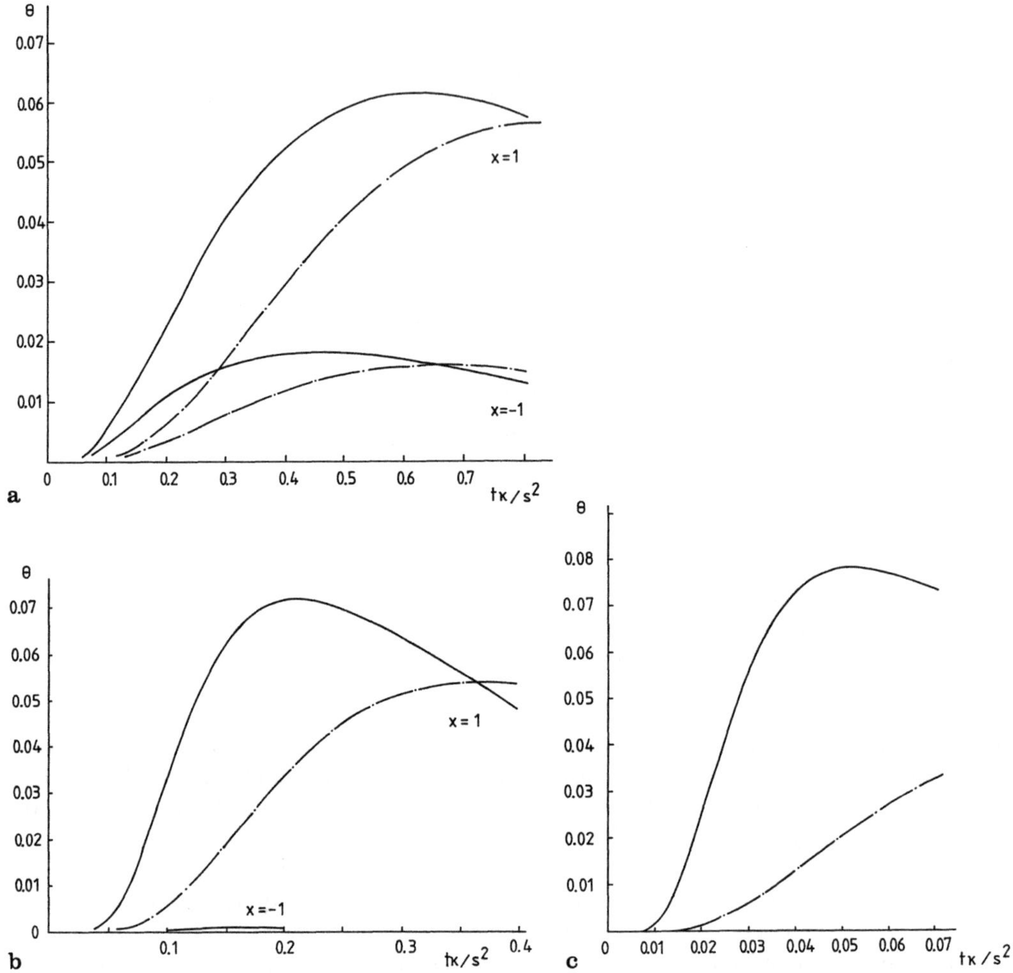

Fig. 3. Flow (——) and wire (-·-·-) temperatures at upstream and downstream sensor wire positions ($y = 0, x = \pm 1$). **a** $P_e = 0.4$; **b** $P_e = 4$; **c** $P_e = 40$

In practice, the two sensor wires are operated differentially, so that it is actually the difference in the voltage across them which is used to derive the heat tracer flight time. At the lower Peclet numbers it was found to be important to include the effect of *upstream* diffusion. Figure 3a (Pe = 0.4), shows that there is a significant rise in flow temperature at the upstream wire. The computation region in this case extended upstream to $x \cong -1.5$; this doubled the necessary computing time, but gave more realistic calibration curves in that they were qualitatively much more similar to the experimental calibrations at the lower Peclet numbers. It is evident from Fig. 3b that once Pe > 4, upstream diffusion was insignificant.

Computed calibrations were obtained from results of this sort by numerically differentiating the sensor signals and applying suitable 'trigger' levels to determine the flight time of the heat tracer. These enabled the flight time ratio, $T_{\text{lin}}/T_{\text{par}}$, to be obtained, where the subscripts refer to linear or parabolic velocity profiles having the same surface stress. Figure 4 shows this ratio, plotted as an 'error' in flight time, as a function of Peclet number, for the particular case $\omega = 0.63$, $\alpha = 10$. The major points to note are, firstly, that the 'error' is generally larger than would be expected on the basis of the difference in convection

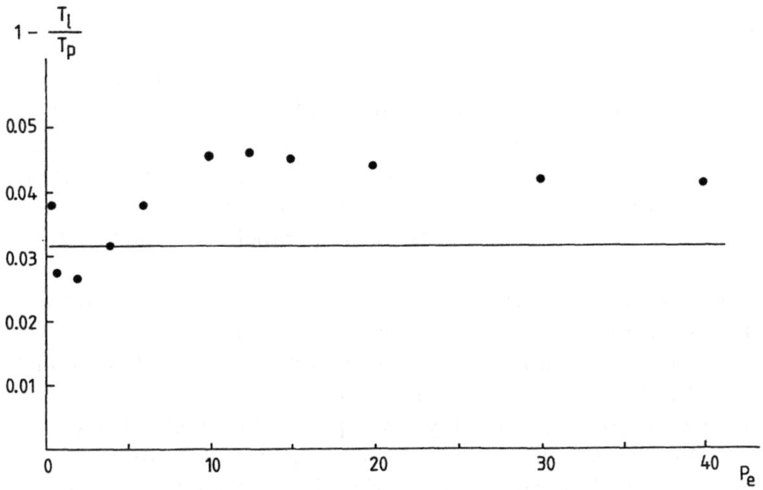

Fig. 4. Time-of-Flight 'Error'. ——, no diffusion ($\omega/2\alpha$); $\omega = 0.63$

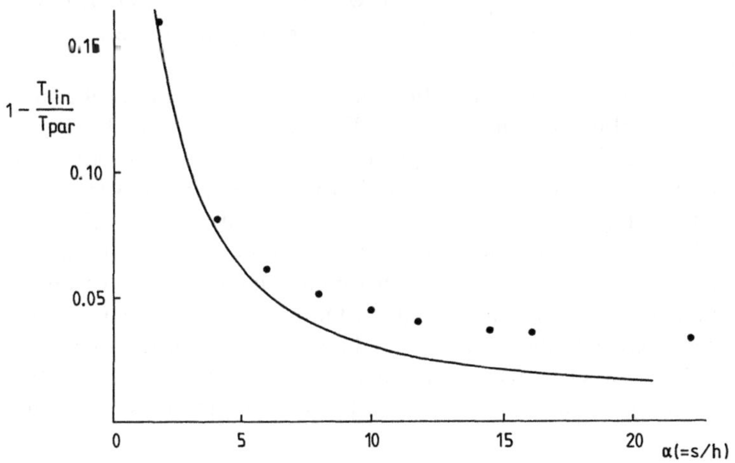

Fig. 5. Time of flight 'error' as a function of wire spacing/height ratio. $\omega = 0.63$ ——, no diffusion ($\omega/2\alpha$)

velocity at the wire height and secondly, that the error is a non-monotonic function of Peclet number. We return to these points later.

Figure 5 shows how T_{lin}/T_{par} varies with α for a fixed value of ω. Note that since the Peclet number is defined as $U_h s/\kappa = h u_\tau^2 s/\kappa\nu$, Pe is not constant in this figure. With $u_\tau^2 = 0.25\ \mathrm{m^2/s^2}$ it is roughly 2α. Our probe geometry typically has $\alpha = 10$. For wires significantly further from the wall (smaller α) the error is dominated by the difference in the convective velocities at the wire height, as expected, and diffusion effects lead to increased errors which become more significant as Pe falls. $\alpha = 10$ does seem a reasonable compromise between the desire to put the wires as close to the wall as possible to ensure that in turbulent flows the instantaneous linear velocity profile assumption is valid at least up to the wire height, and the requirement to minimise errors (arising from calibrations in a *parabolic* flow). Note also that Fig. 5 implies that even for wires mounted on the surface, velocity profile effects would still be significant (as Ginder [6] found).

Experimental Results

The wall probe used for the experiments described here had sensor and pulsed wire lengths of about 2 and 3 mm, respectively, spaced about 0.5 mm apart and 0.05 mm from the surface (so that $\alpha = 10$). It was operated in a manner very similar to that used for a standard pulsed wire anemometer probe, using a micro-computer to perform on-line calibrations and measurements. The calibration relation $U_\tau^2 = A/T + B/T^2 + C/T^3$ was used since this gave a more accurate fit over a wide range of wall shears than the parabolic fits employed for velocity probes.

Individual probe calibrations were performed in a series of three-dimensional rectangular 'laminar flow' channels of various heights ranging between 1 and 6.5 mm. The channel width was 30 mm. Standard analytic solutions for laminar flow in a three-dimensional rectangular channel allowed the surface shear at each flow rate to be determined by a simple measurement of the axial pressure drop in the fully-developed region. Figure 6 shows a typical set of calibrations, all obtained using exactly the same instrument settings (gains, trigger levels, etc.). The expected trend is evident – increasing time of flight (at constant u_τ) for decreasing channel height. Using the values for A, B and C in each case it would be possible, in principle, to cross-plot the data to obtain the 'ideal' calibration that would have been obtained in an infinitely wide channel ($H_c = \infty$). However, it turns out that $1/T$ is a significantly nonlinear function of $1/H_c$ (at constant u_τ) so that extrapolation to the $1/H_c = 0$ case was a rather uncertain procedure.

However, a 'pseudo' $1/H_c = 0$ calibration was obtained by calibrating the probe in a standard, zero-pressure gradient turbulent boundary layer. The results are included in Fig. 6 and in Fig. 7 the data is cross plotted in terms of the flight-time 'error' at constant wall shear vs. Peclet number for the particular case $\omega = 0.5$. The qualitative agreement between these results and those in Fig. 4 is striking. In particular, it should be noted that $(1 - T_{\text{lin}}/T_{\text{par}})$ varies non-monotonically with Pe; the maximum value occurs around Pe = 10, in agreement with the computational results. For Pe > 24 the experimental data is almost certainly too low, because of the influence of unresolvably high instantaneous wall shears in the turbulent flow at high Peclet number (note the relatively high curvature in the turbulent calibration for $1/T > 0.008$). The difference between the measured flight time ratio and the pure convection limit is, again, higher than that indicated by the computations, but there

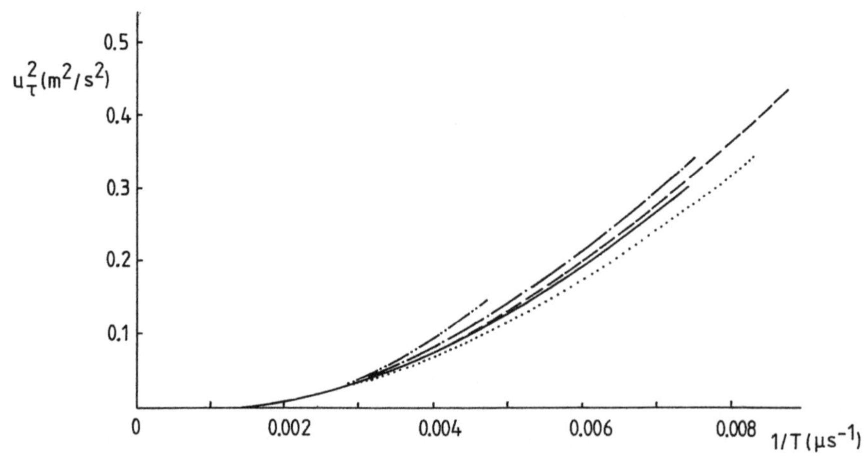

Fig. 6. Calibrations in Laminar Flow Channels. ω = ·· — ··, 1.64; · — ·, 0.86; ---, 0.5; ——, 0.38. ....., Calibration in turbulent boundary layer (effectively, $\omega = 0$)

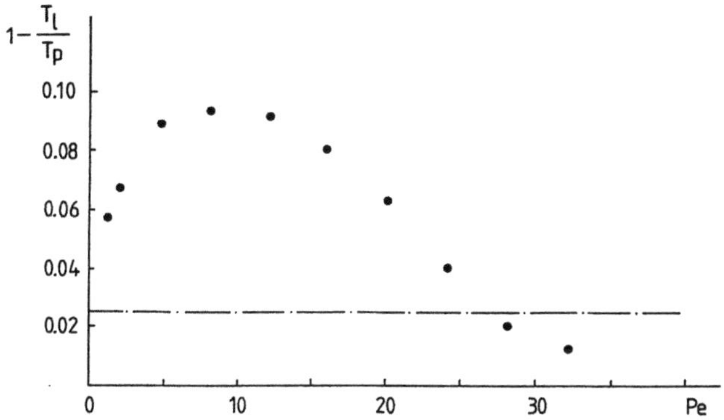

Fig. 7. Experimental Time-of-Flight 'Error'. $\omega = 0.5$. ——, theoretical no diffusion limit $(\omega/2\alpha)$

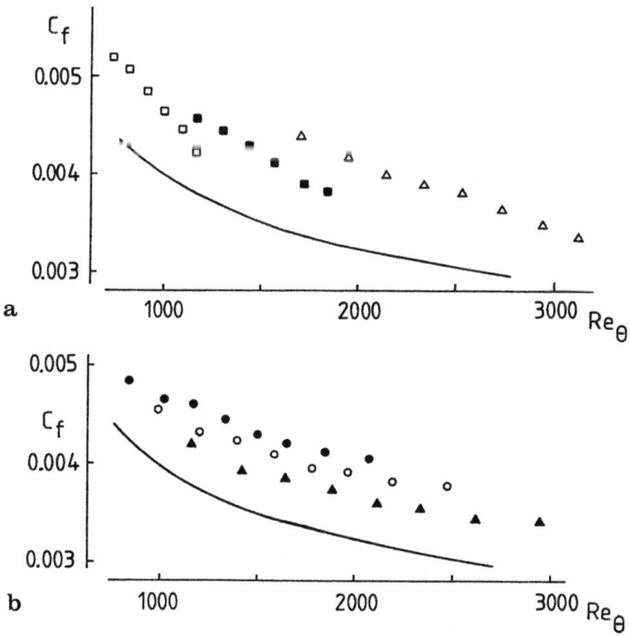

Fig. 8 a and b. Measurement in zero-pressure gradient boundary layer. Probe calibration in channel with $\omega = 0.5$. ——, Preston tube and log-law results; **a** Distance from trip; × (mm): □, 402; ■, 725; △, 1202; **b** $u_\tau^2\,(m^2/s^2)$, Pe: ●, 0.132, 10.6; ○, 0.20, 16.0; ▲, 0.28, 22.4

seems no doubt that the latter have captured the more important physical effects, despite neglect of the viscous wakes of the wires and their influence on the upstream velocity profiles.

Further measurements have emphasized the fact that the errors resulting from a laminar channel flow calibration vary with wall shear. Typical amonst these are the results shown in Fig. 8. Shear stress measurements obtained in a zero-pressure gradient turbulent boundary layer over a range of freestream velocities but at just three axial stations are compared, in Fig. 8 a, with 'standard' results obtained using Preston tubes and log-law plots. The latter showed no more internal scatter than is usual with such techniques (± 1½%) and the smooth line shown is quite close to the classic results in the literature (though there are small

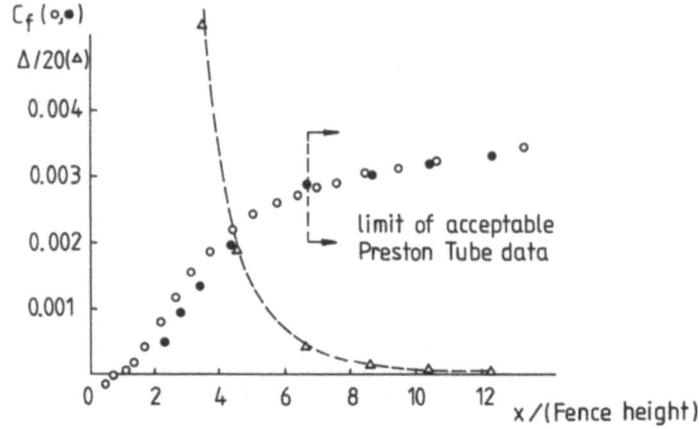

Fig. 9. Mean skin friction in boundary layer approaching a normal fence. ο, pulsed wall probe; •, Preston tube. ▵, $\Delta = (\nu/\varrho u_\tau^3) \cdot dp/dx$ – pressure gradient parameter

differences, probably caused by slight three-dimensional effects). The pulsed wall probe data were obtained using a laminar channel calibration (with $\omega = 0.5$) and not only do not agree with the expected results but do not even seem internally consistent. However, at each axial location the results cover a range of wall shear stress (and hence Peclet number). They are replotted, with some additional data, in Fig. 8b where they are collected as sets of data obtained at nominally the same Peclet number. Evidently, in this Peclet number range, the error reduces as Pe rises, as would be anticipated from the results in Figs. 4 and 7.

Now if the quantitative values of the 'errors' shown in Fig. 7 and evident in Fig. 8 were independent of the particular values of signal gains, trigger level, etc., it would be possible to devise a correction procedure, enabling a single laminar flow channel calibration to be used to derive an equivalent 'infinite channel' ($1/H_c = 0$) calibration. However, both experiment and computations show, not surprisingly, that the errors *do* depend on instrument settings. It must therefore be concluded that laminar flow channel calibration can never be a procedure which will lead to very accurate mean skin friction measurements.

Nevertheless our results do lend some credibility to the procedure of calibrating in a turbulent boundary layer (or channel). Despite the inevitable fluctuation intensities of around 30%, the mean skin friction measured (using on-line linearisation) in such a flow, after probe calibration within it when T is averaged for each calibration point, gives results close to those expected. Actually this can only occur because the various higher order non-linear error terms ignored in the calibration must effectively cancel – C_f (and T) probability distributions are significantly non-Gaussian. The alternative procedure of including the fluctuating part of the time-of-flight directly in the calibration is a more satisfactory approach (Eaton et al., [2]). Figure 9 presents a comparison between Preston tube and pulsed wire mean skin-friction measurements in a separating boundary layer. The flow was produced by mounting a sharp-edged fence spanwise across the false floor in the wind tunnel (as used for the standard boundary layer measurements mentioned above). Upstream of the fence, the (thin) boundary layer enters a region of adverse pressure gradient so that the skin friction falls continually until separation occurs. The pulsed probe was calibrated within the turbulent boundary layer which develops in the absence of the fence. Sufficiently far upstream the data obtained with the fence present agrees well with preston tube measurements. Once the adverse pressure gradient becomes too severe the two sets of measurements deviate and, as expected, Preston tube (or other standard) methods cannot yield accurate

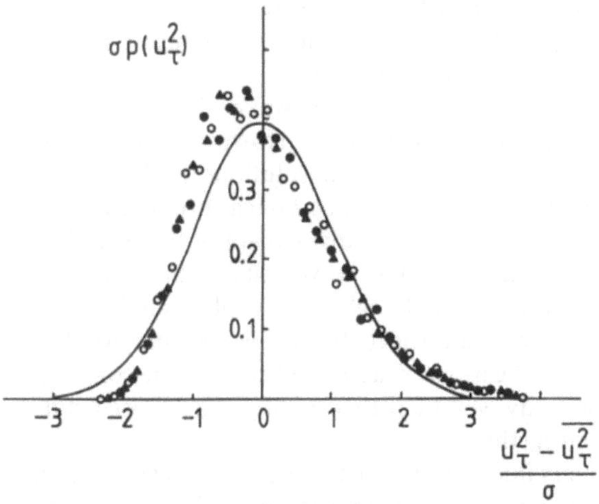

Fig. 10. Skin friction probability density distributions

x, mm	u_τ^2, m^2/s^2	Re_θ	σ/u_τ^2	Skewness
○ 402	0.12	815	0.26	0.66
● 808	0.11	1405	0.27	0.69
▲ 1202	0.11	1915	0.28	0.73

mean values. Patel's [8] criterion for 'too severe' is included in the figure and is consistent with the results.

One of the important features of the pulsed wall probe is, of course, its ability to make measurements of the fluctuating components of wall friction. As an example of its potential Fig. 10 shows probability density distributions of the wall friction in a zero pressure gradient boundary layer at three momentum thickness Reynolds numbers. The data are very similar to the recent hot-wire data of Bhatia et al. [9], for example, although their measurements did not extend to distances closer than about four wall units from the surface ($yu_\tau/\nu = 4$). For the data in Fig. 10, the probe wires were at $yu_\tau/\nu \cong 1.1$ in all three cases shown. It is just as easy to obtain such measurements in highly turbulent, reversing flows; Dianat and Castro [5] contains some examples.

Further Computations

In view of the quite large *quantitative* differences between computed and measured errors (cf. Figs. 1 and 7) it is worth considering the other possible factors that might be relevant. These are of two kinds. Firstly, there are effects related to the parameters in the laminar flow not considered in the numerics, notably the possible effects of heat conduction to the substrate and of the influence of the wires themselves on the flow. Secondly, there may be parameters describing the turbulent motions which have an influence on the probe's response in the boundary layer flows. For example, substrate effects will presumably be different in turbulent flow and, perhaps more importantly, there may be effects on flight time measurements caused by variations in the upstream velocity profile during a typical heat tracer convection time.

Some computations were performed in which the substrate was included directly. At low Peclet numbers it was found that, as expected, the rise in temperature at the sensor position was rather slower, but the substrate did not significantly alter the overall behaviour of the flight time variations discussed earlier. Ginder [6] found much more significant substrate effects, since in his case the probe was surface mounted (thin films).

A complete numerical investigation of the effect of the viscous wake of the wires would be difficult since the whole flow field would need to be computed. However, since the wire Reynolds numbers are always less than unity classical Oseen solutions can be used to estimate the velocity variation behind one of the wires. For our typical probe, the downstream sensor wire is only about 100 pulsed wire diameters downstream of the pulsed wire and it can be shown that (for a uniform upstream profile) the velocity at the sensor wire remains below about 90% of the upstream value over the whole Reynolds number range. Since, in addition, the wake would have significant width by $x/d = 100$ and would be strongly asymmetric because of the upstream shear, the different effects of heat convection for linear, as opposed to parabolic, profiles may well be rather greater than those actually calculated by ignoring the viscous wake altogether.

To estimate the effects of temporal velocity fluctuations during the heat tracer flight time, the time-dependent code was run using a sinusoidally varying velocity profile whose amplitude was arranged to give a 30% local velocity (i.e. wall shear stress) intensity. Consideration of the typical range of wave-numbers in the near wall region (Hinze [10]), compared with typical heat tracer convection times showed that 0(1) cycles of velocity variation can occur during a single shear stress measurement. Such fluctuations will occur randomly so that unless the ratio of the time of flight for a constant shear to that for a fluctuating shear follows phase variations linearly, there will be averaging errors. Our computations showed that this could lead to shear stress over-predictions of at most 4%. Errors will clearly depend on the exact nature of the velocity spectrum and this is certainly an upper bound, since we assumed a 30% intensity for a *single* wave-number component. The only way to reduce the possible significance of this effect in practice is to reduce the wire spacing. However, this leads to deliterious effects from other sources, as discussed previously. Since the errors would anyway seem to be small, this effect can probably be ignored.

Our conclusion from these additional calculations (described in detail in Castro and Dianat, [11]) is that it is the viscous effects which cause the largest part of the difference between the experimentally and computationally obtained differences in laminar and turbulent calibrations.

Final Comments and Conclusions

Whilst the spatial and temporal variations in the instantaneous velocity profile lead to measurement errors which are generally small compared to those that arise if the probe is calibrated in a laminar flow channel, this is not necessarily the case for measurements of the fluctuating component. Figure 11 shows the results of an early experiment in which wall probes with different length sensor wires (1, 2 and 4 mm) were used to make measurements in the standard zero pressure gradient boundary layer. There is a noticeable increase in the measured skin friction intensity (equivalent to u'/U as $y \rightarrow 0$) as the sensor wire length falls. The typical spanwise length scale of the inner layer bursts is about 100 wall units, so these results are not surprising, although they serve to emphasise that unless $l_s u_\tau/\nu < 30$ the skin friction intensity is likely to be significantly underestimated (in agreement with the findings of Willmarth and Sharma, [12]). There was no discernible trend in the mean value results (not shown) although they were all too high since the probe was calibrated in a laminar channel.

The yaw-response of the probe has been found to be accurately cosinusoidal up to at least $\pm 80°$ over the practical Peclet number range (an example is shown in Fig. 12). This is rather

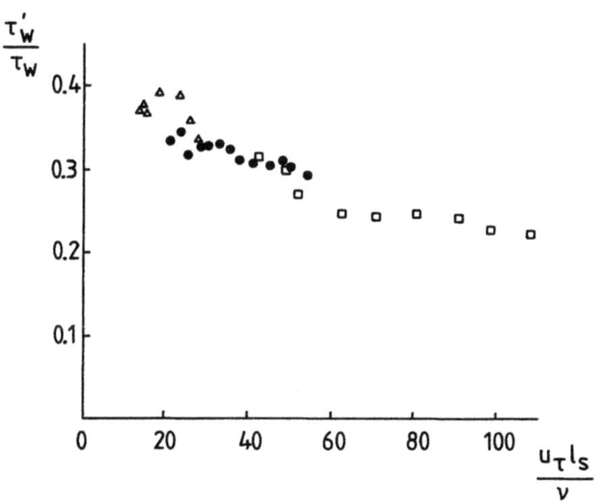

Fig. 11. Wall shear stress intensity as a function of wire length. l_s: ▵, 1.0; •, 2.0; □, 4.0 mm.

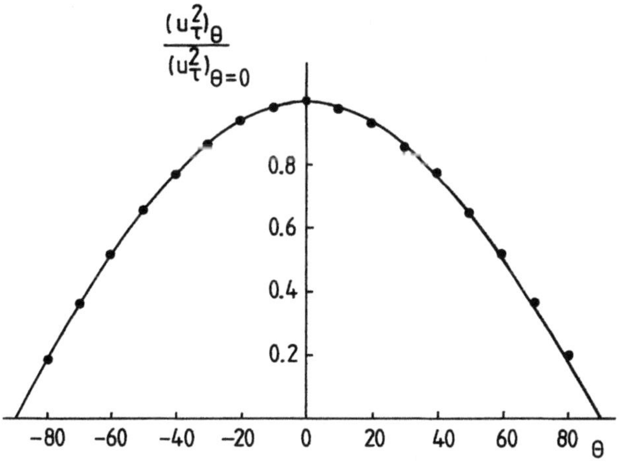

Fig. 12. Probe yaw response. Obtained in laminar channel ($\omega = 0.5$) at $u_\tau^2 = 0.1\ \mathrm{m^2/s^2}$ (Pe ≅ 8). ——, Cos θ

surprising in view of the relatively low Peclet numbers, but gives considerable confidence in the ability of the probe to make accurate measurements of both components of mean and fluctuating skin friction, along with the cross-product $\overline{\tau_u \tau_v}$. Note that there are good reasons for assuming that the latter is, in fact, negligibly small, even in highly turbulent flows (Dianat and Castro, [5]).

Complete details of all our studies are given in Castro and Dianat [11] and the conclusions of the work presented here can be summarised as follows:

1. Differences between calibrations in flows having different upstream velocity profiles are larger than would be expected on the basis of the difference in convection velocity at the probe hight, for the same wall shear. This has been demonstrated by both the numerical computations and the experiments, but the latter results indicate rather larger errors than those suggested by the computations, probably because of exclusion of viscous effects in the former.

2. The resulting errors in skin friction measurements obtained in a turbulent flow with a probe calibrated in a laminar channel are functions of Peclet number (based on u_τ and s, the wire spacing) and s/H_c (where H_c is the channel height) in addition to s/h (where h is the probe height). In particular, both the numerical computations and the experiments have

shown that for given values of the latter two parameters, the errors vary non-monotonically with Peclet number.

3. Since, in addition, the magnitudes of these experimental (and numerical) errors are sensitive to the particular instrument settings (trigger levels, signal gains, etc.), there is no simple way of devising a quantitative correction procedure enabling a single laminar channel flow calibration to be used for accurate skin friction measurements in turbulent flows.

4. However, despite the instrument's inherent non-linearity, calibration in a 'standard' turbulent boundary layer – or, presumably, a turbulent channel flow – does seem viable. This is, in fact, the current practice at Stanford (e.g. Westphal et al., [13]) and Berlin (e.g. Ruderich and Fernholz, [14]). Nonetheless, since fluctuation intensities are at least 30% in such flows, it would seem sensible to pursue the possibility of setting up a *linear* velocity profile laminar flow for calibration purposes. We are currently attempting this using a rotating conical disc rig.

5. In addition to mean flow measurements the technique clearly allows fluctuating data to be obtained. Indeed, we have shown that complete probability density distributions of the instantaneous skin friction can be measured relatively simply. Even in a standard boundary layer, it is difficult to do this in any other way and, for highly turbulent flows, there seems at present no alternative technique. There seems no reason why time-domain measurements could not also be obtained, using techniques like that described by Castro [15] for velocity autocorrelations and spectra from an ordinary pulsed wire anemometer.

Acknowledgmements. This work was undertaken with the financial support of the Procurement Executive of the Ministry of Defence.

References

1. Ginder, R. B., Bradbury, L. J. S. (1973): Preliminary investigation for skin friction measurements in highly turbulent flows. ARC Report 34448
2. Eaton, J. K., Westphal, R. V., Johnston, J. P. (1981): Two new instruments for flow direction and skin friction measurements in separated flows. ISA Trans. **21**, 69
3. Westphal, R., Eaton, J. K., Johnston, J. P. (1981): A new probe for measurement of velocity and wall shear stress in unsteady, reversing flow. ASME J. Fluids Eng. **103**, 478
4. Castro, I. P., Dianat, M. (1984): Surface flow patterns on rectangular bodies in thick turbulent boundary layers. J. Wind Eng. Ind. Aerodyn. **11**, 107
5. Dianat, M., Castro, I. P. (1983): Fluctuating surface shear stresses on bluff bodies. J. Wind Eng. Ind. Aerodyn. **17**, 133
6. Ginder, R. B. (1971): "The Development of a Device to Measure Skin Friction in Highly Turbulent Flows," Ph.D. Thesis, London
7. Bradbury, L. J. S., Castro, I. P. (1971): A pulsed wire technique for velocity measurements in highly turbulent flows. J. Fluid Mech. **49**, 657
8. Patel, V. C. (1965): Calibration of the Preston tube and limitations of its use in pressure gradients. J. Fluid Mech. **23**, 185
9. Bhatia, J. C., Durst, F., Jovanovic, J. (1982): Correction of hot wire anemomentry measurements near walls. J. Fluid Mech. **122**, 411
10. Hinze, J. O. (1975): *Turbulence*, 2nd ed., McGraw-Hill, New York
11. Castro, I. P. and Dianat M. (1985): The measurement of skin friction with a pulsed wire gauge. Mech. Eng. Rep. 85/F1 University of Surrey
12. Willmarth, W. W., Sharma, L. K. (1984): Study of turbulent structure with hot wires smaller than the viscous length. J. Fluid Mech. **142**, 121
13. Westphal, R. V., Johnston, J. P., Eaton, J. K. (1984): Experimental study of flow reattachment in a single-sided sudden expansion. NASA Contracter Report 3765
14. Ruderich, R., Fernholz, H. H. (1986): An experimental investigation of a turbulent shear flow with separation, reverse flow and reattachment. J. Fluid Mech. **163**, 283
15. Castro, I. P. (1985): Time domain measurements in separated flows. J. Fluid Mech. **150**, 183

Part IV

Reacting Flows

Introduction: Turbulence Modeling of Reacting Flows: The State of the Art

N. Peters

Institut für Allgemeine Mechanik, RWTH Aachen, 5100 Aachen, FRG

The interaction of turbulence and chemistry may be classified by two criteria: premixed or non-premixed reactants, slow or fast chemistry. The first criterion is relevant with respect to application: Combustion in large furnaces and gas turbines as well as reactive flows in chemical industry are essentially non-premixed while combustion in the Otto-type reciprocating engine or the reaction of pollutants in the atmosphere occurs in the premixed regime. The second criterion describes the ratio of chemical to convective and/or diffusive time scales. Slow chemistry is not very often of practical interest: there are a few applications like combustion of highly diluted reactants in post-flame regions where chemistry is slow compared to convection and diffusion. However, this case is more easily approached from the turbulence modeling point of view, since it may be treated as a perturbation and extension of the broad fund of knowledge required in non-reacting turbulent flows. On the other hand, reacting flows with fast chemistry present a new challenge to the modeller: The small chemical time scale, multiplied with the molecular diffusion coefficient, introduces a small length scale, typically smaller than the smallest turbulent length scale, the Kolmogorov-scale. If this is the case, chemical activity is confined to thin layers which now are called flamelets. Since flamelets in a turbulent flow field are thin by definition, their structure is one-dimensional normal to the layer and it may be resolved by well developed numerical or asymptotic methods. Therefore, in the flamelet regime chemistry and turbulence may – in principle – be separated.

Among the four regimes that are covered by the two criteria stated above, three are scientifically in quite a good shape, the regimes of unpremixed and premixed reactants with slow chemistry and – at first surprisingly – the regime of unpremixed reactants with fast chemistry. One would think that the fourth regime – premixed reactants with fast chemistry – would be easier than the third one, because premixing prior to reaction eliminates diffusion as a rate limiting step and therefore reduces the number of time scales to be considered. However, premixed reaction introduces new time and length scales of its own by creating a reaction front that moves perpendicular to itself in the turbulent flow. This problem, the interaction of a reaction front – a flame front in combustion – with a turbulence field is far from being well understood. Among the five papers presented in this volume, two, the papers by Jones and Kollmann and by Coupland and Priddin, fall into the regime of non-premixed combustion with fast chemistry, one, the paper by Herring and Wyngaard, into the field of premixed reactants with slow chemistry, while the remaining two, the papers by Moreau, Labbé, Dupoirieux and Borghi and by Escudié and Charnay approach the problem of premixed reactants with fast chemistry from two different view points.

The paper by Jones and Kollmann tries to attack the well known difficulty to predict the CO-level in turbulent diffusion flames correctly by using a *pdf* transport equation. The argument to support *pdf* transport modelling for reacting flows has always been that the chemical source term does not need to be modelled. However, if reaction occurs in thin layers only, reaction and molecular diffusion are closely coupled and the difficulty is simply shifted towards modelling of the molecular diffusion term. The modelling of this term by Jones and Kollmann again employs the turbulent time scale as the only characteristic scale, which is certainly unsatisfactory. Although it seems that chemistry is more realisticly treated in this paper than by the assumption of instantaneous chemical equilibrium, the problem of CO-production in turbulent diffusion remains as a challenge.

The paper by Coupland and Priddin demonstrates the applicability of turbulence modelling even for operating gas turbine combustor chambers. The problem here is not so much the details of the modelling but the development of a sufficiently accurate but cost efficient numerical code, which includes the generation of a body fitted orthogonal mesh for a complex geometry. Comparison of measured and predicted flow patterns of a water flow model are a first step in order to gain confidence in the prediction of the reacting flow field, for which unsufficient data were available. Nevertheless, exit plane temperatures and exit NO_x-levels are well predicted with an adjustment of the turbulent Prandtl number. One may conclude that global answers to industrial problems can be obtained by today's state of the art for non-premixed reacting flows as long as chemical and flow time scales are sufficiently separated. In the gas turbine combustor of Coupland and Priddin combustion chemistry is fast and NO_x-chemistry is slow. Then turbulence is only modified by heat release and this is taken into account by using Favre averaged quantities.

Among the papers on premixed reactants the paper by Herring and Wyngaard addresses the most fundamental question of turbulent transport in presence of chemical reactions. This paper uses direct simulation of a buoyancy driven flow with a linearly reacting passive scalar in a $32 \times 32 \times 32$ box in order to test the validity of eddy transport concepts and their possible generalization. Direct or large eddy simulation is now beginning to become an important tool for the development of turbulence models. Other interesting examples are the papers by Lee and Reynolds, and Moin, Rogers and Moser in this book which address other unsolved problems and provide some new orientation for modellers. Here, as far as scalar transport is concerned, classical mixing length concepts known in thermal convection problems are modified by the slow reaction and a non-local formulation is discussed. A complete invalidation of these models such as by counter-gradient diffusion in premixed flames would require additional interactions between chemistry and turbulence: fast chemistry and/or non-constant density. Then the velocity field will be effected by heat release which is not the case here. Due to the more severe grid size requirements these kind of problems are still resisting the promising tool of direct and large eddy simulation.

The paper by Moreau, Labbé, Dupoirieux and Borghi focusses on an "evergreen" in turbulence modelling, the question of the prediction of recirculation zones. Here several models are compared with experiments for non-reactive flows and flows with combustion. As in previous attempts, the recirculation length is underpredicted by the classical k-ε-model and the turbulent kinetic energy is overpredicted. Calculations with constant viscosity models, where the value of the viscosity has been adjusted, do – not surprisingly – a better job. No scalar measurements or predictions are reported. The claim for more sophisticated turbulence models that would be able to handle this rather complicated premixed turbulent flow is well understood.

The last paper by Escudié and Charnay approaches the important question of the interaction between a premixed flame and a turbulent flow from the quasi-laminar limit by studying the effect of well defined vortices on a V-shaped flame. While vortices of the size of the flame thickness only wrinkle the flame front, larger vortices force it into a quite violent response. This increases the flame front area and thereby the turbulent flame speed. It is certain that more studies of this kind are needed before the interaction between turbulence and combustion is fully understood.

It looks as if the Reacting Flows Session of the Turbulent Shear Flows Symposium is a forum for the assessment of classical models as well as for new ideas. Personally, I would like to see more confrontation with examples where there is more of an interaction – in both directions – between chemistry and turbulence. This requires, of course, comparable historical time scales as well. While the time scale for the development of turbulence models is of the order of twenty years by now, the time scale for finite chemistry models in turbulent flows is only of the order of ten years. Therefore, there is reason to look forward to a second decade in chemically reacting turbulent flows.

Multi-Scalar *pdf* Transport Equations for Turbulent Diffusion Flames

W. P. Jones and W. Kollmann*

Department of Chemical Engineering and Chemical Technology, Imperial College, London SW7 2BY, England

* Department of Mechanical Engineering, University of California, Davis, California 95616, USA

Abstract

Calculations of a turbulent propane air diffusion flame are presented and compared with experiment. The calculation method utilises the $k - \varepsilon$ model to represent turbulent transport and a simple three scalar reaction mechanism to describe the instantaneous formation and consumption rates of propane and carbon monoxide. Turbulent fluctuations in the three scalars are accounted for through the use of a modelled transport equation for the joint *pdf* of mixture fraction, mole number and propane mass fraction which is solved by a Monte Carlo method. In general it is found that CO levels are under-predicted but this is almost certainly a direct consequence of limitations in the reaction scheme used particularly in fuel rich regions of the flame.

Nomenclature

C_D	Molecular mixing model constant	$T(\ )$	Transition probability density function
$C_\mu, C_{\varepsilon_1}, C_{\varepsilon_2}$	$k - \varepsilon$ turbulence model constants		
D	Nozzle diameter	t	Time
D_t	Mean rate of change: $\partial_t + \tilde{U}_\alpha \partial_\alpha$	$\tilde{U}_\alpha$	Density weighted mean component of velocity
f	Mixture fraction		
h	Enthalpy of mixture	u''_α	Fluctuating component of velocity
h_i	Enthalpy of chemical species 'i'	$\boldsymbol{x}$	Position vector
k	Turbulence kinetic energy	Γ	Molecular transport coefficient
N	Total number of species	ε	Turbulence energy dissipation rate
n	Mole number of mixture; number of moles per unit mass	ε_{ij}	Instantaneous scalar 'dissipation' rate of $\phi_i \phi_j$
n_i	Mole number of species 'i'		
p	Pressure	μ_t	'Turbulent' viscosity
$P(\ )$	Probability density function for the set of scalars, ϕ_i	ϱ	Density of mixture
		σ_t	'Turbulent' Prandtl/Schmidt number
R	Universal gas constant		
r	Radius	τ	Turbulence time scale
S_i	Source term; net formation rate of species 'i' per unit mass	ϕ_i	Dependent scalar variable
		ψ_i	Independent scalar variable
T	Temperature, °K	$\langle\ \rangle$	Ensemble average value

Introduction

Calculation methods for turbulent diffusion flames must include a model for describing chemical reactions and the influence of turbulent fluctuations upon them. A technique often adopted is to invoke a 'fast' reaction approximation, e.g. [1] and [2], which when taken together with the assumptions that all molecular transport coefficients are equal and that adiabatic conditions prevail allows the thermochemical state to be determined uniquely in terms of a single strictly conserved scalar variable. Because of the fluctuating nature of turbulent flows, the statistics of the conserved scalar are needed to evaluate mean quantities and these are often described via the introduction of the probability density function (*pdf*)

Turbulent Shear Flows 5

presuming its shape in terms of the mean and variance of the conserved scalar, the values of which are obtained from solution of modelled transport equations. Alternatively and with potentially greater accuracy but at the expense of increased complexity the *pdf* can be obtained directly from solution of its own modelled transport equation, [5] and [6].

While the above approach gives good results under many circumstances no direct kinetic effects can be incorporated and as a consequence various ignition and extinction phenomena and the emission levels of carbon monoxide and unburnt fuel from, for example gas turbines operating at lower power cannot be predicted. The explicit incorporation of finite rate kinetics requires more than one independent scalar. About the simplest mechanism retaining the essential features of the high temperature oxidation of hydrocarbons which can be constructed separates the reaction into two global steps: the burn-up of hydrocarbon to form CO and other products and the oxidation of CO to CO_2. The unique description of such a mechanism can be achieved in terms of a minimum of three independent scalar quantities for which conservation equations must be solved; [7] and [8] provide examples of reaction schemes of this type. For a general treatment of turbulent flows the joint *pdf* of the three scalars is then needed to evaluate mean values. Because of the bounded nature of the scalars the determination of this joint *pdf* via the presumed shape formalism presents considerable difficulties and may well be intractable. In contrast the extension of the *pdf* transport equation approach to handle several scalars is relatively straight forward but because of the resulting increased dimensionality Monte Carlo solution methods become necessary. In the present work this latter approach is adopted. To represent molecular transport in scalar space (molecular 'mixing') a model, based on the closure of [9] for a single scalar is constructed and this is combined with a simple three scalar reaction mechanism to describe the burning of propane in air. The resulting joint *pdf* equation is then applied to the calculation of a turbulent propane-air diffusion flame burning in stagnant surrounds and the results compared with measured concentrations of CO, CO_2 and unburnt fuel and mixture fraction profiles.

Chemical Model

For the calculation of turbulent flames to be tractable, it is necessary that chemical reaction be described in terms of a small number of independent scalar quantities – that is, quantities such as species mass fraction, etc. for which conservation equations must be solved. For this reason a simplified scheme for the oxidation of propane has been constructed using the constrained equilibrium approach of [10]. At any instant the mixture is presumed to comprise the species, H, HO_2, O, OH, H_2, H_2O, CH_3, CH_4, C_2H_2, C_2H_4, C_2H_6, C_3H_8, CO, CO_2, O_2 and N_2, and the oxidation is described by the following rate controlling steps:

$$C_3H_8 + \tfrac{3}{2}O_2 \rightarrow 3\,CO + 4\,H_2, \qquad \text{(i)}$$

$$H + O_2 + M \rightleftharpoons HO_2 + M, \qquad \text{(ii)}$$

$$H + H + M \rightleftharpoons H_2 + M, \qquad \text{(iii)}$$

$$H + OH + M \rightleftharpoons H_2O + M, \qquad \text{(iv)}$$

$$H + O + M \rightleftharpoons OH + M. \qquad \text{(v)}$$

The rate at which propane is consumed via reaction (i) is presumed given by the global expression of [7]:

$$S_{C_3H_8} = -\,8.44 \times 10^9 \exp\{-\,15102/T\}\, \varrho^{0.75}\, n_{O_2}^{1.65}\, n_{C_3H_8}^{0.1}\ \text{kmol/kg sec},$$

and the forward rates for reactions (ii) to (v) are taken from [11]. The reverse rates are then obtained via the appropriate equilibrium constants. The reactions (i) to (v) are also assumed to be the only ones from which a change in the number of moles results. It is further presumed than the diffusion is Fickian, that the values of Prandtl and Schmidt numbers are all equal, that the flow is adiabatic and the pressure is thermodynamically constant. The thermochemical state (including composition, density and temperature) is then obtained via a partial equilibrium assumption in which the free energy is minimised subject to the constraints that the mixture fraction, mole number and mass fraction of C_3H_8 take predetermined values. These predetermined values are those which result from the solution of the appropriate conservation equations. Thus all bimolecular reactions including the main CO oxidation reaction

$$CO + OH \rightleftharpoons CO_2 + H$$

are presumed to be in partial equilibrium. As a consequence the rate of formation and oxidation of carbon monoxide is controlled by the rate at which C_3H_8 and mole number are formed or consumed.

Balance Equations

With the assumption of Fickian diffusion and low Mach number flow the balance equations for mass and energy can be written in the form

$$\varrho D_t \phi_i = \partial_\alpha(\varrho \Gamma_i \partial_\alpha \phi_i) + \varrho S_i(\phi_1, \phi_2 \cdots \phi_N), \tag{1}$$

where the summation convention applies to Greek subscripts only. For the reaction mechanism for the combustion of propane previously described balance equations must be solved for three scalar variables, ϕ_i. They are defined as mixture fraction $\phi_1 \equiv f$, the mole number of the mixture, $\phi_2 \equiv n$ and the mole number of the fuel, $\phi_3 \equiv n_{C_3H_8}$. With the assumption of adiabatic flow the enthalpy at any point in space and time is given by

$$h = (h_0 - h_1)\, f + h_1,$$

where the subscripts '0' and 1 refer to the incoming fuel and air streams. The temperature and density can then be obtained from the ideal gas state equations

$$h = \sum_{i=1}^{N} n_i h_i(T)$$

and

$$\varrho = \frac{p}{RT \sum_{i=1}^{N} n_i}.$$

As a consequence density, temperature and composition can be represented as local functions of the three scalars (ϕ_1, ϕ_2, ϕ_3); the functions turn out to be strongly nonlinear. For turbulent reacting flows the introduction of the probability density function provides an exact means of representing kinetic sources; in the *pdf* balance equation the kinetic source terms appear in closed form. The appropriate single point joint *pdf* $P(\psi_1, \ldots, \psi_N; \mathbf{x}, t)$ of the values ψ_i of the scalars ϕ_i can be obtained from (1) by standard methods [12–14]. For flows with variable density the weighted *pdf* $\tilde{P}$ defined by

$$\tilde{P}(\psi_1 \cdots \psi_N; \mathbf{x}, t) = \frac{\varrho(\psi_1 \cdots \psi_N)}{\langle \varrho \rangle} P(\psi_1 \cdots \psi_N; \mathbf{x}, t)$$

is appropriate. The transport equation for $\tilde{P}$ is given by

$$\langle \varrho \rangle \partial_t \tilde{P} + \langle \varrho \rangle \tilde{U}_\alpha \partial_\alpha \tilde{P} = -\partial_\alpha (\langle \varrho \rangle \langle u''_\alpha | \phi_i = \psi_i \rangle \tilde{P}) + \langle \varrho \rangle \sum_{i=1}^{N} \partial_{\psi_i} [S_i(\psi_i \cdots \psi_N) \tilde{P}]$$
$$- \langle \varrho \rangle \sum_{i=1}^{N} \sum_{j=1}^{N} \partial^2_{\psi_i \psi_j} [\langle \varepsilon_{ij} | \phi_k = \psi_k \rangle \tilde{P}],$$

where all molecular diffusivities Γ_i have been assumed equal and molecular fluxes neglected (high Reynolds number). The scalar 'dissipation' rates defined by

$$\varepsilon_{ij} \equiv \Gamma \nabla \phi_i \cdot \nabla \phi_j$$

determine the rate at which the variances and covariances of ϕ_i are decreased. The source terms S_i incorporate the kinetic mechanism for the C_3H_8 combustion and appear in the *pdf* equation in closed form as convective velocities for the transport of P along the scalar axes.

The *pdf* equation contains two types of term which require closure namely that representing turbulent transport and that describing molecular 'mixing'. The turbulent flux of *pdf* is approximated by means of a gradient-transport model, [9].

$$- \langle \varrho \rangle \langle u''_N | \phi_i = \psi_i, i = 1, \ldots, N \rangle \tilde{P} = \frac{\mu_t}{\sigma_t} \partial_\alpha \tilde{P},$$

where the value of σ_t was chosen to be 0.7 and where

$$\mu_t = C_\mu \langle \varrho \rangle \frac{k^2}{\varepsilon}$$

with k and ε being determined via the $k - \varepsilon$ turbulence model, [15]. The model for the scalar 'dissipation' terms in the *pdf* equation is based on the closure expression developed in [9] for the case of a single conserved scalar variable. The extension to several scalar variables is by no means trivial and depends critically on the topology of the domain of definition D of the scalar variables ϕ_i. The following basic restriction is introduced here: the domain of definition D of the ϕ_i is convex. From the definition of the ϕ_i as mixture fraction and mole numbers it follows that the scalars are bounded and consideration of these bounds shows that the domain of definition of the ϕ_i is a skew tetrahedron. Thus, the domain of definition is a convex and bounded set and satisfies our basic condition. It thus follows that if two points in the domain of definition D are selected then the line connecting them is always inside D and closure models for the scalar dissipation based on the idea of binary interaction of fluid parcels can be applied. To do this consider a subdomain of the flow field centered at a given location which is sufficiently small to be considered homogeneous. The fluid in this subdomain is viewed as a collection of a large number of sufficiently small parcels, which interact pairwise for a given time interval by changing the values of the scalars identifying them. This change is restricted to the intersection of the cube formed by the scalar values of the two points in scalar space (parcels) as diagonally opposed corner points with the domain of definition D. The result of this interaction is governed by the transition *pdf* $T(\psi, \Psi \mid \chi)$ defined as the probability that the interaction of a parcel $\psi_i \leqq \phi_i \leqq \psi_i + d\psi_i$ with a parcel $\Psi_i \leqq \phi_i \leqq \Psi_i + d\omega_i$ produces two parcels with $\chi_i \leqq \phi_i \leqq \chi_i + d\psi_i$ and $\psi_i + \Psi_i - \chi_i \leqq \phi_i \leqq \psi_i + \Psi_i - \chi_i + d\psi_i$ ($i = 1, \ldots, N$ for N scalar variables). If it is furthermore required that the closure model satisfies the properties of the exact term, i.e. it leaves normalization and mean values unaffected but reduces the variances, then the following conditions on T must be satisfied:

$$\underbrace{\int \cdots \int}_{N} d\chi_1 \cdots d\chi_N \, T(\psi, \Psi \mid \chi) = 1$$

and

$$T(\psi, \Psi \mid \chi) = T(\psi, \Psi \mid \psi + \Psi - \chi)$$

and

$$T(\psi, \Psi \mid \chi) = 0 \quad \text{if } \chi \text{ is outside the interval } (\psi, \Psi).$$

The closure approximation for the 'dissipation' term then follows

$$-\sum_{i=1}^{N} \sum_{j=1}^{N} \partial^2_{\psi_i \psi_j} (\langle \varepsilon_{ij} \mid \phi_k = \psi_k = 1, \ldots, N \rangle \, \tilde{P})$$

$$\cong \frac{C_D}{\tau} \left\{ \underbrace{\int \cdots \int}_{N} d\psi \underbrace{\int \cdots \int}_{N} d\Psi \, \tilde{P}(\psi) \, \tilde{P}(\Psi) \, T(\psi, \Psi \mid \chi) - \tilde{P}(\chi) \right\},$$

where C_D is a constant and τ is a turbulence time scale characteristic of the energy containing motions. For the present model the transition *pdf* T is defined as constant in its interval of definition and is thus given by

$$T(\psi, \Psi \mid \chi) = \prod_{i=1}^{N} T_i(\psi_i, \Psi_i \mid \chi_i)$$

and

$$T_i(\psi_i, \Psi_i \mid \chi_i) = \begin{cases} \dfrac{1}{|(\psi_i - \Psi_i)|} & \text{for } \psi_i \leqq \chi_i \leqq \Psi_i \quad \text{or} \quad \Psi_i \leqq \chi_i \leqq \psi_i \\ 0 & \text{otherwise.} \end{cases}$$

The *pdf* equation in closed form is then

$$\langle \varrho \rangle \, \partial_t \tilde{P} + \langle \varrho \rangle \, \tilde{U}_\alpha \partial_\alpha \tilde{P} = \partial_\alpha \left(\frac{\mu_t}{\sigma_t} \, \partial_\alpha \tilde{P} \right) + \langle \varrho \rangle \sum_{i=1}^{N} \partial_{\psi_i} (S_i \tilde{P})$$

$$+ \frac{C_D}{\tau} \left\{ \int \cdots \int d\psi \int \cdots \int d\Psi \, \tilde{P}(\psi) \, \tilde{P}(\Psi) \, T(\psi, \Psi \mid \xi) - \tilde{P}(\chi) \right\}, \quad (1)$$

where C_D is assigned the value 6.0 [1] and where the time scale, τ is obtained from

$$\tau = \frac{k}{\varepsilon}.$$

[1] This value of C_D is selected so that the variance decay rate of a passive scalar in isotropic turbulence is equal to $2 \frac{\varepsilon}{k} \overline{\phi}^2$.

To complete the formulation the balance equations for the mean velocity $\tilde{U}_\alpha$, the turbulence kinetic energy, k and its dissipation rate, ε are needed. These are obtained from solution of

$$\langle\varrho\rangle\,\partial_t\tilde{U}_\alpha + \langle\varrho\rangle\,\tilde{U}_\beta\partial_\beta\tilde{U}_\alpha = -\,\partial_\alpha p + \partial_\beta\{\mu_t(\partial_\alpha\tilde{U}_\beta + \partial_\beta\tilde{U}_\alpha)\}\,, \tag{2}$$

$$\langle\varrho\rangle\,\partial_t k + \langle\varrho\rangle\,\tilde{U}_\alpha\partial_\alpha k = \partial_\alpha\left\{\frac{\mu_t}{\sigma_k}\,\partial_\alpha k\right\} + G - \langle\varrho\rangle\,\varepsilon\,, \tag{3}$$

$$\langle\varrho\rangle\,\partial_t\varepsilon + \langle\varrho\rangle\,\tilde{U}_\alpha\partial_\alpha\varepsilon = \partial_\alpha\left\{\frac{\mu_t}{\sigma_\varepsilon}\,\partial_\alpha\varepsilon\right\} + C_{\varepsilon_1}\frac{\varepsilon}{k}G - C_{\varepsilon_2}\langle\varrho\rangle\,\frac{\varepsilon^2}{k}\,, \tag{4}$$

where $\quad \mu_t = C_\mu\langle\varrho\rangle\,k^2/\varepsilon \quad$ and $\quad G = \mu_t\partial_\alpha\tilde{U}_\beta(\partial_\alpha\tilde{U}_\beta + \partial_\beta\tilde{U}_\alpha)$.

The values of the model constants used were $C_{\varepsilon_1} = 1.44$, $C_{\varepsilon_2} = 1.85$, $C_\mu = 0.065$, $\sigma_k = 1.0$ and $\sigma_\varepsilon = 1.3$; values which correctly reproduce the measured spreading rate and turbulence energy levels in a constant density round jet issuing into stagnant surrounds. All moments of thermodynamic variables (ϱ, T, n_i) are calculated as integrals of the *pdf* $\tilde{P}$.

Solution Method

The closed system of equations was first written in cylindrical polar coordinates in the boundary layer form appropriate to the round jet. For the equations for the mean velocity $\tilde{U}_\alpha$ and turbulence energy, k and its dissipation rate, ε a standard finite-difference procedure, [16] was used and the *pdf* equation was solved using a Monte-Carlo simulation technique. With the coordinates transformed to the $x - \omega$ system of [16] the *pdf* transport equation becomes

$$\frac{\partial\tilde{P}}{\partial x} + (a + b\omega)\,\frac{\partial\tilde{P}}{\partial\omega} = \frac{\partial}{\partial\omega}\left\{c\,\frac{\partial\tilde{P}}{\partial\omega}\right\} + \frac{E + S}{\varrho\tilde{U}}\,,$$

where E and S here represent the molecular 'mixing' and reaction rate terms respectively. This equation is then first approximated by the following explicit operator split difference scheme:

a) Transport in ω-direction (Convection + diffusion):

$$\tilde{P}_j(x^*.\,\psi) = A_j\tilde{P}_{j+1}(x,\psi) + B_j\tilde{P}_{j-1}(x,\psi) + C_j\tilde{P}_j(x,\psi)\,,$$

where $\tilde{P}_j(\ )$ represents the *pdf* at location ω_j and where A_j, B_j and C_j are finite difference coefficients arising from conservative hybrid differencing, e.g. [16], and which depend on the forward step, Δx. The coefficients satisfy

$$A_j + B_j + C_j = 1\,,$$

and for stability Δx must be selected so that the coefficients remain positive.

b) Scalar 'Dissipation':

$$\tilde{P}_j(x^{**},\psi) = P_j(x^*,\psi) + \frac{1}{\bar{\varrho}\tilde{U}}\int_x^{x+\Delta x} E_j(\psi)\,dx\,.$$

c) Reaction:

$$\tilde{P}_j(x + \Delta x, \psi) = P_j(x^{**}, \psi) + \frac{1}{\bar{\varrho}\tilde{U}} \int\limits_x^{x+\Delta x} S_j(\psi)\, dx .$$

If applied as stated the *three* operations (a) to (c) would allow the solution of the *pdf* equation to be advanced from x to $x + \Delta x$ to within a formal accuracy of order Δx. [Note: the locations x^* and x^{**} lie between x and $x + \Delta x$ but the interim solutions at these positions do *not* represent solutions of the equation].

The joint *pdf* at each location ω_j is now represented by an ensemble of particles at each ω-location, with each particle having three scalar properties, and the finite difference scheme (a) to (c) is simulated by a Monte Carlo technique similar to that described in [13]. Transport (convection plus diffusion) is represented by exchanging particles between neighbouring nodes, scalar dissipation is simulated by letting the scalar values of randomly selected particle pairs approach each other by mixing and chemical kinetics by allowing each particle to react over the time interval $\Delta x/\tilde{U}$. The number of particles in each ensemble was 1000 and the forward step size [2], Δx, used for the transport and scalar 'dissipation' simulation was 0.25 of the maximum value permitted by stability considerations. The most time consuming and expensive part of the solution procedure is the treatment required to describe the effect of reaction on the *pdf*. The change in the properties of the particles (representing the *pdf*) due to chemical reaction is given by

$$\frac{\partial \psi_j}{\partial x} = \frac{1}{\bar{\varrho}\tilde{U}} S_j(\psi)$$

and this equation must be integrated over the interval x to $x + \Delta x$ for each particle to obtain the change in the C_3H_8 and mixture mole numbers. It is apparent that this step will be numerically expensive for the stiff equation systems describing chemical reaction and numerical solution within a Monte-Carlo was therefore abandoned as impractical. An efficient way of dealing with finite-rate chemistry is to set up a sufficiently accurate 'look-up' table (once and for all) for the increments, $\Delta\psi_i(\Delta t_1, \ldots)$ of all of the scalar variables effected by reaction and density $\Delta\varrho(\Delta t_1, \Delta t_2, \ldots)$ as a function of selected time intervals $\Delta t_1, \Delta t_2, \ldots$. This requires knowledge of the bounds of the accessible domain in scalar space D_s and discretisation in a regular lattice for efficient interpolation. For the present description of reaction three scalar-variables ψ, $(N = 3)$ suffice to determine the thermo-chemical state and these have been taken to be $\psi_1 \equiv f$ (mixture function), $\psi_2 \equiv n$ (mixture mole number), $\psi_3 \equiv n_{C_3H_8}$ (fuel mole number). Only ψ_2 and ψ_3 are effected by reaction and mixture function ψ_1 is (strictly) conserved. To calculate the bounding surfaces of the accessible domain in the three-dimensional scalar space the following conditions were utilised: (1) Radical concentrations are presumed negligible; (2) Mass fractions must lie within the unit internal and mole numbers are non-negative; (3) The reaction system determines the components in the mixture for minimal and maximal mole numbers. It then follows that the accessible region D_s in scalar space is the skewed tetrahedron, Fig. 1, contained within the unit cube with corner points $A(\psi_1, \psi_2, \psi_3)$ to $D(\psi_1, \psi_2, \psi_3)$ given by

$$A = (0.0,\ 3.4664 \times 10^{-2},\ 0.0),$$

$$B = (6.002 \times 10^{-2},\ 3.5306 \times 10^{-2},\ 0.0),$$

$$C = (0.17549,\ 5.0469 \times 10^{-2},\ 0.0),$$

$$D = (1.0,\ 2.2677 \times 10^{-2},\ 2.2677 \times 10^{-2}).$$

[2] In the Monte Carlo simulation of transport and mixing the computational effort is practically independent of the forward step size.

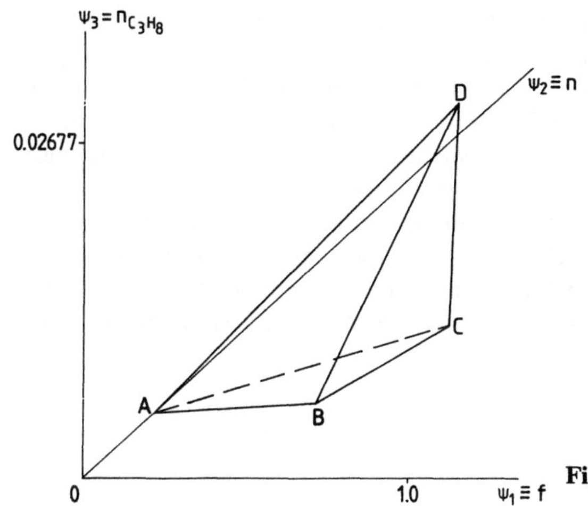

Fig. 1. Accessible domain in the 3-*d* scalar space

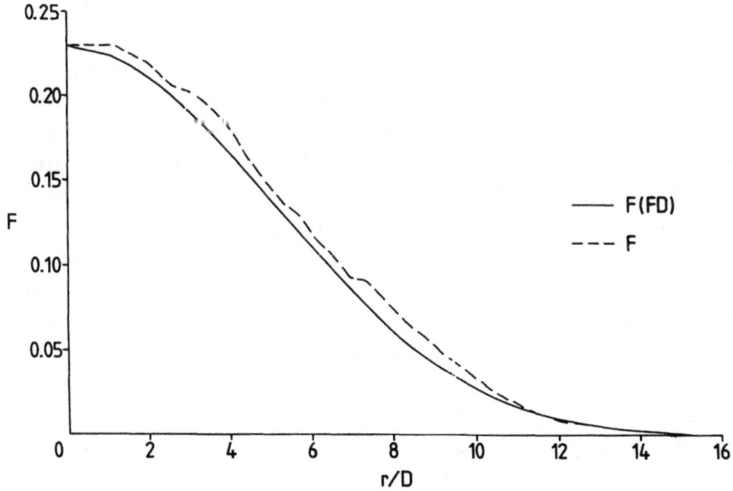

Fig. 2. Radial profile of mixture profile at $x/D = 80$ calculated by finite differences and Monte Carlo Simulation

These values indicate that the volume of D_s is much less than unity. A simple lattice based on the unit cube would thus be very wasteful. The actual interpolation lattice was defined by N_3 planes, $n_{C_3H_8}$ = constant, intersecting D_s with constant spacing. Each intersection forms a triangle, which was discretized by the intersection of N_1 lines parallel to the projection of the corner points A and B (constant spacing) with N_2 lines originating in the projection of corner point C and constantly spaced along $\overline{AB}$. The values chosen for the lattice are $N_1 = 31$, $N_2 = 36$, $N_3 = 81$. Two time increments, $\Delta t_1 = 10^{-5}$ s and $\Delta t_2 = 10\,\Delta t$, were used. The Table consisted of the increments $\Delta\psi_2(\psi_1, \psi_2, \psi_3, \Delta t_1, \Delta t_2)$, $\Delta\psi_3(\ldots)$ and $\Delta\varrho(\ldots)$ and the change in the properties of each particle due to reaction was then obtained by linear interpolation in the Table. The number of finite difference cells in the cross-stream direction was 40 and the total number of particles was 40000.

Testing of the accuracy of the Monte-Carlo simulation technique is limited by the slow convergence of the method with increasing number of particles. However a number of consistency checks were made and these included a comparison of the mean mixture fraction profiles calculated by a straight forward finite difference solution of the mixture fraction balance equation with those extracted from the Monte-Carlo simulation of the joint *pdf* equation. The profiles which arise with the two methods at an axial location of $x/D = 80$ are shown in Fig. 2. The differences in mixture fraction which are evident are a result of truncation errors – because an implicit formulation was used to obtain the finite difference solution whilst the Monte-Carlo method simulates an explicit scheme – and statistical errors associated with the use of a finite number of particles in the simulation. Because of the small forward steps utilised the latter source is felt to provide the major contribution to the discrepancy. While the differences are slightly larger than might be estimated in terms of the number of particles the accuracy of the Monte-Carlo method is considered sufficient for present purposes.

Results and Discussion

The *pdf* transport equation outlined was applied to the calculation of a turbulent propane diffusion flame burning in quiescent air surrounds at atmospheric pressure. The configuration chosen corresponds to the experimental study of [17] in which measurements of the major species, C_3H_8, CO, CO_2, O_2 and N_2 were obtained. Propane issued from a 6 mm diameter nozzle with a nominal velocity of 32 m/s – the Reynolds number based at cold conditions at the nozzle was approximately 40,000 – and the flame was stabilised with a small hydrogen flame. The visible flame length was about 200 nozzle diameters and buoyancy influences were negligible over about the first 100 diameters. To carry out flame calculations some means of ignition must be provided and this was achieved by changing the fuel and mixture mole numbers of particles with mixture fraction values lying between the lean and rich limits to their corresponding fully burnt values. This ignition was carried out at an axial location of $x/D = 0.2$ and at typically five grid nodes close to the outer part of the jet. The results of the calculations are compared with measurements in Figs. 3 to 10. From the measured species and with some approximation mixture fractions could be evaluated and the variation of the computed centreline value of mixture fraction with axial distance is compared with these in Fig. 3. The calculated values lie below the measured values typically by about 5% to 10% and the differences seem to be associated with discrepancies in the initial part of the flame, $x/D \lesssim 20$. However the measured axial variation of propane mole fraction along the centreline is accurately reproduced by the calculations, see Fig. 4. Comparisons of measured and calculated radial composition and mixture fraction profiles at $x/D = 40$ and $x/D = 80$ are shown in Figs. 5 to 10. No smoothing operations have

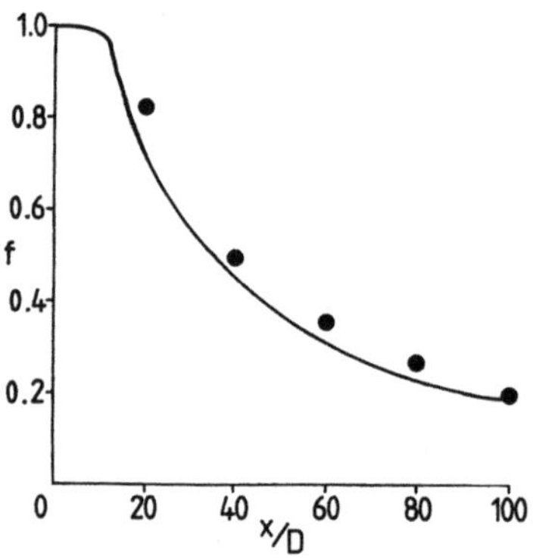

Fig. 3. Axial profile of centre-line mixture fraction; • Measurement [17]

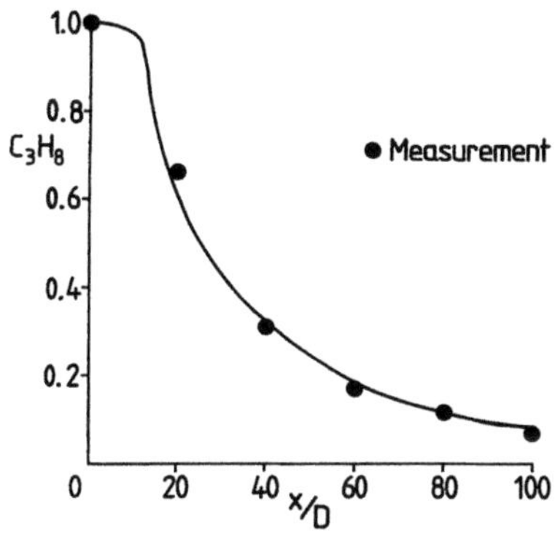

Fig. 4. Axial profile of centre-line C_3H_8 mole fraction; • Measurements [17]

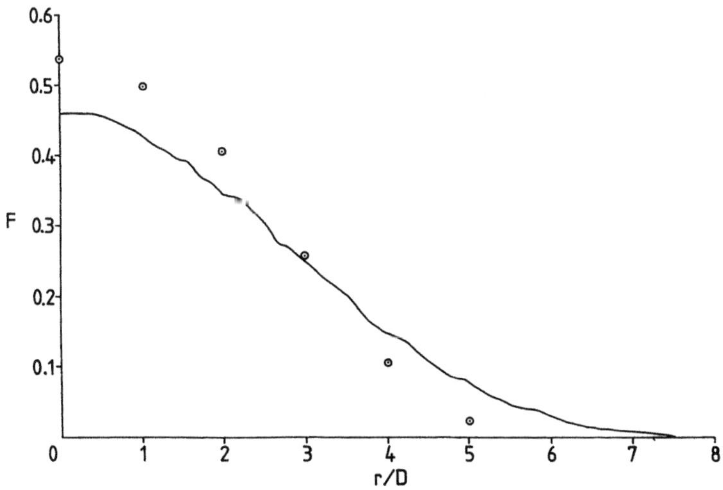

Fig. 5. Radial profile of mean mixture fraction at $x/D = 40$; o Measurement [17]

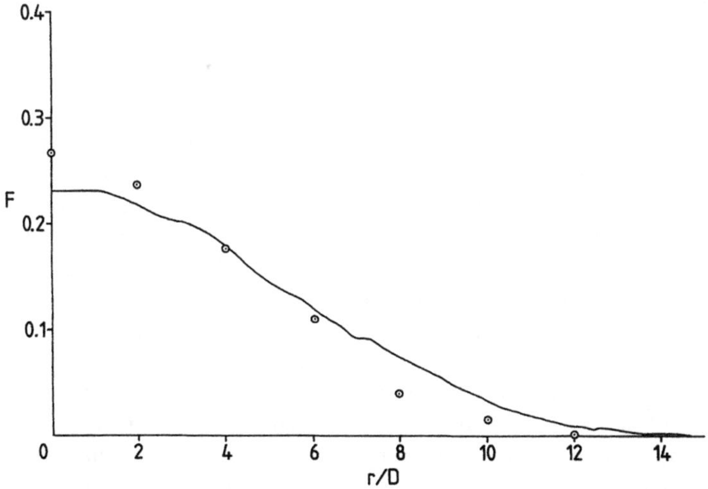

Fig. 6. Radial profile of mean mixture fraction at $x/D = 80$; o Measurements [17]

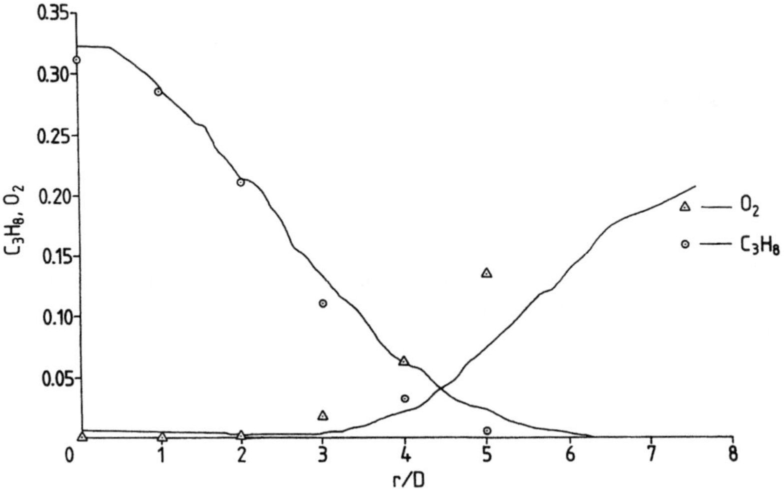

Fig. 7. Radial profiles of C_3H_8 and O_2 mole fractions at $x/D = 40$; ○, △ Measurements [17]

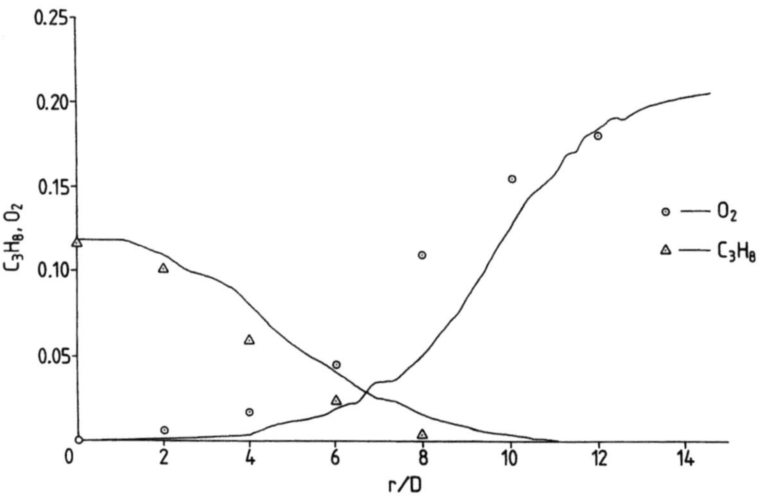

Fig. 8. Radial profiles of C_3H_8 and O_2 mole fractions at $x/D = 80$; ○, △ Measurements [17]

here been performed on the calculated results presented. At both locations the mixture fraction profiles are calculated to have spread slightly more rapidly than measured. However the C_3H_8 profile is extremely well reproduced at $x/D = 40$, Fig. 7; but somewhat overpredicted at $x/D = 80$, Fig. 8, though perhaps not by a significant amount if allowance is made for the mixture fraction profile discrepancies. Given the likely errors in fluid density arising from the discrepancies between measured and calculated CO and CO_2 levels (see below) the level of agreement achieved here can be considered acceptable. The same cannot be said, however for carbon monoxide and carbon dioxide, Figs. 9 and 10. While the calculated profile shapes are similar to those measured the maximum predicted CO level is about 3% compared with 6% measured and there is a corresponding over prediction of the CO_2 levels. To give a perspective of the magnitude of this discrepancy it is worth noting that if full chemical equilibrium had been assumed then the maximum CO level calculated would have been around 15–18%. Nevertheless the level of disagreement is significant and an examina-

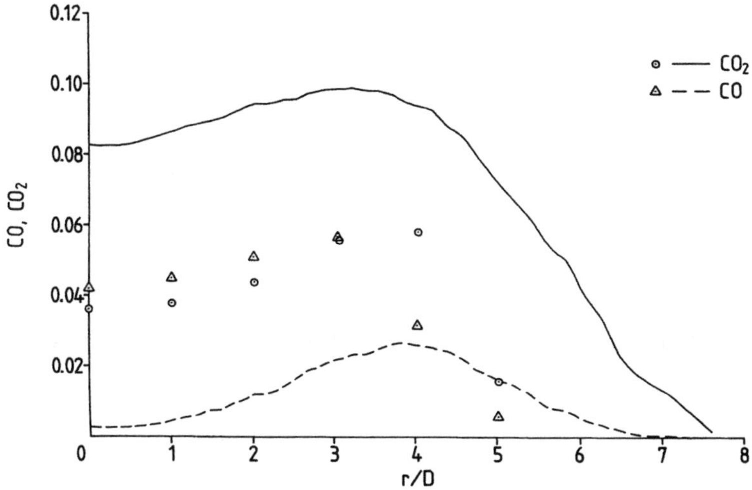

Fig. 9. Radial profiles of CO and CO_2 mole fractions at $x/D = 40$; ○, △ Measurements [17]

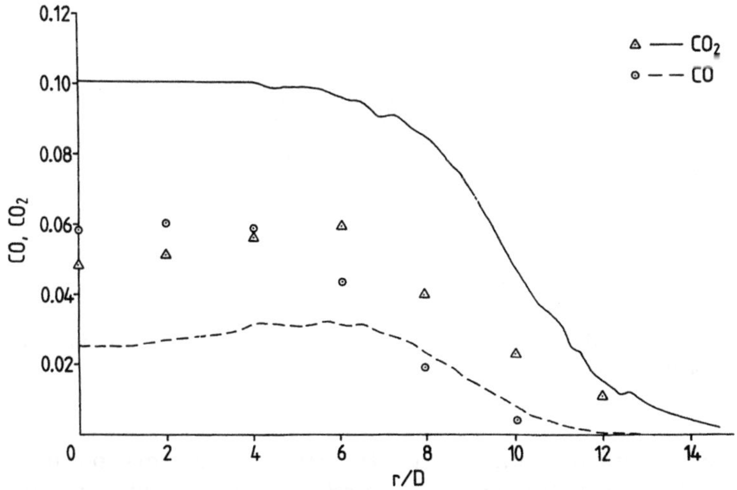

Fig. 10. Radial profiles of CO and CO_2 mole fractions at $x/D = 80$; ○, △ Measurements [17]

tion of the method indicates that it arises at least in the main part from limitations in the reaction kinetic scheme adopted; in fuel rich regions the scheme gives CO consumption rates which are too large[3]. The radial profiles of O_2 are in qualitative agreement with measurements, Figs. 7 and 8, though the calculated concentrations are too low. This feature is without doubt partly associated with the overprediction of CO_2 but perhaps also provides an indication that H_2 concentrations are calculated to be too low; the maximum computed H_2 concentration is about 1%. Radial profiles of mean temperature at the two axial locations are plotted in Figs. 11 and 12; there are no measurements available for comparison but maximum mean temperatures of around 1800 K appear plausible.

The solutions of the three scalar *pdf* equation allow the *pdf* of other scalar quantities to be constructed. To provide an illustration of what can be achieved, Fig. 13 shows the joint

[3] Even larger rates were observed in preliminary calculations using the reaction scheme of [7].

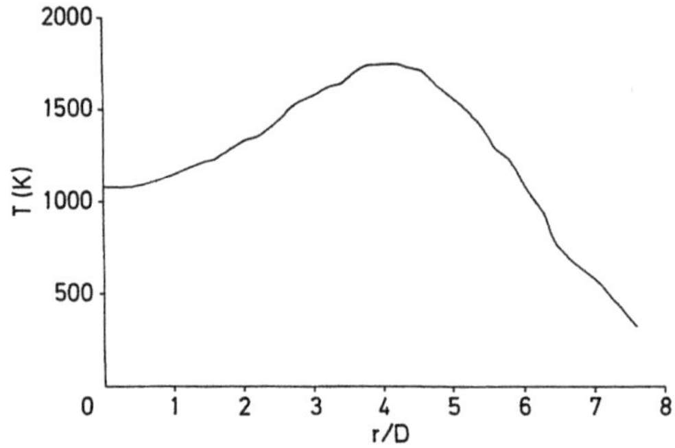

Fig. 11. Mean Temperature profile at $x/D = 40$

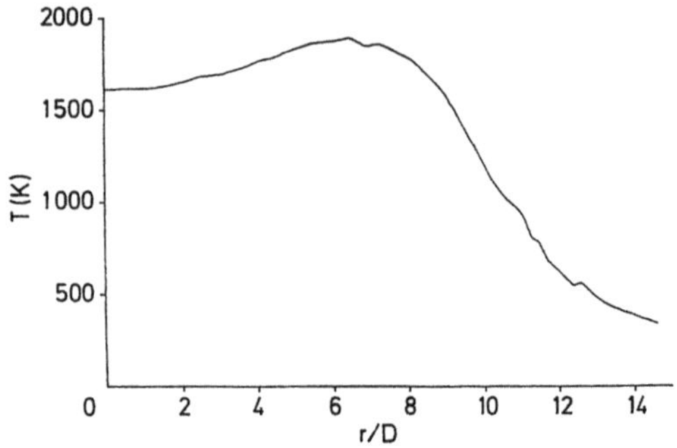

Fig. 12. Mean Temperature profile at $x/D = 80$

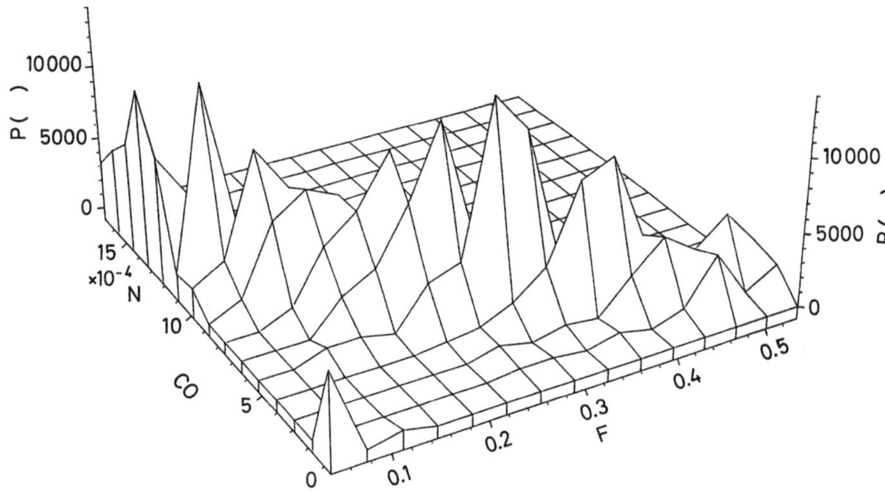

Fig. 13. Joint *pdf* of CO mole number and mixture fraction at $x/D = 40$, $r/D = 2.96$; mean mixture fraction, $\tilde{f} = 0.2534$, CO mole fraction $= 0.02133$

pdf of the CO mole number and mixture fraction at $x/D = 40$ and at a radial position corresponding to $r/D = 3$. It is clear that it cannot be described by any simple function though at this stage it is not possible to draw conclusions about its form.

Conclusions

A method for calculating the properties of turbulent hydrocarbon-air diffusion flames including the levels of CO and unburnt fuel has been described. Chemical reaction is described in terms of the mole number, mixture fraction and the fuel mole number. The *pdf* for these three scalars needed to determine mean values is then obtained from a Monte Carlo solution of its modelled transport equation. This *pdf* transport equation allows nonlinear chemical kinetic source terms to be treated exactly. The procedure has been applied to the calculation of a propane-air flame and the results compared with measurements. From these it appears that the mixture fraction profiles and the rate of disappearance of C_3H_8 are in acceptable agreement with the measurements though the calculated levels of CO and CO_2 are of order 50% in error. It can be concluded with a fair degree of certainty that this arises because of limitations in the reaction scheme used particularly for fuel rich mixtures. A likely explanation of the poor performance of the reaction scheme used is that in fuel rich regions the propane breaks down to form intermediate hydrocarbons which compete with CO to react with the hydroxyl radical so that CO oxidation is suppressed. If this is the case then it is clear that reaction cannot be described adequately by a global fuel breakdown reaction such as (i) involving reaction with O_2. Whether or not this difficiency can be remedied without increasing the number of scalars – by perhaps using alternative global fuel breakdown reactions – remains at this stage an open question. If not then the most obvious extension is to increase the number of independent scalars to four by the addition of, for example, an intermediate hydrocarbon or CO to the formulation.

Acknowledgement. This work was supported by DOE Grant AS03-76SF0034. Thanks are due to Miss J. Poggioli of Imperial College who typed the camera ready manuscript.

References

1. Bilger, R. W. (1976): Turbulent jet diffusion flames. Prog. Energy Comb. Sci. **1**, 87
2. Jones, W. P., Whitelaw, J. H. (1984): "Modelling and Measurements in Turbulent Combustion", in Proc. Twentieth Int. Symp. on Combustion, pp. 233–249, The Combustion Institute, Pittsburgh
3. Jones, W. P. (1980): "Models for Turbulent Flows with Variable Density", in *Prediction Methods for Turbulent Flow* (Ed by W. Kollmann, Hemisphere, New York)
4. Bilger, R. W. (1980): "Turbulent Flows with Non-Premixed Reactions", in *Turbulent Reacting Flows*, ed. by P. A. Libby, F. A. Williams, Topics in Appl. Phys., Vol. 44 (Springer, Berlin, Heidelberg)
5. Janicka, J., Kolbe, W., Kollmann, W. (1978): "The Solution of a *pdf*-Transport Equation for Turbulent Diffusion Flames", Proc. 1978 Heat Trans. Fluid Mech. Inst., Stanford University
6. Nguyen, T. V., Pope, S. B. (1983): Monte Carlo calculations of turbulent diffusion flames. Comb. Sci. Technol. **42**, 13
7. Westbrook, C. K., Dryer, F. L. (1981): Simplified reaction mechanisms for the oxidation of hydrocarbon fuels in flames. Comb. Sci. Techn. **27**, 31
8. Bilger, R. W., Starner, S. H. (1983): A double reaction zone model and perturbation analysis for finite rate kinetics of hydrocarbon fuelled combustion. AIAA Paper 83-0599
9. Kollmann, W., Janicka J. (1982): The probability density function of a passive scalar in turbulent shear flows. Phys. Fluids **25**, 1755
10. Keck, J. C. (1976): "Rate Controlled Constrained Equilibrium for Heating Reactions in Complex Systems", in *Maximum Entropy Formalism*, ed. by R. D. Levine, M. Tribus (MIT Press, Cambridge)
11. Baulch, D. L., Drysdale, D. D., Horne, D. G., Lloyd, A. C. (1972): *Evaluated Kinetic Data for High Temperature Reactions* (Butterworths, England)
12. Lundgren, T. S. (1967): Distribution functions in the statistical theory of turbulence. Phys. Fluids **10**, 969
13. Pope, S. B. (1981): A Monte Carlo method for the *pdf* equations of turbulent reactive flow. Comb. Sci. Technol., **25**, 159
14. O'Brien, E. E. (1980): "The Probability Density Function Approach to Reacting Turbulent Flows", in *Turbulent Reacting Flows*, Springer-Verlag
15. Jones, W. P., Launder, B. E. (1972): The prediction of laminarisation with a two-equation model of turbulence. Int. J. Heat Mass Trans. **15**, 301
16. Pantakar, S. V., Spalding, D. B. (1970): *Heat and Mass Transfer in Boundary Layers* (Intertext Books, London)
17. Godoy, S. (1982): "Turbulent Diffusion Flames"; Ph.D. Thesis, University of London

Modelling the Flow and Combustion in a Production Gas Turbine Combustor

J. Coupland and C. H. Priddin

Rolls Royce plc, P.O. Box 31, Derby DE3 8BJ, U.K.

Abstract

Most papers on flow modelling are concerned with simple laboratory flows. This paper demonstrates the application of current modelling methods to a production gas turbine combustor and demonstrates that the flow field, exit temperature and NOx emissions can be predicted to an acceptable accuracy even in such a complex flow. It is shown how the shape of the exit temperature pattern develops: any proposed modification to improve it can therefore be assessed prior to manufacture and test. Combustion chemistry is represented by a fluctuating equilibrium scheme, and the NOx formation rate is integrated over the fluctuations. The method uses a body-fitted orthogonal coordinate system which can be mapped into any axisymmetric internal flow region. The standard $k-\varepsilon$ model used in the calculations is adequate for the flow field, the only alteration necessary being that the value of the effective Prandtl number for fuel be reduced from the value used in boundary layer flows.

Nomenclature

A	Constant used in mesh generation method
a, b	Exponents in Beta function
$C_\mu, C_{\varepsilon 1}, C_{\varepsilon 2}$	Turbulence model constants
f	Fuel fraction
G	Generation rate of turbulence energy
g^{ij}	Metric tensor
k	Turbulence kinetic energy
$[N_2]$	Nitrogen molecule concentration
n	Number of nodes on combustor outline
$[O]$	Oxygen atom concentration
$P(f)$	Probability density function
p	Static pressure
$R(f)$	Nitric oxide production rate
S	Source term
T	Gas Temperature
U^i	Velocity vector
W	Complex coordinate in combustor plane
z	Complex coordinate in transformed plane
α_i	Interior angle of combustor outline at node i
Γ	Diffusion coefficient
ε	Turbulence energy dissipation rate
μ	Laminar viscosity
μ_T	Turbulent viscosity
ϱ	Density
σ_T	Turbulent Prandtl number
τ^{ij}	Stress tensor
ϕ	Scalar

Introduction

The history of turbulent shear flow calculations has been primarily concerned with a detailed analysis of relatively simple, predominantly two dimensional flows such as jets, wakes and boundary layers. It is understandable and quite right that this should be so, since simple flows are much easier to study, require considerably less computing effort to calculate, less data to measure and are therefore more suitable for turbulence model development. Most practical engineering equipment however is much more complex, particularly the gas turbine combustion chamber, which can contain upwards of 10,000 holes, swirlers, films, contracting ducts, liquid fuel, soot and strongly varying density. Faced with such complexity the tradi-

Turbulent Shear Flows 5

tional engineers view has been largely to abandon theoretical approaches and revert to the cut and try development approach.

Unfortunately this is no longer adequate in todays environment. The requirements placed on the combustor design are becoming increasingly stringent in terms of liner temperatures, exit temperature pattern, emissions, relight and many other criteria. In addition, there are great cost benefits to be gained by reducing the quantity of testing and shortening development timescales. There is therefore a pressing need for methods to be made available to the combustion development engineer by which the complex processes occuring in a combustor can be appreciated and the probable effects of modifications assessed at an early stage in the design.

This paper describes the PACE system now in use at Rolls-Royce for this purpose. Initiated some 8 years ago [1, 2] the system has been developed so that full combustors may be analysed. At the outset it was judged opportune to incorporate state of the art methods, turbulence models and combustion algorithms. For generality of application, a new coordinate system and mesh generator were needed, and in view of the axisymmetry of most current combustors the mesh chosen was an orthogonal body-fitted one in the $X - R$ plane, rotated about the engine axis for the circumferential coordinate.

There are a number of known shortcomings of the $k - \varepsilon$ turbulence model in relatively simple flows. For example, recirculation zone lengths in $2D$ flows are incorrectly predicted [3], different sets of constants for plane jets and round jets are required [4], and highly swirling flows require an anisotropic effective viscosity [5]. All these features (with many others) are present in a gas turbine combustor and it may have been assumed that there is therefore no hope of predicting such complex flows. This is not the case. In practice, simply because there is such a high degree of interaction between the various flow features, the peculiarities of any one is not able to dominate. The flow pattern set up is largely pressure driven, and so the turbulence model assumes a lesser though necessary role.

In addition to its use for complete component design and development the model can be used to investigate small scale features in more detail. The same code as used for the combustor described here has also been used for analysing cooling ring flows, annular ducts with heat transfer, diffusers, and many other flow configurations. A very flexible system therefore exists, of which this paper describes only one application.

The Numerical Model

Equations Solved

An early decision to be made in solving the flow equations is whether to work in Cartesian velocity components and simply transform the grid, an approach favoured by some, e.g. [6, 7], or to transform the velocity components and the grid together. The latter approach has been adopted here, for a number of reasons:

(i) Since the velocities are directed along the mesh lines, there will be regions of the flow where one component will be small or zero, and in these regions there will be little numerical diffusion.

(ii) The form of the resulting equations can still be cast in a control-volume approach with the convective velocities normal to the faces of each cell and stored on the face centres if the usual velocity staggering is employed. Conservation can therefore be guaranteed and proven solution algorithms employed.

(iii) Less storage is required, and the velocity tensor transforms naturally into all three directions.

In this paper general tensor notation is used, repeated indices implying summation. The complete set of Favre averaged conservation equations to be solved are:

$$(\bar{\varrho}\widetilde{U^i})_{,i} = 0 \qquad \text{(mass)}$$

$$(\bar{\varrho}\widetilde{U^i}\widetilde{U^j})_{,j} = -\, g^{ij}(\bar{p} + \tfrac{2}{3}\tilde{k})_{,j} + \tau^{ij}_{,j} \qquad \text{(momentum)}$$

$$(\bar{\varrho}\widetilde{U^i}\phi)_{,i} = [\Gamma_\phi(g^{ij}\tilde{\phi})_{,i}]_{,j} - (\bar{\varrho}\widetilde{u^i\phi})_{,i} + \bar{\varrho}\widetilde{S_\phi} \qquad \text{(scalar)}$$

where

$$\tau^{ij} = (\mu + \mu_T)\,(g^{kj}\widetilde{U^i_{,k}} + g^{ik}\widetilde{U^j_{,k}}).$$

The turbulent viscosity (μ_T) is derived from the $k - \varepsilon$ two-equation model by using

$$\mu_T = C_\mu \bar{\varrho}\tilde{k}^2/\tilde{\varepsilon}$$

with $\tilde{k}$ and $\tilde{\varepsilon}$ derived from their transport equations:

$$(\bar{\varrho}\widetilde{U^i}\tilde{k})_{,i} = [\Gamma_k(g^{ij}\tilde{k})_{,i}]_{,j} + G - \bar{\varrho}\tilde{\varepsilon}$$

$$(\bar{\varrho}\widetilde{U^i}\tilde{\varepsilon})_{,i} = [\Gamma_\varepsilon(g^{ij}\tilde{\varepsilon})_{,i}]_{,j} + C_{\varepsilon 1} G\tilde{\varepsilon}/\tilde{k} - C_{\varepsilon 2}\bar{\varrho}\tilde{\varepsilon}^2/\tilde{k}$$

where

$$G = -\,\bar{\varrho}\widetilde{u_i u_j}(g^{kj}U^i_{,k} + g^{ik}\widetilde{U^j_{,k}}).$$

The standard set of constants for the $k - \varepsilon$ model given in [8] have been used throughout.

Solution Technique

Each equation mentioned above is integrated over its microcontrol volume and the finite difference forms generated therefore satisfy the conservation properties for each solved equation. The result of this process leads to relationships between cell face values. Hybrid differencing is used to derive the algebraic relationships between grid nodes though improved higher order schemes are currently being evaluated.

The resulting linearised equations are solved by a line ADI technique which sweeps in a varied sequence through the solution domain. Each equation is solved for a specified number of complete sweeps before moving to the next equation. A pressure correction technique using the SIMPLE algorithm [9] is used to correct the velocity field so that continuity is satisfied.

Mesh Generation

In order to solve the system of equations in complex geometries a body fitted coordinate system is needed. An orthogonal coordinate system greatly reduces the number of terms in the equations, compared to a non-orthogonal system, but requires more work to generate. There are also problems in generating general $3D$ orthogonal meshes. Fortunately most aircraft gas turbine combustion flow problems are in geometries which are axisymmetric, e.g. main combustors, reheat systems, diffusers and annuli, although the flow itself is definitely three dimensional. It is therefore adequate for such problems to generate an orthogonal mesh in a two dimensional section and form the $3D$ grid by rotation about the engine axis. A typical grid for a segment of a combustor is shown in Fig. 1 a.

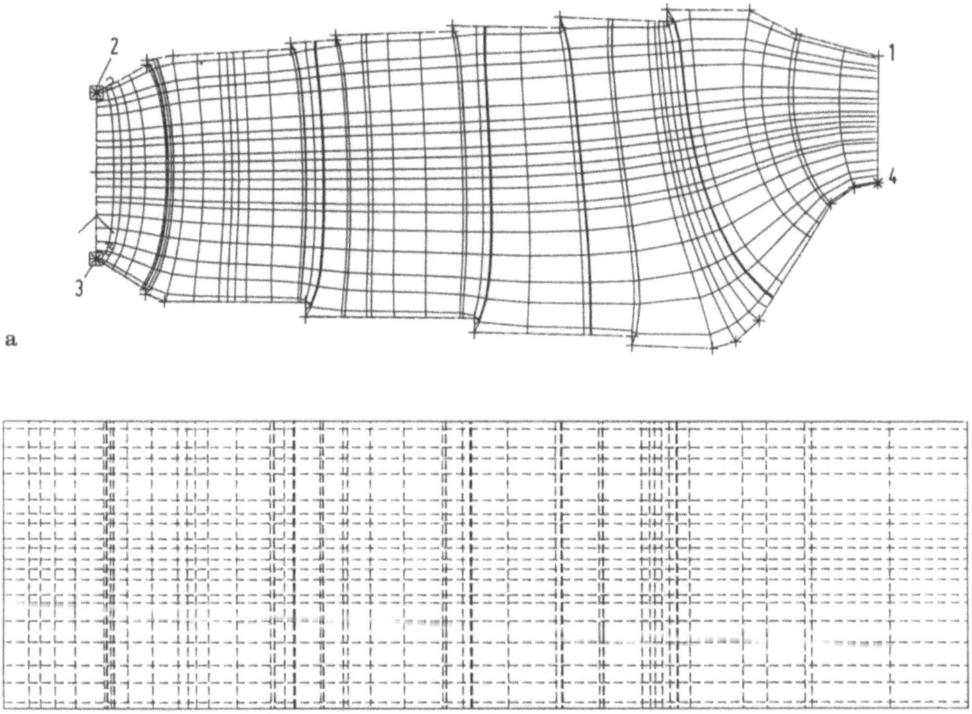

a

b

Fig. 1 a and b. Grid in combustor and transformed plane

To generate the mesh in two dimensions, two applications of the Schwarz-Christoffel transformation [10] are used. The first is used iteratively to find the node locations z_i in the expression below

$$W - W_0 = A \int_{z_0}^{z} \prod_{i=1}^{n} (z' - z_i)^{\alpha_i} \, dz' .$$

This relationship maps the upper half plane onto the interior of the combustor shape, and four of the nodes are selected to map the upper half plane into a rectangle. A rectangular mesh is then fitted inside the rectangle, Fig. 1 b, and this mesh mapped back onto the combustor. This transformation is a conformal mapping, and so the mesh intersection angles are preserved throughout, resulting in an orthogonal mesh in the combustor.

Combustion and Emissions Model

At the high inlet temperature and pressure conditions typical of gas turbine operation, combustion chemistry kinetics are very fast compared with mixing and residence times. It is therefore appropriate to assume that instantaneously, the gas composition is in equilibrium. This equilibrium state will fluctuate with time due to turbulence and therefore, because of the non-linearity of temperature and composition the mean temperature will not be the temperature evaluated at the mean fuel-air mixture.

The equilibrium state can be characterised by any single scalar property of the mixture: in PACE the total fuel fraction f is used. Transport equations are solved for $\tilde{f}$ and its

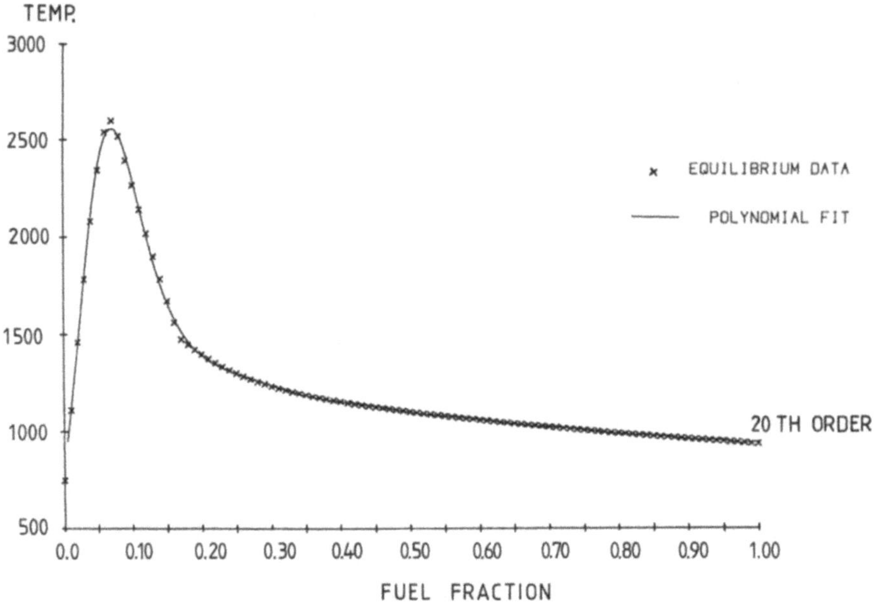

Fig. 2. Variation of temperature with fuel fraction

variance, giving $\widetilde{f^2}$. At any location the probability density function $P(f)$ is generated from $\tilde{f}$ and $\widetilde{f^2}$ assuming that $P(f)$ has a Beta function shape, thus

$$P(f) = f^a(1-f)^b / \int_0^1 f^a(1-f)^b \, df.$$

All other required parameters are determined as a function of f from a program based on the Gordon and McBride equilibrium method [11] and then fitted as a polynomial of high order (typically 20). Figure 2 illustrates how temperature typically varies with fuel fraction.

$P(f)$ is a density-weighted function by definition, since f is a density weighted mean. Any other density-weighted mean (say ϕ) can therefore be found simply from

$$\tilde{\phi} = \int_0^1 P(f)\, \phi(f)\, df$$

and unweighted means where required, by

$$\bar{\phi} = \bar{\varrho} \int_0^1 P(f)\, [\phi(f)/\varrho(f)]\, df.$$

The above is used with $\phi = 1$ to evaluate the mean fluid density.

The benefits of using the Beta function in particular to describe the fluctuating composition are twofold:

(i) Physically, only values of f in the range 0 to 1 have any meaning, and the Beta function ensures this without any artificial clipping.
(ii) Numerically the integrals required can be computed analytically provided polynomial functions are used.

The approach outlined above is adequate for most combustion parameters but kinetic effects must be considered to calculate NO concentration as its rates of formation are much lower than those of the other species. It is assumed in this code that the NO concentration itself is uncorrelated with f, and so only an equation for NO is solved, not its variance. The rate of formation however is a strong function of f, and so, using the extended Zeldovich mechanism

$$N_2 + O \rightleftharpoons NO + N,$$
$$N + O_2 \rightleftharpoons NO + O,$$
$$N + OH \rightleftharpoons NO + H,$$

the source term is calculated as above from

$$S_{NO} = \int_0^1 P(f)\, R(f)\, df,$$

where $R(f)$, the instantaneous formation rate of NO for the above reactions, is derived using a steady state approximation for the nitrogen atom concentration and the rate constants given in [12]. This gives

$$R(f) = 2 * 7.6 * 10^{10} [N_2]\,[O]\, e^{-38000/T}.$$

The backward rate is set so that at equilibrium the net source of NO is zero.

Application to Combustors

Boundary Conditions

Using the model described above, predictions have been made for the annular combustor of one variant of the Rolls-Royce RB 211 gas turbine combustor, Fig. 3. The combustor has 18 fuel injectors and so just a single 20 degree sector of the flame tube was modelled in the calculations. Cyclic symmetry conditions were applied on the circumferential end planes.

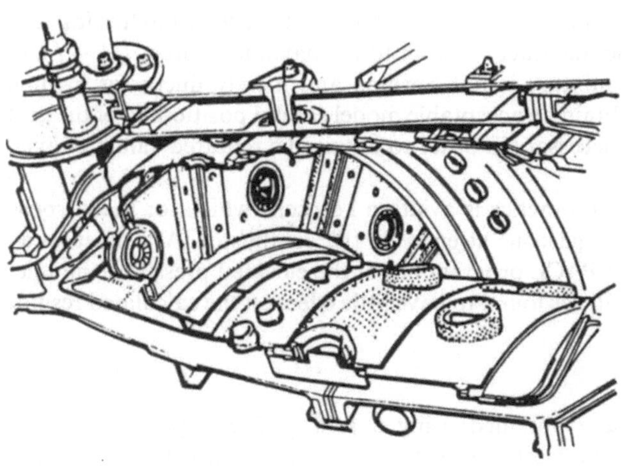

Fig. 3. Gas turbine combustor

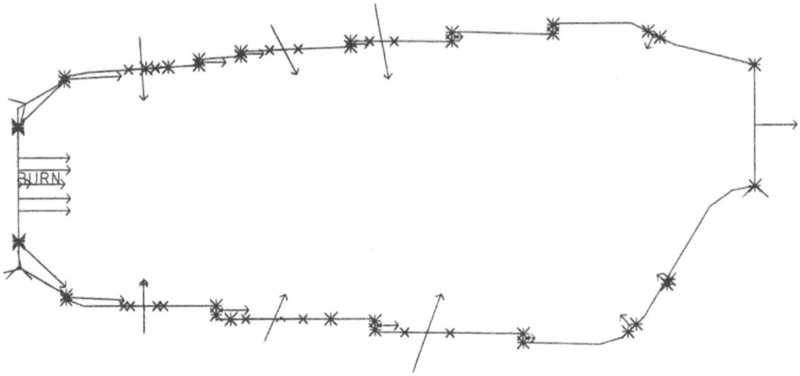

Fig. 4. Combustor boundary flows

Using specified compressor delivery mass flow, temperature, pressure and empirical diffuser loss coefficients a one-dimensional program (CODAS) predicts the feature mass flow splits and injection angles for the specified combustor and annulus geometry.

These mass flows are then used to calculate the velocities on the solution region boundary. A uniform profile of velocity is assumed across each feature. Accurate modelling of the flow through circular holes on the combustor walls, e.g. the primary and dilution zone air flows, is achieved by small adjustments to the gas density in the holes in order to obtain the correct mass and momentum flux.

Figure 4 shows the boundary flows considered in the calculation and illustrates the complexity of the problem. In all there are 42 separate inlet flow features. As will be seen later it is important to model these boundary flows correctly.

Boundary conditions for the turbulence energy were calculated using a turbulence intensity of 1% of the total velocity through a hole and for the dissipation rate using a length scale proportional to the diameter of the hole or the height of the feature. As will be seen later the turbulence generated internally is far greater than that convected in across the boundary, and therefore the results are insensitive to these assumptions. A zero gradient condition is applied at the exit to the combustor.

Fuel

Fuel enters the flame tube through the injector on the baseplate of the combustor. The injector is located centrally in the sector and is modelled by two concentric annuli with specified values of axial velocity, swirl velocity (relative to the injector itself), and fuel fraction in each annulus. The variance of the fuel fraction is set to its maximum value, $f(1-f)$, in the injector. This implies that the mixture is not burning and merely fluctuates in time between pure air and pure fuel. To obtain an acceptable model careful positioning of the grid lines over the fuel injector annuli is necessary in order to obtain the required values of axial and angular momentum flux.

The fuel used by the combustor in practice is aviation kerosene which for the purpose of these calculations was taken to be gaseous at entry to the combustor. A droplet spray evaporation model is included in the PACE program but at the elevated pressures and inlet temperatures mainly of interest here it was found to have little effect on the calculated results.

Grid Dependence

All the results presented below were obtained using a grid of 35322 points (58 axial $\times$ 21 radial $\times$ 29 circumferential). The mesh over the $X-R$ plane is shown in Fig. 1. The

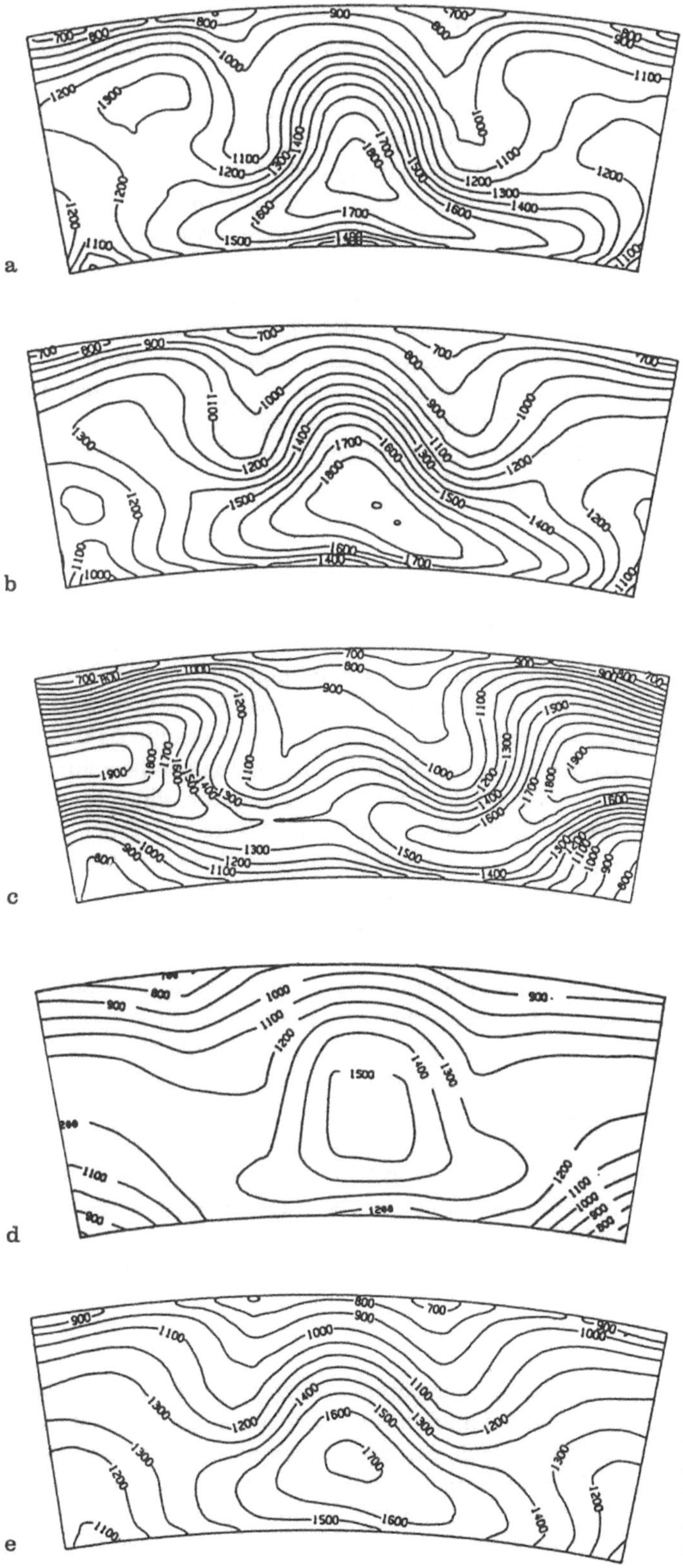

Fig. 5a–c. Exit plane temperature distributions; **a** Fine grid; **b** Coarse grid; **c** Full size dilution holes; **d** Measured; **e** Predicted – $\sigma_T = 0.5$

circumferential mesh lines were evenly spaced except for a slightly finer arrangement around the fuel injector. The mesh line positions were chosen in order to resolve all the boundary flows adequately.

With this mesh a single iteration of the calculation cycle took 69 CPU seconds on a IBM3081 machine. The number of iterations required, starting from a previous solution, to obtain convergence, with residuals of around 0.0005 was typically 150.

A further solution was obtained for the datum condition using a finer grid of 44330 points, most of the mesh refinement being in the axial direction. Figures 5a and b compare the predicted exit temperature distributions obtained using the fine and coarse mesh, respectively. It is seen that the peak temperature at exit is reduced on the fine mesh but that overall the distribution is very similar. It should be noted, however, that the differences observed are also partly due to the different representation of the boundary flows, in particular of the circular holes on the inner and outer walls of the solution region. Jet penetration distances depend to some extent on the grid representation of the hole, and it is therefore most important to choose a grid such that this effect is minimised. Since the coarser mesh was considered acceptable from this viewpoint then, in the interests of economy, all further predictions were made using that mesh.

Computer System

The methods described in this paper are embodied in a computer system called PACE (Prediction of Aerodynamics and Combustor Emissions) developed by Rolls-Royce plc, which comprises five main elements. The preliminary design and analysis suite CODAS is used first to estimate the proportions and directions of flow entering the solution domain, and the geometry defined in this suite is passed to the PACE mesh generator, which produces the meshes seen on Fig. 1. The flows are then read from CODAS into the PACE input processor which converts them into velocity components distributed along the mesh lines, and allows the user to set up the problem definition required (cold/burning, NOx, other variables). It also creates a solution file which is initially empty apart from the boundary values.

The main solution is performed by a batch program which requires no other input – all necessary data has already been provided by the other programs. It is a totally flexible $2D/3D$ code, with dynamic storage allocation for all variables: for the problem reported here, eight primary variables were solved, and all primary and secondary variables are filed for subsequent display.

The display program is the final element in the suite, and this produced the results shown on Figs. 5 to 9. Displays of streaklines, contours and profiles can be drawn on any surface defined by a coordinate value or an X, Y or theta value. Particle tracks in $3D$ and histograms are also available, providing a very flexible interactive facility.

Results

The results to be presented cover the range of experimental and operating combustor pressures. There is unfortunately only a limited amount of experimental information available on any full engine combustor, and this configuration is no exception. Water flow model results are available to give the internal flow pattern, but with combustion the only measurements relevant to this exercise are the flame tube exit temperature pattern and the mean levels of various pollutants, e.g. NOx, carbon monoxide and unburnt fuel. No internal measurements of velocity or turbulence information are available though laser diagnostic methods are currently being applied to verify boundary conditions for these calculations.

Validation Against Water Models

Combuster flow patterns are investigated experimentally using a water flow perspex model of the flame tube/annulus/diffuser system. For comparison with the observed flow patterns a non-combusting atmospheric pressure prediction was obtained. Figures 6a and b give a typical comparison, and the agreement, especially in terms of jet penetration and primary zone flow, is quite good. The predicted flow pattern is visualised using streaklines starting from each mesh point.

Effect of Combustion

If water model testing is to be accepted as a valid tool then it is important that the observed flow patterns on the model should resemble those present in the real combustor at operating pressures. This has been investigated using the numerical model and Fig. 6c shows the flow

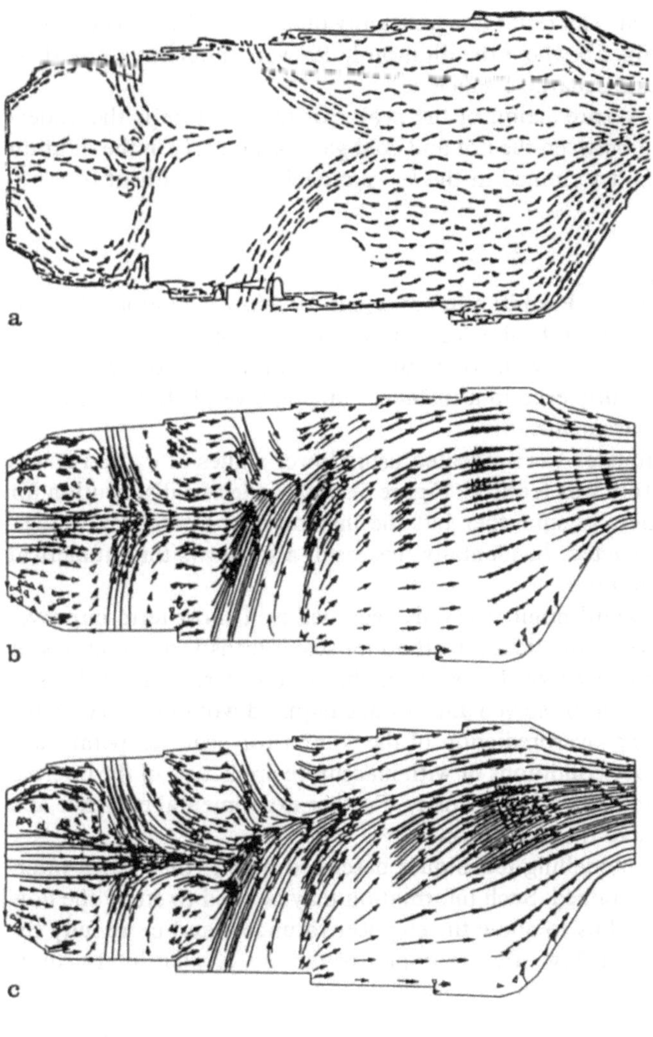

Fig. 6a–c. Combustor flow patterns; **a** Oberseved – Water flow model; **b** Predicted – Cold; **c** Predicted – Combusting

pattern on the same plane obtained from a solution at higher pressure, 10 atmospheres, and with combustion. Comparison of this pattern with Fig. 6b indicates little change due to combustion. This result supports the use of water flow testing as a means of developing gas turbine combustors.

Comparison with Rig Temperature Data

The gas temperature distribution at exit from the combustor is of particular importance, contributing to the design of the turbine blades and vanes for long service life.

Figures 5b and d give the predicted and measured temperature distributions. The measured result is a mean of 18 sectors from a full scale test at 10 atmospheres pressure using a traversing thermocouple rake. Comparison of the two figures shows good qualitative agreement in terms of the position of the peak temperature profile though the maximum temperature point is seen to be too close to the inner wall. Quantitatively the agreement is not so good in that the peak temperature across the whole sector is overpredicted.

To illustrate the importance of modelling the boundary flows correctly Fig. 5c gives the combustor exit temperature distribution for a solution where the dilution holes were modelled using their true geometric area. The result is significantly different to the measurement, Fig. 5d, and to the solution on Fig. 5b where the dilution holes were modelled using their effective areas, i.e. true area times discharge coefficient.

One possible reason for the overprediction of the peak exit temperature is the underprediction of the turbulent mixing between the hot and cold gas, particularly in the dilution zone. The degree of this mixing is determined in the $k - \varepsilon$ model by the gradient of

$$- \mu_T/\sigma_T \, (g^{ij} f)_{,i}$$

where the standard value for turbulent Prandtl number, σ_T, is 0.9. Now in complex flows it is known that the assumption of a constant value for σ_T is not strictly valid. It was suggested in [13] that for recirculating flows a lower value of σ_T than that required in boundary layer flows is more appropriate, and recently even in a turbulent mixing layer, [14], values of 0.4 were deduced from detailed measurements.

The calculations for this combustor were repeated using lower values of σ_T and it was found that the peak exit temperature was reduced. Figure 5e shows the exit distribution for $\sigma_T = 0.5$. The distribution is generally smoother but the maximum temperature, though reduced, is still somewhat overpredicted. Incidentally this result also shows that numerical diffusion is not dominating the results.

The reduction in turbulent Prandtl number is only regarded as a short term improvement to the solution and an improved prediction method for scalar transport in combustor flows is needed. Higher order turbulence model closures are being considered but for this size of problem the additional equations to be solved and storage required would be very costly.

It is possible that the remaining overprediction of the combustor peak temperature is associated with inaccurate boundary conditions or with inaccurate prediction of the internal turbulence levels and the associated turbulent viscosity. Further experimental evidence and modelling work is needed to resolve these questions.

One significant benefit of the modelling techniques described here is that much intermediate information can also be obtained. Such information may be used to study the flow process within the combustor and thus to allow the engineer to understand more fully the effect of changes in geometry and airflow. Two examples of the use of such intermediate information are given.

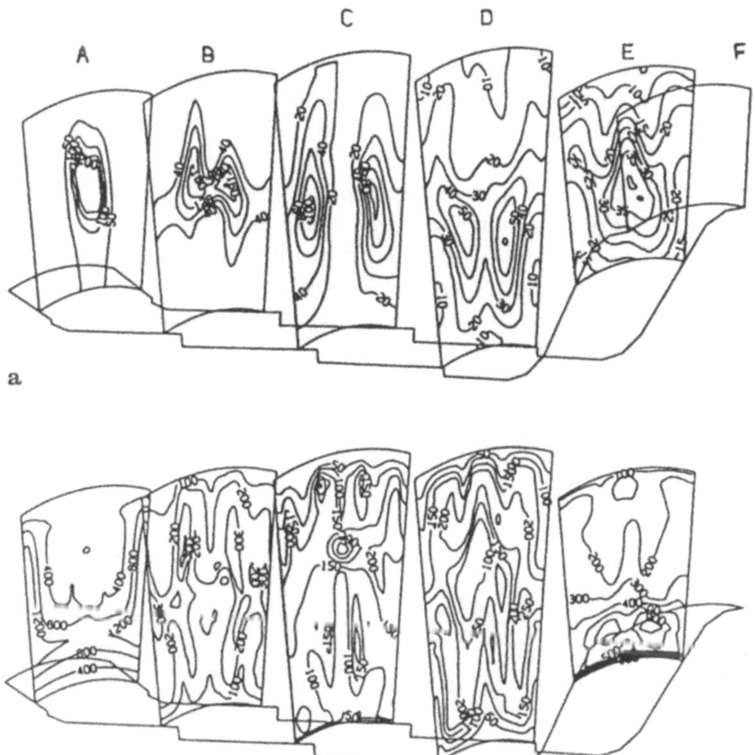

a

b

Fig. 7. Predictions at various axial stations; **a** Fuel fraction * 1000; **b** Turbulence energy (m^2/s^2)

Figure 7 a demonstrates how the exit pattern develops through the combustor by showing contours of fuel fraction, multiplied by 1000, on various axial planes along its length. Fuel fraction rather than temperature is considered as temperature distributions do not distinguish rich and weak zones. On plane A the central fuel fraction peak is associated directly with the fuel injector flow. On plane B the fuel pattern is determined by the flows from the three holes on both inner and outer walls. On plane C the fuel pattern is split by the two central dilution jets on each wall. On plane D the two peaks of fuel fraction are brought together mainly by the inner wall dilution jets on the edge of the solution region. Between planes D and E no further dilution air is introduced and hence the fuel fraction pattern is smoothed out by mixing, a process that continues right up to the exit plane, plane F. Note the similarity between the fuel fraction pattern on plane E and the predicted exit temperature distribution, Fig. 5 e.

Figure 7 b uses the same view of the combustor to illustrate the predicted internal levels of turbulence kinetic energy, k. The high levels seen near the inner wall on plane A are associated with stagnation regions between the cooling flows on the baseplate and the recirculating flow in the primary zone. Note that the primary zone recirculation in this combustor is in a plane normal to Fig. 6 and so is not evident in these flow patterns. The generally high levels seen throughout the combustor indicate the insensitivity of the prediction to the assumed turbulence levels in the boundary flows.

NOx Emissions

The emitted levels of nitrogen oxides (NOx) are greatest at high pressure where the chemical reaction rates are greatest. A high pressure solution, around 27 atmospheres, for the combustor flow has thus been used to obtain a prediction for the NOx emissions from this combustor

Figure 8 shows the predicted growth of mean NOx emission index, expressed as gm NO_2/kg fuel (emission index in EPA/SAE nomenclature), along the combustor length. Interestingly the curve is approximately linear which indicates that nitric oxide is being produced almost uniformly along the combustor until the exit nozzle is reached.

This point is further illustrated on Fig. 9 which shows the distribution of the mean value of the NO formation rate within the combustor.

The measured level of NOx emission index for this combustor at these conditions is 45 which compares favourably with the predicted mean exit value of 46. No adjustments have been made to published rate constants which were taken from [12].

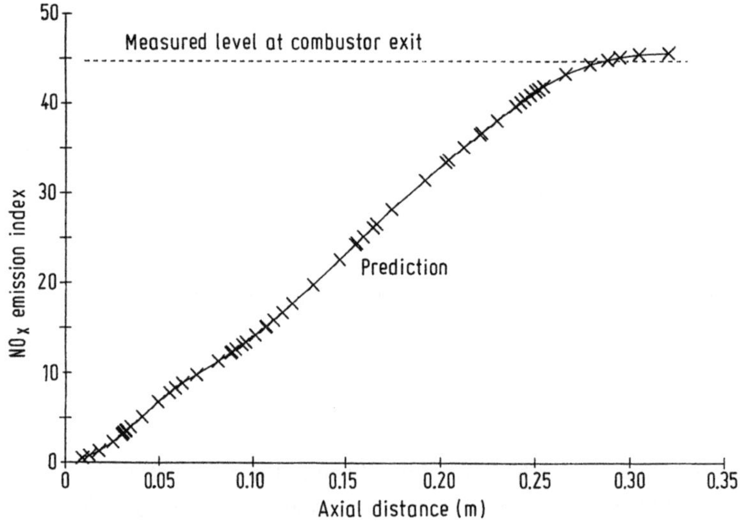

Fig. 8. Axial growth of NOx emission index

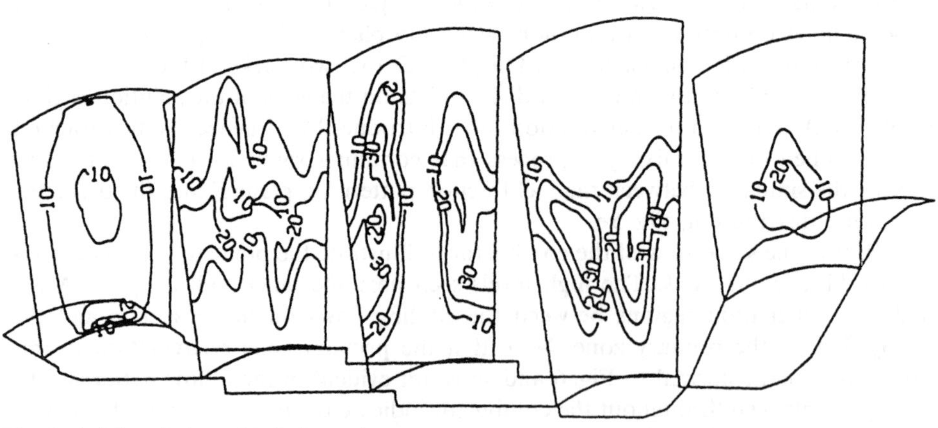

Fig. 9. Axial variation of NO formation rate

Conclusions

From this numerical study of the turbulent flow and combustion in a production gas turbine combustor the following conclusions may be drawn:

(i) Comparisons of predictions with the limited experimental data available are very reasonable.

(ii) A smaller value of turbulent Prandtl number than standard is needed to correctly predict the lateral spread of fuel and hence combustor exit temperature distribution, a value close to 0.4 would seem optimum.

(iii) The assumption of chemical equilibrium for the major species and temperature, together with finite rate effects for pollutant prediction, gives acceptable results.

(iv) The use of a body-fitted finite difference grid provides the ability to predict the flow in many complex flow situations found in practical engineering problems.

(v) The $k - \varepsilon$ model, in spite of its limitations, is acceptable for engineering calculations of practical gas turbine combustion systems.

References

1. Jones, W. P., Clifford, W. C., Priddin, C. H., Chair, R. de (1978): A comparison between predicted and measured species concentrations and velocities in a research combustor. High Temperature Problems in Gas Turbine Engines. AGARD Conf. Proc. **229**
2. Jones, W. P., Priddin, C. H. (1978): Predictions of the Flow Field and Local Gas Composition in Gas Turbine Combustors, Proceedings of the 17th Symposium on Combustion, Combustion Institute, 399–409
3. Sturgess, G. J., Syed, S. A.: (1985): Calculation of Confined Swirling Flows, AIAA 23rd Aerospace Sciences Meeting, AIAA-85-0060
4. Rodi, W. (1972): The Prediction of Free Turbulent Boundary Layers by Use of a Two-Equation Model of Turbulence; Ph.D. Thesis, University of London
5. Boysan, F., Swithenbank, J. (1982): Fundamental mathematical modelling approach to cyclone design. Trans. Inst. Chem. Eng. **60**, 222–230
6. Shyy, W., Correa, M., Tong, S. S. (1984): Demonstration of a New Body-Fitted Coordinate Code for Modelling Gas Turbine Combustor Flows, AIAA/SAE/ASME 20th Joint Propulsion Conference, Cincinnati, Ohio
7. Moore, J., Moore, J. B. (1984): Calculation of a horseshoe vortex flow without numerical mixing. ASME 84-GT-241
8. Launder, B. E., Spalding, D. B. (1974): The numerical computation of turbulent flows. Comp. Meth. Appl. Mech. Eng. **3**, 269–289
9. Patankar, S. V. (1979): *Numerical Heat Transfer and Fluid Flow*, (Hemisphere, London)
10. Anderson, O. L., Davis, R. T., Edwards, D. E., Hawkins, G. B. (1982): Solution of Viscous Internal Flows on Curvilinear Grids Generated by the Schwarz-Cristoffel Transformation, Symposium on the Numerical Generation of Curvilinear Coordinate Systems, Mississippi State University
11. Gordon, S., McBride, B. J. (1971): Computer program for calculation of complex chemical equilibrium compositions. NASA SP273
12. Baulch, D. L., Drysdale, D. D., Horne, D. G., Lloyd, A. C. (1973): *Homogenous Gas Phase Reactions of the H2-N2-O2 System*, Evaluated Kinetic Data for High Temperature Reactions, Vol. 2, (Butterworths, London)
13. Sturgess, G. J., McManus, K. R. (1984): Calculations of Turbulent Mass Transport in a Bluff-Body Diffusion-Flame Combustor, AIAA 22nd Aerospace Sciences Meeting, AIAA-84-0372
14. Chambers, A. J., Antonia, R. A., Fulachier, L. (1985): Turbulent Prandtl number and spectral characteristics of a turbulent mixing layer. Int. J. Heat Mass Trans. **28**, 1461–1468

Convection with a First-Order Chemically Reactive Passive Scalar

J. R. Herring and J. C. Wyngaard

National Center for Atmospheric Research [1], Boulder, C0 80307, USA

Abstract

Convection between horizontal, stress-free, perfectly conducting plates is examined in the turbulent regime for air. Results are presented for an additional scalar undergoing first order decay. We discuss qualitative aspects of the flow in terms of spectral and three-dimensional isosurface maps of the velocity and scalar fields. The horizontal mean profiles of scalar gradients and fluxes agree rather well with simple mixing-length concepts. Further, the mean profiles for a range of the destruction-rate parameter are shown to be nearly completely characterized by the boundary fluxes. Finally, we use the present numerical data as a basis for exploring a generalization of eddy-diffusion concepts that incorporates necessary non-local effects.

Nomenclature

Note: Variables u and p are non-dimensionalized by separation distance between plates and thermal diffusivity. (T, C) are non-dimensionalized by their mean change across the depth. The relation of non-dimensional variables to their dimensional counterparts (indicated by ′) is given in (5).

$C(x, y, z, t)$	Concentration of decaying scalar
$\langle C\rangle$	Horizontal (or ensemble) average of $C = \langle C\rangle (z, t)$
$[C]$	Volume average (over periodic domain) of C
D	Depth of convective layer
E_v	Kinetic energy spectrum
$\boldsymbol{g}$	Acceleration of gravity
$\hat{g}$	Unit vector along $\boldsymbol{g}$
$\boldsymbol{k}$	Wave number vector
k_0	Lowest $\boldsymbol{k}$ in numerical simulation
Nu	Heat flux through lower plate in units $\kappa \Delta T/D$
$Nc(0)$	Scalar flux through lower boundary; units are same as for Nu
$Nc(1)$	Scalar flux through top plate; units are same as for Nu
p	Pressure field
$\boldsymbol{r}$	Position vector
Ra	Rayleigh number $g\alpha(T(0) - T(1))/(\kappa\nu D^3)$
R_c	Critical Ra for slip boundaries $27\pi^4/4$
R_λ	Taylor microscale Reynolds number $= \langle w^2\rangle/\langle(\partial w/\partial z)^2\rangle^{1/2}\nu$
$T(x, y, z, t)$	Temperature
$\langle T\rangle$	Horizontal (or ensemble average) of $T = \langle T\rangle (z, t)$
$T(0)$	Lower plate temperature
$T(1)$	Top plate temperature
$\boldsymbol{u}(x, y, z, t)$	Velocity field (x, y, z)
$u(x, y, z, t)$	x-component of velocity field
$v(x, y, z, t)$	y-component of velocity field
$w(x, y, z, t)$	z-component of velocity field
α	Thermal expansivity
β	Horizontal average of $-\partial T/\partial z$
β_C	Horizontal average of $-\partial C/\partial z$
γ	Ratio of diffusivities for C and T, respectively
ΔT	Temperature excess of lower plate over upper plate
Z	Reacting scalar variance spectrum
ζ	Fluctuation of C from its horizontal average
ε	Fractional destruction rate of C in units of the thermal diffusion time
κ	Thermometric diffusivity
κ_c	Diffusivity for scalar C
ν	Kinematic viscosity
ϕ_1	Toroidal velocity variance; see (6a)

[1] The National Center for Atmospheric Research is sponsored by the National Science Foundation

Turbulent Shear Flows 5

S_w	Velocity derivative skewness $[(\partial w/\partial z)^3]/[(\partial w/\partial z)^2]^{3/2}$	ϕ_2	Poloidal velocity variance; see (6b)
$S_{T,C}$	Scalar mixed skewness $[(\partial A/\partial z)^2 (\partial w/\partial z)]/[(\partial A/\partial z)^2] \cdot [(\partial w/\partial z)^2]^{1/2}$, $A = (C, T)$	σ	Prandtl number (ν/κ)
		θ	Fluctuation of T from its horizontal average
(x, y, z)	Cartesian components of $\boldsymbol{r}$	Θ	Temperature variance spectrum
t	Time	ω	Vorticity $\nabla \times \boldsymbol{u}$

Statement of Problem

We consider a decaying passive scalar $C(\boldsymbol{x}, t)$ in a convective layer, bounded by horizontal infinite conductive plates mounted perpendicular to gravity, $\hat{g}$. The equations of motion are the Boussinesq set, written here in a convenient dimensionless form:

$$\nabla \cdot \boldsymbol{u} = 0, \qquad (\boldsymbol{u} = u, v, w), \tag{1}$$

$$(\partial/\partial t - \sigma \nabla^2)\, \boldsymbol{u} = -\nabla p - \boldsymbol{u} \cdot \nabla \boldsymbol{u} + \sigma \hat{g}\, Ra\, T, \tag{2}$$

$$(\partial/\partial t - \nabla^2)\, T = -\boldsymbol{u} \cdot \nabla T, \tag{3}$$

$$(\partial/\partial t - \gamma \nabla^2 + \varepsilon)\, C = -\boldsymbol{u} \cdot \nabla C. \tag{4}$$

The fields $(\boldsymbol{u}, T, C)$ and coordinates $(\boldsymbol{x}, T)$ are related to their dimensional counterparts $(\boldsymbol{u}', T', C')$ and $(\boldsymbol{x}', t')$ by

$$\{u', T', C', x', t'\} = \{[\kappa/D]\, \boldsymbol{u}, [\Delta T']\, T, [\Delta C']\, C, [D]\, \boldsymbol{x}, [D^2/\kappa]\, t\} \tag{5}$$

$\Delta T = T'(1) - T'(0)$. Boundary conditions at $z = (0, 1)$ are $w(0) = w(1) = 0$; $T(0) = C(0) = 1$; $T(1) = C(1) = 0$. We employ slip boundaries, for which $\partial^2 w/\partial x^2 = 0$, at $z = (0, 1)$. Our goal is to characterize ensemble-mean profiles of temperature $\langle T(\boldsymbol{x}, t)\rangle \equiv \langle T\rangle (z, t)$ and of scalar $(\langle C(\boldsymbol{x}, t)\rangle \equiv \langle C\rangle (z, t))$ fields, and the corresponding fluxes, and their dependence on the dimensionless parameters, $(Ra, \sigma, \gamma, \varepsilon)$. In addition, we inquire as to the validity of eddy transport concepts, and their possible valid generalizations in characterizing the dependence of the fluxes on mean gradients $\{\partial\langle T\rangle (z, t)/\partial z, \partial\langle C\rangle (z, t)/\partial z\}$.

Numerical Results and Interpretations

First we present results of a direct numerical simulation of (1–4). We examine in some detail results for $Ra = (70{,}560)\, Rc$, using $(32 \times 32 \times 32)$ resolution, and an aspect ratio $\sqrt{2}$.

Time series for the fields (w, T, C) at a point near the mid-point of the flow for $Ra = (70{,}560)\, Rc$ are shown in Fig. 1a, b. The higher Ra run was initialized by rapidly increasing Ra at $t = 2.75$ from 70 to 560. We note that both cases are turbulent, with the higher Ra exhibiting more rapid temporal fluctuations. The fields (W, T, C) appear strongly correlated despite rather vigorous temporal activity. Perhaps better measures of turbulent activity are the vertical vorticity ω_z and the velocity-derivative skewness S_w, the latter a volume averaged quantity (see the Nomenclature Section for definitions). These are shown in Fig. 2, this time without a change of time-scale. A value $S_w = -0.5$ is typical of $R_\lambda \cong 40$ experimental flows.

Spectra (computed as isotropic unit-interval wave number shell-averages) for kinetic energy, toroidal component $\phi_1(k)$ poloidal component $\phi_2(k)$, temperature variance spectrum

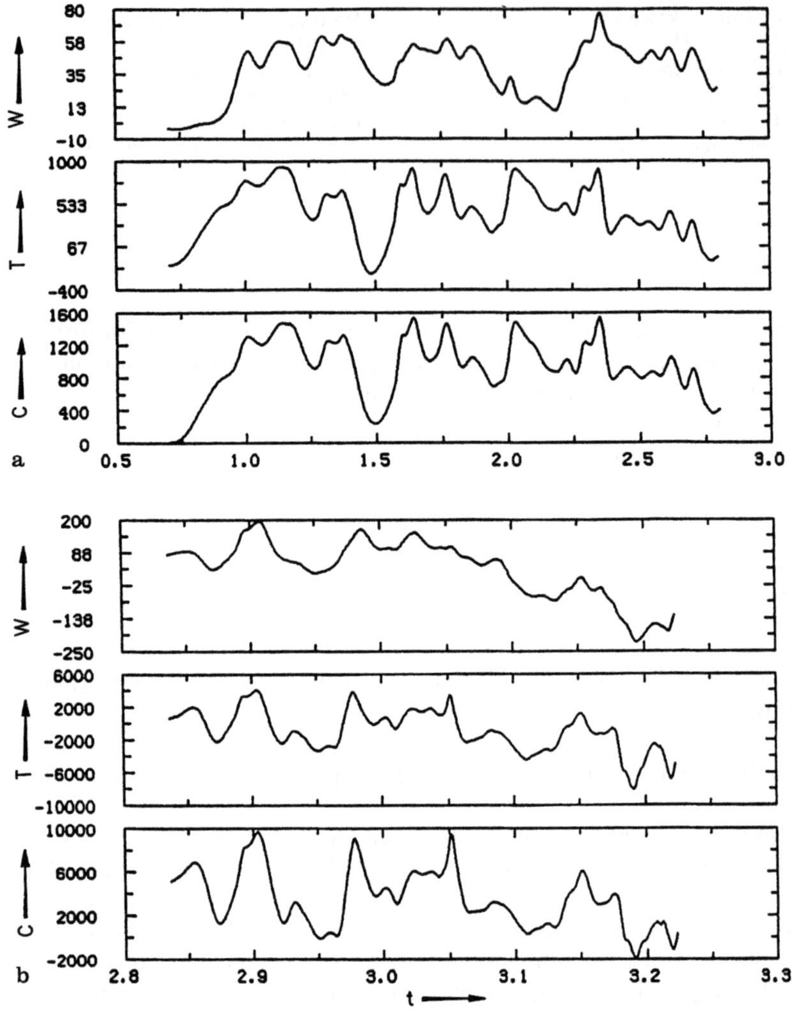

Fig. 1 a and b. Time series for (W, T, C) at a point near of the center of the periodic box, for **a** $Ra = 70\,Rc$, **b** $Ra = 560\,Rc$. Note change of time-scale for **b**, for which only a part of computed time series is shown

$\Theta(k)$, and C-variance $Z(k)$ are shown in Figs. 3 and 4, for the two Ra, and in the stationary state. Spectra ϕ_1, ϕ_2 are defined by

$$\phi_1(k) = \tfrac{1}{2}\,\langle |(\hat{g} \times \boldsymbol{k}) \cdot \boldsymbol{u}(\boldsymbol{k})|^2 \rangle / |\hat{g} \times \boldsymbol{k}|^2, \tag{6a}$$

$$\phi_2(k) = \tfrac{1}{2}\,\langle |\boldsymbol{k} \times (\hat{g} \times \boldsymbol{k}) \cdot \boldsymbol{u}(\boldsymbol{k})|^2 \rangle / |\boldsymbol{k} \times (\hat{g} \times \boldsymbol{k})|^2, \tag{6b}$$

so that

$$\phi_1(k) + \phi_2(k) = E_v(k).$$

$\phi_2(k)$ is the only component directly driven by buoyancy, and $\phi_1(k)$ contains all the vertical vorticity. For the present problem, ϕ_1 is entirely excited by nonlinearity. For isotropic turbulence $\phi_1 = \phi_2$. The turbulence at $Ra = 70\,Rc$ is clearly low Reynolds number ($R_\lambda = 18$), but that at $Ra = 560\,Rc (R_\lambda = 28)$ has a suggestion of a Kolmogorov spectrum.

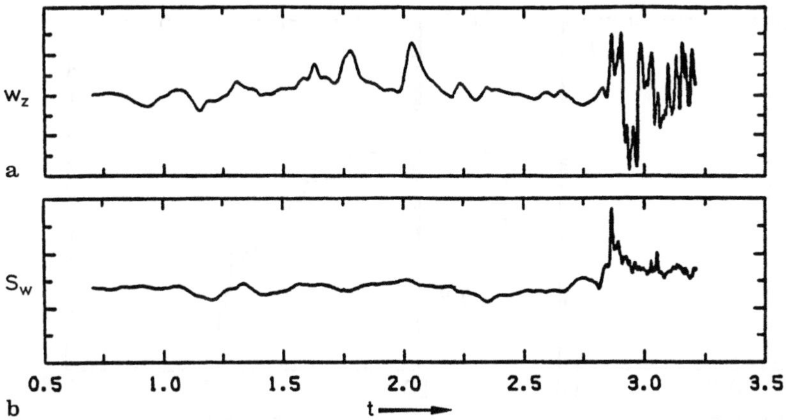

Fig. 2. Time series for (a) $\omega_z(t)$ and (b) skewness factor $-S_w(t)$ (see Nomenclature for definition). At $t = 2.75$, *Ra* is rapidly switched from 70 *Rc* to 560 *Rc*. Only a portion of the computed $Ra = 560\ Rc$ time series is shown

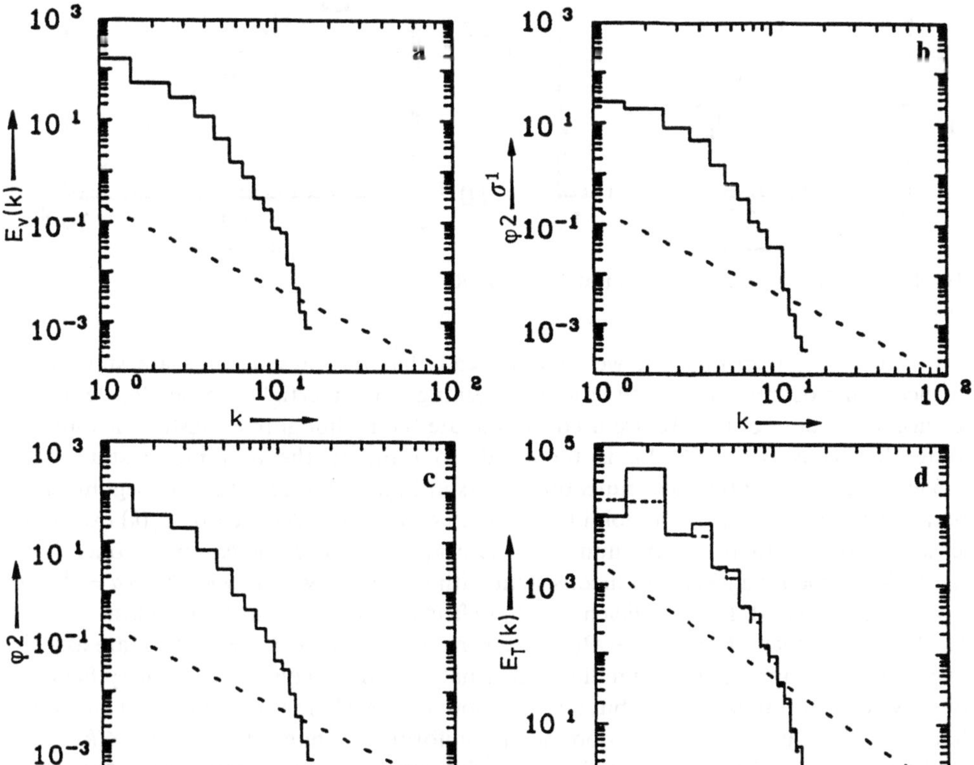

Fig. 3a–d. Spectra **a** $E_v(k)$; **b** $\phi_1(k)$; **c** $\phi_2(k)$; and **d** ($\Theta(k)$, $Z(k)$). Temperature variance and reacting scalar variance spectrum ($Z(k)$ histogram dashed line). Here $Ra = 70\ Rc$. For definition of ϕ_1 and ϕ_2, see (6a) and (6b). Spectra have been time averaged over a stationary portion of the runs. Ordinate scales arbitrary. Dashed line is $k^{-5/3}$

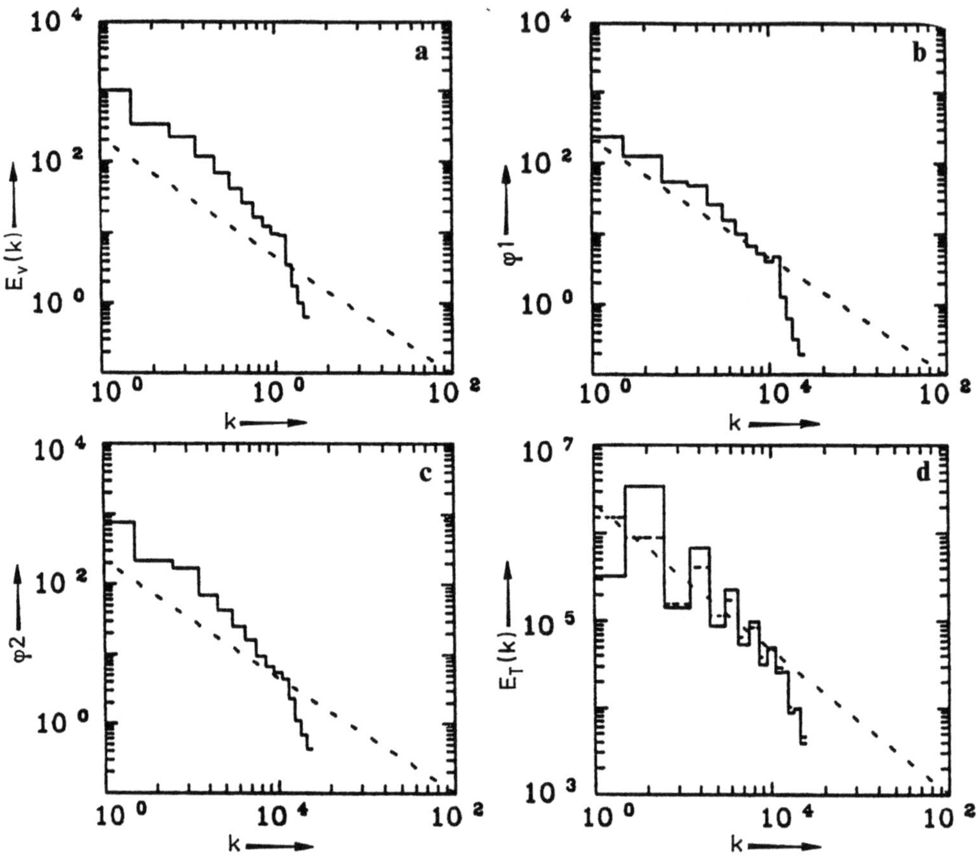

Fig. 4a–d. Same spectra as for Fig. 3 but at $Ra = 560\ Rc$

This is somewhat surprising, but we should recall that the present runs are stationary turbulence, and most numerical simulations pertaining to the inertial range are for decaying turbulence. If the data of Fig. 4a are used to estimate the Kolmogorov constant, a value of ≈ 1.0 results. Thus, $Ra = 560\ Rc$ is too small to estimate the asymptotic state. The $Ra = 560\ Rc$ case has noticeable truncation errors at high k, but also has the beginning of a separate dissipation range. The poloidal component $\phi_2(k)$ clearly exceeds $\phi_1(k)$ except at higher k near the dissipation wave number, where ϕ_1 and ϕ_2 become equipartitioned (isotropy). At $Ra = 560\ Rc$ $\phi_2(k)$ is not inconsistent with $a - 7/3$ law, at low k. The $Ra = 70\ Rc$ case, on the other hand, lacks sufficient inertial effects for isotropy to be manifest.

The large-step fluctuations in the $\Theta(k)$ spectrum, especially for $Ra = 560\ Rc$, are associated with the development of a significant thermal boundary layer. The velocity field, of course, develops a boundary layer, but near the boundaries where w is small, (u, v) are large, so that $\langle u^2 + v^2 + w^2 \rangle (z)$ remains more nearly uniform in z than $\langle \theta^2 \rangle (z)$. The $Z(k)$ spectrum (shown in Fig. 3d and 4d dotted lines) differs noticeably from $\Theta(k)$ only at small k. Of course, for inhomogeneous flows such as that considered here, these spectra are an incomplete characterization of the covariance.

Isosurfaces of the fields (w, T, C) at a given time are shown in Fig. 5 for $Ra = 70\ Rc$. Thermal plumes extending nearly the total vertical depth are clearly visible. Notice the near two-dimensional ridges that form the interior of the lower thermal boundary layer. The top of the (lower) thermal boundary layer is characterized by the pulling of a tuft of warm fluid

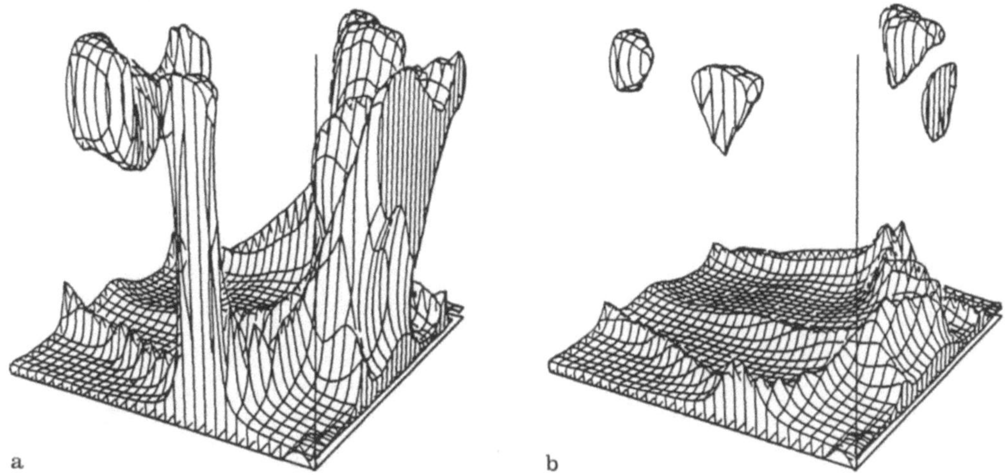

Fig. 5a and b. Isosurfaces bounding regions interior to which **a** $T > 0.7$; and **b** $C > 0.6$ for $Ra = 70\ Rc$ at a time during the stationary phase of the run

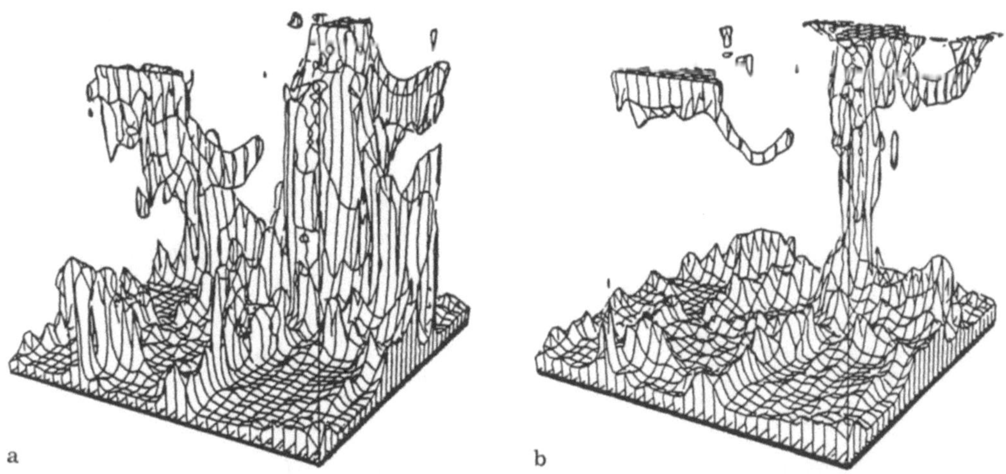

Fig. 6a and b. The same isosurfaces of T **a**, and of C **b** as for Fig. 5, except here $Ra = 560\ Rc$

from the coherent thermal. Once separated from the main body of the thermal the tuft quickly dissipates. Note C-clouds well separated from the lower boundary. The correlation between T and C is clearly seen, despite the rather small values of C in the mid-region. In fact, all calculations reported here (for the full range of ε) have normalized correlation $\langle CT \rangle / \langle C^2 \rangle^{1/2} \langle T^2 \rangle^{1/2} \geqq 0.95$.

As noted above, the thermal boundary layer in Fig. 5 is strongly two-dimensional, with occasional plumes erupting through the boundary-layer ridges, and extending nearly the entire depth of the layer between the plates. The same T- and C-isosurfaces are shown in Fig. 6 for $Ra = 560\ Rc$. Notice that the thermal boundary layer has lost its two-dimensional character; the eruptions are now more frequent, irregular, and the ridges less distinct. The smallest isolated blobs in Fig. 6 are probably numerical truncation errors.

An interesting question here is whether the shapes of Fig. 5 represent in any sense coherent structures frequently thought to be an important component of turbulent flows. As

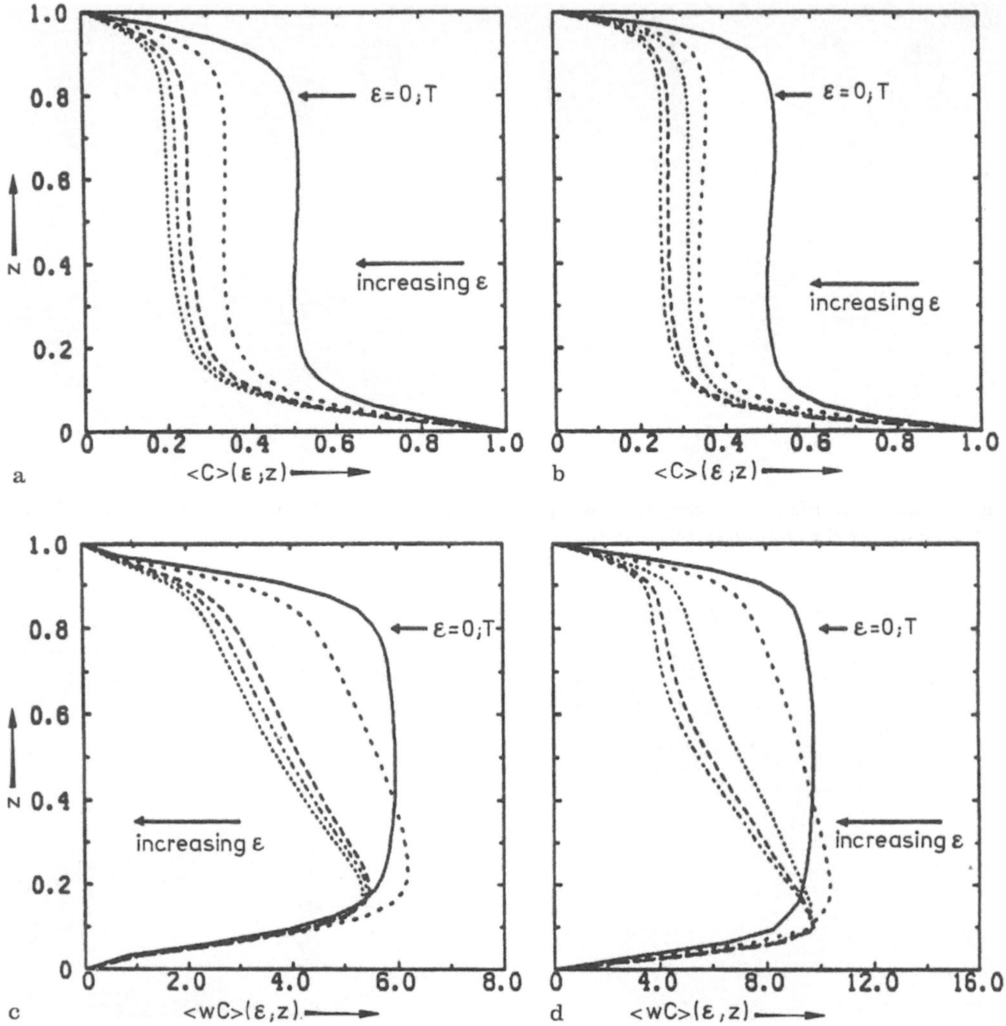

Fig. 7a–d. Horizontal average profiles: **a** $\langle T\rangle$ (*z*) and $\langle C\rangle$ (*z*) for $Ra = 70\ Rc$, $\varepsilon = (0.10, 0.20, 0.25, 0.30)\ \pi^2$; **b** same as **a** except $Ra = 560\ Rc$; **c** $\langle wT\rangle$ (*z*) and $\langle wC\rangle$ (*z*) for $Ra = 70\ Rc$, and same range of ε; and **d** same as **c** except $Ra = 560\ Rc$. For $Ra = 70\ Rc$ these correspond to $(\varepsilon'/\omega_*)\ D = (0.17, 0.26, 0.34, 0.43, 0.51, 0.6)$, and $w_* = w'(1)$ [see (10)]

noted above – and as has been found from a more extensive survey – the blobs that do emerge are rather short-lived. More quantitatively, we examine higher-order statistics of the (*u*, *T*, *C* and *w*) fields, such as their kurtosis. Large kurtosis would mean that small volumes of intense concentrations of fluctuations, although rare events, would nonetheless contribute substantially to the expected value of the fourth moment. A survey of these calculations shows that the kurtosis for all fields is near the Gaussian value of 3. This augurs well for the application of (two-point) closures to this problem (at least for $Ra < 560\ Rc$). We have already noted that the velocity and scalar or temperature mixed skewnesses (S_w, S_T, and S_c) have values typical of homogeneous turbulence (~ -0.48, -0.53, respectively, for $Ra = 70\ Rc$ and $560\ Rc$).

Statistically steady results for the ensemble-mean quantities $\langle(T, C, wT, wC)\rangle(z)$ are shown in Fig. 7 for $\sigma = 0.7$ (air), $Ra = 70\ Rc$ and $560\ Rc$ ($Rc = \pi^4 27/4 = 657.5$), $\gamma = 1$, and

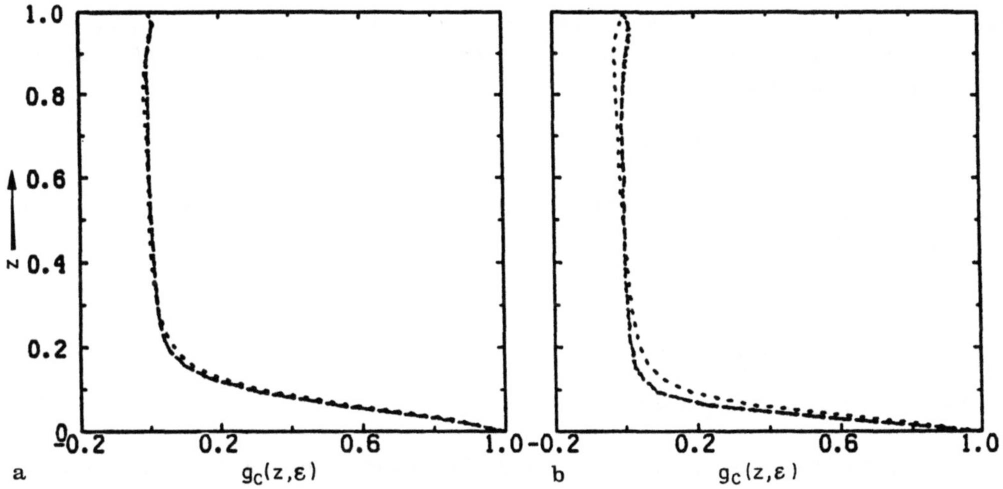

Fig. 8a and b. Flux normalized gradient $g(\varepsilon, z)$ according to (7), and for the range of $\varepsilon = (0, 0.15, 0.25, 0.30, 0.35)\,\pi^2$. Curve **a** $Ra = 70\,Rc$; **b** $Ra = 560\,Rc$

a range of σ. The aspect ratio is $\sqrt{2}$. The profiles $\langle(T, wT)\rangle\,(z)$ are familiar from many earlier studies [1, 2]. The profile $\langle C\rangle$ differs from $\langle T\rangle$ mainly in that its (approximately) constant core value exterior to the boundary layer is smaller than that for $\langle T\rangle$ ($\cong (1/2)$) – a clear result of destruction by ε. It is interesting that $dC/dz \cong 0$ in the core region despite the continuous volumetric destruction of C (for $\varepsilon \neq 0$).

If Ra is very large, it is reasonable to write the profile $\langle C\rangle\,(z)$ as

$$d\langle C\rangle/dz = Nc(0)\,g(z) + Nc(1)\,g(1-z), \tag{7}$$

where

$$Nc(0, 1) = -\,d\langle C\rangle\,(0, 1)/dz \tag{8}$$

is the (conductive normalized) flux at the bottom and top plate. Such a decomposition was profitably introduced by Wyngaard and Brost [3] in their study of scalar diffusion in the convective boundary layer. Since $d\langle C\rangle/dz$ does not possess mid-plane symmetry for $\varepsilon \neq 0$, we may solve (7) for $g(\varepsilon, z)$. Such solutions are shown in Fig. 8a, b for a range of ε and $Ra = (70{,}560)\,Rc$. Notice that the g of (ε, z) (for all ε considered) collapses onto nearly the same curve: thus $g(\varepsilon, z)$ depends on ε primarily through $Nc(0)$ and $Nc(1)$. We may similarly resolve $\langle T\rangle\,(z)$ as the limit of $\varepsilon \to 0$, by L'Hospital's rule.

Simple Estimates of Total Fluxes $\langle wT\rangle$ and $\langle wC\rangle$

The gross features of the above calculation may be understood in terms of simple mixing length concepts [4]. We characterize $\langle T(z)\rangle$ and $\langle C(z)\rangle$ by linear, broken segments with the core regions of essentially zero gradients. The boundary-layer thicknesses (z_T, z_C) for (T, C) are estimated from (3) and (4) as

$$\begin{aligned} [1/z_C^2 + \varepsilon]\,\zeta &= \zeta(z_C)\,w(z_C)/z_C, \\ [1/z_T^2]\,\theta &= \theta w/z_T, \end{aligned} \tag{9}$$

where $C = \langle C \rangle + \zeta$, $T = \langle T \rangle + \theta$, and we estimate $\nabla^2 \theta \sim \theta / z_T^2$, $\nabla^2 \zeta = C / z_C^2$, $u \cdot \nabla \theta \sim w \theta / z_C$, etc. Here, w is interpreted as the rms vertical velocity just exterior to the thermal boundary layer, taken as

$$w(z) = (Ra \, \sigma \, z \, Nu)^{1/3}, \quad 2 Z_T = Nu. \tag{10}$$

Nu is the Nusselt number $= d\langle T \rangle \, (z = 0)/dz$. Using (9–10) permits estimates of $[Nu, Nc(0), Nc(1)]$:

$$Nu \sim (\sigma \, Ra)^{1/3}, \tag{11}$$

$$Nc(0) = 2 \{(\varepsilon z_C + 1)/(\varepsilon z_C + 2)\} \, Nu, \tag{12a}$$

$$Nc(1) = 2 \{(\varepsilon z_C + 1)/(\varepsilon z_C + 2)\} \, Nu. \tag{12b}$$

To derive (12a, b) we use the volume integral of (4) to infer the approximate core value of C:

$$C_0 = 1/(\varepsilon z_C + 2). \tag{13}$$

Finally, combining (9–12) permits an estimate of z_C:

$$2 z_C = (1 - \varepsilon z_C^2)^{4/3} \, Nu. \tag{14}$$

Equation (14) shows the increase of the scalar boundary layer thickness above that for the temperature field stemming from the destruction term, εC in [4]. The coefficients entering these expressions may be estimated from the numerical simulations. We find, for example, $Nu \cong 0.17 \, (\sigma \, Ra)^{1/3}$ from the current 32^3 calculations. Such values may change slightly with increasing resolution and aspect ratio.

General Solution for C by Superposition and Statistical Summary

It is of interest to know statistical properties of a field S which satisfies the more general boundary conditions $S(0) \equiv S_0$, $S(1) \equiv S_1$, but otherwise satisfies (4). We may show that

$$\langle S \rangle \, (\varepsilon, z) = S_0 \langle C \rangle \, (\varepsilon, z) + S_1 \langle C \rangle \, (\varepsilon, 1 - z). \tag{15}$$

Here, $\langle C \rangle \, (\varepsilon, z)$ is the solution to (4). Equation (15) may be established using mid-plane statistical mirror symmetry of (1–4), and linearity of (4). Using these same principles, we compute $\langle S^2 \rangle \, (\varepsilon, z)$ for the general case

$$\begin{aligned} \langle S^2 \rangle \, (\varepsilon, z) &= S_0^2 \langle C^2 \rangle \, (\varepsilon, z) + S_1^2 \langle C^2 \rangle \, (\varepsilon, 1 - z) + 2 S_0 S_1 f(\varepsilon, z), \\ f(\varepsilon, z) &= \langle C(\varepsilon, x, y, z, t) \, C(\varepsilon, x, y, 1 - z, t) \rangle. \end{aligned} \tag{16}$$

A particularly interesting case is that in which C-flux exiting the top boundary is zero. Profiles for that $\langle S \rangle \, (z)$ are shown in Fig. 9.

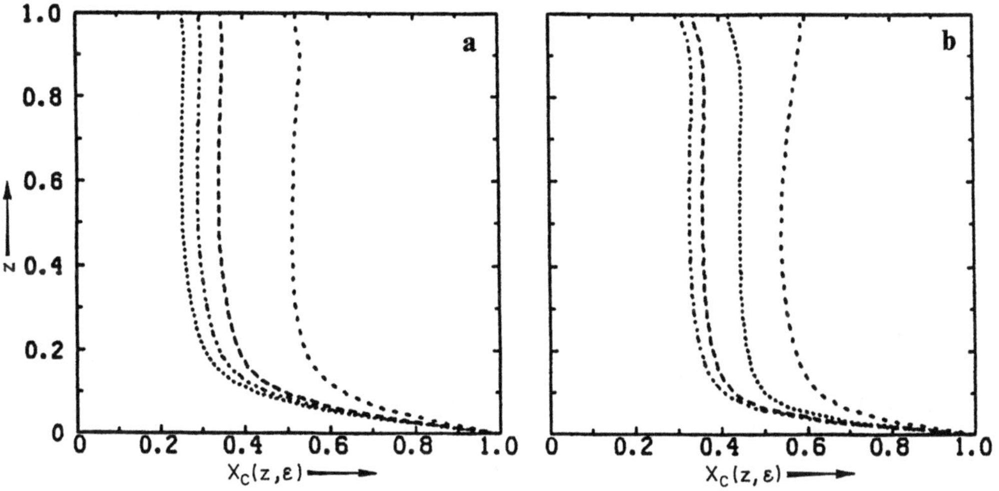

Fig. 9a, b. Profiles $\langle S \rangle$ (z) for $d\langle S \rangle(\varepsilon, 1)/dz = 0$, according to (14) and (15). *Ra* and σ as in Fig. 8

Characterization of Flows via Generalized Eddy Transport Concepts

We may treat the present numerical simulations as a source of experimental information with which to assess theoretical ideas concerning scalar transport in convective regions. We examine-briefly-two related issues: (1) can eddy transport concepts be generalized (possibly along the lines suggested by Fiedler, [5]) so as to comprehend profiles such as those in Fig. 7b for which the core region is stable, and (2) if this is possible, can one then incorporate the scalar destruction, ε, into such a procedure so as to reproduce Fig. 7b, and then carry over the results to $\mathrm{Re} \to \infty$?

An attempt to extract eddy conductivities from the present calculation faces the dilemma of "up-gradient" transport in the core region of the flow (see Fig. 7a, b). Thus, a higher-order-closure model (beyond simple K-theory) is essential to give a valid functional relationship between $\beta(z)$ and $\langle WT \rangle$ (z). Preliminary to a detailed discussion of this issue, we ought to consider several disquieting aspects of this problem.

First, we may consider the possible insight derived from homogeneous closures, e.g., the two-point variety. Such procedures have been suggested as fruitful avenues for eddy viscosity and conductivity in a simpler context [9, 10]. However, a first-order chemical reaction problem may be transformed into a (non-reacting) passive scalar problem (i.e., $C \to \exp(-\varepsilon t)\, C$). Hence, boundaries are essential for the proper ε dependence in the problem. Such procedures may be useful, but only as the limit of inhomogeneous conditions.

Secondly, as to the physics of the stable core, perhaps some insight may be gained by examining those flows for which it is extreme, namely the two-dimensional. Early studies of Moore and Weiss [11] showed that reversed gradients are associated with intensely concentrated convective jets, extending throughout the convective domain. Reversals were more intense at lower Prandtl numbers, for which – contrary to three-dimensional flows – *Nu* increases with Pr. Their solutions were strongly non-Gaussian in the sense that kurtosis of θ and vorticity greatly exceeded 3. Similar results were found in (two-dimensional) cylindrical convection, which shared with the planar case the attribute of no vertical vorticity. Such dynamics may partly survive in three dimensions as two-dimensional sheets (or extensive vertical columns) imbedded in otherwise chaotic flow. This suggests that at least part of the stable core region may not be explained by generalized eddy transport concepts.

Finally, we recall that a $\langle \partial T/\partial z \rangle$ core region is also found in the third cumulant discard (rapid-distortion theory) procedure [12]. The later solutions have somewhat unrealistic stable-zone concentrations just exterior to the (conductive) thermal boundary layer. For this (time-independent) theory, the convective flux-mean gradient relationship is simply

$$\langle WC' \rangle (z) = \int dx' \, G(x, x') \, \langle W(x) \, W(x') \rangle \, \beta_C(z),$$

where

$$(\varepsilon - \kappa_C \nabla^2) \, G(x, t; x', t') = \delta(x, x'). \tag{17}$$

$\delta(x, x')$ and G satisfy the same boundary conditions as θ or ζ. [Note: our discussion here is in terms of C; results for T are obtained by putting $\varepsilon = 0$.] In (17), and the remainder of this section, we have reverted to the physical dimensional form of the equations via (5). Equation (17) displays in embryonic form a "non-locality" relating $\langle WC' \rangle$ to $\beta_C(z)$.

A more complete theory (such as that sketched below) would replace the conductivity in (17) by a suitable eddy conductivity, and in addition would properly include Lagrangian decorrelation effects in the velocity fields $w(x, t)$, and $w(x', t')$. For orientation, we may imagine (17) to be extrapolated to the large R_λ limit in which β_C is much slower in its spatial variation than $\langle w(x) \, w(x') \rangle$ – in reality probably a non-existent limit! This would give

$$\begin{aligned} \langle w\zeta \rangle &= \beta_C(z) \int \langle |w^2(\boldsymbol{k})| \rangle \, d\boldsymbol{k}/(\eta(k) + \varepsilon), \\ \eta(k) &= \kappa_C k^2. \end{aligned} \tag{18}$$

Comparing (18) with closure estimated eddy conductivities [10] suggests that $\eta(k)$ be replaced by

$$\mu(k) \sim \left(\tfrac{2}{3}\right) \int_0^\infty dk \, E(k) \Big/ \left\{ \int_0^k p^2 \, dp \, E(p) \right\}^{1/2}. \tag{19}$$

Proceeding next to a more formal description, we first note that C' satisfies

$$(\partial/\partial t) - K \nabla^2 + \varepsilon + \boldsymbol{u} \cdot \nabla) \, \zeta(x, t) = \beta_C(z) \, \omega. \tag{20}$$

Denoting the (stochastic) Green's function of the lefthand-side operator in (20) by $G(\boldsymbol{x}, t; \boldsymbol{x}', t')$, we may write (in the statistically steady state)

$$\langle w\zeta \rangle (z) = \int d\boldsymbol{x}' \int_0^t dt' \langle w(\boldsymbol{x}, t) \, G(\boldsymbol{x}, t, \boldsymbol{x}', t') \, \beta_C(z) \, w(\boldsymbol{x}, t) \rangle, \tag{21}$$

where we recall that $\langle \; \rangle$ means ensemble average. So far, (21) is exact, and incorporates the heat flux in the limit $\varepsilon \to 0$. We note that the flux-gradient law $\langle WC' \rangle = K'; \beta_c(z)$ is obtained if G is replaced by a local operator, whose timeintegral effect on (18) is a multiplication by a time scale typical of the Lagrangian time scale of the turbulence. Further, in the limit $\varepsilon \to 0$, (18) reduces (far from boundaries) to:

$$\langle W\zeta \rangle \to \beta_C \langle z \rangle / \varepsilon.$$

A valid approach should recover both these limits, and incorporate boundary effects. Making a "quasi-normal" factoring of (18) and a simple but reasonable Ansatz for (G):

$$\langle W\zeta \rangle = \int d\boldsymbol{x} \int_0^t dt' \langle W(\boldsymbol{x}', t') \, W(\boldsymbol{x}', t') \rangle \, \langle G_\varepsilon(x, t; x', t') \rangle \, \beta_C(z'), \tag{22}$$

where

$$\langle(\partial/\partial t - \kappa_C \nabla^2 + \boldsymbol{u} \cdot \nabla + \varepsilon)^{-1}\rangle^{-1} \langle G_\varepsilon\rangle\ (x, t; x' t') = \delta(t - t')\ \delta(\boldsymbol{x}, \boldsymbol{x}'). \tag{23}$$

Since $\langle w(x, t)\ w(x', t')\rangle$ is available from data, we may explore simple approximations which will reproduce the observed profiles for $\langle w C'\rangle$. One approach is to generate $\langle G\rangle$ by using a Gaussian $\boldsymbol{u}$ having the same covariance as the full turbulence simulation, which should give additional insight into the validity of two-point closures for this problem.

Concluding Remarks

The calculations reported here illustrate the extent to which rather low-resolution simulations may be an effective tool in studying inhomogeneous turbulence. The flow appears chaotic at a low value of $R_\lambda(\lesssim 18)$. This is in part a consequence of the slip-boundary conditions. We recall in this connection that Zippelius and Siggia [6] noted that slip-boundary convection becomes chaotic at a much lower Ra than previously supposed. It is also interesting that asymptotic heat flux, i.e., $N_u \sim (\sigma Ra)^{1/3}$, etc., appears to become valid at rather low $R_\lambda(\cong 30)$. An unresolved issue here concerns the effects of including large scales by increasing the aspect ratio. Earlier calculations [7] suggest that at $\sigma = 10$, the presence of larger scales has a profoundly destabilizing effect on the plan-form of the flow. However, a survey of several σ's (10. 1, 0.1) suggested that as $\sigma \to 0$ the need to include very large scales in order to obtain stable statistics is diminished.

The gross energetics and flux budgets for both T and C conform rather well to the simple mixing-length estimates set forth above. Thus we find that the lower and top boundary fluxes (12a, b) and the mid-plane $C(z = 1/2)$ (13–14) are accurately forecast by these simple estimates. However, despite an incipient $-5/3$ law at $Ra = 560\ Rc$, we do not confirm Priestley's [8] law $\langle w^2\rangle \cong z^{2/3}$. The central problem is that $R_\lambda \cong 30$ (the present calculation) is too low to examine asymptotic issues.

We have proposed an extension of eddy-transport concepts by employing an ensemble mean Green's function (with suitable quasi-normal approximations) to write an approximate equation for the scalar fluctuation as a functional of the ensemble mean scalar gradient and the vertical velocity field. As in the analysis of Fiedler [5] this leads to a non-local eddy diffusion operator; however, in our case the two-point second moment of the vertical velocity field is the sole ingredient to complete the calculation. The present $w(x, t)$ statistics may then be used as a data base for testing these ideas.

One interesting stability aspect of these calculations that has a bearing on the predictability of passive scalar flows is elucidated by considering the case $\varepsilon = 0$, and the initial value problem in which C and T are imperfectly correlated. For homogeneous flows, we would expect on general predictability arguments that the fields (C, T) would become, in time, completely decorrelated. However, we have found that in the present inhomogeneous problem (C, T) become perfectly correlated as time proceeds. The correlation-inducing term is, of course, $[\langle w C'\rangle\ \partial\langle T\rangle/\partial z + \langle w T'\rangle\ \partial\langle C\rangle/\partial z]$ which appears to overwhelm the destabilizing effects of turbulence terms $\langle C \nabla \cdot \boldsymbol{u} C\rangle$ in the equations of motion for $\langle C \cdot T\rangle$.

Acknowledgment. All of the graphics and much of the programming for this paper were done by Ms. B. Stankov. We wish to record here our appreciation for her thorough professionalism and gracious generosity.

References

1. Lipps, F. B. (1976): Numerical simulation of three-dimensional Bénard convection in air. J. Fluid Mech. **75**, 113–148
2. Curry, J. H., Herring, J. R., Orszag, S. Z., Loncaric, J. (1984): Order and disorder in two- and three-dimensional turbulence. J. Fluid Mech. **147**, 1–38
3. Wyngaard, J. C., Brost, R. A. (1984): Top-down and bottom-up diffusion of a scalar in the convective boundary layer. J. Atmos. Sci. **41**, 102–112
4. Kraichnan, R. H. (1962): Turbulent thermal convection at arbitrary Prandtl number. Phys. Fluids **5**, 1274–1389
5. Fiedler, B. E. (1984): An integral closure model for the vertical turbulent flux of a scalar in a mixed layer. J. Atmos. Sci. **41**, 674–680
6. Zippelius, A., Siggia, E. D. (1982): Disappearance of stable convection between free-slip boundaries. Phys. Rev. **A26**, 1788–1790
7. Herring, J. R., Jackson, S. (1984): "Thermal Convection: Numerical Experiments near the Onset to Turbulence and the Statistical Theory of Turbulence," in *Turbulence and Chaotic Phenomena in Fluids*, ed. by T. Tatsumi (Elsevier, Amsterdam) 111–116
8. Priestley, C. H. B. (1959): *Turbulent Transfer in the Lower Atmosphere* (Chicago University Press, Chicago)
9. Corrsin, S. (1961): Reactant concentration spectrum in turbulent mixing with a first-order reaction. J. Fluid Mech. **11/3**, 407–416
10. Herring, J. R., Schertzer, D., Lesieur, M., Newman, G. R., Chollet, J. P., Larcheveque, M. (1982): A comparative assessment of spectral closures as applied to passive scalar diffusion. J. Fluid Mech. **42**, 411–437
11. Moore, D. R., Weiss, N. O. (1972): Two dimensional Rayleigh-Bénard convection. J. Fluid Mech. **58**, 289–312
12. Herring, J. R. (1963): Investigation of problems in thermal convection. J. Atmos. Sci. **20**, 325–338

Experimental and Numerical Study of a Turbulent Recirculation Zone with Combustion

P. Moreau, J. Labbé, F. Dupoirieux and R. Borghi*

ONERA, Châtillon Sous Bagneux, France
* Also Laboratoire de Thermodynamique, Faculté de Sciences de Rouen, France

The Problem

Recirculation zones are widely used for the stabilization of combustion in flows with a large velocity. The main example of this is the case of afterburner devices for turbojet engines, where the turbulent flame is stabilized by bluff bodies behind which recirculation zones occur. The prediction of the general shape of the flow field and of the temperatue field of these recirculating flows is of primary importance for the prediction of the "stability domain" of the burners. Two properties of the recirculation zone are of particular interest for combustion stabilization: its volume and the mass flow rate exchanged with the external flow. These two characteristics control the residence times of the fluid particles within the recirculation zone, which has to be large enough to ensure a good stabilization of the combustion.

Without combustion, the prediction of turbulent separated flows behind obstacles is not simple. On one hand the flow is governed by elliptical equations, and on the other hand the turbulence is very complex, non homogeneous, non isotropic and with large curvature of the streamlines. The first attempts to predict by a numerical calculation and a turbulence model such a recirculation zone pointed out some difficulties: a simple and global characterristic, like the recirculation length, has been found always (about 20%) smaller in the calculation than in the experiments (see Pope and Whitelaw [1], Durst and Rasgoti [2], Mobsby and Fussey [3]). It seems, however, that an improvement of the $k - \varepsilon$ turbulence model taking into account the curvature of the streamlines leads to more satisfactory agreement [2]. Such a prediction, even imperfect in its details, particularly concerning the components of the turbulence kinetic energy, can be of large interest for engineering purposes.

What is the situation for recirculation zones behind bluff bodies in combusting flows? It appears that their calculation, including already known turbulence models as well as already known turbulent combustion models, is fully possible. In fact, many examples of such a calculation are existing in the literature ([3] for instance); much more complex flow field have been attacked for numerical prediction also (see [4]). But in simple cases where detailed experimental results are available, the situation is not at all clear. For instance, concerning the recirculation zone behind bluff bodies, it is not clear at the present time whether the combustion increases the size of this zone, or decreases it, with respect to the non burning cases. The first experiments of Winterfeld [5] and Clare et al. [6] have shown an increase of this length, while the ones of Grindley reported in [3], and Pitz and Daily [7] show the opposite. The studies of Taylor and Whitelaw [8], and Fuji and Egushi [9] show both an increase of the length.

Indeed, it is easily understandable that combustion can have opposite influences: i) in the case of a confined duct, the combustion behind the bluff body leads to an increase of the gas specific volume, which, expectedly, produces a flow acceleration, but, less expectedly, compresses the recirculation zone upstream and decreases its length; ii) on the contrary, the dilatation of gases due to higher temperature within the recirculation zone itself can increase

Turbulent Shear Flows 5

its size; iii) the turbulence itself can be increased at the boundaries of the recirculation zone, due to the higher velocity gradient created here by the acceleration due to the heat release, increasing consequently the effective eddy viscosity, and decreasing the length of the recirculation; iv) but, at the contrary, the increase of the laminar viscosity due to the higher temperature can produce a drop of the turbulence kinetic energy, and then of the eddy viscosity, lengthening the recirculation.

All these facts can explain that opposite conclusions have been drawn; in addition, concerning the turbulence itself, very few is known at this time about the influence of the fluctuations of density due to the heat release (see Borghi and Escudié [10], and Vandromme et al. [11]).

Another interesting point is the presence of coherent structures, or quasi periodical temporal fluctuations, within the recirculation zone itself or in the shear layers surrounding it; an experimental evidence of these can be found in [7], and very dramatic unsteady regimes have been found associated to these, and extensively studied by Ganji and Sawyer in [12].

The purpose of the study presented here is to bring new detailed experimental results on this type of flow as well as to discuss it with respect to the available computation methods.

Experimental Study

The Experimental Set Up

It consists in a two dimensional flow behind a downstream facing step; the duct behind the step is of section 100 × 100 mm, and the step height is 35 mm. This device is fed by a premixed Air-Methane flow at about 50 m/s at the level of the step, preheated at 530 K; when the combustion is operated, the equivalence ratio has been chosen to 0.8; comparative tests have been performed also without combustion.

The methane is injected 50 cm upstream, and the flow is conveniently homogeneous in the section of the step. Due to the absence of a settling chamber upstream, the turbulence level in this section is rather high: the turbulence has been found almost isotropic, and $\bar{u}'^2$ has been measured as 10 $(m/s)^2$, with respect to 50 m/s for $\bar{u}$.

Visualizations by a shadowgraph technique and high speed (4000 Fr/s) cinematography have been performed. They revealed the strongly turbulent nature of the flame, with an apparently thick undulated flame. The undulations are sometimes very large with respect to the apparent thickness of the instantaneous flame, and a quasi periodic structure is seen. But these undulations never take the form of classical coherent structures seen in two dimensional mixing layers and clearly put into evidence in [12].

The Laser Doppler

The flow field has been studied with a two colors Laser Doppler Anemometer, which can be displaced in the 3 axis; the blue fringes have been oriented at − 45° with respect to the longitudinal axis of the device, the green ones being oriented at + 45°, simultaneously. Bragg cells have been used to determine the sign of the velocity. The measuring device is the same as that used in [13], and enables us to measure simultaneously the two longitudinal and lateral components of the velocity. Some additional tests have been made to measure the spanwise velocity component, which has been found always close to zero, as expected.

The seeding has been performed with submicron particles of zirconium oxide, 1.5 m upstream of the step, in order to obtain a very homogeneous and dense seeding.

Experimental Results

The mean flow field, with and without combustion, has been studied in details, and the first conclusion is that the recirculation length is larger without combustion than with combustion: the ratio length/step height is 15/3.5 without combustion and 11/3.5 with combustion. These values compare very well with the ones of the similar study of Pitz and Daily [7]; of course this ratio depend on several parameters as the blockage ratio (height of the step/height of the duct), here 3.5/10, and also on the shape of the step (see Davies and Beer [14]). Figure 1 shows the velocity field obtained (x is the longitudinal coordinate, one curve is displayed for each centimeter behind the step, y is the transverse one).

Large velocity fluctuation has been found within the recirculation zone: $(\bar{u}'^2 + \bar{v}'^2)^{1/2}$ is of the order of 7 m/s without combustion and 10 to 15 m/s with combustion. The maximum value is in the shear layer without combustion but is displaced inside the recirculation zone with combustion. With combustion a stronger anisotropy of the turbulence is clearly seen. Figure 2 gives the detailed curves. Figure 3 shows that the $\overline{u'v'}$ Reynolds stress component has been clearly increased by the combustion (100% global increase); it seems to behave similarly, but not proportionally, with: $\partial\bar{u}/\partial y + \partial\bar{v}/\partial x$. Comparing the curves of Figs. 2 and 3, one can see that the maximum of $\overline{u'^2}$, $\overline{v'^2}$ and $\overline{w'^2}$ corresponds to the one of

a

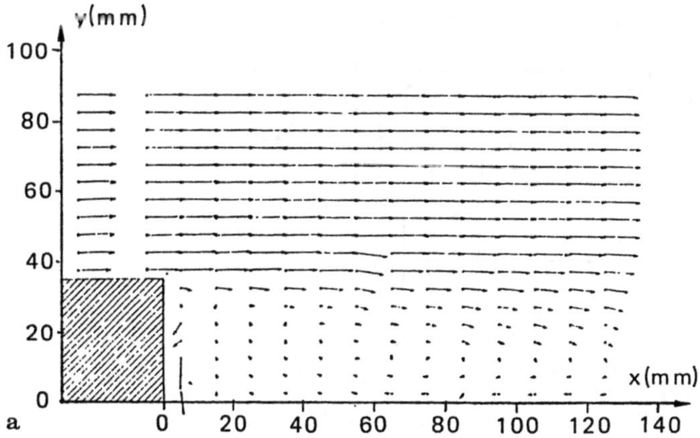

Fig. 1a. Velocity field without combustion

b

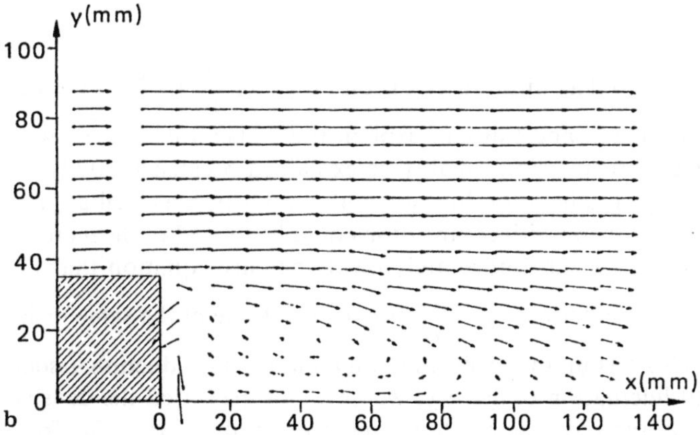

Fig. 1b. Velocity field with combustion

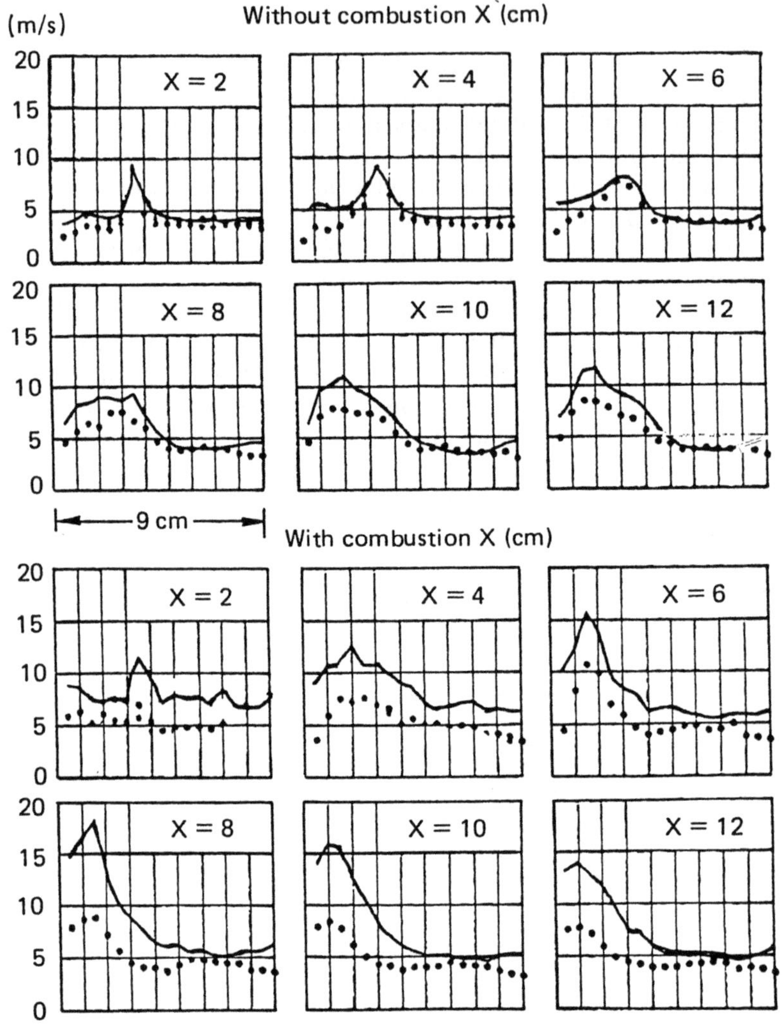

Fig. 2. Profiles of $\sqrt{\overline{u'^2}}$, $\sqrt{\overline{v'^2}}$

$\overline{u'v'}\left(\frac{\partial \bar{u}}{\partial y}+\frac{\partial \bar{v}}{\partial x}\right)$, with combustion as well as without. With combustion $\left(\frac{\partial \bar{u}}{\partial y}+\frac{\partial \bar{v}}{\partial x}\right)$ is weaker than without, but $\overline{u'v'}$ is much stronger, and this can explain the higher value of $\overline{u'^2}$ with combustion than without. The anisotropy can probably be explained because $\partial \bar{u}/\partial y$ is the strongest component of the velocity gradient, in flow without combustion as well as with combustion, and the increase of anisotropy with combustion could be related to the increase of $\overline{u'v'}$. Concerning $\overline{u'v'}$, it does not appear immediately on its balance equation that the combustion could increase its level: the term $-\varrho \overline{v'^2}\frac{\partial \bar{u}}{\partial y}$ is expected to be of the same order with and without combustion. This sets up the problem of an additional production associated with pressure gradient and fluctuation of density; we have not measured the density, but the calculations that will be presented later show that the combustion is far from completed in a large part of the recirculation zone, and particularly in the shear layer; on

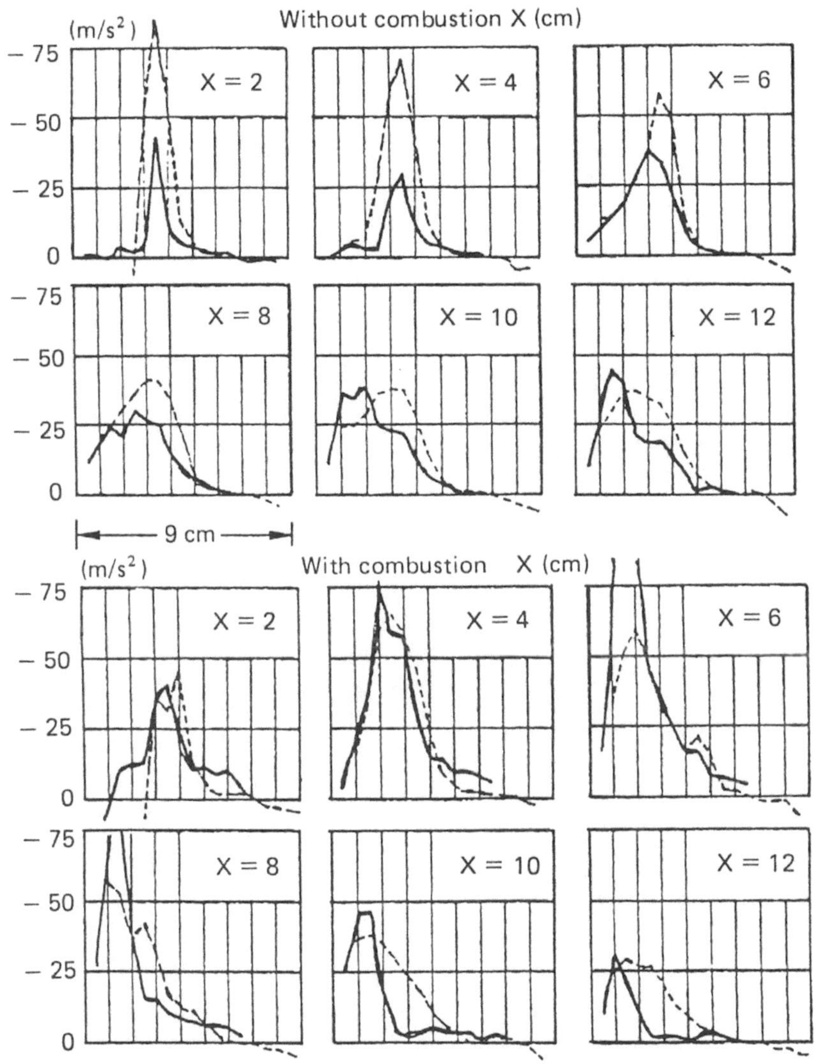

Fig. 3. Profiles of $\overline{u'v'}$ and $d\bar{u}/dy + d\bar{v}/dx$ —— $\overline{u'v'}$ --- $(d\bar{u}/dy + d\bar{v}/dx) \times 0.02$

the other hand $\partial\bar{p}/\partial y$ is largely positive in this region, in such a way that $-\overline{u''}\dfrac{\partial\bar{p}}{\partial y}$ is a good candidate as an additional production term for $\overline{u'v'}$.

Finally, it can be remarked that $\overline{u'^2}$, $\overline{v'^2}$, $\overline{w'^2}$ are very high also outside the shear layer. Within the recirculation zone $\overline{u'v'}$ is high, even if $\left(\dfrac{\partial\bar{u}}{\partial y}\right)$ is well below its maximum value, and this could explain that turbulence production occurs also there; but one can also think that turbulence is accumulated within the recirculation zone by the convection terms. Outside the recirculation zone, in the external flow, one can remark in addition that the presence of combustion clearly increases the $\overline{u'^2}$ level at the entrance of the combustor ($x = 2$ or 4); but this increase is very strongly dissipated and, downstream, the level is just slightly higher than without combustion; no clear explanation of this fact appears possible at the moment.

The joint probability density function of u' and v' has been obtained also. Figure 4 shows an example of such p.d.f. obtained in a cross section of the recirculation, at corresponding

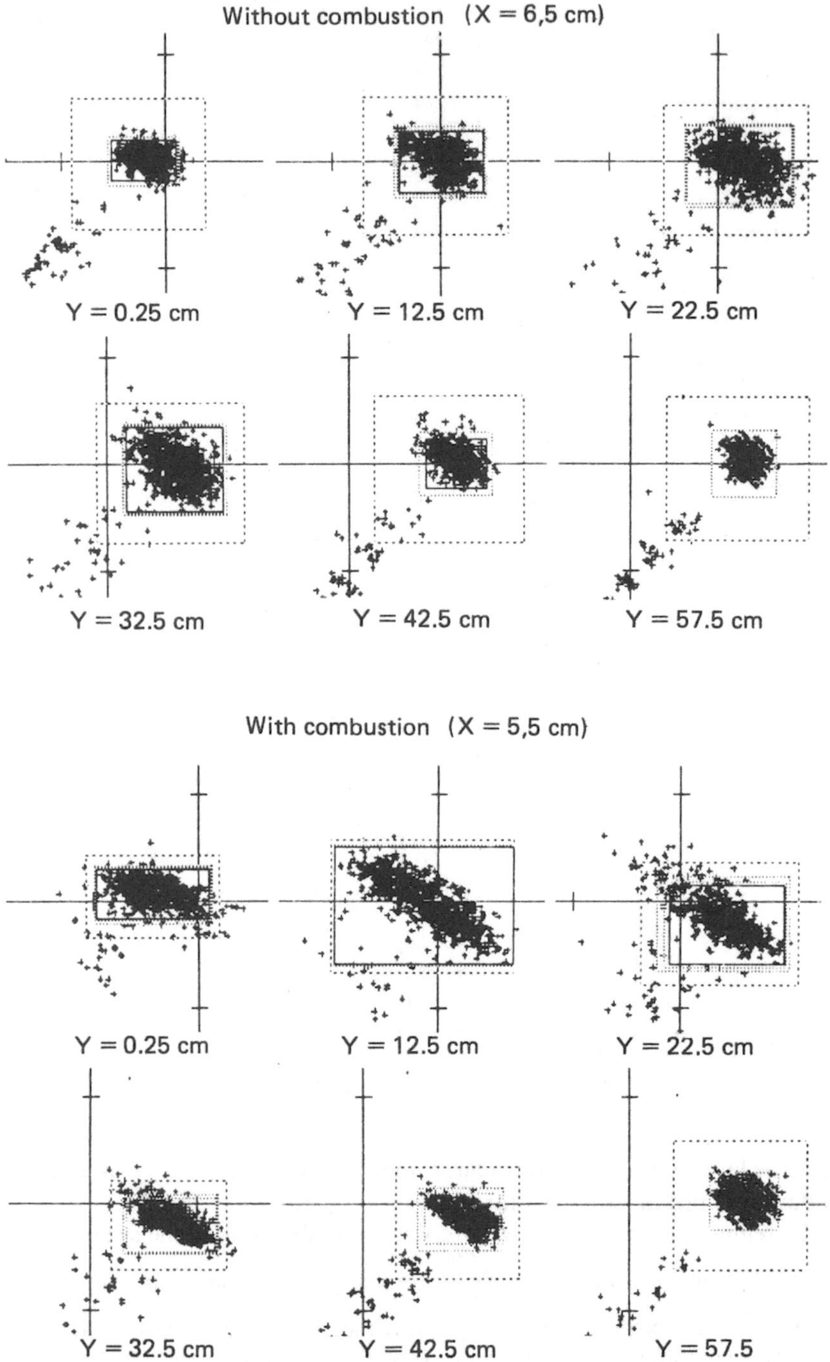

Fig. 4. Instantaneous velocity distribution for different points of a given cross section

stations without and with combustion. The p.d.f. appears almost everywhere very classical, except that the strong correlation $\overline{u'v'}$ is seen. Except in the reattachment zone, the p.d.f. does not show any bimodal shape.

Numerical Study

The numerical study has been attempted with a calculation method of the classical balance equations of the aerothermochemistry. Dutoya and Michard [15] describe with more details the numerical method used. A first calculation has been made with a pseudo-laminar turbulence model, that is with a closely related constant "eddy viscosity". A value of 0.02 (m^2/s) has been found which agrees relatively well, in the whole field, with the experimental results for the mean values of the velocities with and without combustion. Figure 5 shows the agreement obtained, with and without combustion. In particular the shortening and even the change of shape of the recirculation are surprisingly well reproduced by this very simple model. The calculation domain begins 6 cm upstream to the step.

In addition to the turbulence model, the numerical method includes a model for the mean turbulent combustion rate; the model used is based on a very fast reactions assumption, giving the so called "Eddy Break up" formula:

$$\mathring{W} = \frac{Y_F}{Y_{F_0}} \left(\frac{Y_{F_0} - Y_F}{\tau_{\mathrm{EBU}}} \right),$$

where $\mathring{W}_F$ is the mean combustion rate, Y_F the local mean value of the fuel mass fraction, and Y_{F_0} its entrance value. τ_{EBU} is a constant (with the dimension of a time), which has to be adjusted empirically; the value used here is 2.2 10^{-5} s, in agreement with other applications of the same calculation method to the development of a turbulent premixed flame in a duct [16, 17]. Within these experiments a characteristic value of the turbulence integral time scale has been estimated to $3 \cdot 10^{-5}$ s.

Further calculations have been attempted with other models. We have used a "subgrid scale model", where the eddy diffusivity ν_t is given by a mixing length type formula:

$$\nu_t = l^2 \left(2 \left(\frac{\partial \bar{u}}{\partial x} \right)^2 + 2 \left(\frac{\partial \bar{v}}{\partial y} \right)^2 + \left(\frac{\partial \bar{u}}{\partial y} + \frac{\partial \bar{v}}{\partial x} \right)^2 \right)^{1/2}$$

with $l^2 = c \, \Delta x \, \Delta y$, where Δx is the grid size in x direction and Δy in y direction.

The best fit with the experiment results was obtained with $c = 0.02$, but the agreement was not better than for the previous calculation with $\nu_t =$ cste.

This S.G.S. model has to be completed to give us a turbulence time scale to be used with the "Eddy Brak up" model; we have taken here

$$\tau_{\mathrm{EBU}} = 0.4 \left(2 \left(\frac{\partial \bar{u}}{\partial x} \right)^2 + 2 \left(\frac{\partial \bar{v}}{\partial y} \right)^2 + \left(\frac{\partial \bar{u}}{\partial y} + \frac{\partial \bar{v}}{\partial x} \right)^2 \right)^{-1/2}.$$

The $k - \varepsilon$ model has then been implemented in the calculation method; the L.D.A. Measurements at the entrance of the combustion chamber have been taken into account to prescribe inlet conditions for k, the turbulence kinetic energy, and the entrance conditions for ε have been deduced from a given integral length scale in this section. The influence of realistic measurement errors of these entrance conditions on the recirculation length has

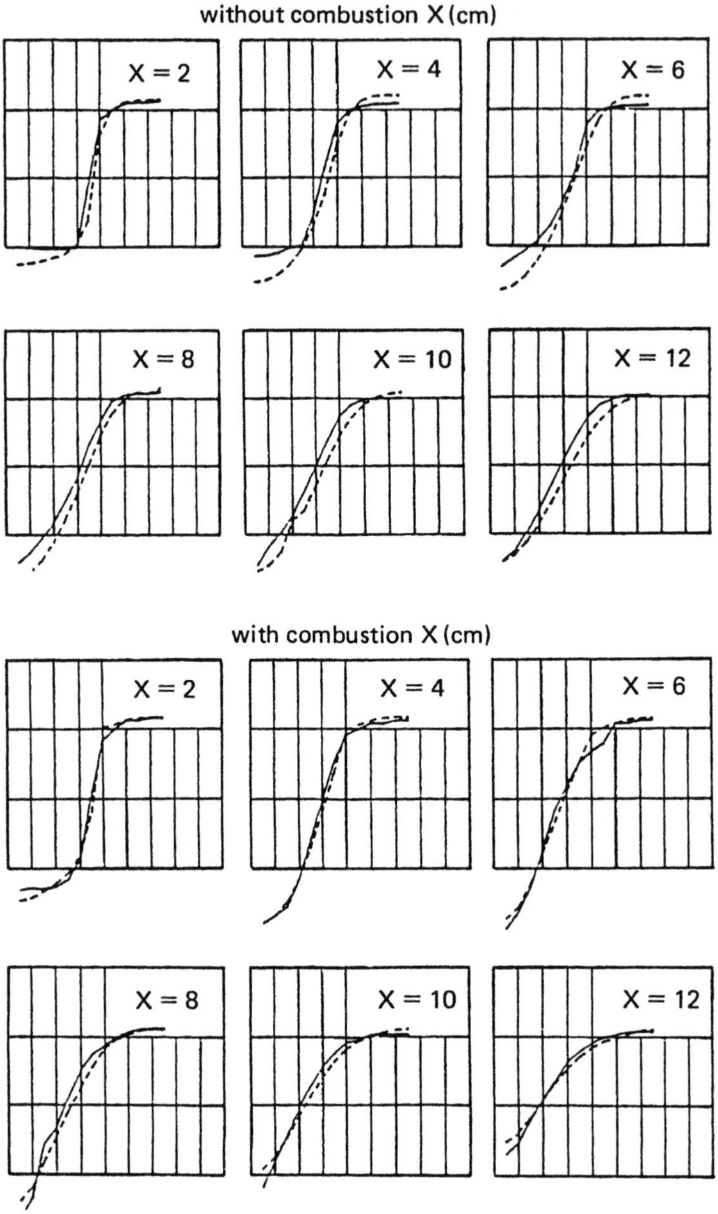

Fig. 5. Axial velocity profiles

been found very poor. With the classical values of the coefficients of the $k - \varepsilon$ model, the recirculation length has been underpredicted: 12 cm instead of 15 cm without combustion and 7 cm instead of 9 cm with combustion. It appears, however, that the shortening due to the combustion is very well given by the calculations.

Figure 6 compares the transverse profiles of $(\bar{u}'^2 + \bar{v}'^2)/2$ and k obtained by the calculation and by measurements in the section $X = 4$ cm, respectively. The computed value is underestimated in the non combusting case, but closer to the experiments with combution. With a very rough point of view, one can say, however, that the k profiles and the effect of

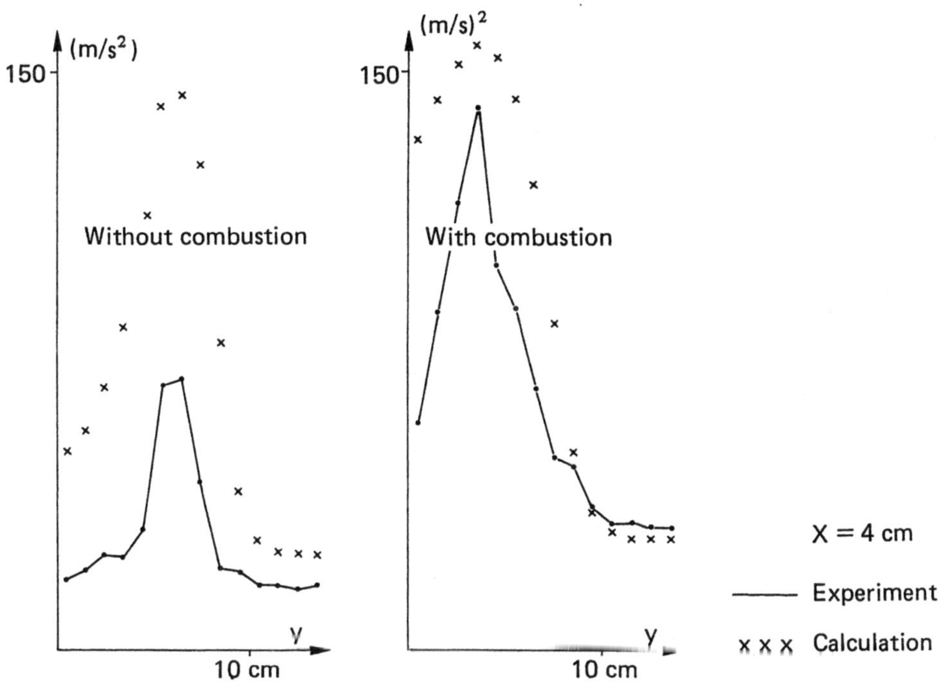

Fig. 6. Profiles of turbulent kinetic energy

the combustion predicted within 25%. The achievement of a closer agreement necessitates probably to take into account both effects of anisotropy and density fluctuations.

Concluding Remarks

The present conclusions that can be pointed out from our experimental and computational study can be summarized as follows:

i) It has been confirmed that the recirculation zone behind bluff body in a confined flow is shortened when the combustion is stabilized by this bluff body; ii) This shortening is well reproduced by the calculation results. It seems probable that the effect is related to the increase of massic volume of combusting gases behind the recirculation zone, and not to the turbulence modification, because all turbulence models used here lead to similar effects; iii) The turbulence within the recirculation zone is clearly modified when combustion occurs: the anisotropy of the turbulence is enhanced, the Reynolds stress is strongly increased, and the location of the maximum of turbulence kinetic energy is displaced; iv) The length of the recirculation region, and the mean velocity profiles can be computed in good agreement with a constant eddy viscosity model (without and with combustion). The classical $k - \varepsilon$ model leads to worse agreement; the prediction of the turbulence kinetic energy is achieved only within 25%, on a global (and not point to point) basis.

More detailed predictions of turbulence characteristics need the use of a more sophisticated turbulence model, including the calculation of the anisotropy as well as the effect of density fluctuation and curvature.

References

1. Pope, S. B., Whitelaw, J. H. (1976): The calculation of near-wake flows. J. Fluid Mech. **73**, 9–32
2. Durst, F., Rastogi, A. K. (1979): "Turbulent Flow over Two-Dimensional Fences", in Turbulent Shear Flows 2, ed. by L. J. S. Bradbury et al. (Springer, Berlin, Heidelberg) 218
3. Mobsby, J. A., Fussey, D. E. (1979): "A Three-Environment Mixing Model for Turbulent Flame Studies", 17th Int. Symp. on Combustion, 411
4. Jones, W. P., Mac Guirk, J. J. (1980): Mathematical modelling of gas turbine combustion chambers. AGARD Conf. Proc. **275**
5. Winterfeld, G. (1965): "On Process of Turbulent Exchange Behind Flame Holders", 10th Int. Symp. on Combustion, The Combustion Institute
6. Clare, H., Durao, D. F. G., Melling, A., Whitelaw, J. H. (1976): Investigation of a *V*-gutter stabilized flame by laser anemometry and Schlieren photography. AGARD Conf. Proc. **193**
7. Pitz, R. W., Daily, J. W. (1981): Experimental study of combustion in a turbulent free shear layer formed at a rearward facing step. AIAA 19th Aerospace Sciences Meeting (St. Louis)
8. Taylor, A. M. K. P., Whitelaw, J. H. (1980): Velocity and temperature measurements in a premixed flame within an axisymmetric combustor, AGARD Conf. Proc. **281**
9. Fuji, S., Egushi, K. (1981): A comparison of cold and reacting flows around a bluff body flame stabilizer. ASME J. **103**, 328–334
10. Borghi, R., Escudié, D. (1984): Assessment of a theoretical model of turbulent combustion by comparison with a simple experiment. Combust. Flame **56**, 149–164
11. Vandromme, D., Ha Minh, H., Viegas, J. R., Rubesin, M. W., Kollmann, W. (1983): Second-order closure for the calculation of compressible wall bounded flows with an implicit Navier-Stokes solver, 4th turbulent shear flow (Karlsruhe)
12. Garyi, A. R., Sawyer, R. F. (1979): An experimental study of the flow field and pollutant formation in a two dimensional, premixed, turbulent flame. AIAA 17th Aerospace Sciences Meeting (New Orleans).
13. Labbe, J., Magre, P., Collin, G. (1983): Flow measurements in an experimental combustion chamber. AIAA 18th Thermophysics Conference (Montreal)
14. Davies, T. W., Beer, J. M. (1969): The turbulence characteristics of annular wake flow. International Seminar on Heat and Mass Transfer, Herceg-Noir
15. Dutoya, D., Michard, P. (1980): An implicit finite volume method for the calculation of elliptic flows. La Recherche Aérospatiale, 1980-2
16. Dupoirieux, F. (1983): "Prédiction d'écoulements turbulents réactifs", AGARD Conf. Proc. **353**
17. Dupoirieux, F. (1985): "A new model of combustion for the numerical computation of turbulent reactive flows", to be presented at the 10th International Colloquium on Dynamics of Explosions and Reactive Systems (Berkeley)

Experimental Study of the Interaction Between a Premixed Confined Laminar Flame and Coherent Structures

D. Escudié and G. Charnay

Ecole Centrale de Lyon – Laboratoire de Mécanique des Fluides 36, Avenue Guy de Collongue – B.P. 163, 69131 Ecully Cedex, France

Abstract

The present work deals with the experimental study of a premixed laminar *V*-shaped flame, expanding in a confined combustion chamber, and interacting with coherent structures.

A tomographic technique is used to visualize the flame front and a detailed description of the instantaneous velocity (Laser Doppler Anemometry) and mean temperature field (coated thermocouple) are used.

The results are interpreted in the light of preliminary flow characterization without combustion, regarding the vortex scales compared with the flame thickness and the position of the vortex generating cylinder

Introduction

Despite important efforts in the field of turbulent combustion, a few particular points such as the concept of flame generated turbulence or the interaction between turbulence and combustion are not completely understood. Although many numerical or analytical predictions exist for such flows, experimental studies are lacking specifically in the case of confined combustion.

Investigations of turbulent premixed flames begin in 1940 with the theoretical and experimental studies of Damköhler [1]. He first sets into evidence the fact that large-scale turbulence wrinkles a premixed laminar flame without modifying its internal structure, whereas small-scale turbulence primarily affects the transport processes inside the flame. Some years later, Karlovitz et al. [2] improved the later analyses and put forward the idea of turbulence created by the flame.

These necessary steps for the comprehension of turbulent combustion are now supplemented by studies of the interaction between turbulence and combustion. It is known that turbulence influences combustion and vice-versa; hence it seems difficult to dissociate both phenomena. Visualization techniques tend to explain qualitatively the effect of turbulence on combustion, and detailed descriptions of the aerodynamical field give information of the effects of the flame on turbulence.

The purpose of this study is to describe the interaction between a premixed confined laminar flame and coherent structures. Many questions arise at the present time:

- How is the flame front modified by the vortex, which is a form of simplified two-dimensional turbulence?
- Are the flame velocity, the thickness and the mean position of the flame front changed?
- Does the flame have an effect on the vortex upstream the front?
- Is the vortex "consumed" by combustion or can it be observed on the other side of the *V*-shaped flame?

Turbulent Shear Flows 5

• What happens if the characteristic dimension of the vortex is of the same order of magnitude as the flame front thickness?

Experimental works has been done in swirling flows [3–7], but only one (Namer [8]) is devoted to a detailed description of a non-confined premixed laminar flame (C_2H_4) in a Karman vortex street. The most important results of this study are a deflexion of the vortex street upstream the flame and a combination of dilatation and increased dissipation which "consumes" the vortices.

For the sake of easier numerical assessment, the chemical reaction considered here is a premixed lean hydrogen-air flame. This laminar *V*-shaped premixed flame is expanding into a confined quartz chamber where pressure gradient can take a prominent part in the kinetic energy balance [9, 10]. A vortex generating cylinder is placed upstream in the flow and the Reynolds number can vary from small values where regular vortex streets are formed, to large values (irregular range) in which turbulent velocity fluctuations accompany the periodic formation of vortices. This variation depends on the cylinder diameter, so it corresponds to a coherent structure size changing from a small scale having the same order of magnitude as the flame front thickness to a large scale which can probably cause the flame extinction in a particular case (vortex generating cylinder close to the stabilizing wire).

First, the flame front is visualized by means of a tomographic technique [11] which allows as to consider qualitatively the influence of the vortex upon combustion. A shadowgraph technique is also used to control the combustion process since the flame is not luminous.

In the second part, a detailed description of the velocity field including mean and rms profiles of the axial velocity component is presented, with and without combustion. Measurements of the mean thermal field are also provided; this allows the mean position of the flame front to be evaluated in terms of the characteristic cylinder eccentricity.

Experimental Set Up and Conditions

Experiments are performed in a quartz cylindrical chamber the diameter and length of which are respectively 114 mm and 200 mm. The experimental conditions are summarized in Fig. 1. The lean premixed flame (6.5% H_2 volumic concentration in air) is stabilized on a 0.4 mm diameter catalytic platinum wire. The average upstream flow velocity is 5 m/s.

Two vortex generating cylinders are used to perturb the flame front. Their axial distance upstream of the catalytic wire is prescribed and is equal to 10 mm. The eccentricity $E(Y/D)$ can vary from 0 mm (right upstream of the platinum wire) to 12 mm. The following Table presents the Reynolds numbers (Re $= U_D/\nu$), Strouhal numbers (St) and Strouhal frequencies ($f_S = S\,U/D$), for the two cases under consideration.

D (mm)	U (m/s)	Re	St	f_S
0.4	5	133	0.18	2250
1	5	333	0.20	1000

The downstream development of the wake behind a cylinder has been well known since the fifties, particularly owing to Roshko's [11] and Kovasznay's [12] works. Vortex-street patterns are stable and well-defined for long distances downstream, occur for only a small range of cylinder Reynolds numbers (Re $= 40$ to Re $= 150$). For the first cylinder ($D_1 = 0.4$ mm) the scale (Λ) of the vortices is of the same order as the flame front thickness ($\Lambda \sim \delta$). When the cylinder diameter is larger ($D_2 = 1$ mm) the vortex street is turbulent and

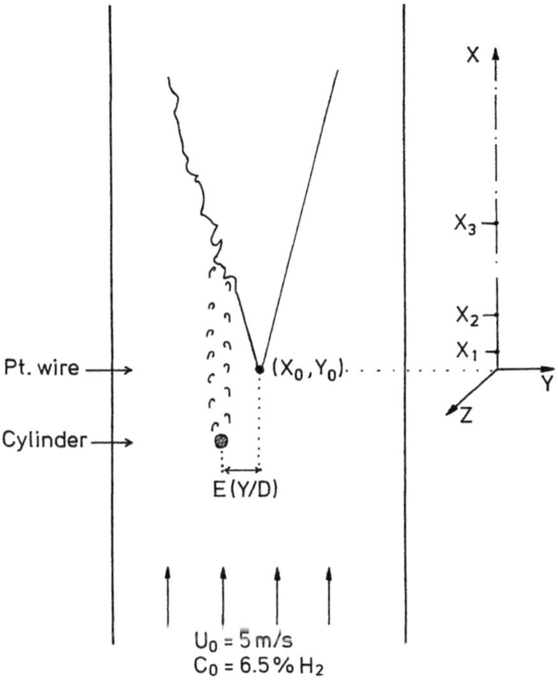

Fig. 1. Experimental conditions

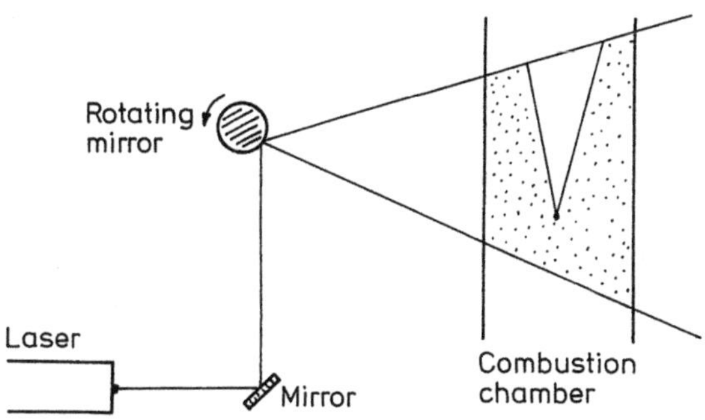

Fig. 2. Laser tomography apparatus

degenerates into fully turbulent flow beyond approximately 50 diameters downstream the cylinder. In this case the vortex scale is greater than the flame front thickness ($\Lambda > \delta$).

The mean temperature measurements are achieved with a 100 μm diameter coated Chromel-Alumel thermocouple. For mean velocity and turbulence measurements a classical forward scatter dual beam Laser Doppler Velocimetry technique is used (2 W Spectra Physics Argon ion laser, 514 nm, and a Bragg cell), and the whole flow is seeded with an Alumina (Al_2O_3) – Silica (SiO_2) mixture.

It can be noted that a preliminary study has been carried out by means of hot-wire anemometry. A turbulent residual rate of about 0.3% was obtained and values of Strouhal frequencies were confirmed by FFT analyzer.

The laser tomographic study was originated by Boyer [13] and is used for the flow visualization. In this case, the premixed lean mixture of hydrogen and air is seeded with incense which does not have any perturbating effect upon the flame, as shown by preliminary experiments and calculations. As the particles pass through the flame front they evaporate and burn. Because of the illumination with a sheet of laser light (cf. Fig. 2), and of light scattering due to the particles, the flame appears as the interface between the bright and dark areas (burned regions remain dark). This type of 2-D visualization can explain clearly the effects of coherent structures on combustion, whereas shadowgraph visualization, which integrates information along the optical path, is used to control the combustion process and allows the flame angle to be approximated.

Results and Discussion

Shadowgraph Visualization

This is one of the oldest optical techniques [14] for the visualization of flows with density gradients, and is appropriate to flames where density variation is generated by temperature increase. When the flame front is perturbed by the cylinder, it is difficult to interpret the interaction because of integration effects. However, for a laminar flow the V-shaped flame angle is rather good and corresponds to the value obtained according to the mean temperature profiles (cf. temperature measurements), i.e. $2\alpha = 4°$. So the flame studied is little open but this is not a problem, as one could expect; on the contrary this permits us to see an effect of the vortex – flame front interaction from one V-branch to the other one.

Laser Tomography

In this part of the study, the light sheet ($\sim$ 0.7 mm thickness) is created by a 5 W Argon laser. The mirror rotation frequency and the exposure time are such that each small part of the photograph corresponds to 1/6000° s. Hence the two extremities (up to down) are 1/300° s shifted in time. Three typical cases are presented in this paper. The first one is concerned with interaction between cylinder D_1 and the flame front (Fig. 3); the two others are devoted to cylinder D_2 (Figs. 4, 5).

A laser tomography study of a laminar flame in a Karman vortex was performed by Hertzberg et al. [15]. They found by means of a high-speed camera that the wavelength and the frequency of the perturbed flame motion correspond to that of the vortex street. In our experiment as will be seen, the phenomena are more complex.

For the case of laminar combustion, a perfect linear frontier between burned and unburned gas is observed. It is important to note that this lean flame presents neither wrinkle nor instabilities. When the cylinder $D_1 = 0.4$ mm is put upstream in the flow and for a weak eccentricity $E(Y/D_1) = 2.5$, small structures ($\Lambda_1 \sim 1$ mm) immediately appear behind the catalytic wire in a well defined zone and damp down 4 cm downstream. Besides, an instability expands sometimes at 10 cm over the Platinum wire towards the top of the combustion chamber. So a small perturbation with a characteristic scale Λ_1 of the same order as the thickness flame front would not only increase transfers [16] but also make the flame unstable.

The effect on the flame front is very sensitive, when the dynamical perturbation is generated by a 1 mm diameter rod. Even for large eccentricity [$E(Y/D_2) = 8$] combustion reacts to the perturbation wrinkling as if it were a turbulent flow; this is easy to understand since the interaction occurs at a X-distance greater than 40 D_2 (turbulent spectrum). The flame front is more and more wrinkled and burned packs flake away as the cylinder approaches. For $E(Y/D_2) = 4$ (Fig. 4), the cylinder influence is obvious on the other side of the V-flame, from the lower part (in which no wakes interference exists) to the top: strangely enough the combustion front loses its linear aspect by waving. This effect is emphasized when the cylinder is drawn towards the catalytic wire: behind the rod, the flame front is

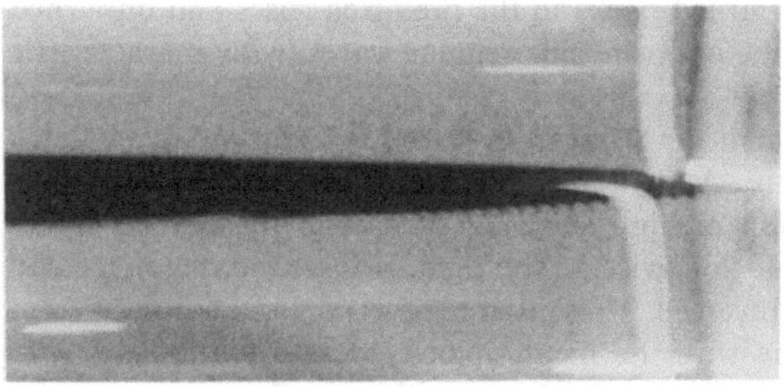

Fig. 3. Tomography of the little cylinder-flame interaction $E(Y/D_1) = 2.5$

Fig. 4. Tomography of the large rod-flame interaction $- E(Y/D_2) = 4$

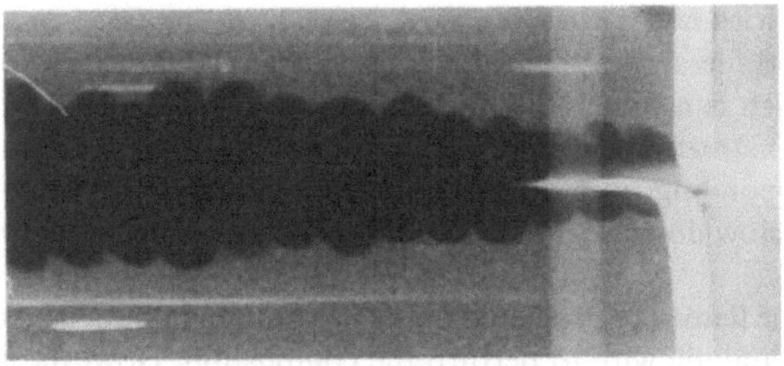

Fig. 5. Tomography of the large rod-flame interaction $- E(Y/D_2) = 0$

always wrinkled, vortex carrying away burned packs in the premixed gases and unburned packs in burned combustion gases; on the opposite side regular waves (with characteristic length ~ 7 mm) still persist.

If the cylinder is immediately upstream the wire, i.e. $E(Y/D_2) = 0$ (Fig. 5), the *V*-flame presents a regular flame front curvature. A particular structure, in which eddies travel downstream, clearly shows a first region of high shearing stresses giving rise to eddies rotating counter clockwise on the left and clockwise on the right. In a second region located upper, the direction of eddy rotation is reversed. These two regions correspond to different mean velocity profiles: right behind the wire the mean velocity decays without or with combustion: while as the wake effect decreases, the velocity within the flame increases far away from the wire.

It is not so easy to provide an explanation for this organized arrangment. Computations of the mean gradient $\partial U/\partial y$ (from L.D.A. measurements) in the mean flame front position always show a very small variation of this quantity. So, this phenomenon is not only due to shearing stresses, and it was impossible to distinguish any wire vibration. Perhaps some instability may occur even if it is contrary to Blevin's results [17]; he showed experimentally and theoretically (linear analysis) that no instability is possible when $E(Y/D) = 0$.

In fact we feel that the flame front can have a regular structure only if the periodic perturbation remains the same locally; this perturbation propagating along the flame front upstream and downstream. When a small part of this flame front is submitted to vortices ($\Lambda_2 < \delta$) rotating in opposite directions ($E(Y/D_2) < 4$), instantaneous shearing stresses are important: the flame front is wrinkled, sometimes broken, and local extinction may take place. But when interaction occurs between the flame front and the same eddy [$4 > E(Y/D_2) > 0$ on the other *V*-branch of the flame] a regular motion is created. It seems proved, particularly in the $E(Y/D_2) = 4$ eccentricity case, that the vortex is not "consumed" by the flame contrary to Namer's results [8].

Velocity Measurements

The previous visualizations lead us to investigate three typical stations downstream the flameholder: $X_1 = 13$ mm (in the wire wake), $X_2 = 30$ mm (wake supposed to be balanced by combustion acceleration) and $X_3 = 80$ mm (combustion acceleration).

The experimental study has been carried out for the two vortex generating cylinders (D_1 and D_2), without and with combustion, and for various eccentricity values: $E(Y/D) > 10$ (without cylinder – Pt wire interaction), $4 < E(Y/D) < 2$ (interaction in the X_3 station) and $E(Y/D) = 0$ (cylinder upstream the Pt wire).

Up to now only instantaneous axial velocity has been obtained, so it must be kept in mind that tangential velocity measurements will be necessary for a better final understanding.

All the figures present the axial mean velocity values versus *Y*-position on the combustion chamber diameter, for the three *X* stations and a chosen eccentricity.

Without combustion the velocity profile across the wake of a cylinder is characterized by the defect in the mean velocity which is associated with the drag on the cylinder. For example the mean velocity profiles downstream the cylinder $D_1 = 0.4$ mm and the platinum wire are plotted in Fig. 6a. The corresponding ones downstream the cylinder $D_2 = 1$ mm and the wire are shown in Fig. 6b; the slow down region is more or less large depending on the cylinder diameter.

The same profiles obtained across the flame are presented in Figs. 7a and b. In both cases the cylinder is too far away from the catalytic wire to perturb the combustion. However a global acceleration of the flow due to confinement and associated with the important

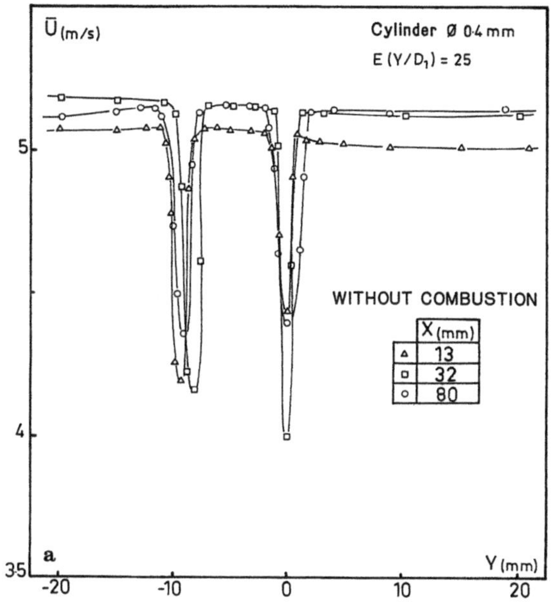

Fig. 6a. Mean axial velocity without combustion $D_1 = 0.4$ mn, $E(Y/D_1) = 25$

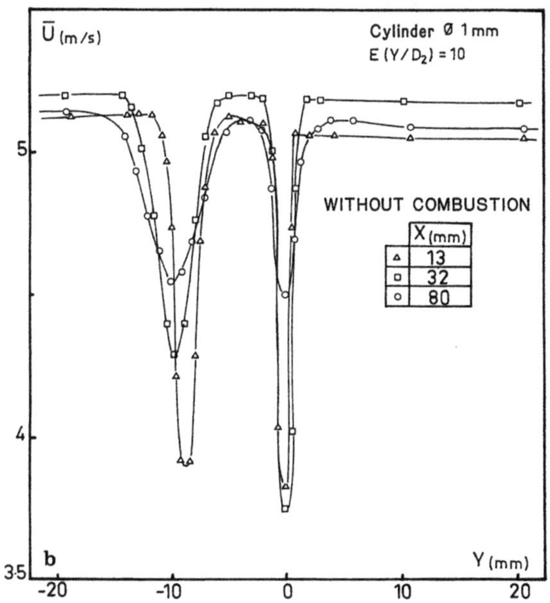

Fig. 6b. Mean axial velocity without combustion $D_1 = 1$ mn, $E(Y/D_2) = 10$

decrease of density (related to temperature rise) is noted. Moreover these first curves justify the choice proposed for the stations: in the X_1 station the wake is still important despite of combustion, but in the X_2 section this wake effect is nearly balanced by combustion energy production, and further on the velocity significantly increases.

Even though the cylinder does not interact with combustion, it plays a secondary role: it helps to know the action of the flame front on the upstream flow. It is easy to see the mean

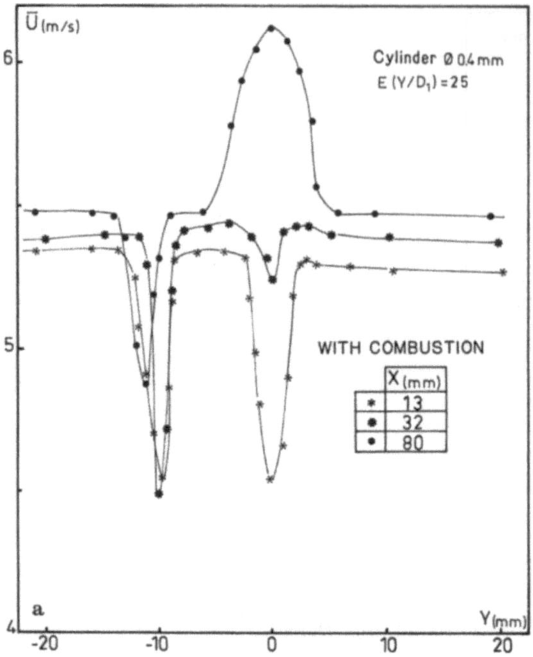

Fig. 7a. Mean axial velocity with combustion $D_1 = 0.4$ mn, $E(Y/D_1) = 25$

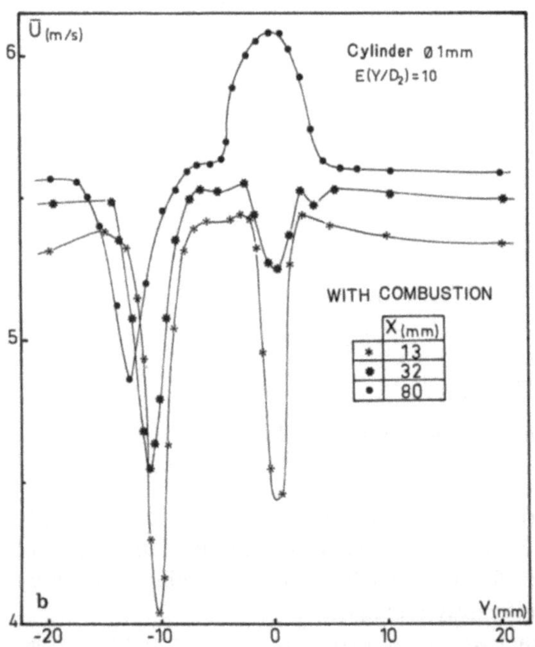

Fig. 7b. Mean axial velocity with combustion $D_2 = 1$ mn, $E(Y/D_2) = 10$

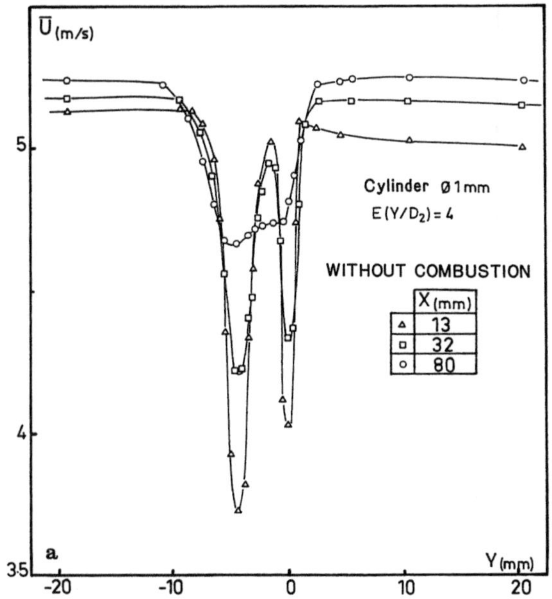

Fig. 8a. Mean axial velocity without combustion $D_2 = 2$ mn, $E(Y/D_2) = 4$

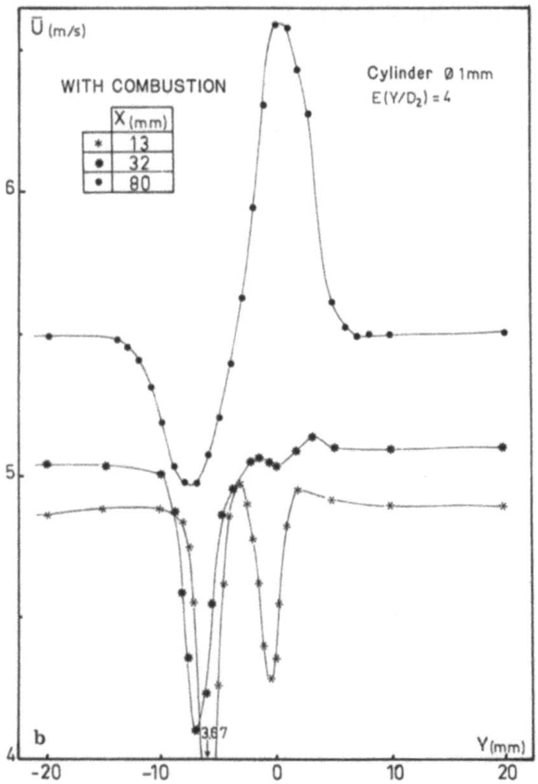

Fig. 8b. Mean axial velocity with combustion $D_2 = 1$ mn, $E(Y/D_2) = 4$

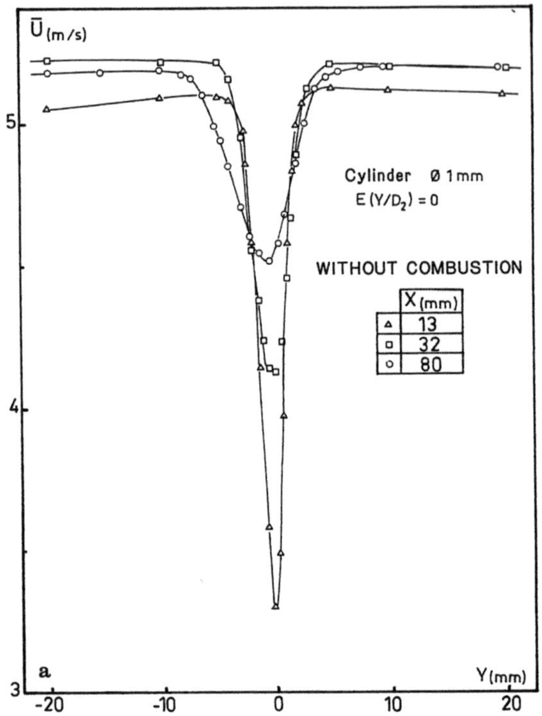

Fig. 9a. Mean axial velocity without combustion $D_2 = 1$ mn, $E(Y/D_2) = 0$

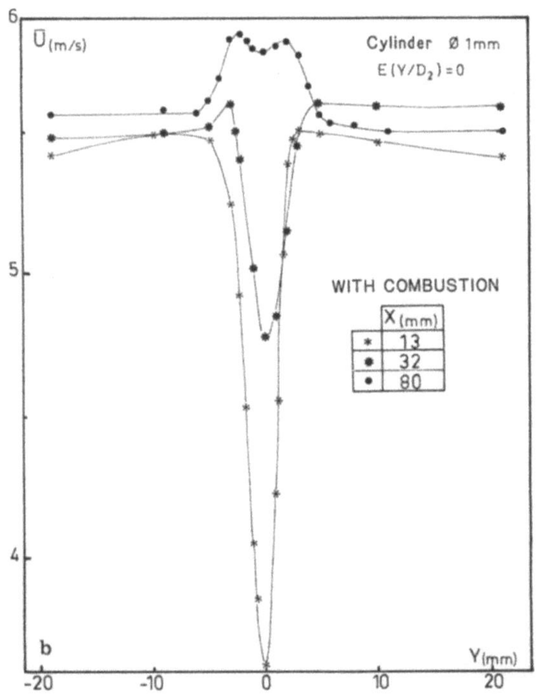

Fig. 9b. Mean axial velocity with combustion $D_2 = 1$ mn, $E(Y/D_2) = 0$

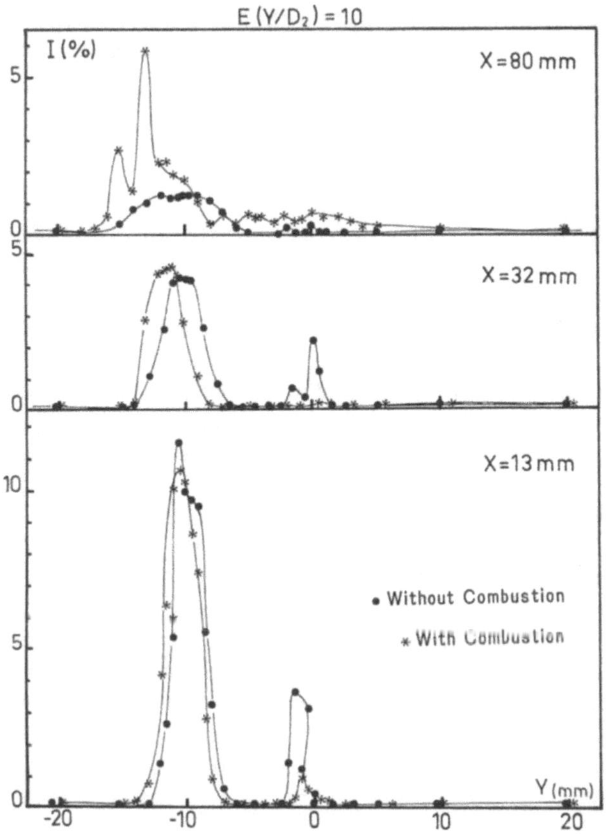

Fig. 10. Rms velocity profiles with and without combustion $D_2 =$ 1 mn, $E(Y/D_2) = 10$

deflection of the streamlines, compared with the non combustion case (Figs. 6a and b), and this deflection of the wake and of the flow upstream the flame is probably due to the pressure drop across the flame [$\tau = (\varrho u/\varrho b) - 1 = 3$]. Of course, this effect of combustion on the upstream flow field must be taken into account since the flow will in turn affect the flame development.

When the cylinder is drawn near the platinum wire [$E(Y/D) < 10$] the two wakes interfere with each other little by little. The mean velocity profiles corresponding to two characteristic cases [$E(Y/D_2) = 4$ and $E(Y/D_2) = 0$] are presented in Figs. 8 and 9, respectively. Without combustion (Figs. 8a and 9a) the wake interaction occurs only in the X_3 station for $E(Y/D_2) = 4$; the slowing down zone is then wider than if $E(Y/D_2) = 0$: in this case the profiles are identical to those obtained for a larger diameter cylinder ($\sim$ 1.2 mm). In the same way for $E(Y/D_1) = 0$ (small rod situated right upstream the Pt wire) the wake is equivalent to a 0.55 mm diameter cylinder. This is confirmed by results in combustion (Fig. 9b) where the important velocity slow down in the wake of this cylinder (with a larger apparent diameter) is hardly balanced; the flame looks like being stabilized on a greater flame holder.

With combustion the eccentricity $E(Y/D) = 4$ remains important. Fig. 8b emphasizes the axial mean velocity increase in the flame and the constant velocity of the unburned gases compared with the non-combustion curves. These features have to be refined and we are presently improving experiments to investigate tangential and axial velocities on the whole Y axis. Up to that point all the interpretations should be dismissed.

Rms values of the axial velocity component always vanish inside the flame. For example rms profiles, with and without combustion, in the case of $E(Y/D_2) = 10$ are shown in Fig. 10.

In the X_1 and X_2 sections the maxima in turbulent intensity without combustion are due to the velocity gradients in the wakes of the cylinder and wire. With combustion they do not exist because of the dissipation terms increase due to temperature and consequently a relaminarization appears in the burned gases.

Mean Temperature Measurements

A typical temperature profile without vortex generating cylinder is plotted in Fig. 11. Because of the small value of the Lewis number (Le = 0.3) the maximum temperature is greater than the computed adiabatic temperature.

When the flame is expanding into the combustion chamber, i.e. from the X_1 to the X_3 station, the maximum remains nearly the same but the curves widen out. Thus the mean position of the flame front can be obtained from a point located on the steepest temperature gradient and not on the maximum velocity gradient as often used in experimental studies [18]. Actually the latter leads to a mean flame front position inside the combustion zone, i.e. too small a flame angle value.

Results for the different mean flame positions versus the D_2 diameter cylinder eccentricity are presented in Fig. 12. Results for the larger rod are chosen because in the small cylinder case, measurements are not sufficiently accurate to set the phenomenom clearly into evidence.

The mean flame front position is not modified when $E(Y/D) \geqq 10$ but it is affected when the vortex street generated by the cylinder is introduced. The mean flame position is shown to move from its undisturbed location; it becomes much thicker and is shifted to the side probably owing to the complex nature of the flow in this region. However, this results will have to be specified in the future, particularly after the tangential velocity measurements and an estimation of the vortex velocity with respect to the flame burning velocity ($\bar{U}_L = \bar{U}_0 \sin\alpha = 18$ cm/s in our study).

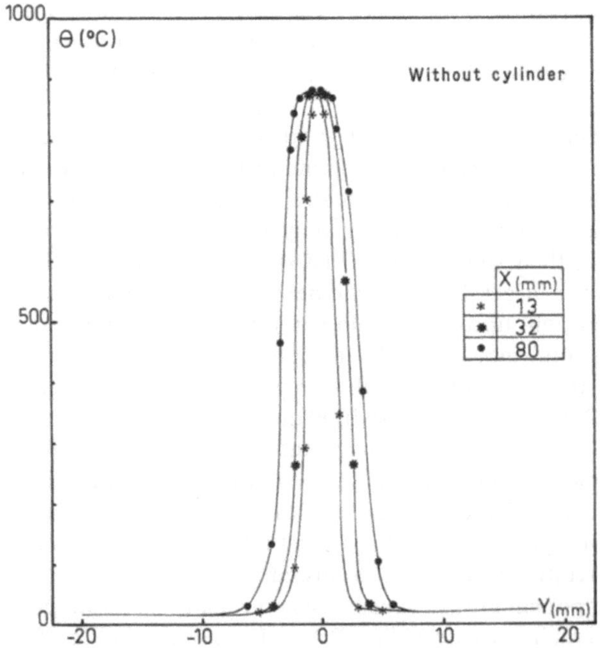

Fig. 11. Mean Temperature profile without vortex generating cylinder

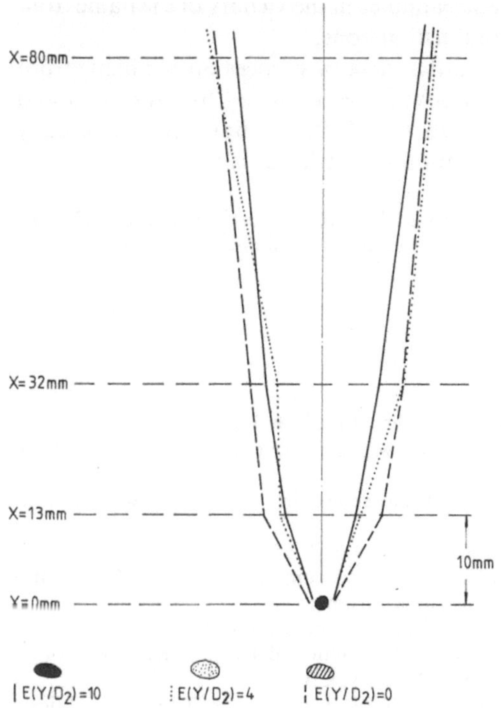

Fig. 12. Mean flame position

Lastly, when the rod is immediately upstream the flame holder, combustion looks really stabilized as if it were on a larger cylinder with the same angle and thickness.

Summary and Conclusion

This study permits us to supply several experimental results for better understanding of Combustion-Turbulence interaction and especially the premixed *V*-shaped flame modified by a Karman vortex street.

Two types of visualization (Laser Tomography and Shadowgraph) associated with instantaneous measurements of the axial velocity field and the mean temperature field are used to make a detailed description of this perturbed combustion.

The main results are:

- The existence of a limited organized region near the flameholder and the emergence of an unstable wave in the case of a small perturbation (D_1, $\Lambda_1 \sim \delta$).
- A wrinkled flame front with packets of burned and unburned gases travelling downstream if the generating cylinder is larger (D_2, $\Lambda_2 > \delta$).
- For a particular eccentricity value [$E(Y/D) \sim 4$] the mean position of the flame front is shifted and a characteristic structure appears on the opposite *V*-branch showing that vortices are probably "not consumed" by combustion.
- When the rod is right upstream the flame holder, combustion looks stabilized as on a larger equivalent cylinder and the flame front has a regular pattern because of the same local disturbance created by vortices rotating in the same direction.

Moreover one could notice a deflection of the streamlines in the vicinity of the flame front and a relaminarization of the flow inside the combustion zone.

In fact most of these results are concerned with the effect of vortices on the flame front and many questions remain. Thus this study has now to be improved by tangential and phase-locked velocity measurements which will lead to a better comprehension of eddy transformation beyond the flame and of the generation of particular waves.

Acknowledgement. The research reported here is supported by a CNRS (Centre National de la Recherche Scientifique) grant. We also wish to thank Pr. R. Borghi for helpful discussions.

References

1. Damköhler, G. (1940): Z. Electrochem. **45**, 11
2. Karlovitz, B., Denniston, D. W., Wells, D. E. (1951): J. Chem. Phys. **19**, 541
3. Mestre, A., Benoit, A. (1973): "Combustion in Swirling Flow", in 14th *Symposium on Combustion*, 719
4. Chigier, N. A., Dvorak, K. (1975): "Laser Anemometer Measurements in Flames with Swirl", in 15th symposium on combustion, 573
5. Lilley, D. (1977): Swirl flows in combustion: A review. AIAA J. **5**, 1063
6. Gouldin, F. C., Leibovich, S. (1979): Effect of swirl on premixed combustion. NASA Conference Publication, premixed prevaporized combustor technology forum
7. Fujii, S., Eguchi, K., Gomi, N. (1981): Swirling jets with and without combustion. J. **19**, 1438
8. Namer, I. (1980): "An Experimental Investigation of the Interaction Between a Karman Vortex Street and a Premixed Laminar Flame"; Ph.D. Thesis Berkeley, CA
9. Escudié, D., Trinité, M., Paranthoën, P. (1983): Modification of turbulent flow field by an oblique premixed hydrogen-air flame. Prog. Astron. Aeron. **88**, 147
10. Borghi, R., Escudié, D. (1984): Assessment of a theoretical model of turbulent combustion by comparison with a simple experiment. Combust. flame **56**, 149
11. Roshko, A. (1953): On the development of turbulent wakes from vortex streets. NACA Technical Note, 2913
12. Kovasznay, L. S. G. (1949): Hot-wire investigation of the wake behind cylinders at low Reynolds numbers. Proc. Roy. Soc. Ser. A, **198**, 174
13. Boyer, L. (1980): Laser tomographic method for flame front movements studies. Combust. Flame, **39**, 321
14. Dvorak, V. (1880): Ann. Physik, **3**, 502
15. Hertzberg, J. R., Namazian M., Talbot, L. (1984): A laser tomographic study of a laminar flame in a Karman vortex street. Combust. Sci. Technol. **38**, 205
16. Petersen, R. E., Emmons, H. W. (1961): Stability of laminar flames. Phys. Fluids **4**, 456
17. Blevins, R. D. (1977): Instabilities of tube rows and tube arrays. Flow-induced vibrations. V.N.R. Company, 88
18. Cheng, R. K., Ng, T. T. (1984): On defining the turbulent burning velocity in premixed *V*-shaped turbulent flames. Combust. Flame **57**, 155

Index of Contributors

Acharya M. 176
Adrian R. J. 79
Aupoix B. 253

Bertoglio J.-P. 19
Boisson H. C. 82
Borghi R. 337
Bradbury C. J. S. 278
Bradshaw P. 171
Braza M. 82

Castro I. P. 278
Charnay G. 347
Chassaing P. 82
Coupland J. 310
Cousteix J. 253

Dakos T. 7
Dianat M. 278
Dupoirieux F. 337
Durst F. 197

Eaton J. K. 266
Erm L. P. 186
Escudié D. 347
Escudier M. P. 176

George W. K. 134
Gibson M. M. 7
Glauser M. N. 134

Ha Minh H. 82
Haworth D. C. 44
Hayakawa M. 110
Helland K. N. 67
Herring J. R. 324
Hussain A. K. F. 110

Itsweire E. C. 67

Jeandel D. 19
Jones W. P. 296
Joubert P. N. 186
Jovanovic J. 197

Kanevce Lj. 197
Kawall J. G. 98
Keffer J. F. 98
Kim J. 221
Kollmann W. 296
Kourta A. 82

Labbé J. 337
Lasheras J. C. 124
Lee M. J. 54
Leib St. J. 134
Leonard A. 234
Liandrat J. 253
Louchez P. R. 98

Maxworthy T. 124
Menon S. 110
Metcalfe R. W. 110
Moreau P. 337

Pauly W. R. 266
Perry A. E. 3
Peters N. 293
Pope S. B. 44
Priddin C. H. 310

Reynolds W. C. 54
Rohr J. J. 67

Saripalli K. R. 146
Simts A. J. 186
Spalart P. R. 234

Van Atta C. W. 67
Veeravalli S. 31

Warhaft Z. 31
Westphal R. V. 266
Wyngaara J. C. 324

Springer Series in Computational Physics

Editors: **H. Cabannes, M. Holt, H. B. Keller, J. Killeen, S. A. Orszag, V. V. Rusanov**

Computational Methods for Kinetic Models of Magnetically Confined Plasmas

By **J. Killeen, G. D. Kerbel, M. C. McCoy, A. A. Mirin**
1986. 77 figures. VIII, 199 pages. ISBN 3-540-13401-8

Y. N. Dnestrovskii, D. P. Kostomarov

Numerical Simulation of Plasmas

Translated from the Russian by N. V. Deyneka
1986. 97 figures. XIV, 304 pages. ISBN 3-540-15835-9

R. Peyret, T. D. Taylor

Computational Methods for Fluid Flow

2nd corrected printing. 1985. 125 figures. X, 358 pages
ISBN 3-540-13851-X

R. Gruber, J. Rappaz

Finite Element Methods in Linear Ideal Magnetohydrodynamics

1985. 103 figures. XI, 180 pages. ISBN 3-540-13398-4

C. A. J. Fletcher

Computational Galerkin Methods

1984. 107 figures. XI, 309 pages. ISBN 3-540-12633-3

R. Glowinski

Numerical Methods for Nonlinear Variational Problems

1984. 82 figures. XV, 493 pages. ISBN 3-540-12434-9
(Originally published as "Glowinski, Lectures on Numerical Methods...", Tata Institute Lectures on Mathematics, 1980)

M. Holt

Numerical Methods in Fluid Dynamics

2nd revised edition. 1984. 114 figures. XI, 273 pages
ISBN 3-540-12799-2

O. G. Mouritsen

Computer Studies of Phase Transitions and Critical Phenomena

1984. 79 figures. XII, 200 pages. ISBN 3-540-13397-6

O. Pironneau

Optimal Shape Design for Elliptic Systems

1984. 57 figures. XII, 168 pages. ISBN 3-540-12069-6

M. Kubíček, M. Marek

Computational Methods in Bifurcation Theory and Dissipative Structures

1983. 91 figures. XI, 243 pages. ISBN 3-540-12070-X

Y. I. Shokin

The Method of Differential Approximation

Translated from the Russian by K. G. Roesner
1983. 75 figures, 12 tables. XIII, 296 pages. ISBN 3-540-12225-7

Finite-Difference Techniques for Vectorized Fluid Dynamics Calculations

Editor: **D. L. Book**
With contributions by J. P. Boris, M. J. Fritts, R. V. Madala, B. E. McDonald, N. K. Winsor, S. T. Zalesak
1981. 60 figures. VIII, 226 pages. ISBN 3-540-10482-8

D. P. Telionis

Unsteady Viscous Flows

1981. 132 figures. XXIII, 408 pages. ISBN 3-540-10481-X

F. Thomasset

Implementation of Finite Element Methods for Navier-Stokes Equations

1981. 86 figures. VII, 161 pages. ISBN 3-540-10771-1

F. Bauer, O. Betancourt, P. Garabedian

A Computational Method in Plasma Physics

1978. 22 figures. VIII, 144 pages. ISBN 3-540-08833-4

Springer-Verlag
Berlin Heidelberg New York
London Paris Tokyo

Springer

MIX
Papier aus verantwortungsvollen Quellen
Paper from responsible sources
FSC® C105338

If you have any concerns about our products,
you can contact us on
ProductSafety@springernature.com

In case Publisher is established outside the EU,
the EU authorized representative is:
Springer Nature Customer Service Center GmbH
Europaplatz 3, 69115 Heidelberg, Germany

Printed by Libri Plureos GmbH
in Hamburg, Germany